Aspects of the life record

Major events	Dominant forms[b]		Systems
	Homo		Quaternary[c]
			Tertiary[c]
Grasses become abundant	Mammals		
		Flowering plants	
Horses first appear			
Extinction of dinosaurs			Cretaceous
Birds first appear			Jurassic
Dinosaurs first appear	Reptiles	Conifer and cycad plants	Triassic
			Permian
	Amphibia		
Coal-forming swamps			Pennsylvanian
			Mississippian
		Spore-bearing land plants	Devonian
	Fish		Silurian
Vertebrates first appear (fish)			
			Ordovician
	Marine invertebrates		
		Marine plants	
First abundant fossil record (marine invertebrates)			Cambrian
	Primitive marine plants and invertebrates		Precambrian
	One-celled organisms		

[c]Some geologists prefer to use the term Cenozoic for the Quaternary and the Tertiary.
[d]In most European and some American literature Pennsylvanian and Mississippian are combined in a period called the Carboniferous.
[e]Subdivisions not firmly established.

PHYSICAL GEOLOGY

L. DON LEET LATE HARVARD UNIVERSITY

SHELDON JUDSON PRINCETON UNIVERSITY

MARVIN E. KAUFFMAN FRANKLIN AND MARSHALL COLLEGE

PHYSICAL GEOLOGY FIFTH EDITION

PRENTICE-HALL, INC.   ENGLEWOOD CLIFFS, NEW JERSEY 07632

L. Don Leet, Sheldon Judson, and Marvin E. Kauffman: **PHYSICAL GEOLOGY**, FIFTH EDITION

© 1978 by Prentice-Hall, Inc., Englewood Cliffs, New Jersey 07632

Printed in the United States of America

10 9 8 7 6 5 4 3 2 1

LIBRARY OF CONGRESS CATALOGING IN PUBLICATION DATA

Leet, Lewis Don, 1901–1974.
    Physical geology.

    Includes bibliographies and index.
    1. Physical geology. I. Judson, Sheldon, 1918–
joint author. II. Kauffman, Marvin Earl, 1933–
joint author.
QE28.2.L43 1978   551   77-13432
ISBN 0-13-669739-9

Prentice-Hall International, Inc., London
Prentice-Hall of Australia Pty. Limited, Sydney
Prentice-Hall of Canada, Ltd., Toronto
Prentice-Hall of India Private Limited, New Delhi
Prentice-Hall of Japan, Inc., Tokyo
Prentice-Hall of Southeast Asia Pte. Ltd., Singapore
Whitehall Books Limited, Wellington, New Zealand

PHYSICAL GEOLOGY, FIFTH EDITION, has been film-composed in different styles of Cairo, a version of various so-called antique, or "Egyptian," typefaces introduced into Great Britain during the first quarter of the nineteenth century.

David R. Esner was responsible for production, with the assistance of Mary Helen Fitzgerald; design and layout are by Betty Binns Graphics; Betty Binns and David R. Esner art-directed; and the line illustrations were executed by Vantage Art, Inc.

The cover photograph is by David Muench: Differential erosion has produced this spectacular formation in Arches National Park, Utah. The La Sal mountains are in the background.

PREFACE

Since *Physical Geology* first appeared 24 years ago, knowledge of the earth and its near neighbors has expanded in spectacular fashion; moreover, the remarkable contemporary discoveries in the earth sciences are more than ever perceived to be founded on data, principles, and techniques studied over a long time. It is proper, then, that this fifth edition, like its predecessors (particularly the third edition), attempts to integrate new knowledge with traditional knowledge in a balanced and concise synthesis, carefully clarified for those students who will not make geology their life's work. We have once again paid special attention to users' comments and introduced a number of changes that we hope will make *Physical Geology* even more efficient and attractive than its four previous versions.

Chapter 1 sketches out the grand themes of geological understanding, themes that will be expanded upon and illuminated by other information throughout the textbook: time, the cycling of rock materials, and the powerful theories that explain how ocean basins are born, how they grow, and how they disappear—that is, the process of sea-floor spreading and the demonstration that continents drift and move as elements of the large plates composing the earth's crust. And in Chapter 8 we discuss how the earth's magnetic field has reversed itself many times in the past—a discovery that not only helped establish the reality of continental drift but also allowed us to measure its rate and to establish a new way of telling geologic time.

Research of the last 15 years has contributed to an explanation of why most volcanoes and earthquakes occur where they do, and Chapters 3 and 6 indicate that we are closer than ever before to understanding how these major earth forces work: There is now real hope, for instance, that before long we shall be able to predict earthquakes with some certainty.

This edition, especially in Chapters 14 to 16, deals extensively with the current issues of sources and utilization of energy; and we discuss how people can modify the natural processes of earth and atmosphere to affect the physical and biological environment, often to human disadvantage.

And in a new Chapter 17 we have briefly surveyed the exploration of near neighbors in our planetary system.

The foregoing innovations and consolidations of related material do not mean, however, that the standard geological fare has been neglected: The highly accepted treatment in former editions of such topics as atoms and minerals (now Chapter 2), weathering and sedimentary rocks (Chapter 4), deformation (Chapter 7), and glaciation (Chapter 12) has been continued in this edition, sometimes expanded where necessary to accommodate findings, sometimes combined to throw into higher relief the unity of geology.

Finally we call attention to a new and contemporary format, enabling the book to be kept to around 500 pages, to completely redrawn illustrations in a crisper and less sophisticated style than that of the fourth edition, and to clear black-and-white photographs. We hope that these devices will encourage students in their acquisition of geological knowledge.

We are indebted to the many persons who have helped with previous editions, an assistance that has proved invaluable in the preparation of this edition. In addition helpful comments and reviews of the manuscript were received from several colleagues, whom we wish to thank: George T. Farmer, Jr., Madison College, Harrisonburg, Virginia; William R. Farrand, University of Michigan, Ann Arbor; Martin H. Link, Los Angeles Harbor College, Wilmington, California; John F. Shroder, Jr., University of Nebraska, Omaha; John Stolar, Cheyney State College, Cheyney, Pennsylvania; and Charles P. Walters, Kansas State University, Manhattan.

The staff at Prentice-Hall has continued to provide professional competence, amicable support, and a great deal of patience; we particularly thank David R. Esner and Logan M. Campbell. Sheldon Judson has had editorial assistance from Joan Wyckoff, for which he is grateful. Marvin Kauffman has enjoyed comfort and assistance from family and friends; and he recognizes this aid with love and gratitude.

We owe a large debt to L. Don Leet, an active author of the first four editions of this book. His untimely death occurred before work on this edition had begun. But we gladly acknowledge that we have relied heavily on his earlier contributions.

SHELDON JUDSON
MARVIN E. KAUFFMAN

SUMMARY CONTENTS

COMPLETE CONTENTS

3

IGNEOUS ACTIVITY AND IGNEOUS ROCKS

3.1 VOLCANOES 38

3.2 BASALT PLATEAUS 52

3.3 IGNEOUS ACTIVITY AND EARTHQUAKES 52

3.4 THE EARTH'S HEAT 53

PHYSICAL GEOLOGY

1

TIME AND A CHANGING EARTH

Geology is the science of the earth, an organized body of knowledge about the globe on which we live—about the mountains, plains, and ocean deeps, about the history of life from slime-born amoeba to humanity, and about the succession of physical events that accompanied this orderly development of life (Figure 1.1).

Geology helps us unlock the mysteries of our environment. Geologists explore the earth from the ocean floors to the mountain peaks to discover the origins of our continents and the encircling seas. They try to explain a land surface so much varied that in less than an hour a traveler can fly over the Grand Canyon of the Colorado River, over the dammed waters of Lake Meade, over Death Valley, the lowest point in the contiguous United States, and over Mount Whitney, the highest point. They probe the action of glaciers that crawled over the land and then melted away over half a billion years ago and of some that even today cling to high valleys and cover most of Greenland and Antarctica, the remnants of a recent but presently receding Ice Age.

Geologists search for the record of life from the earliest one-celled organisms of ancient seas to the complex plants and animals of the present. This story,

from simple algae to seed-bearing trees and from primitive protozoans to highly organized mammals, is told against the ever-changing physical environment of the earth.

For the earth has not always been as we see it today, and it is changing (but slowly) before our eyes. The highest mountains are built of materials that once lay beneath the oceans. Fossil remains of animals that swarmed the seas millions of years ago are now dug from lofty crags. Every continent is partially covered with sediments that were once laid down on the ocean floor, evidence of an intermittent rising and settling of the earth's surface.

In this chapter we take a preliminary look at some of the important concepts in the study of our changing earth. In subsequent chapters we shall discuss at greater length the subjects touched on here. Therefore you must read this chapter with the understanding that the assertions we make will be more fully explained at appropriate later points. This chapter, however, is intended to provide a framework within which to organize your thinking about much of that more detailed material: It serves as a kind of map by which to chart our exploration of physical geology.

FIGURE 1.1

The earth seen from the *Apollo 8* spacecraft during the first circumlunar voyage. The atmosphere, which partially obscures the earth here, makes possible not only life but oceans, running water, glaciers, winds, and the familiar processes of weathering. [NASA.]

1.1 TIME

In 1654 a biblical scholar, James Ussher, archbishop of Armagh, declared that the earth had been created at 9:00 A.M. on October 26, 4004 B.C.—an event and a time presumably synonymous in his mind with the birth of the universe. Since his announcement, over 300 years, or nearly 6 percent of all the time assigned by the archbishop to earth history, have passed, and our notion of earth time has changed dramatically. We now know that the earth is about 4.5 billion years old—nearly a million times the age attributed to it in the seventeenth century. We are less certain about the universe, of which our earth is a part. But most current evidence suggests that the universe is more than three times the age of the earth.

In the following paragraphs we briefly discuss some of the ways in which geologic time is divided and measured, deferring until Chapter 18 a more detailed consideration of the subject.

ABSOLUTE AND RELATIVE TIME

An initial and casual reaction to the notion of time is that we can mark it off without much difficulty, even though we recognize, as Thomas Mann wrote in *The Magic Mountain*, that "Time has no division to mark its passage, there is never a thunder-storm or blare of trumpets to announce the beginning of a new month or year. Even when a new century begins it is only we mortals who ring bells and fire off guns."[1]

Nevertheless, our modes—seconds, years, millennia, and the rest—are all we have to work with; and consequently we can consider geologic time from two points of view: as relative or as absolute. **Relative time**—that is, whether one event in earth history came before or after another event—disregards years. On the other hand, whether a geologic event took place a few thousand years ago, a billion years ago, or at some date even farther back in earth history is reported in **absolute time.**

Relative and absolute time in earth history have their counterparts in human history. In tracing the history of the earth, we may wish to know whether some event, such as a volcanic eruption, occurred before or after another event, such as a rise in sea level, and how these two events are related in time to a third event, perhaps a mountain-building episode. In human history, too, we try to determine the relative position of events in time. In studying United States history, we find

[1]Thomas Mann, *The Magic Mountain*, trans. H. T. Lowe-Porter, p. 225, Modern Library, Inc., New York, 1955.

it important to know that the Revolutionary War preceded the Civil War and that the Canadian-American boundary was fixed some time between these two events.

Sometimes events in both earth history and human history can be established only in relative terms. Yet our record becomes increasingly precise as we fit more and more events into an actual chronological calendar: If we did not know the date of the United States-Canadian boundary treaty—if we knew only that it was signed between the two wars—we could place it between 1783 and 1861. (Recorded history, of course, provides us with the actual date, 1846.)

Naturally, we should like to be able to date geologic events with similar precision. But so far this has been impossible, and the accuracy in determining the dates of human history—that is, written human history—will likely never be achieved in geologic dating. Still, we can determine approximate dates for many geologic events, which are probably of the correct order of magnitude. We can say that dinosaurs became extinct about 63 million years ago and that about 11,000 years ago the last continental glacier began to recede from New England and the area bordering the Great Lakes.

Radioactivity Radioactive elements (those whose nuclei spontaneously emit particles to produce new elements, as discussed in Section 18.1) have provided the most effective means of measuring absolute time. The rate at which a given radioactive element decays is (so far as we have been able to determine) unaffected by changes in physical conditions or by time. So if we know the amount of original radioactive material (the *parent*) that remains, the rate of radioactive decay, and the amount of new elements (the *daughters*) that has formed, then we can calculate the time elapsed since radioactive decay began. Of course, the calculation is not quite so simple, as we discuss in Section 18.1. Nevertheless, from the time that the first radioactive age determinations were made (in 1907) to now, we have learned enough about the techniques and pitfalls of radioactive dating to be confident in the thousands of dates now available, particularly those made during the last two decades. A variety of elements has proved useful, with a time range from that of carbon 14, which can be used to date events that occurred a few hundred to a few tens of thousands of years ago, to that of an element such as uranium 238, which has the potential of dating events several times greater than the age of the earth.

DIVISIONS OF GEOLOGIC TIME

The application of radioactive elements to the measurement of geologic time became useful only after geologists had already constructed a calendar of geologic events. This calendar, still in use, was based on the ages of rock units relative to each other. Such relative ages were determined by conclusions drawn from a number of phenomena, including the superposition of younger rocks on older rocks, the cutting of older rocks by more recently formed rocks, and the progressive evolutionary stages of plant and animal life, as represented by remains in some rocks. The determination methods are discussed in detail in Section 18.2. It suffices to say here, however, that the arrangement of rock units and the earth events they record in the geologic calendar, as determined before the twentieth century, have been confirmed by the absolute dates of later radioactive dating.

The rock units in their proper chronological order make up the **geologic column,** which is reproduced on the front endpaper of this book.

UNIFORMITARIANISM

Modern geology was born in 1785, when James Hutton (1726–1797), a Scottish medical man, gentleman farmer, and geologist (Figure 1.2), formulated the principle

FIGURE 1.2

Many concepts of present-day geology stem directly from observations by James Hutton. [From F. D. Adams, *Birth and Development of the Geological Sciences,* reprinted by permission of Dover Publications, Inc., New York, 1938.]

now known to geologists as the **doctrine of uniformitarianism.** This principle simply means that the physical processes operating in the present to modify the earth's surface have also operated in the geologic past, that there is a uniformity of processes past and present.

Here is an example. We know from observations that modern glaciers deposit a distinctive type of debris made up of rock fragments that range in size from submicroscopic particles to boulders weighing several tons. This debris is jumbled, and many of the large fragments are scratched and broken. We know of no agent other than glacier ice that produces such a deposit. Now suppose that in the New England hills or across the plains of Ohio or in the deep valleys of the Rocky Mountains we find deposits that in every way resemble glacial debris but find no glaciers in the area. We can still assume that the debris was deposited by now-vanished glaciers. On the basis of evidence like this geologists have worked out the concept of the great Ice Age (Chapter 12).

Such an example can be multiplied many times. Today most earth features and rocks exposed at the earth's surface are explained as the result of past processes similar to those of the present; we shall find many conclusions of physical geology based on the conviction that modern processes have also operated in the past.

Armed with Hutton's concept of uniformitarianism, nineteenth-century geologists were able to explain earth features on a logical basis. But the very logic of the explanation gave rise to a new concept for students of the earth. Past processes presumably operated at the same slow pace as those of today. Consequently, as we have implied in the preceding sections, very long periods of time must have been available for those processes to accomplish their tasks: It was apparent that a great deal of time was needed for a river to cut its valley or for hundreds or thousands of feet of mud and sand to be deposited on an ocean bottom, hardened into solid rock, and raised far above the level of the sea.

The concept of almost unlimited time in earth history is thus a necessary outgrowth of the application of the principle that **the present is the key to the past.** For example, geologists know that mountains as high as the modern Rockies once towered over what are now the low uplands of northern Wisconsin, Michigan, and Minnesota. But only the roots of these mountains are left, the great peaks having long since disappeared. Geologists explain that the ancient mountains were destroyed by rain and running water, creeping glaciers and wind, and landslides and slowly moving rubble and that these processes acted essentially as they do in our present-day world.

Now think of what this explanation means. We know from firsthand observation that streams, glaciers, and winds have some effect on the surface of the earth. But can such feeble forces level whole mountain ranges? Instinct and common sense tell us that they cannot. This is where the factor of time comes into the picture. True, the small, almost immeasurable amount of erosion that takes place in a human lifetime has little effect; yet when the erosion during one lifetime is multiplied by thousands and millions of lifetimes, mountains can be worn away. Time makes possible what seems impossible.

The distinguished American geologist Adolph Knopf has written:

If I were asked as a geologist what is the single greatest contribution of the science of geology to modern civilized thought, the answer would be the realization of the immense length of time. So vast is the span of time recorded in the history of the earth that it is generally distinguished from the more modest kinds of time by being called "geologic time."[2]

1.2 EARTH MATERIALS AND THE ROCK CYCLE

Geology is based on the study of rocks, and we seek to know their composition, their distribution, how they are formed and destroyed, and why they are lifted up into continental masses and depressed into ocean basins.

Rock is the most common of all the materials on earth. It is familiar to everyone. We may recognize it as the gravel in a driveway, the boulders in a stream, the cliffs along a ridge. And it is common knowledge that firm bedrock is exposed at the earth's surface or lies beneath a thin cover of soil or loose debris (Figure 1.3).

The solid rock and the soil and debris above it are

[2]Adolph Knopf, *Time and Its Mysteries*, ser. 3, p. 33, New York University Press, New York, 1949.

FIGURE 1.3 (LEFT)

Firm rock lies beneath a thin cover of soil or unconsolidated material or crops out at the surface, as it does here near the mouth of the Fremont Canyon in Utah. The snow-capped Henry Mountains are in the background. [U.S. Geological Survey, J. R. Stacy.]

FIGURE 1.4 (RIGHT)

Rock, debris, and soil particles are composed of minerals. Each type of mineral exhibits its own chemical composition and physical characteristics. Here is a cluster of crystals of the mineral pyrite, FeS_2. The larger crystals measure approximately 1 cm across. [Willard Starks.]

made up of minerals (Figure 1.4). Each type (and over two thousand different minerals have been described) has its own chemical composition and physical characteristics to identify it (see Chapter 2). But a bare handful, a dozen more or less, makes up the bulk of the minerals that we see in the upper layers of the earth. If we were to examine the rock material of these layers from place to place, we should begin to see differences among different samples, including, for instance, the types of mineral in each rock; and on the basis of simple observation we should begin to place rocks into different classes.

THE THREE ROCK FAMILIES

Observations, then, have led geologists to divide the earth's rocks into three main groups based on mode of origin: **igneous**, **sedimentary**, and **metamorphic**. Later on, we shall discuss each type in detail, but here is a short explanation of all three.

Igneous rocks, the ancestors of all other rocks, take their name from the Latin *ignis*, "fire." These "fire-formed" rocks were once a hot, molten mass known as **magma**, which subsequently cooled into firm, hard rock. Thus the lava flowing across the earth's surface from an erupting volcano soon cools and hardens into an igneous rock (Figure 1.5). There are other igneous rocks exposed at the surface that actually cooled some distance beneath it. We see such rocks today only because erosion has stripped away the rocks that covered them during their formation.

Most sedimentary (Latin *sedimentum*, "settling") rocks are made up of particles derived from the breakdown of preexisting rocks. Usually these particles are transported by gravity, water, wind, or ice to new loca-

A small volcanic cone in the crater of Vesuvius near Naples, Italy, belches out an ash-laden cloud of hot gases. This volcano is one of a chain of volcanoes, both active and extinct, that reaches from Mount Etna in Sicily to north of Rome, far up the Italian boot. They all represent features built by molten material periodically extruded through the earth's crust. [Vincenzo Carcavallo.]

Layering is a characteristic of sedimentary rocks. Here, in Pueblo County, Colorado, beds of limestone alternate with layers of shale, slightly recessed because they weather more rapidly. [G. K. Gilbert.]

This aerial photograph shows the twisted pattern of deformed and metamorphosed rocks. The beds, which show as bands 100 to 300 m wide, were once horizontal sedimentary rocks. They were deeply buried beneath the surface, tilted and folded by earth forces, and then exposed to view by subsequent erosion. [Royal Canadian Air Force.]

tions, where they are deposited in new arrangements. For example, waves beating against a rocky shore may provide the sand grains and pebbles for a nearby beach. If these beach deposits were to be hardened, we should have sedimentary rock. One of the most characteristic features of sedimentary rocks is the layering of the deposits that go to make them up (Figure 1.6).

Metamorphic rocks compose the third large family of rocks. Metamorphic (from the Greek words *meta*, "change," and *morphē*, "form") refers to the fact that the original rock has been changed from its primary form to a new form. Earth pressures, heat, and chemically active fluids beneath the surface may all be involved in changing an originally sedimentary or igneous rock into a metamorphic rock (Figure 1.7).

THE ROCK CYCLE

We have suggested that there are definite relationships among sedimentary, igneous, and metamorphic rocks: With time and changing conditions, any one of the rock types may be changed into some other form. These relationships form a cycle, as shown in Figure 1.8, which is simply a way of tracing out the various paths that earth materials follow. The outer circle represents the complete cycle; the arrows within the circle represent shortcuts in the system that can be, and often are, taken. Notice that the igneous rocks are shown as having been formed from a magma and as providing one link in a continuous chain. From these parent rocks, through a variety of processes, all other rocks can be derived.

First, weathering attacks the solid rock, which either has been formed by the cooling of a lava flow at the surface or is an igneous rock that was formed deep beneath the earth's surface and then exposed by erosion. The products of weathering are the materials that will eventually go into the creation of new rocks—sedimentary, metamorphic, and even igneous. Landslides, running water, wind, and glacier ice all help to move the materials from one place to another. In the ideal cycle, this material seeks the ocean floors, where layers of soft mud, sand, and gravel are consolidated into sedimentary rocks. If the cycle continues without interruption, these new rocks may in turn be deeply buried and subjected to pressures caused by overlying rocks, to heat, and to forces developed by earth movements. The sedimentary rocks may then change in response to

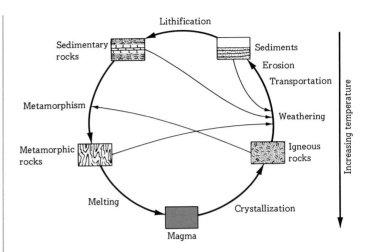

FIGURE 1.8

The rock cycle, shown diagrammatically. If uninterrupted, the cycle will continue completely around the outer margin of the diagram, from magma through igneous rocks, sediments, sedimentary rocks, and metamorphic rocks and back again to magma. The cycle may be interrupted, however, at various points along its course to follow the path of one of the arrows through the interior of the diagram.

these new conditions and become metamorphic rocks. If these metamorphic rocks undergo continued and increased pressure and heat, they may eventually lose their identity and melt into a magma. When this magma cools, we have an igneous rock again; we have come full circle.

But notice too that the complete **rock cycle** may be interrupted. An igneous rock, for example, may never be exposed at the surface and hence may never be converted to sediments by weathering. Instead, it may be subjected to pressure and heat and converted directly into a metamorphic rock without passing through the intermediate sedimentary stage. Other interruptions may take place if sediments, or sedimentary rock, or metamorphic rock are attacked by weathering before they continue to the next stage in the larger, complete cycle.

This concept of the rock cycle was probably first stated in the late eighteenth century by James Hutton, of whom we have already spoken:

We are thus led to see a circulation in the matter of this globe, and a system of beautiful economy in the works of nature. This earth, like the body of an animal, is wasted at the same time that it is repaired. It has a state of growth and augmentation; it has another state, which is that of diminution and decay. This world is thus destroyed in one part, but it is renewed in another; and the operations by which this world is thus constantly renewed are as evident to the scientific eye, as are those in which it is necessarily destroyed.[3]

We can consider the rock cycle to be a kind of outline of physical geology, as a comparison of Figure 1.8 with the Contents of this book will show.

1.3 PLATE TECTONICS, SEA-FLOOR SPREADING, AND CONTINENTAL DRIFT

To generations of geologists it has been clear that the earth is a dynamic, changing body: As we have pointed out, new rocks, sedimentary rocks, are made from the weathered debris of older rocks, and they can be crumpled, metamorphosed, and lifted into high mountain chains; or old rocks can be melted and the resulting magma cooled to form igneous rocks. Indeed, these changes can be traced in nature and pictured as we have in the rock cycle in Figure 1.8. But only recently have we been able to fit the rocks and their alterations into a worldwide, integrated system and to explain in a general way the origin of continents and ocean basins and of mountain ranges and continental plains and the location of volcanoes and earthquake belts. Two of the processes involved are referred to as **plate tectonics** and **sea-floor spreading.** Although we shall later extensively discuss these processes—particularly in Chapter 8—we take a preliminary look at them here: They include the movement of several large plates that, fitted together, form the rigid rind of the earth. This movement causes the growth as well as the closure of ocean basins and the creation of earthquakes, volcanoes, and mountain building along the plate boundaries. The movement also accounts for the shifting of continents (**continental drift**) over the last several hundred million years. The processes focus on the outer 200 km (kilometers) of the earth, a subject to which we shall also return, and a brief sketch of what we know about this zone is presented below.

LITHOSPHERE, ASTHENOSPHERE, CRUST, AND MANTLE

The outer 50 to 100 km of the earth is a rigid shell of rock, called the **lithosphere** (from the Greek *lithos,* "rock," and "sphere"). Yet as observations from deep mines tell us, the temperature of the earth increases by around 15°C (Celsius) with each kilometer of depth. So at a depth of about 70 km the temperature averages 1000°C, at which rock will slowly flow if pressure is applied. This "soft" zone from 70 to 100 km is called the **asthenosphere** (Greek *asthenēs,* "weak"). As we shall see, its existence helps explain some of the earth's major movements, both vertical and horizontal.

The lithosphere and asthenosphere are distinguished by temperature, but we can also divide the outer part of the earth into shells on the basis of composition. We speak, therefore, of the skin of the earth as the **crust** and of the bulk of the earth beneath the crust as the **mantle.** A wide variety of rock types constitutes the earth's crust, and we can make direct observations of most of the types. A discontinuous cover of sedimentary rocks, a few meters to a few kilometers thick, overlies igneous and metamorphic rocks. Beneath the continents the crust is thicker (averaging about 35 km) than that beneath the oceans (about 5 km). A dark-colored, relatively heavy igneous rock called **basalt** dominates the crust beneath the oceans, and we quite naturally call it **oceanic crust.** Its density, about 3 t/m^3 (tonnes per cubic meter), contrasts with the lighter-weight **continental crust,** with a density of about 2.6 t/m^3. A large portion of the continental crust is composed of the igneous rock called **granite,** which not only is less dense than basalt but also is light gray to pink in color. Beneath the rocks of the crust at a depth of 5 to 50 km lies the mantle, made of rocks with a density of about 3.3 t/m^3; the asthenosphere lies within the upper mantle (Figure 1.9).

VERTICAL MOVEMENTS AND ISOSTASY

Precise surveying shows that, when a sufficiently large lake forms behind a dam, it will depress the earth's

[3]James Hutton, *Theory of the Earth,* vol. 2, p. 562, Edinburgh, 1795. Hutton's theory of the earth was first presented as a series of lectures before the Royal Society of Edinburgh in 1785. These lectures were published in book form in 1795. Seven years later, Hutton's concepts were given new impetus through a more readable treatment, called *Illustrations of the Huttonian Theory,* by John Playfair.

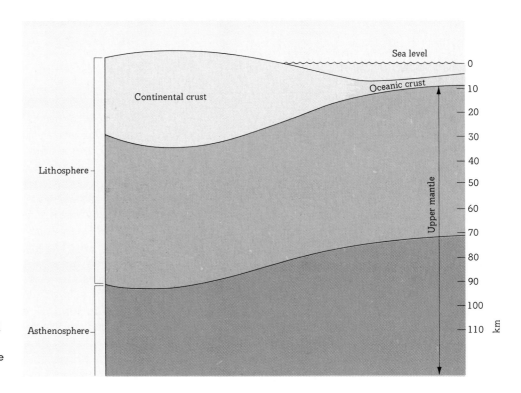

FIGURE 1.9

The relationships among continental crust, oceanic crust, upper mantle, lithosphere, and asthenosphere. (See also Figure 6.44.)

crust slightly. We also know that, when glaciers expanded during the Ice Age, their weight depressed large areas of the crust. Conversely, when the glaciers waned and disappeared, the land rose, recovering its earlier elevation; and even today some areas of Scandinavia and Canada are still rising in response to the melting of the last glaciers. The ancient geologic record provides other examples: Studies of thick sequences of sedimentary rocks show numerous situations in which thousands of meters of sediments accumulated in shallow marine environments. There is no way to explain a continuous pile of shallow-water sediments unless the basin in which they had accumulated was shallow from the beginning and kept sinking slowly as additional sediments were added.

In all these examples of **vertical movement** portions of the earth's crust behave as if they were floating in a soft, slowly flowing zone. And that is what we believe happens. The crust and the uppermost mantle that lies above the asthenosphere have some strength. But they cannot resist the pull of gravity, and they respond to the addition or removal of a load. So if material such as water, ice, or sediments are added at the surface, that overloaded area (and the column of rigid rock beneath) sinks slightly into the asthenosphere; conversely, if a load is removed, an area of the crust floats upward. This floating balance of the crust is called **isostasy** (meaning "equal standing"; Figure 1.10). It explains why the thick, lightweight continental crust stands high in relation to the ocean basins underlain by thinner, heavier oceanic crust.

HORIZONTAL MOVEMENTS

The geologic record also reveals that large portions of the earth's crust have moved horizontally. And if we accept the idea of continental drift, then we can say that the entire surface of the earth is subject to lateral movements, measured in thousands of kilometers. This notion brings us back to our preliminary look at plate tectonics and sea-floor spreading.

Plates of the earth If we plot earthquakes on a world map, we find them concentrated in narrow, well-defined belts (as in Figure 1.11). These belts are also marked by extensive volcanic activity. Between the belts are large areas, both on the continents and in the ocean basins, where earthquakes and volcanoes, although not entirely absent, are infrequent. We now know that the zones of intense activity mark the bound-

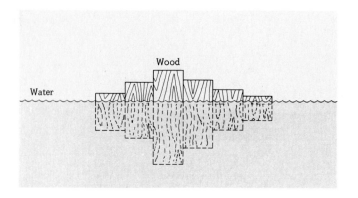

FIGURE 1.10

Isostasy is a condition of balance, of floating equilibrium, illustrated here by blocks of wood of varying height floating in water. Each block sinks to a different depth in the water, but the amount above the surface is proportionally the same for each—in this example about one-third. For each block the amount above the water is balanced by the part below the water. Large blocks of the earth's lithosphere behave similarly, floating in the asthenosphere.

aries between rigid **plates** of the earth's crust—plates forming a jigsaw puzzle whose pieces constantly jostle one another. It is this movement that is associated with volcanoes and earthquakes.

We are now able to demonstrate (as discussed in Section 8.3) that plates move at rates between 1 and 9 cm/year (centimeters per year). The relative motion along plate margins can be any of three types: Two plates may slide by, and parallel with, each other; two plates may diverge from each other; or two plates may converge. In fact each of these movements is found along the margin of any given plate.

What goes on along the boundary of diverging or converging plates?

If two plates are pulling apart, it must mean that something is being added to their separating margins. For example, the broad rise that runs down the center of the Atlantic Ocean—the Midatlantic Ridge—marks the axis of separation between the Old and New Worlds, with the North and South American plates moving generally westward as the Eurasian and African plates drift eastward. The crest of the Midatlantic Ridge is marked by a precipitously walled *rift valley* along its entire length. Along the valley new volcanic material (basalt) wells upward from the mantle. The

newly formed oceanic crust is then slowly transported laterally away from the ridge (sea-floor spreading), allowing room for still younger material to be added along the axial valley.

If the plates grow along a diverging boundary, they must be destroyed elsewhere to allow space for the new material. This happens along the converging boundaries of plates. Look, for example, at the western boundary of the South American plate, where that plate, which includes both the southwestern Atlantic Ocean basin and the South American continent, collides along most of its western margin with a large Pacific Ocean plate, the Nazca plate. The convergence of the two plates is marked by the Andes Mountains, intense earthquakes, and volcanic activity; in addition a deep, narrow trench characterizes the ocean floor just west of the South American shore (see Figure 1.12).

The leading edge of the South American plate is riding over the leading edge of the Nazca plate. The horizontal movements of the crust are piling up an excess of rock material where the two plates collide, and isostasy operates to ensure that this thickened column of the earth's crust not only sinks lower into the asthenosphere but stands higher above sea level.

Now look at the trench just off South America. The ocean floor here is depressed much below the average deep ocean floor. The leading edge of the Nazca plate is being shoved downward and eastward beneath the South American plate, pulling part of the oceanic crust and floor with it. Along the dipping zone of contact between the two plates, earthquakes occur as the plates slip by each other. The bulk of the Nazca plate continues downward under the leading edge of the South American plate until temperature and pressure are so great that, along its front edge, it begins to lose its rigidity and is reincorporated into the mantle in the asthenosphere. During this process, igneous activity goes on. The thermal energy needed to generate molten material comes from the friction generated between the plates, from the heat of radioactivity in the thickened rock pile, and from the heat generated by the weight of the thickened crust.

A hypothesis first seriously introduced in 1912 by Alfred Wegener, a German meteorologist, continental drift received little acceptance until the mid-1960s, when the reality of plate tectonics and sea-floor spreading began to emerge. Wegener had pictured the continents as moving about in individual units; but after the concept of continental drift had become scientific-

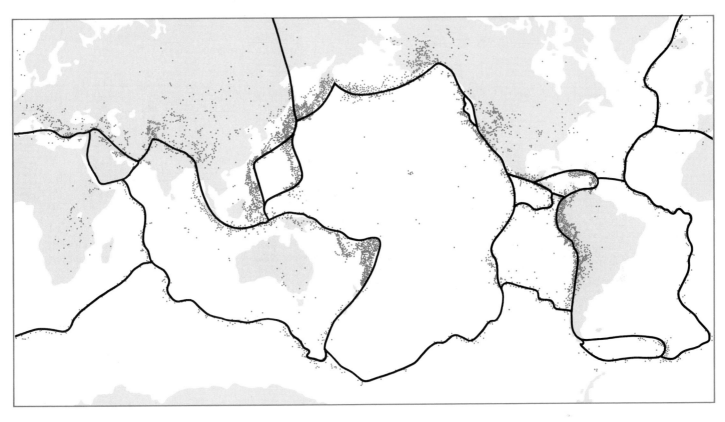

FIGURE 1.11

Earthquakes are concentrated in narrow bands around the earth. The boundaries of
the earth's major plates follow these bands. (See also Figure 8.19.)

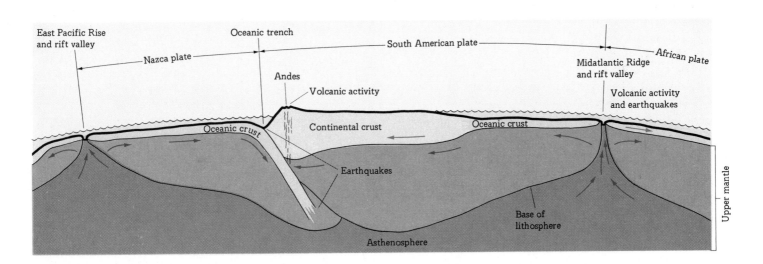

FIGURE 1.12

The relationships among the African, South American, and Nazca plates. See text
for discussion.

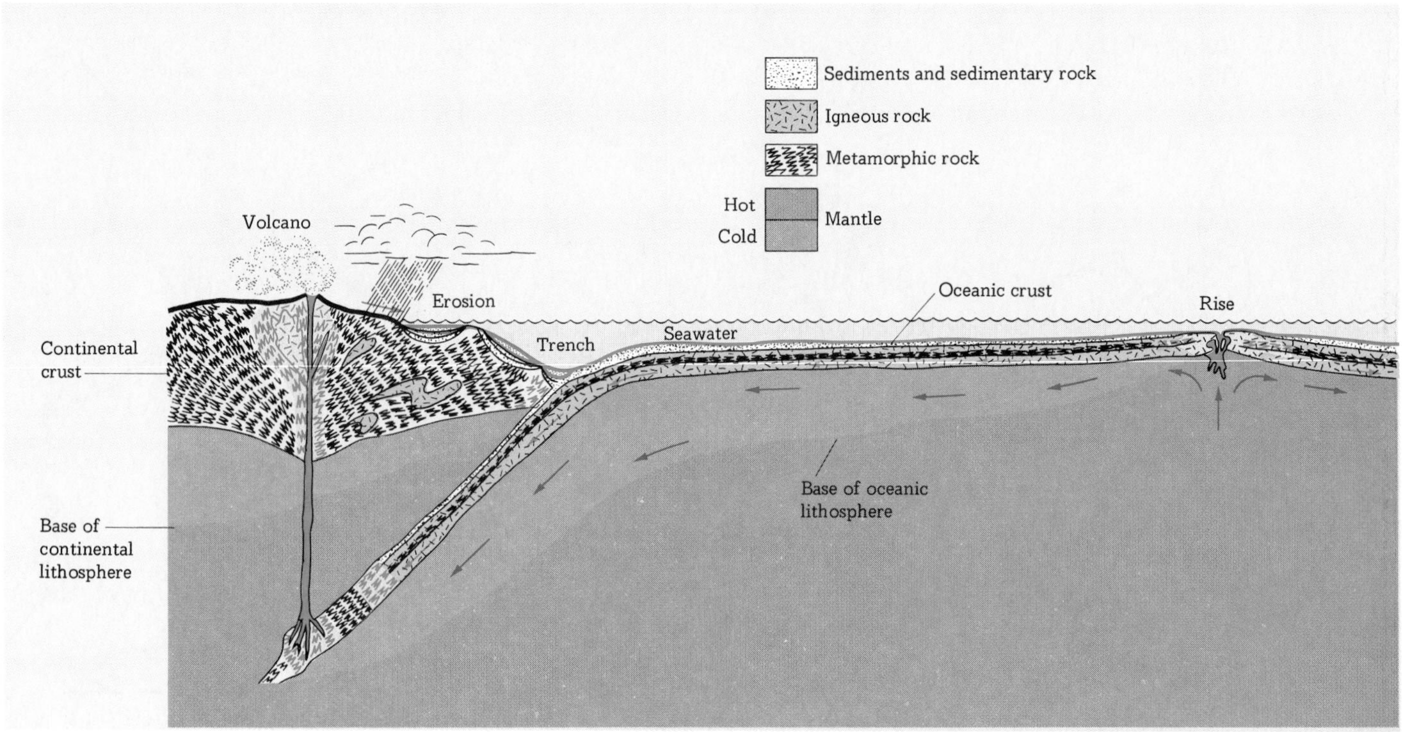

FIGURE 1.13

A cross section through a growing oceanic plate and its disappearance beneath the edge of a continent shows that the igneous, sedimentary, and metamorphic components of the rock cycle are present in both continental and oceanic areas. At locations where the rock cycle has converted one rock type into another, the newly formed rock is shown in color. [Modified from Sheldon Judson, Kenneth S. Deffeyes, and Robert B. Hargraves, *Physical Geology*, p. 28, Prentice-Hall, Inc., Englewood Cliffs, N.J., 1976.]

ally respectable, it was demonstrated that the moving units included not only continents but portions of ocean basins: It is a plate that moves, and any earth feature—whether continent or ocean—that it includes moves with it.

THE ROCK CYCLE AGAIN

In later sections we shall return to the subject of plate tectonics and sea-floor spreading, expanding the bare outline presented above and laying out the data that have led us to this model of crustal movement. For the moment, however, accept the reality of plate tectonics,

and apply it to the concept of the rock cycle that we introduced in Section 1.2 and Figure 1.8.

In describing the rock cycle, we pictured it without reference to world geography and without providing a general mechanism for the cycle. The plate-tectonic model furnishes some of this detail. We can visualize sediments moved by water, wind, and ice from the continents to the oceans, only to be brought back to the continents by the slowly moving ocean floors. Again, plate movements, particularly along plate boundaries, create temperatures and pressures that lead to metamorphism and igneous activity; and especially along converging boundaries, continents grow in thickness and elevation, allowing erosion to eat deeper and deeper into the thickened wedge of continental rocks and create more sediments for transfer back into the oceans (see Figure 1.13).

The chapters that follow deal with specific parts or processes that are involved in the rock cycle. In Chapter 2 we begin our more detailed examination of the earth with a discussion of minerals, the building blocks from which the earth's lithosphere is made.

OUTLINE

Geology is the science of the earth and deals with the materials that compose it, the history of life, and the physical events of earth history.

Geologic time encompasses the age of the earth, about 4.5 billion years.

Absolute time and relative time are used to divide geologic time. Relative time—whether an event in earth history came before or after another event—disregards years. Absolute time measures the age of a geologic event in years.

Radioactivity provides most of the absolute dates in geology and depends on the breakdown, at a constant rate, of unstable elements to stable ones.

Divisions of geologic time are listed in a **geologic-time table**, and the rock units representing these time divisions are shown in a **geologic column**, as reproduced on the front endpaper of this book.

Uniformitarianism is the principle that holds that the earth processes now going on also operated in the past.

Earth materials are composed of minerals, which in turn make up rocks.

The three rock families are **igneous**, **sedimentary**, and **metamorphic**.

The rock cycle traces the changes as rocks are transformed from one type into another.

Plate tectonics, **sea-floor spreading**, and **continental drift** describe the motion of large pieces of the outer 50 to 100 km of the earth.

Lithosphere, asthenosphere, crust, and **mantle** are terms applied to certain portions of the outer 200 km of the earth.

Vertical movements of the crust often reflect the floating balance reached by portions of the earth's lithosphere as it sinks or rises in the asthenosphere, a process called **isostasy**.

Horizontal movements are often the result of plate tectonics and sea-floor spreading.

Plates of the earth are defined by narrow bands of earthquakes and volcanic activity. Plates move relative to each other in converging, diverging, and parallel directions. Plates grow at diverging boundaries and are destroyed at converging boundaries.

The rock cycle can be related to the events accompanying plate tectonics and sea-floor spreading.

SUPPLEMENTARY READINGS

Eicher, Donald L. *Geologic Time*, Prentice-Hall, Inc., Englewood Cliffs, N.J., 1968. *A short book that describes the various ways by which we measure geologic time.*

Marvin, Ursula B. *Continental Drift*, Smithsonian Institution Press, Washington, 1973. *An easy-to-read but authoritative account of the history of the idea of continental drift.*

Sullivan, Walter *Continents in Motion*, McGraw-Hill Book Company, New York, 1974. *The story of our ideas about a changing earth, which leads us to our present concepts of plate tectonics, sea-floor spreading, and continental drift.*

2

MINERALS

The surface of the earth, the part on which we live, is composed of rocks and minerals. Rocks are made from minerals; minerals, from elements; and elements, from other, smaller particles. Minerals and rocks are basic to our concern with geology, but a brief look at the origin of the elements and particles that form minerals will help our understanding.

The word **mineral** has many different meanings in everyday usage; but in our discussion it will be used to refer to a **naturally occurring element or compound of elements that has been formed by inorganic processes.** Minerals are everywhere about us. Almost any even small plot of ground will offer numerous samples. They can occur in several forms, such as rocky ledges, the soil of plowed fields, or the sands of a river bank. Even some of the most common types are valuable enough to make mining them commercially worthwhile; and the rarer minerals of gold and silver have provided the basis of wealth and power since the dawn of civilization.

The physical universe is composed of what we call **matter;** yet one of the most elusive problems in science is to define matter precisely. It has long been customary to refer to the states of matter as solid, liquid, or gas, and we say that matter has physical properties, such as color or hardness, as well as chemical properties, which govern its ability to change or to react with other bits of matter. But all this simply tells us *about* matter, not what matter is.

Many centuries ago, Greek philosophers speculated on this problem, arguing over whether the smallest pieces into which anything could be divided were just miniatures of the original—microscopic drops of water, extremely small grains of sand, infinitesimal pieces of salt—or whether at some point down the line certain particles might be found that were joined together differently to form water, sand, salt, and all the other substances that make up what we think of as the material world.

Today, we know that the second explanation is in a general way the true one. These particles are called **atoms.** So if we are to deal with matter, we must first learn about the atoms and their combinations. This information will then give us a basic understanding of what rocks actually are, why they differ, and how they can be changed.

2.1 PARTICLES AND ELEMENTS

ELECTRIC CHARGE

All matter appears to be essentially electrical in nature. Some of the earliest ideas about what we now call **electricity** sprang from a very simple experiment—the rubbing together of a piece of amber and a piece of fur. After the rubbing both the fur and the amber were found to be capable of picking up light pieces of other materials, such as feathers or wool. The interesting thing was that materials *attracted* by the amber were *repelled* by the fur. So scientists decided that there must be two kinds of electricity: One we now call **positive,** and the other **negative.** This fact was first discovered by the Greeks about 600 B.C. Then, late in the sixteenth century, William Gilbert, personal physician to Elizabeth I, proposed that the power responsible for this phenomenon be called electricity, from the Greek word ēlektron, meaning "amber." And in technical terms we say that **like electric charges repel each other and unlike charges attract each other** (see Figure 2.1).

ATOMS

Just what happens when amber is rubbed with fur? Particles pass from the fur to the amber, and the amber becomes charged—we say "negatively." So we reason that the particles that bring about this condition must also be negatively charged. These negatively charged particles are called **electrons,** and they are fundamental constituents of atoms, of which all matter is composed.

Atoms cannot be assembled from electrons alone, however; for all electrons are negative and would not stick together by themselves. So scientists reasoned that there had to be some positively charged particles. At last they were found and were called **protons.** In addition to having a positive charge, a proton differs from an electron in another respect: It acts like a much heavier unit of matter. Because any quantity of matter is arbitrarily described by a number called its **mass,** the proton is said to have greater mass than the electron. In fact, *we define the fundamental particles of atoms in terms of mass and electric charge* (see Table 2.1).

Scientists continued to chip away at atoms until finally they turned up a third particle, with a mass about equal to that of the proton but with no electric

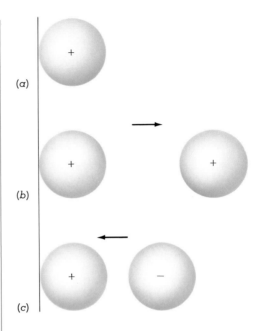

(a)

(b)

(c)

FIGURE 2.1

Like charges of electricity repel each other; unlike charges attract each other. The positively charged sphere in (a) repels another positive charge (b) but attracts a negative charge (c).

TABLE 2.1

Fundamental particles

	Electric charge	Mass, u[a]
Electron	−1	0.00055
Proton	+1	1.00760
Neutron	0	1.00890

[a] Atomic mass unit.

FIGURE 2.2

Electron shells around a nucleus. In true scale the diameter of the shells would be 20,000 to 200,000 times the diameter of the nucleus. (If the sun were the nucleus, the electron shells would embrace more space than does the entire solar system.) Yet the nucleus contains 99.95 percent of the mass of the entire atom.

charge. This electrically neutral particle they called the **neutron** (Table 2.1).

Protons, neutrons, and electrons are the fundamental particles that combine to form atoms.

Nobody has yet come forward with a completely adequate explanation of exactly what these fundamental particles are, but we do know that **energy** is intimately involved in their makeup and in their combinations.

Atomic structure All the information that we have about the structure of atoms has been established by indirect observation. Using doubly charged positive particles called **alpha particles** (two protons and two neutrons bound together) to bombard atoms, physicists have made some interesting discoveries: For example, if these particles are shot at a target, such as a piece of metal, made up of billions of atoms, no more than 1 alpha particle in 10,000 hits anything inside the target.

It has been concluded, therefore, that the target must be largely open space.

Repeated tests have shown that an atom contains a nucleus of protons and neutrons surrounded by a cloud of electrons. They have also revealed that a normal atom has as many electrons as it has protons. The number of protons plus the number of neutrons in an atom constitute its **mass number** (the neutrons contribute to the mass of the atom, but they do not affect its electrical charge). The electrons, being negatively charged, spin very rapidly around the nucleus (like planets about the sun); otherwise they would be pulled in by the attraction of the positive protons. In fact, so small is the atom, they complete several quadrillion round trips per second, at a speed of hundreds of kilometers per second. Again like the planets revolving around the sun, electrons revolve around the nucleus at different distances.

Atomic size The electrons form a protective shield around the nucleus and define the size of the atom (see Figure 2.2). In describing atomic dimensions, we use a special unit of length, the *angstrom* (Å or sometimes A), which is a hundred-millionth of a centimeter, that is, 0.00000001 cm, usually written 1×10^{-8} cm. Atomic nuclei have diameters that range from a ten-thousandth to a hundred-thousandth of an angstrom—that is, from 1×10^{-4} to 1×10^{-5} Å. Atoms of the most common elements have diameters of about 2 Å, which is roughly 20,000 to 200,000 times the diameter of the nucleus. Again we see that matter consists mostly of open space. If the sun were truly the nucleus of our atomic model, the diameter of the atom would be greater than the diameter of the entire solar system.

Atomic mass Although the nucleus occupies only about a trillionth of the volume of an atom, it contains 99.95 percent of the atom's mass. In fact, if it were possible to pack a cubic centimeter with nothing but protons, this small cube would weigh more than 100 million t at the earth's surface.

Ions An ion is an electrically unbalanced form of an atom or group of atoms. An atom is electrically neutral. But if it loses an electron from its outermost shell, the portion that remains behind has an extra unmatched positive charge. This unit is known as a positively charged ion. If the outermost shell gains an electron, the ion has an extra negative charge and is known as a negatively charged ion. As we shall see later, more than one electron may be lost or gained, leading to the formation of ions with two or more units of electrical charge.

ELEMENTS

An atom is the smallest unit of an element. **Each element is a special combination of protons, neutrons, and electrons.** The distinguishing feature of each element is the number of protons in its nucleus. An element has an **atomic number** corresponding to this number, and it is given a name and a symbol.

Element 1 is a combination of one proton and one electron (see Figure 2.3). Long before its atomic structure was known, this element was named hydrogen, or "water former" (from the Greek roots *hydōr*, "water," and *-genēs*, "born") because water is formed when hydrogen burns in air. The symbol of hydrogen is H. Because it has a nucleus with only one proton, hydrogen assumes place 1 in the table of elements.

Element 2 consists of two protons (plus two neutrons in the most common form) and two electrons (see Figure 2.4). It was named helium, with the symbol He, from the Greek *hēlios*, "the sun," because it was identified in the solar spectrum before being isolated on the earth. Because of the two protons in its nucleus, helium takes place 2 in the table of elements.

Each addition of a proton, with a matching electron to maintain electrical balance, produces another element. Neutrons seem to be included more or less indiscriminately, although there are about as many neutrons as protons in the common form of many of the elements.

There are three classes of elements: metals, nonmetals, and metalloids, of which 77 are classed as metals, 17 as nonmetals, and 9 as metalloids. These are keyed in the list of elements in Appendix B. A metal typically shows a peculiar luster, called *metallic luster*, is a good conductor of electricity or heat, is opaque, and may be fused, drawn into wire, or hammered into sheets. A nonmetal lacks some or all of these properties. A metalloid has some metallic and some nonmetallic properties.

The elements appearing in nature begin with the lightest, hydrogen (place 1), and end with the heaviest, the metal uranium (place 92). Technetium (43), promethium (61), astatine (85), and francium (87) are not readily found in nature; they have been produced only for an instant during radioactive decay.

Isotopes   Every element has forms that, although essentially identical in chemical and physical properties, have different masses. Such forms are called **isotopes**— from the Greek *isos*, "equal" or "the same," and *topos*, "place"—because each form has the same num-

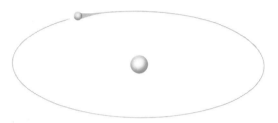

Hydrogen

FIGURE 2.3

Diagram of a hydrogen atom, which consists of one proton and one electron. This is the simplest atom.

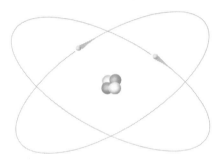

Helium

FIGURE 2.4

Diagrammatic representation of an atom of helium. The nucleus consists of two protons and two neutrons, and accordingly has a mass number of 4. There are two electrons (negative charges) to balance the positive charges on the two protons.

Since there are two protons in the nucleus, this atom is number 2 in the table of elements. The nucleus of helium (two protons + two neutrons) without any accompanying electrons is an alpha particle.

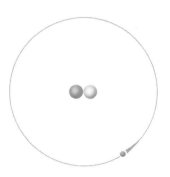

FIGURE 2.5

Schematic of deuterium, an isotope of hydrogen formed by the addition of a neutron to the nucleus. It has a mass number of 2.

ber of protons in the nucleus and occupies the same position in the table of elements. Isotopes show differences in mass as a result of differences in the number of neutrons in their nuclei. For example, hydrogen with one proton and no neutrons in its nucleus has a mass number of 1. When a neutron is present, however, the atom is an isotope of hydrogen with a mass number of 2, called deuterium (Figure 2.5). All elements have isotopes.

Information about atomic structure is commonly expressed symbolically by a subscript number prefixed to the element's symbol to indicate the atomic number and by a superscript for the mass number. Thus hydrogen would be represented by $_1^1H$, and its isotope deuterium by $_1^2H$.

Many of the elements as they occur in nature are mixtures of isotopes. In general, regardless of the source of the element, the proportion of isotopes com-

posing it is constant unless special steps have been taken to separate them. Thus chlorine, whether occurring naturally or prepared in the laboratory, has an atomic weight of 35.457, which represents the average weight of a mixture of about three parts of $_{17}^{35}Cl$ to one part of $_{17}^{37}Cl$.

ORGANIZATION OF MATTER SUMMARIZED

Matter is the substance of the physical universe. It is composed of atoms. Atoms are combinations of the fundamental particles protons, neutrons, and electrons (except for the common isotope of hydrogen, which is a combination of one proton and one electron). An atom is the smallest unit of an element. In nature 88 elements have been found; 4 others have been identified as short-lived transients; and 11 have been artificially prepared.

Atoms are rarely found alone in nature. They are found in combination with each other and with other atoms. Atoms of different elements combine by exchanging or sharing electrons to form compounds. The smallest unit of a substance is a molecule. The 88 naturally occurring elements have been found combined in nature, forming numerous compounds. They have also been combined in the laboratory to form the millions of compounds we use today.

Water is the only substance to occur naturally in the three states of solid, liquid, and gas.

2.2 DISTRIBUTION OF ELEMENTS IN NATURE

We have said that all the elements consist of just two essential nuclear particles, or building blocks: protons and neutrons. The nucleus of the simplest element—hydrogen—is a single proton. One of the earliest notions concerning elements was that they consist of varying combinations of hydrogen atoms. Though considerably more complicated, most modern theories on the origin of elements are based on this essential idea.

BIG-BANG THEORY

One of the more popular theories about how elements began was first proposed by George Gamow, then at George Washington University and later at the Univer-

sity of Colorado. He and his fellow workers developed the **big-bang theory**. Astronomical observations that the entire universe is expanding led to the conclusion that about 10 to 20 billion years ago the cosmos began in a big bang, or explosion, from a very dense "core" made up mainly of neutrons (electrons would be pressed into protons under the tremendous pressure). As this large ball expanded, some of the neutrons decayed to protons and electrons. Each proton captured a neutron, and this mass captured another neutron, thus building up the masses. By an extremely rapid series of events all the elements were formed in this first stage of the universe's history; indeed, it is suggested that the whole process of element formation occurred within a few minutes after the initial explosion. The material thus created continued to expand and eventually clumped into galaxies, stars, and planets.

Although there are a number of problems with Gamow's concept, many contemporary theories are based on this same model modified in various technical ways.

STEADY-STATE THEORY

A second group of theories on the origin of the elements relies on a more complex series of events. Fred Hoyle of Cambridge University, among others, has concluded that the elements were built in the hot interior of stars rather than in a primordial explosion. Nuclear reactions and transformations are known to be occurring constantly in the stars. The long-lasting energy of the sun and other stars is explained by two chains of nuclear reactions, which together produce tremendous amounts of energy and build new nuclei: **proton-proton fusion** and the **carbon cycle.**

In outline the steady-state theories propose that an initial potential universe consisted of cold but turbulent hydrogen atoms in a dilute gaseous state. Part of the gas condensed into stars because of gravitational attraction. Under this gravitational force the stars contracted and their interiors grew very hot and dense. At temperatures in the millions of degrees the protons moved so fast as to fuse and form deuterons (hydrogen 2), which combined with other protons to form helium 3. Pairs of helium 3 nuclei fused to produce helium 4, releasing two surplus protons for each fusion. The overall result of these proton-proton chains was the formation of single atoms of helium from sets of four atoms of hydrogen.

Two helium nuclei may combine temporarily at very high temperatures to form a nucleus of mass 8, which is very unstable (none is found in nature: one of the stumbling blocks in the Gamow theory of continuous buildup of all the elements). However, beryllium 8 has been produced temporarily in the laboratory and will form in the hot, dense interior of a star. Thus a beryllium 8 nucleus during its very brief existence may fuse with a helium 4 nucleus, resulting in a nucleus of carbon 12.

Once carbon 12 nuclei have been formed in the hydrogen-helium core of a star, they may combine with more and more helium nuclei to complete the carbon cycle, with the formation of oxygen 16, neon 20, and perhaps magnesium 24 and others. Intervening elements, including nitrogen 14, apparently form by secondary processes rather than by main-line buildup of the elements.

ELEMENT ABUNDANCE

The relative abundance of elements found on the earth contrasts markedly with the abundance of elements detected in the universe.

About 90 percent by weight of the earth consists of four elements: iron (35 to 40 percent), oxygen (25 to 30 percent), silicon (13 to 15 percent), and magnesium (about 10 percent). Four other elements are present amounting to about 1 percent: nickel, calcium, aluminum, and sulfur. Another seven elements amount to from 0.1 to 1 percent: sodium, potassium, chromium, cobalt, phosphorus, manganese, and titanium. The earth is made up almost entirely of these fifteen elements, with all the remaining elements contributing to less than 0.1 percent of its weight.

Now let us look at the abundances in the universe. Hydrogen and helium respectively make up around 76 and 23 percent of the weight of the universe. Thus *all* the remaining elements amount to only a little more than 1 percent of the weight of the universe. The sun, like the universe as a whole, is mostly hydrogen and helium although it has slightly lower relative amounts of the remaining common elements.

Apparently, in the formation of the earth and similar planets of the solar system, the lighter gases—hydrogen and helium especially—were lost very quickly. This process left the earth with a residue of unusual relative abundances of elements compared with the abundances of elements in the universe as a whole.

2.3 A FIRST LOOK AT MINERALS

At the beginning of this chapter we defined a mineral as a natural element or compound formed by inorganic processes. Now we shall see specifically what this definition means.

MINERAL COMPOSITION

More than 2,000 minerals are known. Some are composed of only one element, such as diamond (C), copper (Cu), silver (Ag), gold (Au), mercury (Hg), and sulfur (S). Others are made up of relatively simple compounds of elements. For example, the mineral halite is composed of two elements, sodium (Na) and chlorine (Cl), in equal amounts. The chemical symbol for halite, NaCl, indicates that every sodium ion present is matched by one chlorine ion. The mineral pyrite, sometimes known as "fool's gold," is also composed of two elements, iron (Fe) and sulfur, but in this mineral there are two atoms of sulfur for each atom of iron, a relationship expressed by the chemical symbol FeS_2. Other minerals have a more complex composition.

A mineral's composition can vary slightly: An occasional replacement by other elements whose atoms are of similar size does not create a new mineral. We can therefore say that **every mineral is composed of elements in definite or slightly varying proportions.**

CHARACTERISTICS OF MINERALS

All the properties of minerals are determined by the composition and internal atomic arrangement of their elements. We can identify minerals on the basis of their chemical properties, but physical properties are the ones most often used. Such physical properties include crystal form, hardness, specific gravity, cleavage, color, streak, and striations.

Crystal form When a mineral grows without interference or obstacle it will be bounded by plane surfaces symmetrically arranged and will acquire a characteristic **crystal form,** which is the external expression of its internal crystalline structure. The faces of crystals are defined by surface layers of atoms. These faces lie at angles to one another that have definite characteristic values, the same value for all specimens of the same mineral. The size of the faces may vary from specimen to specimen, but the angles between the faces remain constant.

Every crystal consists of atoms arranged in a three-dimensional pattern that repeats itself regularly. A combination of atoms whose repetition can produce a given mineral is called its **unit cell.** In a crystal of copper all the atoms are alike and are arranged in a cube. This arrangement is limited by the size of the copper atoms. By repetition the entire crystal is built up (see Figure 2.6). The principal surface layers correspond to the faces of a cube; the smaller surface layer obtained by cutting off a corner of a cube is called an *octahedral face.* Native copper found in deposits of copper ore is often found in the form of crystals with cubic and octahedral faces. Remember that the atoms in the figure are greatly enlarged: If the crystal had sides only 0.1 mm (millimeter) long, there would be about 400,000 atoms in a row along each edge.

In crystals of halite there are ions of two different kinds arranged in the regular pattern shown in Figure 2.7. The smaller ones are those of sodium and the larger ones are those of chlorine. The atomic arrangement is limited by the size of the ions. The unit cell of halite is a cube made up of six chlorine ions surrounding one sodium ion, with all equidistant. The mineral is held together by ionic bonds. Repetition of the unit cell in three dimensions forms the cubic halite crystal.

The mineral quartz occurs in many rocks as irregular grains because its growth has been constricted. Even in these irregular grains, however, the atoms are arranged according to their typical crystalline structure. And where conditions have permitted the mineral to develop freely, it forms crystals that are always six-sided prisms. Whether an individual crystal of quartz is only 1 mm long or 25 cm, the faces of the prism always meet at the same angle (see Figure 2.8). This is an example of the constancy of interfacial angles. Although faces of a crystal may be of different sizes and shapes because of malformation, their similarity is frequently evidenced by natural striations, etchings, or growths. On some crystals the similarity of faces of a form can be seen only after etching with acid.

There are six different systems of crystals, classified according to the angles between their faces and described in Appendix A.

Some elements or compounds may develop several different crystal forms, producing different minerals.

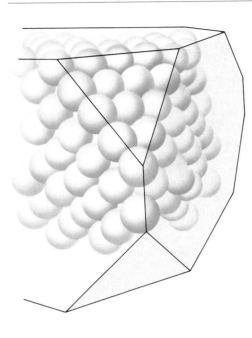

FIGURE 2.6

An atomic view of a copper crystal, showing small octahedral faces and large cubic faces. [After Linus Pauling, *General Chemistry*, p. 23, W. H. Freeman and Company, San Francisco, copyright © 1953.]

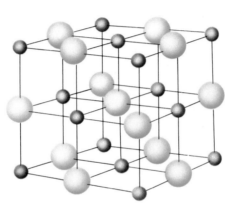

FIGURE 2.7

Arrangement of sodium ions with positive electrical charge, Na^+, and chlorine ions with negative electrical charge, Cl^-, to form the ionic compound NaCl, common salt. Na^+ has a radius of 0.97 Å, and Cl^- a radius of 1.81 Å.

FIGURE 2.8

Quartz crystals. Regardless of the shape or size of crystals, the angles between true crystal faces remain the same. Transverse striations on prism faces are most clearly seen on the two large crystals, which also carry blotches of foreign matter; but striations are present on the faces of the other crystals, too. The large, stubby crystal, from Dauphiné, France, is about 35 cm tall; the other crystals are from Brazil. [Harvard Mineralogical Collection, Walter R. Fleischer.]

FIGURE 2.9

A large, perfect crystal of diamond (uncut), which is the external expression of the orderly internal arrangement of atoms of carbon. Weight, 84 carats; from Kimberley, South Africa. (Shortly after this picture was taken, the specimen was stolen from an alarm-guarded safe of the Harvard Mineralogical Collection and has probably been cut up.) [Harry Groom.]

This assumption of two or more crystal forms by the same substance is called **polymorphism.** Each results from the conditions under which it was formed. For example, carbon's two common forms are graphite and diamond: The crystal form of diamond is an eight-sided solid, an octahedron (Figure 2.9); the crystal form of graphite is a flat crystal with six sides. Although both minerals are composed of carbon, the difference in their crystal forms comes from the arrangement of their carbon atoms—one pattern in diamond, another in graphite (see also Figure 2.11). The crystal form of pyrite is a cube (Figure 2.10), and that of marcasite is a flattened tubular shape. Both are composed of FeS_2.

Here again the reason for the difference in crystal form lies in the internal arrangement of the atoms.

Every mineral has a characteristic crystal form, which is the external shape produced by its crystalline structure.

Cleavage *Cleavage* is the tendency of a mineral to break in certain preferred directions along smooth plane surfaces. Cleavage planes are governed by the internal arrangement of the atoms; they represent the directions in which the atomic bonds are relatively weak. Because cleavage is the breaking of a crystal between atomic planes, it is a directional property, and any plane throughout a crystal parallel to a cleavage surface is a potential cleavage plane. Moreover, it is always parallel to crystal faces or possible crystal faces because faces and cleavage both reflect the same crystalline structure.

Cleavage is a *direction* of weakness, and a mineral sample tends to break along planes parallel to this direction. This weakness may be due to a weaker type of atomic bond, as we have indicated, to greater atomic spacing, or frequently to a combination of the two. Graphite has a platy cleavage because of relatively weak bonds between the carbon layers. Diamond has but one bond joining all its carbon atoms; its cleavage takes place along planes having the largest atomic spacing.

Cleavage may be perfect, as in the micas, more or less obscure, as in beryl and apatite, or entirely lacking, as in quartz.[1]

Striations A few common minerals have parallel, threadlike lines or narrow bands, called *striations*, running across their crystal faces or cleavage surfaces. These lines can be clearly seen on crystal faces of quartz and pyrite, for example (Figure 2.10).

Once again, this property is a reflection of the internal arrangement of the atoms and the conditions of growth of the crystals.

Hardness *Hardness* is another physical property governed by the internal atomic arrangement of the mineral elements. (The stronger the binding forces between the atoms, the harder the mineral.) The degree of hardness is determined by the relative ease or difficulty with which one mineral is scratched by another or by a fingernail or knife. It might be called the mineral's "scratchability." Hardness (H) ranges from 1 through 10:

---

[1]See Appendix A and later parts of this chapter for certain specific mineral names and characteristics.

Range	Scratchability
$H < 2.5$	Will leave mark on paper; can be scratched by fingernail
$2.5 < H < 3$	Cannot be scratched by fingernail; can be scratched by penny
$3 < H < 5.5$	Cannot be scratched by penny; can be scratched by knife
$5.5 < H < 7$	Cannot be scratched by knife; can be scratched by quartz
$7 < H$	Cannot be scratched by quartz

To illustrate, if you pick up a piece of granite and try to scratch one of its light-colored grains with a steel knife blade, this mineral simply refuses to be scratched. But if you drag one of these grains across a piece of glass, the glass is easily scratched. Clearly, then, these particular mineral grains in granite are harder than either steel or glass. But if you have a piece of topaz handy, you can reveal their vulnerability; for although they are harder than steel or glass, they are not so hard as topaz.

Minerals differ widely in hardness (see Appendix A): Some are so soft that they can be scratched with a fingernail; some are so hard that a steel knife is required to scratch them; diamond, the hardest mineral known, cannot be scratched by any other substance.

Specific gravity Every mineral has an average weight per unit volume. This characteristic weight is usually described in comparison with the weight of the same volume of water. The number that represents this comparison is called the **specific gravity** of the mineral and is given in *grams per cubic centimeter* or *tonnes per cubic meter*.

The specific gravity of a mineral increases roughly with the mass of its constituent elements and with the closeness with which these elements are packed together in their crystalline structure. Most rock-forming minerals have a specific gravity of around 2.7 although the average specific gravity of metallic minerals is 5. Pure gold has the highest specific gravity, 19.3.

It is not difficult to acquire a sense of relative weight by which to compare specific gravities. We can learn to distinguish between two bags of equal size, one filled with feathers and one filled with lead, and experience in hefting stones has given most of us a sense of the "normal" weight of rocks.

Color Although *color* is not a reliable property in identifying most minerals, it is strikingly characteristic for a few. These include the intense azure blue of azur-

ite, the bright green of malachite, and the pale yellow of sulfur. Magnetite is iron black, and galena is lead gray. The color of other minerals, such as quartz, can be quite variable because of slight impurities.

Minerals containing iron are usually "dark-colored." In geologic usage *dark* includes dark gray, dark green, and black. Minerals that contain aluminum as a predominant element are usually "light-colored," a term that includes purples, deep red, and some browns.

Streak The *streak* of a mineral is the color it displays in finely powdered form. The streak may be different from the color of the hand specimen. Although the color of a mineral may vary between wide limits, the streak is usually constant.

One of the simplest ways of determining the streak of a mineral is to rub a specimen across a piece of unglazed porcelain known as a *streak plate*. The color

FIGURE 2.10

Cubic crystals of pyrite. Striations are clear on the large specimen. Note that those in adjacent faces are perpendicular to each other. The small specimen, about 5 cm wide, consists of three intergrown cubes. [Harvard Mineralogical Collection, Walter R. Fleischer.]

of the powder left behind on the streak plate helps to identify some minerals. Because the streak plate has a hardness of 7, it cannot be used to identify minerals with greater hardness.

Hematite, Fe_2O_3, may be reddish brown to black; its streak, however, is light to dark bloodred, which becomes black on heating. Limonite, $FeO(OH) \cdot nH_2O$, sometimes known as "brown hematite" or "bog-iron ore," has a color that is dark brown to black but a streak that is yellowish brown. Cassiterite, SnO_2 (tin-stone), is usually brown or black, but it has a white streak.

Other physical properties Minerals have other physical properties that may be helpful in identification: magnetism, electrical properties, luster, fluorescence, fusibility, solubility, fracture, and tenacity (see Appendix A).

2.4 MINERAL STRUCTURE

In the section on crystal form we have seen that atoms of elements in a mineral assume an orderly pattern, its **crystalline structure.** (For example, in halite the ions of sodium are in alternation with the ions of chlorine.)

Each mineral has a unique crystalline structure that will distinguish it from another mineral even if the two are composed of the same element or elements. Let us again consider the minerals diamond and graphite. Each is composed of only one element, carbon (Figure 2.11). In diamond each atom of carbon is bonded (see next section) to four neighboring carbon atoms. This complete joining of all its atoms results in very stable bonds and is the reason why diamond is so hard. In graphite each atom of carbon is bonded in a plane to three neighboring atoms. This bonding forms sheets, or layers, of carbon piled one on another, but the sheets can easily be separated. Thus graphite is a soft substance. Pyrite and marcasite are another pair of minerals with identical composition, FeS_2, but with different crystalline structures. In pyrite ions of iron are equally spaced in all directions. In marcasite they are not equally spaced. The difference in spacing accounts for their being two different minerals (Figure 2.12).

Other minerals may have more complicated crystalline structure: They may contain more elements and have these joined together in more complex patterns. The color, shape, and size of any given mineral may vary from one sample to another, but the internal atomic arrangement of its component elements is *identical* in all specimens of a particular mineral.

So we find it necessary to include in our definition of a mineral not only that it is an inorganic solid element or compound with a diagnostic chemical composition but also that it has a unique orderly internal atomic arrangement of its elements.

COMPOUNDS

Compounds are combinations of atoms of different elements. **Organic** compounds are those in which the carbon atom (or carbon and oxygen) plays a large part—that is, those mostly formed by life processes. Others are **inorganic** compounds. The method by which the atoms are held together (or "bonded") varies, as described in the next three sections.

IONIC BOND

The number of electrons in the outer shell of an atom determines the manner and ease with which the atom can join with other atoms to form compounds. Compounds may be formed by different atoms' losing or gaining electrons. For example, an atom of sodium has only one electron in its outermost shell but eight in the next shell, whereas chlorine has seven in its outermost shell. When a sodium atom and a chlorine atom join, the sodium's outermost electron slips into the vacancy in the chlorine's outermost shell. Chlorine gains the electron and becomes a negative ion, Cl^-, whereas sodium, by losing it, becomes a positive ion, Na^+ (see Figure 2.13). The result is an **ionic compound,** the mineral *halite*, or common table salt, a most abundant substance. (NaCl means that one atom of sodium is joined with one atom of chlorine.) The unit is a **molecule** of salt. A molecule is the smallest unit of a compound.

When electrons are lost or added to form an ion, electrical forces between the nucleus and the electrons are thrown out of balance, and this affects the radius of the ion. For example, if sodium loses the electron from its shell 3, it has only two shells of electrons left. This is the same electronic supply that Ne (neon) has, but the

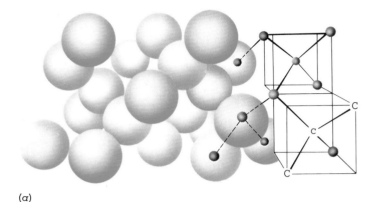

(a)

FIGURE 2.11

Different arrangements of atoms of carbon produce the crystalline structure of (a) diamond and (b) graphite. [After Linus Pauling, *General Chemistry*, W. H. Freeman and Company, San Francisco, copyright © 1953.]

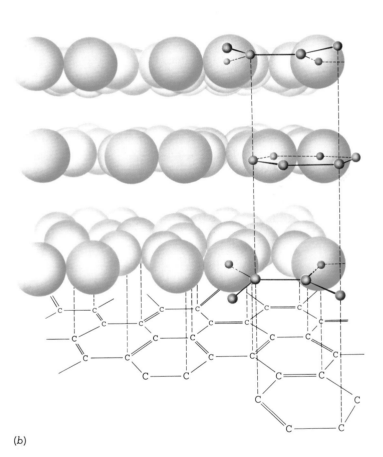

(b)

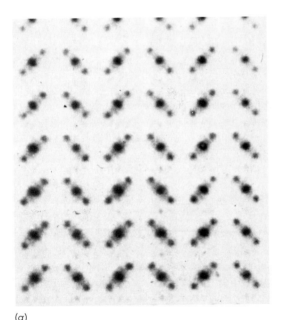

(a)

(b)

FIGURE 2.12

X-ray photographs: (a) pyrite, showing orderly arrangement characteristic of crystalline structure. The mineral is composed of iron and sulfur, FeS_2. Large spots are atoms of iron; small ones, atoms of sulfur. Each atom of iron is bonded to two atoms of sulfur, and spacing of iron atoms is the same in both directions of the plane of the photograph. Magnification, approximately 2.2 million ✕. (b) Marcasite, FeS_2. Note difference between horizontal and vertical spacing of iron atoms. Compare with pyrite. Magnification, about 2.8 million ✕. [Martin J. Buerger.]

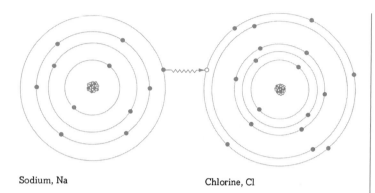

Sodium, Na   Chlorine, Cl

FIGURE 2.13

In the formation of the compound halite an atom of sodium, which has only one electron in its outermost shell, joins an atom of chlorine, which needs one electron to fill its outermost shell. The sodium's lone outermost electron slips into the vacant place to fill the chlorine's outermost shell.

radius of Na^+ is 0.97 Å, whereas that of an atom of Ne with the same number of electrons and shells is 1.60 Å. The excess of unbalanced positive charge in Na^+ pulls in its remaining electrons and shrinks the apparent radius of the ion. In fact, if two full shells of electrons always meant the same radius, the positive ions of the geologically important elements 11 through 17 would have about the same radius after all shell 3 electrons were lost. However, as more and more outer electrons are removed, the increasing relative power of the cen-

tral positive charge pulls in the remaining electrons until the radii decrease progressively from 0.97 Å for Na^+ to 0.27 Å for Cl^{7+} (see Table 2.2).

COVALENT BOND

Atoms can combine in other ways to form compounds, as, for example, in the formation of a water molecule. The oxygen atom has six electrons in its outermost shell and needs two more to achieve the stable number of eight. If two hydrogen atoms, each with its single electron, approach an oxygen atom, the hydrogen electrons in effect slip into the vacant slots in the outermost shell of the oxygen atom (see Figure 2.14). So the hydrogen and oxygen nuclei are sharing their electrons in a way called a **covalent bond**. Again, the result is a compound that is totally different from the elements themselves. This compound is water, whose symbol, H_2O, represents the elements that make it up and the proportions in which they are present—a combination so perfect that water is one of nature's most stable compounds.

Because the oxygen atom has essentially gained two electrons, it becomes negatively charged. And each hydrogen atom, acting as though it had lost its electron, takes on a positive charge. As a result, a molecule of water acts like a small rod, one end positively charged and the other negatively (see Figure 2.15). These ends are referred to as a positive pole and a negative pole because of the molecule's similarity to

TABLE 2.2
Radii of atoms and ions[a]

At. no.	Element	Atom		Ion	
		Symbol	Radius, Å	Symbol	Radius, Å
8	Oxygen	O	0.60	O^{2-}	1.32
10	Neon	Ne	1.60		
11	Sodium	Na	1.86	Na^+	0.97
12	Magnesium	Mg	1.60	Mg^{2+}	0.66
13	Aluminum	Al	1.43	Al^{3+}	0.51
14	Silicon	Si	1.17	Si^{4+}	0.42
15	Phosphorus	P	1.08	P^{5+}	0.35
16	Sulfur	S	1.04	S^{6+b}	0.30
17	Chlorine	Cl	1.07	Cl^{7+b}	0.27

[a] Radii from Cornelius S. Hurlbut, Jr., *Dana's Manual of Mineralogy*, 18th ed., John Wiley & Sons, Inc., New York, 1971.
[b] S^{6+} and Cl^{7+} are not the usual ionic forms. The elements would normally gain electrons to become S^{2-} and Cl^-.

FIGURE 2.14 (TOP)

Two hydrogen atoms and one oxygen join to form water, H_2O, by a covalent bond. In this bond the hydrogen electrons do double duty in a sense, filling the two empty places in the outer shell of oxygen yet remaining at normal distance from their nuclei. The result is the formation of a molecule of water, the smallest unit that displays the properties of that compound.

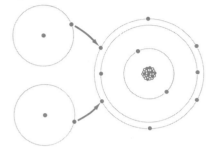

Hydrogen, H

Hydrogen, H   Oxygen, O

FIGURE 2.15 (MIDDLE)

The dipolar character of water: The oxygen has, in effect, gained two electrons, hence a double negative charge, whereas the hydrogen atoms have each lost the effective service of an electron and represent positive charges. Accordingly, the water molecule acts like a small rod positively charged on one end and negatively charged on the other. Some possible combinations of water molecules are suggested here.

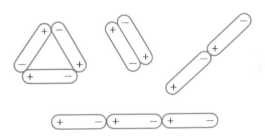

FIGURE 2.16 (BOTTOM)

How water dissolves salt. Water dipoles attach themselves to the ions that compose the salt and overcome the ionic attractions that hold the salt together as a solid. Each sodium and chlorine ion is then convoyed by a number of water dipoles into the body of the liquid.

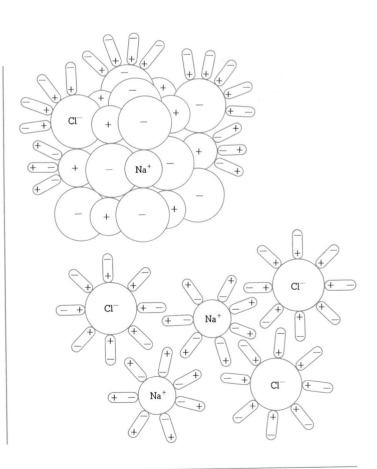

a bar magnet. The molecule, then, is a *dipole* ("two-pole"); and water is known as a **dipolar compound.** This polarity gives water special properties that make it an extremely important agent in geological processes. The mechanism by which water dissolves salt (see Figure 2.16) is an illustration of the ease with which water dissolves various substances and participates in weathering and other geological activities.

METALLIC BOND

The atoms of metallic elements have a special kind of bonding, which is responsible for metals' being good conductors of heat and electricity. In a piece of metal the atoms of a single element are closely packed together; their outermost electrons are not shared or exchanged but are free to move around and connect with any atoms in the solid. These wandering electrons, shared by any of the atoms in the metal, create the **metallic bond.**

The relative freedom of movement of the electrons in this relationship accounts for the high level of electrical conductivity. When there is no current, electrons jump randomly from atom to atom. When an electrical potential is applied to a good conductor, such as a copper wire, electrons flow through the wire without stopping to attach themselves to any particular atoms. They thus convey electricity (or heat) throughout the whole wire.

2.5 THE MOST COMMON MINERALS

Although there are more than 2,000 minerals known, only a limited number comprise most of the rocks of the earth's crust. These can be grouped into closely related families of minerals. As we pointed out in Section 2.3, minerals are homogeneous crystalline materials but not necessarily pure substances. Most **rock-forming minerals,** or minerals found most abundantly in the rocks of the crust, have variable compositions caused by ionic substitution of some elements for other elements. These substitutions are distributed at random throughout the crystalline structure. Such replacement is called **solid solution.** The resultant mineral is homogeneous and has the same physical properties throughout, but its composition is slightly variable.

SILICATES

More than 90 percent of the rock-forming minerals are silicates, compounds containing silicon, oxygen, and one or more metals. Each silicate mineral has as its fundamental unit, the **silicon-oxygen tetrahedron** (see Figure 2.17). This is a combination of one "small" silicon ion, with a radius of 0.42 Å, surrounded as closely as geometrically possible by four "large" oxygen ions, each with a radius of 1.40 Å, forming a tetrahedron. The oxygen ions contribute an electric charge of 8 − to the tetrahedron, and the silicon ion contributes 4 +. Therefore the silicon-oxygen tetrahedron is a complex ion with a net charge of 4 −. Its symbol is $(SiO_4)^{4-}$.

The most common of the rock-forming silicates are olivine, augite, hornblende, biotite, muscovite, feldspars, and quartz. They are listed and classified in Table 2.3.

Ferromagnesians In the first four of these rock-forming silicates—olivine, augite, hornblende, and biotite—the silicon-oxygen tetrahedra are joined by ions of iron and magnesium. These silicate minerals are known as **ferromagnesians,** from the joining of the Latin *ferrum,* "iron," and "magnesium." All four ferromagnesians are very dark or black and have a higher specific gravity than do the other rock-forming minerals.

Olivine is a nesosilicate because it is composed of isolated silicon-oxygen tetrahedra held together by positive ions of magnesium and iron. Its elements are so firmly held together by their ionic bonds that it exhibits no cleavage and is a relatively hard mineral, 6.5 to 7. Because there are no planes of weakness, olivine fractures when struck a blow.

Olivine is an example of a mineral that undergoes compositional changes. Its formula is $(Mg, Fe)_2SiO_4$. The proportions of magnesium and iron vary; so their symbols are in parentheses. The proportion of silicon and oxygen remains constant. Iron and magnesium substitute for one another quite freely in olivine's crystalline structure as they each have two electrons in the outer shell and their ionic radii are almost identical, 0.97 and 0.99 Å. This tendency for two or more ions to substitute for one another in a continuous series is known as **isomorphism.** Because of the varying amount of magnesium and iron, olivine is a solid-solution mineral series. The end members of the series are forsterite and fayalite: When the positive ions are all magnesium, the mineral is forsterite, Mg_2SiO_4; when they are all iron, the mineral is fayalite, Fe_2SiO_4.

Olivine is geologically important as it makes up

FIGURE 2.17

The silicon-oxygen tetrahedron, $(SiO_4)^{4-}$: (a) from above, (b) from the side. This is the most important complex ion in geology because it is the central building unit of nearly 90 percent of the minerals of the earth's crust.

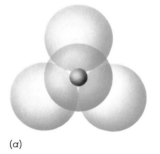

(a)

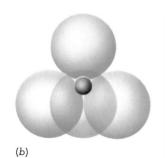

(b)

TABLE 2.3
Silicate classification[a]

Class[b]	Arrangement of SiO₄ tetrahedra	Rock-forming minerals
Nesosilicates	Isolated	Olivine
Sorosilicates	Double	Epidote (epidote family of hydrous calcium-aluminum-iron silicates)
Cyclosilicates	Rings	Beryl (beryllium aluminosilicate)
Inosilicates	Chains (single)	Augite (pyroxene family)
	Chains (double)	Hornblende (amphibole family)
Phyllosilicates	Sheets	Biotite (black mica) Muscovite (white mica)
Tectosilicates	Frameworks	Orthoclase (potassium feldspar) Plagioclase (calcium-sodium feldspar) Quartz

[a]From Cornelius S. Hurlbut, Jr., *Dana's Manual of Mineralogy*, 18th ed., John Wiley & Sons, Inc., New York, 1971.
[b]The prefixes are from the Greek: *nesos*, "island"; *sōros*, "group"; *kyklos*, "ring"; *inos*, "chain" or "thread"; *phyllon*, "sheet"; *tektōn*, "builder."

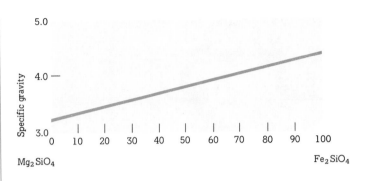

FIGURE 2.18

Specific gravity of olivine.

(a)

several percent of the crustal rocks at the surface and is believed to predominate in the heavier and deeper-seated rocks. Its specific gravity is 3.27 to 4.37, increasing with iron content (Figure 2.18). Its color ranges from olive to grayish green, sometimes brown. This mineral usually occurs in grains or granular masses.

Augite, an inosilicate, has a crystalline structure based on single chains of tetrahedra,[2] as shown in Figure 2.19, joined by ions of iron, magnesium, calcium, sodium, and aluminum. It is dark green to black, with a

[2]In the remainder of this chapter "tetrahedra" is equivalent to "silicon-oxygen tetrahedra."

FIGURE 2.19

Single chain of tetrahedra viewed (a) from above and (b) from an end. Each silicon ion (small black sphere) has two of the four oxygen ions of its tetrahedron bonded exclusively to itself, and it shares the other two with neighboring tetrahedra fore and aft. The resulting individual chains are in turn bonded to one another by positive metallic ions. Because these bonds are weaker than the silicon-oxygen bonds that form each chain, cleavage develops parallel to the chains.

(b)

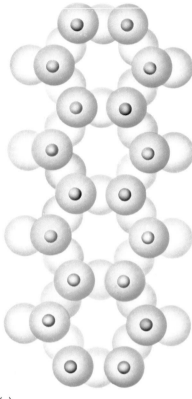

(a)

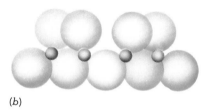

(b)

FIGURE 2.20

Double chain of tetrahedra viewed (a) from above and (b) from an end. The doubling of the chain of Figure 2.19 is accomplished by the sharing of oxygen atoms by adjacent chains.

colorless streak. Its hardness is 5 to 6, and its specific gravity ranges from 3.2 to 3.4. It has rather poor cleavage along two planes almost at *right* angles to each other. This cleavage angle is important in distinguishing augite from hornblende. (A good way to recall it is to remember that "augite" rhymes with "right.")

Augite is the commonest variety of a group of minerals designated by the family name **pyroxenes** (see Appendix A) and characterized by a fundamental crystalline structure built on single chains of tetrahedra. Pyroxenes crystallize and form a series whose members are closely analogous chemically to members of the amphibole group (see below).

Hornblende has a crystalline structure based on double chains of tetrahedra, as shown in Figure 2.20, joined by the iron and magnesium ions common to all ferromagnesians and by ions of calcium, sodium, and aluminum. It too is an inosilicate. Hornblende's color is dark green to black, like that of augite; its streak, colorless; its hardness, 5 to 6; and its specific gravity, 3.2. Two directions of good cleavage meet at angles of approximately 56° (degrees) and 124°, which help distinguish hornblende from augite (see Figure 2.21).

Hornblende is an important and widely distributed rock-forming mineral. It is the commonest mineral of a group called the **amphiboles,** which closely parallel the pyroxenes in composition but contain hydroxyl, OH. Amphiboles are characterized by double chains of tetrahedra. The minerals classified as hornblendes are in reality a complex solid-solution series from anthophyllite to hornblende (see Appendix A).

Biotite, black mica, is a potassium-magnesium-iron-aluminum silicate, $K(Mg,Fe)_3[AlSi_3O_{10}(OH)_2]$. (It was named for a distinguished French physicist, J. B. Biot.) Like other micas, it is a phyllosilicate constructed of tetrahedra in sheets, as shown in Figure 2.22. Each silicon ion shares three oxygen ions with adjacent silicon ions to form a pattern like wire netting. The fourth, unshared oxygen ion of each tetrahedron stands above the plane of all the others. The basic structural unit of mica consists of two of these sheets of tetrahedra with their flat surfaces facing outward and their inner surfaces held together by positive ions. In biotite the ions are magnesium and iron. These basic double sheets of mica, in turn, are loosely joined together by positive ions of potassium.

Layers of biotite or any of the other micas can be peeled off easily (see Figure 2.23) because there is perfect cleavage along the surfaces of the weak potassium bonds. In thick blocks biotite is usually dark green

(a)   (b)

FIGURE 2.21

Cleavage of hornblende (a) compared with that of augite (b). The top "roof" of the hornblende specimen and the top and perpendicular left-hand faces of the augite are cleavage surfaces. Throughout each specimen easiest breaking is parallel to these surfaces. On the front face of the augite are some "steps" outlined by cleavage planes. Such steps are the most common manifestation of cleavage, which seldom produces pieces as large as this one 5-cm wide. [Harvard Mineralogical Collection, Walter R. Fleischer.]

FIGURE 2.22

Tetrahedral sheets. Each tetrahedron is surrounded by three others, and each silicon ion has one of the four oxygen ions to itself, sharing the other three with its neighbors.

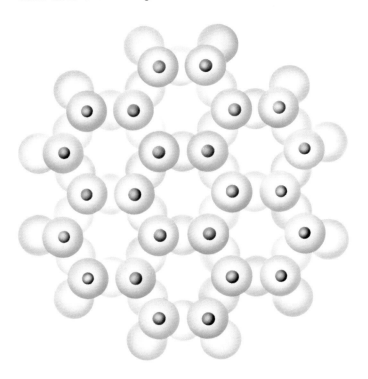

FIGURE 2.23

Mica cleavage. The large block (or "book") is bounded on the sides by crystal faces. Cleavage fragments lying in front of the block are of different thicknesses, as indicated by their degrees of transparency. The reference grid is made of 5-cm squares. [Walter R. Fleischer.]

or brown to black. Its hardness is 2.5 to 3, and its specific gravity is 2.8 to 3.2.

Nonferromagnesians   The other common rock-forming silicate minerals are known as **nonferromagnesians** simply because they do *not* contain magnesium or iron. These minerals are muscovite, feldspars, and quartz. They are all marked by their light colors and relatively low specific gravities, ranging from 2.6 to 3.0.

Muscovite is white mica, so named because it was once used as a substitute for glass in old Russia (Muscovy). It has the same basic crystalline structure as biotite, but in muscovite each pair of tetrahedra sheets is tightly cemented together by ions of aluminum. As in biotite, however, the double sheets are held together loosely by potassium ions, along which cleavage readily takes place. It is a phyllosilicate with a formula of, essentially, $KAl_2[AlSi_3O_{10})(OH)_2]$. In thick blocks the color of muscovite is light yellow, brown, green, or red. Its hardness ranges from 2 to 2.5, with specific gravity 2.8 to 3.1.

Feldspars are the most abundant rock-forming silicates. Their name comes from the German *feld*, "field," and *spath*, a term used by miners for various nonmetallic minerals. The name reflects the abundance of these minerals: "field minerals," or minerals found in any field. Feldspars make up nearly 54 percent of the minerals in the earth's crust.

The feldspars are silicates of aluminum with potassium, sodium, or calcium. Crystals of feldspars of different systems resemble each other closely in angles and crystal habit. They all show good cleavages in two directions, which make an angle of 90° or close to 90° with each other. Their hardness is about 6, and their specific gravity ranges from 2.55 to 2.76.

Feldspars are tectosilicates because all the oxygen ions in the tetrahedra are shared by adjoining silicon ions in a three-dimensional network. However, in the centers of one-quarter to one-half of the tetrahedra aluminum ions with a radius of 0.51 Å and an electric charge of 3+ have replaced silicon ions with their radius of 0.42 Å and electric charge of 4+. The nega-

tive electric charge resulting from this substitution is corrected by the entry of K^+, Na^+, or Ca^{2+} into the crystalline structures.

The feldspars are **orthoclase** with potassium and **plagioclase** with sodium or calcium (see Table 2.4). Orthoclase is named from the Greek *orthos*, "straight," and *klasis*, "a breaking," because the two dominant cleavages intersect at a right angle when a piece of orthoclase is broken (see Figure 2.24). Aluminum replaces silicon in every fourth tetrahedron, and positive ions of potassium correct the electrical unbalance. The introduction of aluminum in the tetrahedra cannot be regarded as solid solution: Aluminum is not a vicarious constituent of orthoclase whose percentage varies from sample to sample or which may be wholly replaced by silicon; it is an essential constituent and is not replaceable by silicon without breakdown of the crystalline structure. Orthoclase is white, gray, or pinkish; its streak is white; and its specific gravity is 2.57.

Plagioclase ("oblique-breaking") feldspars are so named because they have cleavage planes that intersect at about 86°. One of the cleavage planes is marked by striations. Plagioclase feldspars may be colorless, white, or gray although some samples show a striking play of colors called opalescence.

The plagioclase feldspars albite and anorthite are end members of a solid solution series. The amount of aluminum varies in proportion to the relative amounts of calcium and sodium to maintain electrical neutrality; the more calcium, the greater the amount of aluminum. Here the variation in amount of aluminum may be properly regarded as ionic substitution. The number of ions of calcium that replace sodium is equaled by the number of aluminum ions replacing silicon in the silicon-oxygen framework. Sodium can substitute quite easily for calcium in a mineral's crystalline structure because their ionic radii are almost identical, 0.97 and 0.99 Å, even though there is a slight difference in electrons present in the outer shell—one in sodium and two in calcium.

In albite aluminum replaces silicon in every fourth

TABLE 2.4 **The feldspars**	Diagnostic ion	Name	Symbol	Formula[a]
	K^+	Orthoclase (potassium feldspar)	Or	$K(AlSi_3O_8)$
	Na^+	Albite (sodium feldspar)	Ab	$Na(AlSi_3O_8)$
	Ca^{2+}	Anorthite (calcium feldspar)	An	$Ca(Al_2Si_2O_8)$

[a]In these formulas the symbols inside the parentheses indicate the tetrahedra. The symbols outside the parentheses indicate the diagnostic ions that are worked in among the tetrahedra.

tetrahedron, and positive ions of sodium correct the electrical unbalance. The specific gravity of albite is 2.62. In anorthite aluminum replaces silicon in every second tetrahedron, and positive ions of calcium correct the electrical unbalance. The specific gravity of anorthite is 2.76.

Although the plagioclase feldspars have different compositions, they have the same crystal form and for that reason are called **isomorphous**. At the same time they can be thought of as solutions of albite in anorthite in varying proportions, or the reverse. For that reason, they have been called examples of solid solution. This term, however, does not imply that in the intermediate minerals there are molecules of either albite or anorthite as there would be of, say, sodium and chlorine in a solution of salt in water. Intermediate minerals simply have compositions that are conveniently described as mixtures of the pure end members of the series.

Quartz is the most common rock-forming silicate mineral that is composed "exclusively" of tetrahedra. It is a tectosilicate. Every oxygen ion is shared by adjacent silicon ions, which means that there are two ions of oxygen for every ion of silicon. This relationship is presented by the formula SiO_2. Because of its crystalline structure, quartz is a relatively hard mineral, 7. Because there are no planes of weakness in its crystalline structure, it does not exhibit cleavage but fractures along curved surfaces when struck a blow. Being composed of fairly light elements, its specific gravity is 2.65.

Of all the rock-forming minerals, quartz is most nearly a pure chemical compound. However, spectrographic analyses show that even in its most perfect crystals there are traces of other elements. Quartz usually appears smoky to clear in color, but less-common varieties include purple or violet *amethyst*, massive rose-red or pink *rose quartz*, smoky yellow to brown *smoky quartz*, and *milky quartz*. These color differences are caused by the other elements, present as minor impurities. They are not caused by, and do not affect, the crystalline structure of quartz.

OXIDE MINERALS

Oxide minerals are formed by the direct union of an element with oxygen. These have relatively simple formulas compared to the complicated silicates. The oxide minerals are usually harder than any other class except the silicates, and they are heavier than others except the sulfides. Within the oxide class are the chief

FIGURE 2.24

Feldspar cleavage in specimens of orthoclase. The large block on the right and the small fragment on the black box (a 5-cm cube) show the cleavage planes at nearly 90°, a characteristic of feldspars. [Walter R. Fleischer.]

ores of aluminum, iron, tin, chromium, and manganese. So we have the common oxide minerals ice, H_2O, corundum, Al_2O_3, hematite, Fe_2O_3, magnetite, Fe_3O_4, and cassiterite, SnO_2.

SULFIDE MINERALS

Sulfide minerals are formed by the direct union of an element with sulfur. The elements that occur most commonly in combination with sulfur are iron, copper, lead, zinc, silver, and mercury. Some of these sulfide minerals occur as valuable ores, such as pyrite, FeS_2, chalcocite, Cu_2S, galena, PbS, and sphalerite, ZnS.

CARBONATE AND SULFATE MINERALS

In discussing silicates, we found them to be built around the complex ion $(SiO_4)^{4-}$, that is, the silicon-oxygen tetrahedron. Two other complex ions are also of great importance in geology. One of these consists of a single carbon ion with three oxygen ions packed around it, the complex ion $(CO_3)^{2-}$. Compounds in which this ion appears are called **carbonates**. For example, the combination of a calcium ion with a carbon-oxygen ion produces calcium carbonate, $CaCO_3$, known in its mineral form as *calcite*. This mineral is the principal component of the common sedimentary rock limestone. The other complex ion is $(SO_4)^{2-}$, a combination of one sulfur ion and four oxygen ions. This complex ion combines with other ions to form **sulfates,** for example, $CaSO_4$, the mineral *anhydrite*.

HALIDES

A variety of minerals results from the combination of certain elements (positive ions) with the halogen elements (chlorine, iodine, bromine, and fluorine): the very common rock salt, *halite*, NaCl, its poisonous partner, *sylvite*, KCl, and many other chlorides, iodides, bromides, and fluorides. Halides occur as precipitates from evaporating ponds, salt flats, and natural brines.

MINERALOIDS

Some substances do not yield definite chemical formulas upon analysis and show no sign of crystallinity. These are said to be **amorphous** and have been called **mineraloids.** A mineral may exist in a crystalline phase with a definite composition and crystalline structure, or under certain conditions of formation practically the same substance may occur as a mineraloid. Mineraloids are formed under conditions of low pressure and temperature and are commonly substances originating during the process of weathering of the materials of the earth's crust. They characteristically occur in mammillary, botryoidal, stalactitic, and similar-shaped masses. Their ability to absorb other substances accounts for their wide variations in chemical composition. Bauxite, limonite, and opal are examples of mineraloids.

2.6 ORGANIZATION AND ASSOCIATION OF MINERALS

We know that minerals are special combinations of elements or compounds (see summarizing Table 2.5), and now we can complete our definition of a mineral: (1) It is a naturally occurring element or inorganic compound in the solid state;[3] (2) it has a diagnostic composition; (3) it has a unique crystalline structure; and (4) it exhibits certain physical properties as a result of its composition and crystalline structure.

ASSOCIATIONS OF MINERALS

Certain groups of minerals are most commonly found associated in specific geological settings. For example, some minerals formed at high temperatures commonly occur in igneous rocks that crystallize deep within the earth. Other mineral assemblages reflect their origin under the high temperatures *and* increased pressures associated with metamorphic events. Finally, there are some mineral groupings that usually develop under conditions approaching the atmospheric temperatures and pressures found at the earth's surface. Such mineral associations are useful indicators of past conditions at the time of either their origin or their recombination to form new minerals, where temperature, pressure, or chemical environment has changed.

2.7 RESHUFFLING THE ELEMENTS AND MINERALS

As we noted in Chapter 1, rocks continuously undergo changes during their cyclic passage through time, and the minerals making up rocks are thus changed. Likewise, the elements constituting these minerals are rearranged in combinations with new elements in varying proportions.

One location where much of this reorganization occurs is along plate boundaries within the earth's lithosphere (see Section 1.3). For example, at the trailing edge of a plate new material is welling up from the deeper portion of the earth's mantle, solidifying and being added as new plate material.

On the other hand, at the leading edge of a lithospheric plate other possibilities exist. The plate may be moving down into the mantle along a subduction zone.

[3]Some mineralogists do not restrict the definition to the solid state but include such substances as water and mercury.

TABLE 2.5

Summary of the organization of some common minerals[a]

| Elemental minerals | Compound minerals | | | | | |
	Silicates [elements + $(SiO_4)^{4-}$]	Oxides (elements + O)	Sulfides (elements + S)	Carbonates [elements + $(CO_3)^{2-}$]	Sulfates [elements + $(SO_4)^{2-}$]	Halides (elements + Cl^-, I^-, Br^-, or F^-)
Copper	Ferromagnesian	Cassiterite	Chalcocite	Calcite	Anhydrite	Halite
Diamond	Augite	Corundum	Galena	Dolomite	Gypsum	Sylvite
Gold	Biotite	Hematite	Pyrite	Magnesite		
Graphite	Hornblende	Ice	Sphalerite			
Platinum	Olivine	Magnetite				
Silver	Nonferromagnesian					
Sulfur	Feldspars					
	Orthoclase					
	Plagioclase					
	Albite					
	Anorthite					
	Muscovite					
	Quartz					

[a] Minerals may be either elements or compounds, but not all elements or compounds are minerals.

Here the increased temperature and pressure cause the minerals in the rocks to undergo changes into other minerals and even to be melted if carried deep enough.

A second type of leading edge of lithospheric plate may occur. The plate may carry a continent upon its back, and this continental mass may encounter another, similar mass carried by a plate approaching from the opposite direction. The resulting continent-to-continent collision produces both uplift of mountains and downwarp of "roots" under those mountains (see Chapter 7). Elements, minerals, and rocks found in the mountains will undergo changes during uplift, includ-ing weathering and erosion. Similarly, those materials found in the roots of the mountains will undergo changes resulting from increased temperature and pressure: melting or remelting of the constituents.

In this way earth materials are constantly under-going a reshuffling in terms of their combinations, relative proportions, and associations. New minerals are formed from old, new concentrations of elements and minerals develop where they did not exist previously, and old minerals are broken down into their constituent parts (elements or combinations of elements) and re-combined in differing arrangements of elements, minerals, and rocks.

OUTLINE

Minerals are naturally occurring inorganic elements or compounds.

Matter, the substance of the physical universe, is described in terms of atoms with mass and electric charge.

Electric charge is of two kinds, arbitrarily called **positive** and **negative.**

Atoms are composed of protons, neutrons, and electrons.

Atomic structure is expressed by a nucleus of protons and neutrons surrounded by a cloud of electrons.

For the most common elements **atomic size,** determined by the protective shield of electrons, is about 2 Å in diameter around a nucleus of 10^{-5} to 10^{-4} Å.

Atomic mass is 99.95 percent in the nucleus.

Ions are electrically unbalanced forms of atoms or groups of atoms.

Elements are particular combinations of protons, neutrons, and electrons.

Isotopes are forms of elements with the same number of protons in the nucleus but different mass because of their different number of neutrons.

Gamow's **big-bang theory** assumes that all elements developed very rapidly in an explosive event about 10 to 20 billion years ago.

Hoyle's **steady-state theory** concludes that elements are constantly being formed in the hot interior of stars from the hydrogen and helium fuels.

Mineral identification is accomplished by examination of chemical and physical properties, including crystal form, cleavage, striations, hardness, specific gravity, color, and streak.

Crystal form is the external shape produced by a mineral's crystalline structure.

Cleavage is the tendency of a mineral to break in certain preferred directions along smooth plane surfaces.

Striations are parallel, threadlike lines or narrow bands running across crystal faces or cleavage surfaces.

Hardness is governed by the internal atomic arrangement of the elements of a mineral.

Specific gravity is a number that compares the weight of a given volume of a mineral with the weight of the same volume of water.

Color is not a reliable property for identifying most minerals, but minerals containing iron are usually dark-colored and those containing aluminum are light-colored.

Streak is the color of a mineral in finely powdered form.

Other physical properties include magnetism, electrical properties, luster, fluorescence, fusibility, solubility, fracture, and tenacity.

Mineral structure is the internal orderly arrangement of atoms, which is unique for each mineral.

Compounds are combinations of atoms of different elements.

Molecule is the smallest unit of a compound.

Ionic compounds occur when electrons are lost or added to atoms.

Covalent bonding results from sharing electrons rather than from gaining or losing them.

Metallic bonds are responsible for metals' being such good conductors of heat and electricity.

Rock-forming minerals are 90 percent silicates, based on a complex ion called the silicon-oxygen tetrahedron.

Ferromagnesians olivine, augite, hornblende, and biotite contain iron and magnesium.

Nonferromagnesians include muscovite, the feldspars, and quartz.

Quartz is the only rock-forming silicate mineral composed exclusively of tetrahedra.

Oxide minerals are formed by the direct union of an element with oxygen.

Sulfide minerals are formed by the direct union of an element with sulfur.

Carbonate minerals are built around the complex ion $(CO_3)^{2-}$.

Sulfate minerals are built around the complex ion $(SO_4)^{2-}$.

Halides form from combinations of positive ions with chlorine, iodine, bromine, and fluorine.

Mineraloids do not have definite compositions or crystalline structure.

Organization of minerals lies in naturally occurring combinations of elements or compounds in the solid state, each with diagnostic composition and unique crystalline structure as well as certain physical properties.

Associations of minerals commonly occur in specific geologic settings and reflect the conditions of temperature, pressure, or chemical environment at the time of their origin or when recombined into new minerals.

New combinations of reshuffled elements and compounds produce new minerals, especially along edges of lithospheric plates.

SUPPLEMENTARY READINGS

Fowler, William A. "The Origin of the Elements," *Sci. Am.*, vol. 195, no. 5, 1956. *Discusses Gamow and Hoyle theories.*

Frye, Keith *Modern Mineralogy*, Prentice-Hall, Inc., Englewood Cliffs, N.J., 1974. *Somewhat advanced treatment from chemical fundamentals to a complete survey of mineralogy.*

Hurlbut, Cornelius S., Jr. *Minerals and Man*, Random House, Inc., New York, 1969. *For the layman a lively, balanced presentation of expert information on the nature, origin, and properties of 150 of the world's principal minerals, together with stories of how they have been used throughout the ages; with 217 illustrations, including 160 in color.*

Hurlbut, Cornelius S., Jr. *Dana's Manual of Mineralogy*, 18th ed., John Wiley & Sons, Inc., New York, 1971. *Latest edition of the textbook that has dominated this field for most of this century.*

Loeffler, Bruce M., and Roger G. Burns "Shedding Light on the Color of Gems and Minerals," *Am. Sci.*, vol. 64, no. 6, 1976. *Gives first good explanation for variations in color of minerals due to selective absorption of different wavelengths of light.*

3

IGNEOUS ACTIVITY AND IGNEOUS ROCKS

From time to time throughout geologic history molten rock, working its way to the surface through covering rock, has poured out or been blown onto the ground, there to solidify again into rock. Igneous activity consists of movements of molten rock inside and outside the earth and includes the effects associated with them.

In some areas molten rock has been extruded through extensive fissures in the earth's surface. This activity, called **fissure eruptions,** has built large plateaus. In other places molten rock has escaped to the surface through vents; around these vents the ejected material has accumulated to build up landforms that we know as **volcanoes.**

As noted in Chapter 1, much of the earth's igneous activity is concentrated along the margins of lithospheric plates. Molten material rises along the midoceanic-ridge zones of ocean basins and the rifted margins of continents. Other igneous masses are associated with the convergent edges of moving plates along subduction zones (see Figure 1.12). Later we shall look at the character of these igneous intrusions and extrusions as they relate to plate boundaries. In this chapter we shall also examine the movement of plates over zones of hot rising material and consider the resulting surface features.

3.1 VOLCANOES

A volcano may grow in size until it becomes a mountain (Figure 3.1). Normally cone-shaped, it has a pit at the summit, which may be either a crater or a caldera. A **crater** is a steep-walled depression out of which volcanic materials are ejected. Its floor is seldom over 300 m in diameter; its depth may be as much as 100 m. A crater may be at the top of a volcano or on its flank.

The much larger **caldera** is a basin-shaped depression, more or less circular, with a diameter many times greater than that of the included volcanic vent or vents. Most calderas, in fact, are more than 1,500 m in diameter; some are several kilometers across and several hundred meters deep.

Between eruptions a volcano's vent may become

choked with rock congealed from a past eruption. Sometimes small jets of gas come out through cracks in this rock plug. A volcano is built by, and remains active because of, materials coming from a large deep-seated reservoir of molten rock. While in the ground, this molten rock is called **magma.** When extruded on the surface, it is called **lava.**

The largest volcano on earth is Mauna Loa, part of the island of Hawaii. It is 600 km around the base, and its summit towers nearly 10 km above the surrounding ocean bottom. This and the rest of the island represent accumulations from eruptions that have gone on for more than 1 million years. The largest volcano yet found anywhere is Olympus Mons ("Mount Olympus") on Mars. Photographed originally from the *Mariner 9* spacecraft, this volcano may be over 23 km high, with a caldera 65 km across (see also Chapter 17).

EVOLUTION OF VOLCANOES AND TYPES OF VOLCANISM

Many volcanoes start their eruptive cycles with **basaltic**[1] lava flows and then **andesitic**[2] lava flows, and then for thousands of years they produce only fragmental materials that build up their cones. In their final stages a cone-disrupting explosion occurs with ejection of only fragmental materials. The cycle ends when acidic lava congeals in the volcanic vent. First eruptions consist of basic low-viscosity lava, which flows readily. With succeeding eruptions lava becomes more acidic and more viscous. Eventually, lava becomes so viscous that even gases escape with difficulty, great pressures building up in the magma with eruptions of fragmental pyroclastic debris resulting.

There exists a distinction between volcanoes on continents and those in oceanic areas. Although highly fluid basalt occurs in both regions, associated volcanic rocks are markedly different.

Highly viscous magmas are commonly associated with continents and islands near continental borders. These magmas also tend to have more gas and therefore tend to produce much more explosive volcanic

[1]Basalt is a fine-grained igneous rock composed mainly of plagioclase feldspar and iron-magnesium (ferromagnesian) minerals; this and other rock terms are defined later in this chapter and appear in the classification chart, Figure 3.28.
[2]Andesite is a volcanic rock composed essentially of the plagioclase mineral andesine, with varying amounts of pyroxene, hornblende, or biotite.

FIGURE 3.1

Pavlof Volcano on the Alaska Peninsula. [U.S. Navy.]

eruptions than do the less viscous oceanic magmas.

Oceanic magmas characteristically contain more silicon in relation to the alkalies (sodium and potassium) and are called **tholeiitic basalts.** The initial bases of most of the Hawaiian volcanoes consist of this material. Most young eruptions in oceanic basins consist of basalt, whereas most young eruptions on continents are andesitic (hence the "andesite line," separating typical oceanic from typical continental geology; see Figure 6.42), showing increases in alkali content.

Nearly all samples of lava dredged from the deep ocean floor are tholeiitic basalt. Most of the samples taken from the volcanic islands along the Midatlantic Ridge are **alkalic basalt.**

Volcanic eruptions on stable portions of continental masses tend to be like those of the ocean basins although more explosive. The volcanism in folded mountain belts, on the other hand, tends to begin with basalts, sodium-rich basalts called *spilites*, and *serpentinized peridotites* (ultrabasic rocks). Volcanic activity may be associated with periods of mountain building and often also occurs after the folding ceases.

VOLCANIC ERUPTIONS

In its reservoir far below the earth's surface magma is composed of elements in solution. Some are vaporized as magma approaches the surface. These volatile com-

FIGURE 3.2

Pahoehoe lava. [U.S. Geological Survey.]

ponents play an extremely important role in igneous activity and are the primary agents in producing a **volcanic eruption.** As the magma nears the surface, volatiles tend to separate from the other components and migrate through them to the top of the moving mass. The gases accumulate if the volcanic vent is blocked. Pressure builds up until it can no longer be confined. Then it pushes out. If its temperature is 1000° C or higher, the gases expand several thousandfold as they escape, shattering rock that blocks the vent and throwing it and magma into the air. After the explosion, the magma still in the ground is left poorer in volatile components but may be fluid enough to pour out.

Volcanic eruptions are of different types, from relatively quiet outpourings of lava to violent explosions accompanied by showers of volcanic debris. The type of eruption depends to a great extent on the magma's composition and its volatile content. Lavas rich in ferromagnesians tend to flow relatively quietly, whereas those high in silica, SiO_2, are viscous and explosive.

COMPOSITION OF LAVAS

Lavas are of three principal types, classified primarily by their proportions of silica: acid, intermediate, and basic. The acid lavas have 70 percent or more of silica; intermediate lavas, 50 to 70 percent; and basic lavas, less than 50 percent.

All lavas have some composition similarities, but no two volcanoes erupt lavas of exactly the same composition. In fact, the composition may vary from one eruption to another in the same volcano. Generally, however, as we have indicated, over the course of its history, a volcano starts with basic lava, grades into intermediate, and ends its activity with dominantly acidic lava. Volcanoes erupting primary, basic lavas are confined to ocean basins. In the Pacific Ocean four are on the Hawaiian Islands, and three are on the Galápagos. In the Atlantic Ocean they range from Iceland along the Midatlantic Ridge to Tristan da Cunha. Intermediate-type lavas are erupted by volcanoes ringing the Pacific and constituting the island arcs. In the Mediterranean Sea a number of Italian volcanoes erupt lavas believed to have been contaminated by the ingestion of quite large quantities of limestone.

The composition of volcanic eruptions will therefore vary, depending upon the source of the magma and the nature of the rocks through which it passes on its upward journey to the earth's surface. Magma generated entirely within material of uniform composition will be composed of that single rock type (such as basic types along the midoceanic ridges). Those magmas resulting from melting of lithospheric crust that has been carried down along subduction zones will produce intermediate-type rocks (such as along the perimeter of the Pacific Ocean).

The surfaces of lava flows commonly have one of two contrasting shapes, for which the Hawaiian names of **pahoehoe** and **aa** are used. In the pahoehoe type, sometimes known as "corded" or "ropy," the surface is smooth and billowy and frequently molded into forms that resemble huge coils of rope (Figure 3.2). In the aa type the surface of the lava is covered with a random mass of angular, jagged blocks. The surface that develops depends on the viscosity of the lava, with viscosity increasing as lavas become more acidic. Pahoehoe is the shape produced on the surface of a basic lava, and aa is the shape produced when the flow becomes sluggish because of an increase in silica content.

Temperature of lavas Measurements of lava temperature have been made at some volcanic vents. These show that basic lavas have an average temperature of around 1100°C. The hottest lavas known are the Hawaiian lavas measured at Kilauea. Measurements were made in the lava lake Halemaumau by the use of Seger cones (ceramic material constructed to melt at various temperatures). The cones were placed in pipes, and then the pipes were lowered into the lake of lava. At a depth of 13 m the lava had a temperature of 1175°C. This decreased to 860°C at a depth of 1 m and increased again to 1000°C at the lava surface. Temperatures as high as 1350°C were recorded above the lava surface, where volcanic gases were reacting with atmospheric oxygen.

In 1910 Frank A. Perret recorded temperatures between 900 and 1000°C from the lava at Mount Etna in Sicily and in 1916 to 1918 observed lavas in Mount Vesuvius from 1015 to 1040°C. Minimum temperatures observed on lavas are around 750°C.

VOLCANIC GASES

As one might expect, taking an accurate sampling of volcanic gases is not an easy job. It is also difficult to decide whether the gases have come exclusively from the magma or partly from the surrounding rocks. We can, however, make a few generalizations from measurements at Kilauea.

Close to 70 percent of the volume of gases collected directly from a molten lake of lava was steam. Next in abundance were carbon dioxide, nitrogen, and sulfur gases, with smaller amounts of carbon monoxide, hydrogen, and chlorine. Even when gases other than steam make up only a small percentage of the total volume, their absolute quantities may be large. For example, in 1919 during the cooling of material erupted in 1912 from Mount Katmai in Alaska, the total amount of hydrochloric acid released was estimated at 1,250,000 t, and the total amount of hydrofluoric acid was approximately 200,000 t.

When any igneous rock is heated, it yields some quantity of gases. Water vapor predominates, and measurements indicate that it constitutes about 1 percent of fresh—that is, unweathered—igneous rocks. Estimates of the average water content of actual magma range from about 1 to 8 percent, with the weight of opinion centering around 2 percent. A silicate melt will not hold more than about 11 percent of volatiles under any circumstances.

PYROCLASTIC DEBRIS

Fragments blown out by explosive eruptions and subsequently deposited on the ground are called **pyroclastic debris**. The finest of these constitute **dust**, which is made up of pieces of the order of 10^{-4} cm in diameter. When volcanic dust is blown into the upper atmosphere, it can remain there for months, traveling great distances. The following fragments settle around or near the volcanic crater:

Ash Fragments consisting of sharply angular glass particles. Smaller than cinders.
Cinders Small, slaglike, solidified pieces of magma 0.5 to 2.5 cm across. Cinder **cones** may develop from the accumulation of these fragments (see Figure 3.3).
Lapilli Pieces about the size of walnuts.
Blocks Coarse, angular pieces of the cone or masses broken away from rock that blocks the vent.

FIGURE 3.3

A cinder cone (approximately 0.5 km in diameter) with basaltic lava spreading from its base. Some older cones and flows can also be seen. San Francisco Mountains area, Coconino County, Arizona. [From Harvard University, Kirk Bryan Library of Geomorphology.]

FIGURE 3.4

Two examples of pumice: fine-textured and coarse-textured.

Bombs Rounded masses that congeal from magma as it travels through the air.

Pumice Pieces of magma up to several centimeters across that trap bubbles of steam or other gases as they are thrown out (see Figure 3.4). After these solidify, they are honeycombed with gas-bubble holes, which give them sufficient buoyancy to float on water.

Fiery clouds During the Katmai eruption, a great avalanche of incandescent ash mixed with steam and other gases was extruded. Heavier than air, this highly heated mixture rolled down the mountain slope. Masses of such material are called **fiery clouds** (sometimes referred to by the French equivalent, *nuées ardentes*). The volume of material extruded was so great that it covered a valley of 140 km² to a depth of 30 m. For the next 10 years the steam and gases kept erupting from this extruded material through a great number of holes, called **fumaroles,** and the area was given the name Valley of Ten Thousand Smokes.

Fiery clouds have characterized eruptions at Mount Pelée on Martinique Island in the West Indies to such an extent that they have come to be known as Peléan types of eruption. Further characteristics are magma expelled as pumice and ash and in the final stage of eruptions a mass of viscous lava accumulated in a domelike form when the gas content is so reduced that it no longer shatters the magma on reaching the surface.

At a few minutes before 8:00 A.M. on May 8, 1902, a gigantic explosion occurred through one side of Pelée. A fiery cloud at temperatures around 800°C swept down the mountainside and engulfed the city of Saint

Pierre, wiping out its 25,000 inhabitants and many refugees from other parts of the island who had gathered there during the preceding days, when the eruption was building up with minor explosions and earthquakes (Figure 3.5); estimates of the death toll ran as high as 40,000. By the middle of October, a dome of lava too stiff to flow had formed in the crater, and from it a spine was extruded like a great blunted needle, with a diameter of 100 to 200 m and at its maximum a height of 300 m above the crater floor.

The eruptions at Pelée were apparently similar to those at Mount Mayon in the Philippines on April 27, 1968 (see Figure 3.6).

Worldwide effects In 1783 Asama in Japan and Laki in Iceland had explosive volcanic eruptions. Large quantities of dust were blown into the upper atmosphere. The sun's effectiveness in heating the earth's surface was so reduced that the winter of 1783 and 1784 was one of the severest on record. Benjamin Franklin was the first person to connect the unusual weather with the volcanic eruptions, publishing his ideas in May, 1784.

During 1814 and 1815 the earth's temperature was reduced, following volcanic eruptions of Mayon and Mount Tambora on Sumbawa island, east of Java. The eruption of Tambora threw so much dust into the air that for 3 days there was absolute darkness over a distance of 500 km. With this dust and the dust erupted from Mayon in the atmosphere, the amount of the sun's heat reaching the earth's surface was significantly reduced. The year 1815 became known as the "year without a summer," marked throughout the world by

FIGURE 3.5

Ruins of Saint Pierre, Martinique, shortly after its destruction by a fiery cloud on the morning of May 8, 1902. [Underwood and Underwood.]

FIGURE 3.6

Fiery clouds of eruption of Mount Mayon, Philippine Islands, April 27, 1968. [Smithsonian Institution.]

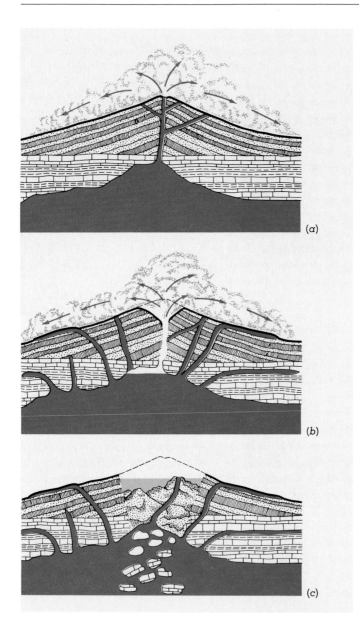

FIGURE 3.7

Sequence of events proposed by one hypothesis for the formation of a caldera. (*a*) An eruption begins with fiery clouds and dust clouds distributing materials on slopes and surrounding country. (*b*) Eruption continues; part of cone is blown away; and lava flows join in draining the magma reservoir. (*c*) Most of cone collapses into the reservoir; later activity forms cinder cone in caldera. [After H. Williams, "Calderas and Their Origin," *Bull. Univ. Calif. Dept. Geol. Sci.*, vol. 25, pp. 239–346, 1941.]

long twilights and spectacular sunsets also caused by the dust in the stratosphere.

FORMATION OF CALDERAS

Calderas may be formed by explosion, by collapse, or by a combination of both (see Figure 3.7). It is often difficult to determine just which mechanism is responsible.

The caldera on Bandai (37.58°N, 140.05°E), on the island of Honshu, Japan, is one example of a caldera formed by explosion. After 1,000 years of dormancy, Bandai exploded on July 15, 1888, blowing off its summit and part of its northern slope. After the violent explosion had subsided, a caldera was discovered over 1.5 km in diameter, with walls 360 m high.

The caldera on Kilauea was probably formed by the collapse of the summit. As great quantities of magma escaped from the reservoir beneath the volcano, support for the summit was withdrawn, and large blocks of it fell in, forming the caldera.

Crater Lake, in southern Oregon, lies in a basin that is an almost perfect example of the caldera shape. This is circular, with a diameter of a little more than 8.5 km and a maximum depth of 1,200 m, and it is surrounded by a cliff that rises 750 to 1,200 m. Crater Lake itself is about 600 m deep. The caldera was formed when the top of a symmetrical volcanic cone, Mount Mazama, vanished during an eruption. Geologists have studied the deposits on the slopes and tried to piece together its history.

First, a composite cone was slowly built to a height of around 3,600 m. Then glaciers formed, moving down from the crest and grooving the slopes as they traveled. An explosive eruption occurred 6,600 years ago, and the caldera was formed. Later activity built up a small cone inside the caldera, which now protrudes above the surface of Crater Lake as Wizard Island.

Not all observers agree on the origin of the Crater Lake caldera itself, however. The question is whether all or nearly all the missing material from the cone was actually blown out during an eruption or whether the caldera was created when the summit collapsed. The answer to the mystery should be provided by an analysis of the unconsolidated material found in the vicinity. Does this material consist of pyroclastics formed during an eruption, or does it consist of the broken remnants of Mazama's blown-off summit? The problem is that the summit itself originally included pyroclastics and con-

FIGURE 3.8

Profile of one of the world's nearly perfect composite cones: **Mayon, on Luzon.** [Harvard University, Gardner Collection.]

gealed lavas from earlier eruptions, and it is difficult to distinguish between the two. One investigator has concluded that, of the 70 km³ of Mazama that disappeared, only 8 km³ are represented in the materials now lying on the immediate slopes and that the rest of it was dropped into the volcano when the roof of an underlying chamber collapsed.[3] This chamber may have been partially emptied by the ejection of large volumes of material during an eruption. H. Williams found evidence that ash spread over a radius of nearly 50 km. Others have reported Mount Mazama ash as far east and northeast as Alberta and eastern Montana, and some of the magma may have worked its way beneath the surface into adjoining areas. This explanation of the Crater Lake caldera has been challenged on the grounds that it is based on an invalid distinction between "old" and "new" pyroclastics in the debris that covers the area.

CLASSIFICATION OF VOLCANOES

Volcanoes are classified according to the materials that have accumulated around their vents. Thus we have shield volcanoes, composite volcanoes, and cinder cones.

When the extruded material consists almost exclu-

[3]H. Williams, "Calderas and Their Origin," *Bull. Univ. Calif. Dept. Geol. Sci.*, vol. 25, pp. 239–346, 1941.

sively of lava poured out in quiet eruptions from a central vent or from closely related fissures, a dome builds up that is much broader than it is high, with slopes seldom steeper than 10° at the summit and 2° at the base. Such a dome is called a **shield volcano.** The five volcanoes of the island of Hawaii are shields.

Sometimes a cone is built up of a combination of pyroclastic material and lava flows around the vent. This form is called a **composite volcano** and is characterized by slopes of close to 30° at the summit, tapering off to 5° near the base. Mayon, on Luzon, is one of the finest examples of a composite cone (Figures 3.6 and 3.8).

A single volcano may develop as a shield volcano during part of its history and as a composite volcano later. Mount Etna is an example of such a volcano (see Figure 3.9).

Finally, small cones consisting mostly of pyroclastic debris, particularly cinders, are called **cinder cones.** They achieve slopes of 30° to 40° and seldom exceed 500 m in height. Many cinder cones have flows of basalt issuing from their base (Figure 3.3). Parícutin, in Mexico, is an example of a cinder cone that has developed in modern times, as we shall describe.

HISTORY OF SOME VOLCANOES

A volcano is considered **active** if there is some historical record of its having erupted. If it has not done so but

FIGURE 3.9

Mount Etna viewed from the sea near Catania, Sicily. The flat slopes to the left are those of a shield volcano. When Etna changed its eruptive habit late in its history, the explosive ejection of fragmental material built a 300-m pyroclastic cone on the summit of the broad shield of lava flows. These pyroclastics form the irregular and steepened slopes nearest the smoking vent. [Vittorio Sella.]

FIGURE 3.10

Vesuvius in eruption, 1906. The snow-clad slopes mark the modern volcano. To the left of this and in the foreground is the jagged remnant of Mount Somma. [A. & C. Caggiano.]

shows notable lack of erosional alteration, pointing to eruption within quite recent geologic time, it is considered **dormant,** or merely "sleeping," and capable of renewed activity. If a volcano not only has not erupted within historic time but also shows wearing away by erosion and no signs of activity (such as escaping steam or local earthquakes), it is considered **extinct.**

Vesuvius Vesuvius, on the shore of the Bay of Naples, has supplied us with a classic example of the reawakening of a dormant volcano (Figure 3.10). At the time of Christ, Vesuvius was a vine-clad mountain, Mount Somma, a vacation spot in southwest Italy favored by wealthy Romans. For centuries it had given no sign of its true nature. Then, in A.D. 63, a series of strong earthquakes shook the area; and around noon on August 24, 79, Somma started to erupt. The catastrophe of

that August day lay silent for nearly 17 centuries. When the remains of Herculaneum and Pompeii—cities that had been buried by the eruption—were uncovered, a story to grip the world's imagination transpired: Roman sentries had been buried at their posts; family groups, in the supposed safety of subterranean vaults, had been cast in molds of volcanic mud cemented to a rocklike hardness, along with their jewels, candelabra, and the food they had hoped would sustain them through the emergency.

Somma is believed to have erupted first about 10,000 years ago as a submarine volcano in the Bay of Naples. It then emerged as an island and finally filled in so much of the bay around it that it became a part of the mainland. It is the youngest volcano found in that vicinity.

There must have been an exceedingly long interval of quiet before the eruption of the year 79 because no earlier historical records of volcanic activity exist. But in 79 part of the old Somma cone was destroyed, and the cone now called Vesuvius started. During this eruption Pompeii was buried by pyroclastic debris; people were asphyxiated by gases from the ash and suffocated from the dust. Herculaneum was overwhelmed by mud flows of water-soaked ash deposited to a considerable depth.

Eruptions of pyroclastic debris occurred at intervals after 79. The longest period of quiescence lasted 494 years and was followed by an eruption in 1631, which poured out the first lava in historic time. This rejuvenation is attributed to chemical changes accompanying the magmatic ingestion of great quantities of limestone in a reservoir with a roof about 5 km below sea level (see Figure 3.11).

Krakatoa   One of the world's greatest explosive eruptions took place in 1883 at Krakatoa, in Sunda Strait between Java and Sumatra. Krakatoa had once been a single island, a volcanic mountain built up from the sea bottom. At a remote period in the past it split apart during an eruption. By 1883, after a long period of rebuilding, three cones had risen above sea level and had merged. These cones, named Rakata, Danan, and Perboewatan, and various unnamed shoals made up the outline of Krakatoa.

On August 26, 1883, a series of explosions began. The next day, at 10:20 A.M., a gigantic explosion blew the two cones Danan and Perboewatan to bits. A part of the island that had formerly stood 800 m high was left covered by 300 m of water. The noise of the eruption was heard on Rodrigues Island, 5,000 km across the Indian Ocean, and a wave of pressure in the air was recorded by barographs around the world. A great flood of water created by the activity drowned 36,500 persons in the low coastal villages of western Java and southern Sumatra. Columns of ash and pumice soared kilometers into the air, and fine dust rose to such heights that it was distributed around the globe and took more than 2 years to fall. During that time sunsets were abnormally colored all over the world. A reddish-brown circle, known as "bishop's ring," which was seen around the sun under favorable conditions, gave evidence not only of the continued presence of dust in the upper air but of the approximate size of the pieces—just under 0.002 mm. Since 1883, Krakatoa has revealed from time to time that it is in the process of actively rebuilding.

Parícutin   About 320 km west of Mexico City (19.50°N, 102.05°W) Parícutin sprang into being on February 20, 1943. Nine years later, it had become quite inactive, but during its life it was studied more closely than any other newborn vent in history.

Many stories of the volcano's first hours have been told. According to the version now generally regarded as the most reliable, Parícutin began about noon as a thin wisp of smoke rising from a cornfield that was being plowed by Dionisio Pulido. By 4 P.M. explosions

FIGURE 3.11

Schematic reconstruction of supposed relationships in the Somma-Vesuvius volcano. [After J. H. F. Umbgrove, *The Pulse of the Earth*, 2nd ed., p. 74, Martinus Nijhoff, The Hague, 1947.]

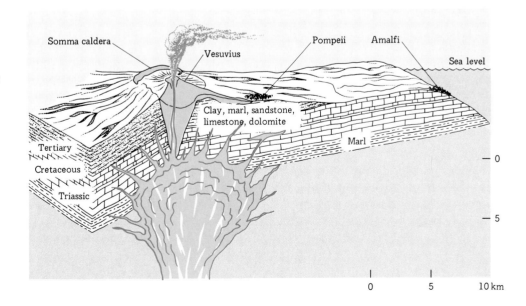

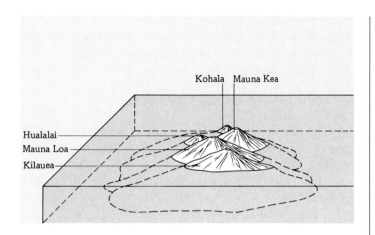

FIGURE 3.12

Schematic drawing of the five volcanoes that have been built up from the sea floor to merge and form the island of Hawaii (viewed here from the southeast).

TABLE 3.1

Volcanoes of the island of Hawaii[a]

Name	Area, km²	Portion of island, %	Summit elevation, m
Mauna Loa	5,290	50.5	4,100
Mauna Kea	2,390	22.8	4,135
Kilauea	1,430	13.7	1,225
Hualalai	755	7.2	2,475
Kohala	610	5.8	1,650

[a] From H. T. Stearns and G. A. Macdonald, "Geology and Groundwater Resources of the Island of Hawaii," *Hawaii Div. of Hydrog. Bull. 9*, p. 24, 1946.

were occurring every few seconds, dense clouds of ash were rising, and a cone had begun to build up. Within 5 days the cone was 100 m high, and after 1 year it had risen to 425 m. Two days after the eruption began, the first lava flowed from a fissure in the field about 300 m north of the center of the cone. At the end of 7 weeks this flow had advanced about 1.5 km. Some 15 weeks after the first explosion, lava had also begun to flow from the flanks of the cone itself.

After 9 years of activity, Parícutin abruptly stopped its eruptions and became just another of the many small "dead" cones in the neighborhood. The histories of these other cones, parasites of Toncítaro or of neighboring major volcanoes, undoubtedly parallel the story of Parícutin.

Other new volcanoes that have developed during historic time are Jorullo (18.85°N, 101.82°W) and Monte Nuovo (40.83°N, 14.10°E). Jorullo broke out in the middle of a plantation about 70 km southeast of Parícutin, in 1759. Monte Nuovo erupted in 1538, just west of Vesuvius.

Hawaii The Hawaiian Islands are peaks of volcanoes projecting above the ocean and strung out along a line running 2,400 km to the northwest. The Marquesas, Society, Tuamotu, Tubuai, Samoan, and other volcanic groups of the South Pacific form lines roughly parallel to the Hawaiian Islands.

At the northwestern end of the Hawaiian chain are the low Ocean and Midway islands. At the southeastern end is Hawaii, the largest of the group, 140 km long and 122 km wide, and the tallest deep-sea island in the world. It is composed of five volcanoes—Kohala, Hualalai, Mauna Kea, Mauna Loa, and Kilauea (see Figure 3.12). Each has developed independently, and each has its own geologic history: Lava from Mauna Kea has buried the southern slope of Kohala, and lava from Mauna Loa has buried parts of Mauna Kea, Hualalai, and Kilauea. The dimensions of the volcanoes are listed in Table 3.1 and represent portions of the volcanoes above sea level at the present time; they do not take into account buried slopes.

Kohala has been extinct for many years, but Mauna Kea shows evidence of having been active in the recent geological past although not within recorded history. Hualalai last erupted in 1801, and Mauna Loa was active about 6 percent of the time from 1832 to 1977. During the same interval, Kilauea was active about 66 percent of the time. Prior to that, extending back to A.D. 140, native legend tells of 40 to 50 eruptions of Kilauea.

FIGURE 3.13 (ABOVE)

Schematic section and plan view of a plume similar to that postulated for producing the volcanoes of the Hawaiian Islands. As the lithospheric plate moves slowly over a cylindrical upwelling of magmatic material, a new volcano develops from time to time, older volcanoes becoming inactive as the plate moves away from this hot spot.

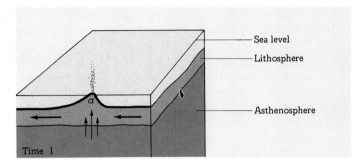

FIGURE 3.14 (BELOW)

Bend in seamount line (Emperor and Hawaiian chains) marks change in movement of Pacific plate over hot spot in mantle. Volcanoes of progressively younger age developed as the plate moved first in a northwesterly direction and later in a more westerly direction. An age of 70 million years has been obtained from rocks near the northwestern extremity of the Emperor chain (*a*), 46 million years for Koko seamount (*b*), 20 million years for Midway Island (*c*), and less than 1 million years for the volcano Kilauea on Hawaii (*d*).

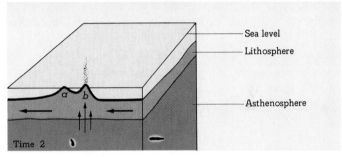

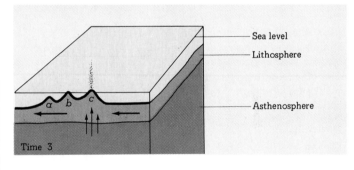

Investigation of the Hawaiian Island chain suggests that it is probably the result of the relative northwesterly motion of the Pacific plate over a zone of melting in the earth's mantle (a **plume**), which causes crustal extension and produces a "hot spot" in the earth's crust (according to J. Tuzo Wilson,[4] and W. Jason Morgan[5]). Lavas of basalt composition pour out at these **hot spots**, forming islands, which in turn become inactive as motion of the plate continues (Figure 3.13). Radiometric ages of the Hawaiian Island basalts tend to support this hypothesis, with the ages becoming younger toward the hot spot (the southeastern end of the chain) and older away from it. The Emperor seamount chain may be a continuation of the Hawaiian Islands, and indicates even older ages and a shift in the direction of plate migration over this hot spot (see Figure 3.14).

Measurements made on tilt meters show that both Mauna Loa and Kilauea swell up during the period when magma is rising from below. The uplift reaches a maximum just before an eruption, but after the eruption the mountains shrink back again. Such measurements

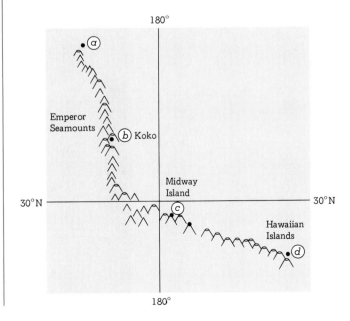

[4]"Continental Drift," *Sci. Am.*, vol. 208, no. 4, 1963.
[5]"Hot-spots, Plate Motions, and Polar Wandering," *Geol. Soc. Am. Abstr*, vol. 4, no. 3, p. 202, 1972.

on Kilauea[6] show that slow swelling over a period of months, about 16 on the average, produces a maximum uplift of 40 to 135 cm. Flank eruptions of the volcano allow sudden collapse of the summit over a period of days, amounting to 20 to 50 cm of subsidence per eruption. The swelling extends over an area about 10 km in diameter centered on, or just south of, Kilauea caldera. From the area and the amount of swelling involved, it was computed that in the 1960 eruption each centimeter of uplift or subsidence represented deformation of 1 million m³ of earth materials. Total volumes of 20 to 30 million m³ have been computed for larger eruptions.

[6]R. W. Decker, D. P. Hill, and T. L. Wright, "Deformation Measurements on Kilauea Volcano, Hawaii," *Bull. Volcanol.*, vol. 29, pp. 721–732, 1966.

Records of tilt can also be used to forecast an impending eruption.

When Kilauea is active, magma rises up within the mountain and floods out as lava into a pit in the floor of the caldera. Occasionally the lava flows out over the rim of the pit onto the floor and gradually raises its level. Usually, however, the lava is confined to the pit, forming what is termed a **lava lake**. This lava lake may last for years and then disappear completely for equally long periods. The level of the lake falls when lava flows from the flanks of the volcano, both above and below sea level. From time to time the system is drained, and the caldera floor collapses. Then the magma rises again, lava floods into the caldera, and the process is repeated.

Disappearing islands   Submarine volcanoes, like

FIGURE 3.15

A temporary island was formed at Metis Shoal of the Tonga Archipelago during a submarine eruption that began December 11, 1967. The island stayed above the surface for 58 days. [From Smithsonian Institution, Center for Short-lived Phenomena.]

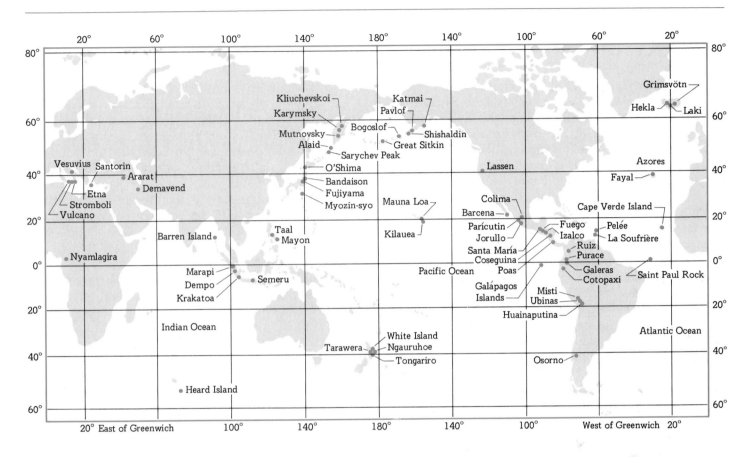

FIGURE 3.16

Location of some active volcanoes [Various sources.]

Krakatoa, that build themselves up above sea level, blow off their heads, and then rebuild produce the so-called disappearing islands of the Pacific Ocean. In 1913, for example, Falcon Island (20.4°S, 175.6°W), in the South Pacific, suddenly disappeared after an explosive eruption. On October 4, 1927, accompanied by a series of violent explosions, it just as suddenly reappeared. The island of Bogosloff (about 56°N, 168°W), in the Aleutians, was first reported in 1826 and has been playing hide-and-seek with mapmakers ever since.

On December 11, 1967, a volcanic eruption started at Metis Shoal of the Tonga Archipelago. After 20 days a new island about 700 m long and 100 m wide had poked up 20 m above the sea (Figure 3.15). From the start of the activity the island stayed above the surface for a total of 58 days. By February 1, it was a few "jagged rocks and water washing across." On February 19, there were "very high breakers on subsurface rocks" at the site; and by April 1, the shoal was completely under water, and there were no observed breakers.[7]

DISTRIBUTION OF ACTIVE VOLCANOES

We find evidence of volcanic eruptions in rocks of all ages. Apparently, igneous activity has been going on throughout geologic time and has occurred in the highest mountain ranges, on the bottom of the ocean, and on open plains. There are more than 500 active volcanoes in the world today. Some are located in Figure 3.16.

[7]Event Report, "The Submarine Volcanic Eruption and Formation of a Temporary Island at Metis Shoal, Tonga Islands," Center for Short-lived Phenomena, Smithsonian Institution, Washington, June 15, 1968.

3.2 BASALT PLATEAUS

On June 11, 1783, after a series of violent earthquakes on Iceland near Mount Skapta, an immense outpouring of lava began along a 16-km line, the Laki Fissure. Lava poured into the Skapta River, drying up the water and overflowing the stream's channel, which was 150 to 200 m deep and 60 m wide in places. Soon the Skapta's tributaries were dammed, and many villages in adjoining areas were flooded. The lava flow was followed by another 1 week later and a third on August 3. So great was their volume that they filled a former lake and an abyss at the foot of a waterfall. They spread out in great tongues 20 to 25 km wide and 30 m deep. As the lava flow diminished and the Laki Fissure began to choke up, 22 small cones formed along its length, relieving the waning pressures and serving as outlets for the final extrusion of debris.

This is one of the few authenticated instances within historical time of the mechanism known as fissure eruption, or lava flood. There is strong evidence, however, that floods of this sort occurred on a gigantic scale in the geological past. The rocks produced by lava floods are known as **flood basalts,** or **plateau basalts,** because of the tendency to form great plateaus. The low viscosity required for lava to flood freely over such great areas is characteristic only of lavas that have basaltic composition.

Iceland itself is actually a remnant of extensive lava floods that have been going on for over 50 million years and that have blanketed 0.5 million km². The congealed lava is believed to be at least 2,700 m thick in this area. The Antrim Plateau of northeastern Ireland, the Inner Hebrides, the Faeroes, and southern Greenland are also remnants of this great North Atlantic, or Britoarctic, Plateau.

Of equal magnitude is the Columbia Plateau in Washington, Oregon, Idaho, and northeastern California (see Figure 3.17). In some sections, more than 1,500 m of rock have been built up by a series of fissure eruptions. Individual eruptions deposited layers ranging from 3 to 30 m thick, with an occasional greater thickness. In the canyon of the Snake River, Idaho, granite hills from 600 to 750 m high are covered by 300 to 450 m of basalt from these flows. The Columbia Plateau has been built up during the past 30 million years. The principal activity took place a million years ago in northeastern California and Idaho, but some flows in the Craters of the Moon National Monument in southern Idaho, probably the most recent of United States fissure eruptions, are believed to have occurred within the last 250 to 1,000 years.

Other extensive areas built up by fissure eruptions include north-central Siberia, the Deccan Plateau of India, Ethiopia, around Victoria Falls on the Zambezi River in Africa, and parts of Australia.

3.3 IGNEOUS ACTIVITY AND EARTHQUAKES

Most volcanic eruptions are associated with earthquakes. In recent years, records of this seismic activity have given warnings of impending eruptions. The relationship between local earthquakes and eruptions is important enough to be described in detail in the following case.[8]

Raoul Island, a volcanic island, is 8 km across and is the largest of the Kermadec group, located approximately 1,000 km northeast of New Zealand. It lies in a very active zone of earthquake activity extending from Tonga Island to New Zealand.

Unusual seismic activity began on November 10,

[8]R. D. Adams and R. R. Dibble, "Seismological Studies of the Raoul Island Eruption, 1964," *New Zealand J. Geol. Geophys.*, vol. 10, no. 6, 1967.

1964, when the first of a swarm of local earthquakes were recorded. In the 10 weeks before this there had been only 1 earthquake. Then, within 4 h (hours), over 80 earthquakes per hour were being recorded. On November 11 earthquakes became less frequent, but the ground began to shake continuously, a phenomenon called **volcanic tremor.** Within 1 day the volcanic tremors were large enough to mask records of small individual earthquakes. By November 13 the tremor dropped off to about one-half its maximum intensity, and earthquakes were again being recorded at about 30 to 40 per hour. At 21 h 57 min (minutes) Greenwich time the largest earthquake of the series occurred. Raoul's seismograph was temporarily out of order, but this earthquake was large enough to be recorded on other seismographs, some as far away as North America. The United States Coast and Geodetic Survey determined it to be a moderate-sized earthquake and located its center at 20 km west of the island at a depth of 77 km. Volcanic tremor and frequency of earthquakes decreased then, until on November 20 only 10 to 15 per hour were being recorded.

Suddenly, on November 20, just before 18 h Greenwich time, the amplitude of volcanic tremor increased, and there was a "sound of a big landslide" accompanied by a cloud of steam, followed by "another roaring noise," which heralded the appearance of a great column of black mud that shot up to 1,000 m in the center of the cloud, with rocks flying out of the column and falling back into the crater. During the eruption a new crater 100 m in diameter was blown out within the main crater. Within 1 h the level of volcanic tremor settled down, with about 10 earthquakes per hour.

The Raoul eruption followed the not unusual pattern of occurring when the associated earthquake activity was on the wane after an extremely sudden rise. Seismographs are being watched carefully in the hope that they will provide forewarning of any future outbursts of volcanic activity.

DEPTH OF MAGMA SOURCE

Magma supplying eruptive materials of active volcanoes comes from deep within the earth. The depths of earthquakes and earth tremors associated with eruptions are giving us some clues to the depths of the magma.

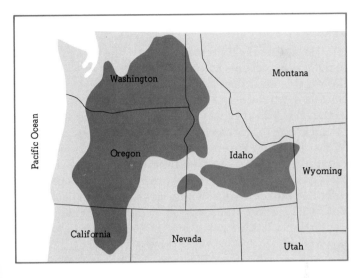

FIGURE 3.17

Map of Columbia Plateau, showing areas built by fissure eruptions.

Swarms of earthquakes and "harmonic tremor" with foci 60 km below the summit of Kilauea preceded the 1959 eruptions, which suggests that 60 km may be the depth at which magma enters the "plumbing system" of Kilauea.

The 1964 eruption on Raoul Island was preceded by an earthquake centered 20 km west of the island at a depth of 77 km. Later quakes were nearer and shallower, but there is a strong suggestion that the chain of events really started with magma at a depth of 77 km moving into the volcano's feeding conduit.

3.4 THE EARTH'S HEAT

The principal requirement for igneous activity is heat, and evidence has accumulated that there is a great deal of heat in the earth's interior. That the earth's temperature increases with depth was first noted as early as the seventeenth century, when German miners were reporting a steady rise in temperature with increasing depth in their mine shafts. Further proof has now come from the temperature measurements made in other mines and from deep drill holes in the earth.

The increase in temperature with depth is called the **thermal gradient.** Reported rates of increase have varied from place to place even in areas far removed from igneous activity. The thermal gradient can range from less than 10°C/km to as much as 50°C/km. Excluding seasonal temperature changes that affect only the top few meters of the earth's surface, a thermal gradient of 30°C/km seems to be about average. These figures get us "into hot water," so to speak, very quickly. Within 50 km of depth, a thermal gradient of 30°C/km would result in a temperature of 1500°C, believed to be high enough to melt any known rock, and at the earth's center it would produce a temperature of 192,000°C, more than 40 times the temperature of the sun's atmosphere.

Heat travels extremely slowly through soil and rock. Measurements and calculations have shown that the earth's **heat flow,** which is the product of the temperature gradient and the thermal conductivity of earth materials, is reasonably constant. Its average value is 1.2 μcal (microcalories)/cm^2-s (second), that is, 1.2 $\times$ 10^{-6} cal/cm^2-s. Present indications are that there is little difference in the heat flow from a number of different land and sea areas. The average heat flow represents about 40 cal/cm^2-year, which does not seem like much. But in its steady, quiet way it totals more energy in a year than all that released by seismic and igneous activity.

CAUSE OF IGNEOUS ACTIVITY

Igneous activity requires magma, and magma requires that rocks be melted. Where does this melted rock come from? Earthquake records indicate that most magma comes from the mantle. The heat to melt solid rock is apparently produced in two ways: Radioactive decay generates heat energy as a product of the transformation of one element into another; movement of rock masses past one another produces enormous amounts of frictional energy manifested as heat. These two mechanisms can account for the total heat necessary to generate magmas.

The ultimate causes of igneous activity are related to the same internal forces that elevate mountains, cause earthquakes, and result in metamorphism. Volcanic outpourings of lava and gases originate near the mantle's upper boundary. Hot magma squeezes upward, driven by the internal forces that are seemingly in constant motion within the earth (see Chapter 8 for more discussion of convection cells and convective overturn within the mantle). This magma may remain for relatively long periods in shallow chambers within the crust, or it may erupt at once through fissures in the earth's surface. The complex interaction of many factors determines when and how an eruption is triggered. Geologists are just now learning more about these phenomena through geophysical and geochemical studies combined with detailed field examination of critical regions.

3.5 MASSES OF IGNEOUS ROCKS

Igneous rocks are formed from the solidification of magma. From time to time, molten rock within the earth's crust works its way out to the surface, as we have discussed. There it may pour or be blown forth from vents or fissures and solidify into rock, or it may remain trapped within the crust, where it slowly cools and solidifies. Rocks of igneous origin compose 95 percent of the outermost 10 km of the globe. During our discussion of igneous activity, we dealt primarily with the extrusion of lava and pyroclastic debris and with some of the landforms that result: basalt plateaus and volcanoes. These are surface masses of igneous rock.

When magma within the crust loses its mobility and ceases activity, it solidifies in place, forming igneous rock masses of varying shapes and sizes. Today, such rocks can be seen at the surface on continents where previously overlying rocks have been worn away by erosion. For example, the first internal part of a volcano to be exposed by erosion is the plug that formed when magma solidified in the vent. Revealed next are the channels through which the magma moved to the surface. Finally, in some regions of ancient activity, the crust has been so elevated (see Chapter 7) and eroded that the reservoir that once stored the magma can now be seen as a solid rock mass at the surface. Solidified offshoots of magma from the reservoir, having intruded themselves into other rocks within the crust, are included in these masses, which were not necessarily connected with the eruption of a volcano.

PLUTONS

All igneous rock masses that were formed when magma solidified within the earth's crust are called **plutons.** When rocks have a definite layering, we may speak of the magma that invades them as **concordant** if its boundaries are parallel to the layering or **discordant** if its boundaries cut across the layering.

Plutons are classified according to their size,

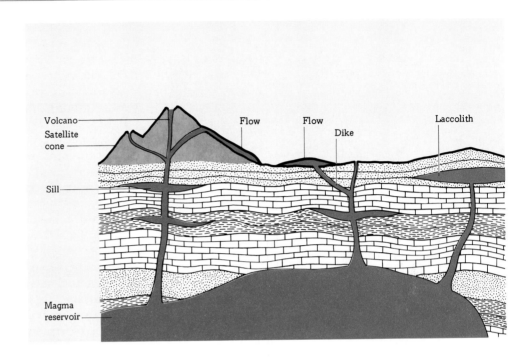

FIGURE 3.18

Plutons and landforms associated with igneous activity.

shape, and relationship to surrounding rocks. They include sills, dikes, lopoliths, laccoliths, and batholiths (Figure 3.18).

Tabular plutons A pluton with a thickness that is small relative to its other dimensions is called a **tabular pluton**.

Sills A tabular concordant pluton is called a **sill**. It may be horizontal, inclined, or vertical, depending on the attitude of the rock structure with which it is concordant.

Sills range in size from sheets less than 1 cm in thickness to tabular masses hundreds of meters thick. Because a sill is an intrusive form—that is, it has forced its way into already existing rocks—it is always younger than the rocks that surround it. A sill must not be confused with an ordinary lava flow that has been buried by other rocks later on. There are fairly reliable ways of distinguishing between the two types: A buried lava flow usually has a rolling or wavy top pocked by the scars of vanished gas bubbles and showing evidence of erosion, whereas a sill has a more even and unweathered surface; also, a sill may contain fragments of rock that were broken off when the magma forced its way into the surrounding structures.

The Palisades along the west bank of the Hudson River near New York are the remnants of a sill that was several hundred meters thick. Here the magma was originally intruded into flat-lying sedimentary rocks. These are now inclined at a low angle toward the west.

Dikes A tabular discordant pluton is called a **dike** (see Figure 3.19). Dikes originated when magma forced its way through the fractures of adjacent rocks.

The width of individual dikes ranges from a few centimeters to many meters. The Medford dike near Boston, Massachusetts, is 150 m wide in places. Just how far we can trace the course of a dike across the countryside depends in part on how much of it has been exposed by erosion. In Iceland dikes 15 km long are common, and many can be traced for 50 km; at least one is known to be 100 km long.

As magma forces its way upward, it sometimes pushes out a cylindrical section of the crust. Today, as a result, we find exposed at the surface some roughly circular or elliptical masses of rock that outline the cylindrical sections of the crust. These solidified bodies of magma are called **ring dikes.** Large ring dikes may be many kilometers around and hundreds or thousands of meters deep. Ring dikes have been mapped with widths of 500 to 1,200 m and diameters ranging from 2 to 25 km.

Some dikes occur in concentric sets. These originated in fractures that outline an inverted cone, with

FIGURE 3.19

Basalt dike (approximately 1 m thick) cutting through granite at Cohasset, Massachusetts. [John A. Shimer.]

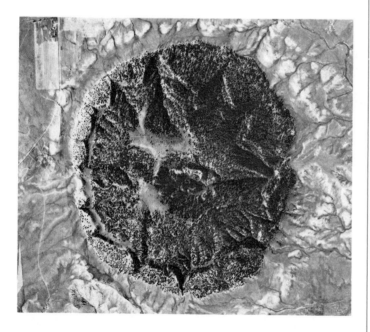

FIGURE 3.20

Laccolith with igneous core still covered: Green Mountain Dome, Sundance Quadrangle, Wyoming. (Dark core about 0.5 km in diameter.) [From Harvard University, Kirk Bryan Library of Geomorphology.]

the apex pointing down into the former magma source. These dikes are called **cone sheets.** In Scotland, the dip of certain cone sheets suggests an apex approximately 5 km below the present surface of the earth. Dikes are also found in approximately parallel groups called **dike swarms.**

Lopoliths **Lopoliths** are tabular concordant plutons shaped like a spoon, with both the roof and the floor sagging downward.

A well-known example is the Duluth lopolith, which crops out on both sides of Lake Superior's western end and appears to continue beneath the lake. It has been computed to be 250 km across and 15 km deep, with a volume of 200,000 km³.

Most lopoliths are composed of rock that has been differentiated into alternating layers of dark and light minerals, presenting the appearance of thinly bedded sedimentary rock.

Massive plutons Any pluton that is not tabular in shape is classified as a **massive** pluton.

Laccoliths A massive concordant pluton that was created when magma pushed up the overlying rock structures into a dome is called a **laccolith** (see Figure 3.20). If the ratio of the lateral extent of a pluton to its thickness is less than 10, the pluton is arbitrarily classed as a laccolith; if this ratio is more than 10, the pluton is classed as a sill. Obviously, because it is extremely difficult to establish the lateral limits of a pluton, in many cases it is best to use the term **concordant pluton,** supplemented by whatever dimensional details we can observe.

A classic development of laccoliths is found in the Henry, La Sal, and Abajo Mountains of southeastern Utah, where their features are exposed on the Colorado Plateau, a famous geological showplace.

Batholiths A large discordant pluton that increases in size as it extends downward is called a **batholith.** The term "large" in this connection is generally taken to mean a surface exposure of more than 100 km². A pluton that has a smaller surface exposure but exhibits the other features of a batholith is called a **stock.**

Batholiths are exposed thousands of meters above sea level, where they have been lifted by forces operating in the earth's crust. Thousands of meters of rock that covered the batholiths have been stripped away by the erosion of millions of years. We can observe these roots of mountain ranges in the White Mountains of New Hampshire and in the Sierra Nevada (Figure 3.21).

Although batholiths provide us with some valuable data, they also raise a host of unsolved problems. All these problems bear directly on our understanding of igneous processes and the complex events that accompany the folding, rupture, and eventual elevation of sediments to form mountains. (We shall discuss these mountain-forming processes in Chapters 6 and 7.)

We can summarize what we know about batholiths as follows:

1 Batholiths are located in mountain ranges. Although in some mountain ranges no batholiths are exposed at the present time, we never find batholiths that are not associated with mountain ranges. In any given mountain range the number and size of the batholiths seem to be directly related to the intensity of the folding and crumpling that have taken place. This does not mean, however, that the batholiths caused the folding and crumpling. Actually, there is convincing evidence to the contrary, as we shall see in some of the following features.

2 Batholiths usually run parallel to the axes of mountain ranges.

3 Batholiths have been intruded across the folds, indicating that they were formed after the folding of the mountains—although the folding may have continued after the batholiths were formed.

4 Batholiths have irregular dome-shaped roofs. This characteristic shape is related to **stoping,** one of the mechanisms by which magma moves upward into the crust. As the magma moves upward, blocks of rock are broken off from the structures into which it is intruding. At low levels, when the magma is still very hot, the stoped blocks may be melted and assimilated by the magma reservoir. Higher in the crust, as the magma approaches stability and its heat diminishes, the stoped blocks are frozen in the intrusion as **xenoliths,** that is, "strange rocks" (see Figure 3.22).

5 Batholiths are primarily composed of granite or granodiorite (a plutonic rock consisting of quartz, calcic plagioclase, orthoclase, and some mafic constituents such as pyroxene, hornblende, and/or biotite).

6 Batholiths give the impression of having replaced the rocks into which they have intruded instead of having pushed them aside or upward. But if that is what really took place, what happened to the great volumes of rock that the batholiths appear to have replaced? Here we come up against the problem of the origins of batholiths—in fact, against the whole mystery of igneous activity. Some observers have been

FIGURE 3.21

Weathering has exposed this portion of the Sierra Nevada batholith in Yosemite National Park, California. [W. C. Bradley.]

FIGURE 3.22

Xenolith in Mount Airy granite, North Carolina, which provides evidence in support of magmatic origin of this granite because the inclusion could not have retained its sharp edges and separate identity during granitization. [T. M. Gathright II.]

led even to question whether granitic batholiths were formed from true magmas at all. The suggestion has been made that the batholiths may have been formed through a process called **granitization,** in which solutions from magmas move into solid rocks, exchange ions with them, and convert them into rocks that have the characteristics of granite but have never actually existed as magma. We shall return to this highly controversial proposal in Chapter 5.

7 Batholiths contain a great volume of rock. The Sierra Nevada batholith of California is 650 by 60 to 100 km, and a partially exposed batholith in southern California and Baja California is probably 1,600 by 100 km.

8 Gravity measurements have been interpreted[9] as indicating that the downward extent, or "thickness," of many batholiths is some 10 to 15 km.

Figures 3.23 and 3.24 summarize the conventional representation of the concordant and discordant plutons.

3.6 FORMATION OF IGNEOUS ROCKS

Igneous rocks at the surface today have been formed from magma. As we pointed out earlier in this chapter molten rock in the ground is magma, and when it is extruded on the surface, it is lava. And when solidified pieces of magma are blown out, they are pyroclastic debris.

Pyroclastic debris eventually becomes hardened into rock through the percolation of ground water. In one sense, rocks formed in this way could be classified as sedimentary; but because they consist of solidified pieces of magma, we shall include them in our discussion of igneous rocks. Volcanic ash that has hardened into rock is called **tuff.** If many relatively large angular blocks of congealed lava are embedded in a mass of ash and then hardened to rock, the rock is called **volcanic breccia.** If such included pieces are mainly rounded fragments, the rock is called **volcanic conglomerate.**

Magma, extruded as lava at the surface, cools and solidifies to form igneous rocks. The offshoots of magma that work their way into surrounding rock below the surface cool more slowly and solidify. Even the magma reservoir eventually cools and solidifies, but it takes much longer because it is a larger mass. **All igneous rocks have been formed from the solidification of magma.**

CRYSTALLIZATION OF MAGMA

Magma solidifies through the process of *crystallization.* At first, magma is a **melt,** a liquid solution of elements at high temperature. After a decrease occurs in the heat that keeps the magma liquid, the melt starts to solidify. Bit by bit, mineral grains begin to grow. As this growth goes on, gases are released. Now we no longer have a complete liquid but rather a liquid mixed with solid and gaseous materials.[10] As the temperature continues to fall, the mixture solidifies until igneous rock is formed.

FIGURE 3.23

Concordant plutons.

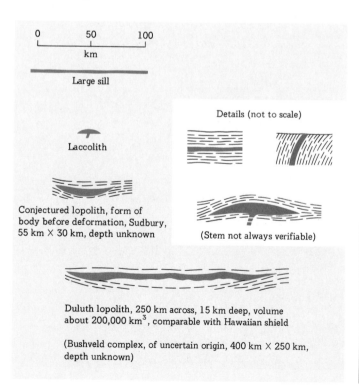

Large sill

Details (not to scale)

Laccolith

Conjectured lopolith, form of body before deformation, Sudbury, 55 km × 30 km, depth unknown

(Stem not always verifiable)

Duluth lopolith, 250 km across, 15 km deep, volume about 200,000 km³, comparable with Hawaiian shield

(Bushveld complex, of uncertain origin, 400 km × 250 km, depth unknown)

0   50   100
km

---

[9]M. H. P. Bott and Scott B. Smithson, "Gravity Investigations of Subsurface Shape and Mass Distribution of Granite Batholiths," *Bull. Geol. Soc. Am.,* vol. 78, pp. 859–878, 1967.

[10]To reflect this changing picture, we define a magma as any naturally occurring silicate melt, whether or not it contains suspended crystals or dissolved gases.

Igneous rocks may consist of interlocking grains of a single mineral or a mixture of several. As pointed out in Section 2.5, the most common rock-forming minerals are the silicates olivine, augite, hornblende, biotite, muscovite, orthoclase, plagioclase, and quartz.

BOWEN'S REACTION PRINCIPLE

Magma is a solution of elements, but it does not crystallize in the way ordinary solutions do. Most solutions of a given composition always crystallize into a solid of the same composition. This happens regardless of conditions during solidification. If crystallization of a magma were similar, it would always yield rock of the same composition. However, magma of a given composition may be able to crystallize into a number of different kinds of rock.

In 1922, N. L. Bowen[11] of the Geophysical Laboratory of the Carnegie Institution of Washington proposed that the differences in end products depend on the rate at which the magma cools and on whether early-formed minerals remain in or settle out of the remaining liquid during its crystallization. He suggested that, as a magma cools, the first-formed minerals undergo continuous modification with the liquid remaining after they crystallized. He called this process **reaction**. Reaction is the key to magmatic crystallization.

Based on a number of laboratory experiments with silicate melts, Bowen was able to arrange the rock-forming minerals of igneous rocks into **reaction series**. He found that an important characteristic of a silicate melt is that, as one mineral develops at a certain temperature, it will be converted upon further cooling into a different mineral by reaction with the liquid around it. The reaction series are of two different types, continuous and discontinuous. In the *continuous* reaction series some early-formed minerals are converted into new minerals by continuously changing their composition but not their crystalline structure. In the *discontinuous* reaction series some early-formed minerals react with the melt to change to new minerals with different compositions and a different crystalline structure.

Taking a magma with olivine basaltic composition, Bowen arranged the rock-forming minerals into a discontinuous reaction series of the ferromagnesian minerals and a continuous reaction series of the feldspars.

[11]N. L. Bowen, "The Reaction Principle in Petrogenesis," *J. Geol.*, vol. 30, pp. 177–198, 1922.

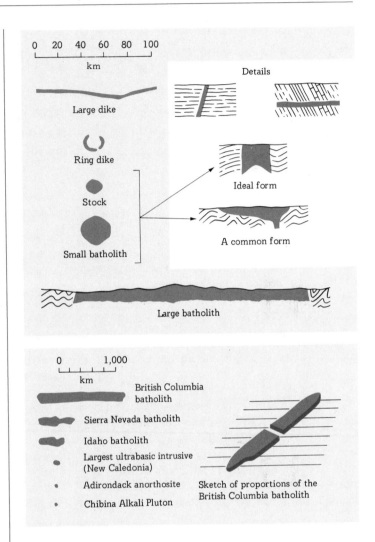

FIGURE 3.24

Discordant plutons.

As this magma cools, olivine and the calcium feldspar, anorthite, are the first minerals to form. If no minerals settle out, the melt may solidify to basalt or gabbro. If some of the early-formed minerals settle out, however, in a process called **fractionation**, the reaction process will continue further and the remaining minerals react with the remaining melt. Augite becomes hornblende, and the calcium-sodium feldspar becomes the sodium feldspar, albite. The greater the degree of fractionation, the more extensive the reaction process. With a high degree of fractionation the whole reaction series is gone through and later-formed minerals will be rich in silica (see Figure 3.25). Thus, according to Bowen's

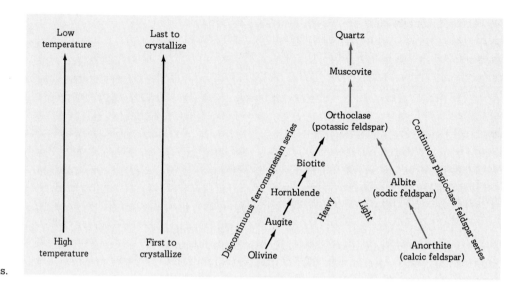

FIGURE 3.25

Bowen's reaction series.

reaction principle, the fractional crystallization of an olivine basaltic magma can lead to the formation of successively more siliceous rocks until finally a rock of granitic composition is reached.

Bowen's reaction principle explains how an olivine basaltic magma may solidify as one rock type or may produce several rock types. It might crystallize to a gabbro that is composed largely of ferromagnesian minerals, or it might produce a granite that is composed predominantly of sialic minerals. The rock formed depends on the extent to which early-formed

TABLE 3.2

Magma crystallization and depth[a]

Thickness, m	Time required, yr
1	0.033
10	3
100	300
1,000	30,000
10,000	3,000,000

[a]R. A. Daly, *Igneous Rocks and the Depths of the Earth*, p. 63, McGraw-Hill Book Company, New York, 1933.

minerals were removed from further reaction with the melt and on the rate of cooling.

Rate of crystallization The rate at which a magma crystallizes influences the extent to which fractionation and reaction take place. When magma cools rapidly, there is no time for minerals to settle or to react with the remaining liquid. This occurs when a partially crystallized magma is extruded onto the surface or injected into thin dikes or sills. But when a large body of magma cools slowly, deep within the crust, a high degree of fractionation or chemical reaction may take place. So the rate of crystallization varies with depth. For example, a magma consisting largely of nonferromagnesians at 1100°C, exposed to the air on top and ranging in thickness from 1 to 10,000 m, would solidify as shown in Table 3.2.

There are some objections to Bowen's hypothesis. According to his reaction principle, only about 5 or 10 percent of an original basic magma could solidify into granite or granodiorite (a composition between that of granite and of diorite). This extremely small percentage cannot account for the large masses of these rocks found on the continents. Where are the earlier-formed, more basic rocks? We shall examine this problem further in Chapter 5 under the topic of granitization.

3.7 TEXTURE OF IGNEOUS ROCKS

Texture, a term derived from the Latin *texere*, "to weave," is a physical characteristic of all rocks. The term refers to the general appearance of rocks. In referring to the texture of igneous rocks, we mean specifically the size, shape, and arrangement of their interlocking mineral grains.

GRANULAR TEXTURE

If magma has cooled at a relatively slow rate, it will have had time to develop grains that the unaided eye can see in hand specimens. Rocks composed of such large mineral grains are called **granular** (see Figure 3.26).

The rate of cooling, however, though important, is not the only factor that affects the texture of an igneous rock. For example, if a magma is of low viscosity—that is, if it is thin and watery and flows readily—large, coarse grains may form even though the cooling is relatively rapid; for in a magma of this sort the ions can move easily and quickly into their rock-forming mineral combinations.

APHANITIC TEXTURE

The rate at which a magma cools depends on the size and shape of the magma body, as well as on its depth below the surface. For example, a small body of magma with a large surface area—that is, a body that is much longer and broader than it is thick—surrounded by cool, solid rock loses its heat more rapidly than would the same volume of magma in a spherical reservoir. And because rapid cooling usually prevents large grains from forming, the igneous rocks that result have **aphanitic** textures. Individual minerals are present but are too small to be identified without a microscope.

GLASSY TEXTURE

If magma is suddenly ejected from a volcano or a fissure at the earth's surface, it may cool so rapidly that there is no time for minerals to form at all. The result is a **glass,** which by a rigid application of our definition is not really a rock but is generally treated as one. Glass is a special type of solid in which the ions are not

FIGURE 3.26

Enlarged photograph of a piece of granular igneous rock, taken through a slice that has been ground to translucent thinness (known as a *thin section*). The photograph shows the rock to be composed of interlocking crystals of different minerals. (Field of view approximately 0.5 cm in diameter.)

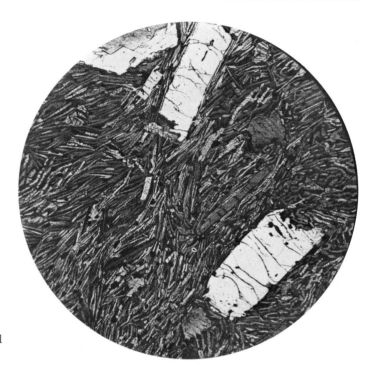

FIGURE 3.27

Enlarged photograph of a thin section of porphyritic igneous rock. (Largest crystal approximately 0.125 cm by 0.25 cm.)

arranged in an orderly manner. Instead, they are disorganized, like the ions in a liquid. And yet they are frozen in place by the quick change of temperature.

PORPHYRITIC TEXTURE

Occasionally, a magma cools at variable rates—slowly at first, then more rapidly. It may start to cool under conditions that permit large mineral grains to form in the early stages, and then it may move into a new environment where more rapid cooling freezes the large grains in a **groundmass** of finer-grained texture (see Figure 3.27). The large minerals are called **phenocrysts.** The resulting texture is said to be **porphyritic.** *Porphyry,* from the Greek word for "purple," was originally applied to rocks containing phenocrysts in a dark-red or purple groundmass.

In rare cases magma may be suddenly expelled at the surface after large mineral grains have already formed. Then the final cooling is so rapid that the phenocrysts become embedded in a glassy groundmass.

3.8 TYPES OF IGNEOUS ROCK

Several systems have been proposed for the classification of igneous rocks. All are artificial in one detail or another, and all rely on certain characteristics that cannot be determined in the field or from hand specimens. For our present purposes we shall emphasize texture and composition. Such a classification is entirely adequate for an introductory study of physical geology and even for many advanced phases of geology.

This classification appears in tabular form in Figure 3.28, together with a graph that shows the proportions of silicates in each type of igneous rock. The graph demonstrates the *continuous progression from rock types in which light-colored minerals predominate to rock types in which dark-colored minerals predominate.* The names of rocks are arbitrarily assigned on the basis of average mineral composition and texture. Sometimes intermediate types are indicated by such names as granodiorite. There are many more igneous rocks than are shown in this figure.

FIGURE 3.28

General composition of igneous rocks is indicated by a line from the name to the composition chart: Granite and rhyolite consist of about 50 percent orthoclase, 25 percent quartz, and 25 percent plagioclase feldspars and ferromagnesian minerals. Granite is the most important granular rock, and basalt the most important aphanitic rock. [Composition chart modified from L. Pirsson and A. Knopf, *Rocks and Rock Minerals*, p. 144, John Wiley & Sons, Inc., New York, 1926.]

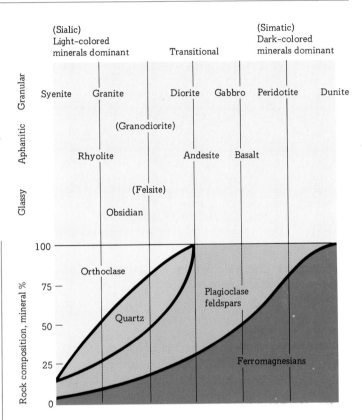

LIGHT-COLORED IGNEOUS ROCKS

The igneous rocks on the left side of Figure 3.28 are light both in color and specific gravity. They are sometimes referred to as **sialic** rocks.

It has been estimated that granites and granodiorites together comprise 95 percent of the igneous rocks of the continents that have solidified from magma. The origin and history of some granites are still under debate, but we use the term here to indicate composition and texture, not origin.

Granite is a granular rock (see Figure 3.29). Its mineral composition is as follows:

2 parts orthoclase feldspar + 1 part quartz + 1 part plagioclase feldspars + small amount of ferromagnesians = GRANITE

Rocks with the same mineral composition as granite but with an aphanitic rather than granular texture are called **rhyolite.**

The glassy equivalent of granite is **obsidian** (see Figure 3.30). Although this rock is listed near the left side of the composition chart, it is usually pitch black in appearance. Actually, though, pieces of obsidian thin enough to be translucent turn out to be smoky white against a light background.

DARK-COLORED IGNEOUS ROCKS

Of the total volume of rock formed from magma that has poured out onto the earth's surface, it is estimated that 98 percent is basalts and andesites.

A popular synonym for basalt is **trap rock,** from a Swedish word meaning "step." This name refers to the tendency of certain basalts to form columns that look

FIGURE 3.29

Granite. (Specimen approximately 5 cm by 6 cm.) [Robert Navias.]

FIGURE 3.30

Obsidian, the glassy equivalent of granite. (Specimen approximately 10 cm across.) [Robert Navias.]

like stairways in some outcrops (see Figure 3.31). These are a product of the cooling process and become evident after weathering.

Basalt has aphanitic texture. Its mineral composition is as follows:

1 part plagioclase feldspars + 1 part ferromagnesians = BASALT

The granular equivalent of basalt is **gabbro**. **Peridotite** is a granular igneous rock that is composed largely of ferromagnesian minerals.

INTERMEDIATE TYPES

Composition Igneous-rock compositions blend continuously from one to another as we go from the light to the dark side of the classification chart. *Andesite* is the name given to aphanitic igneous rocks that are intermediate in composition between granite and basalt. These rocks were first identified in the Andes mountains of South America—hence the name. Andesites are mostly found in areas around the Pacific Ocean. The granular equivalent of andesite is **diorite**.

Texture Going from the top to the bottom of the chart in Figure 3.28, we find that the rock textures grade continuously from granular to aphanitic, whereas the composition remains the same. For example, if we read down along the first vertical rule, we find that granite,

rhyolite, and obsidian become progressively finer-grained although all three have essentially the same composition. The same is true of gabbro and basalt.

In addition to these textures, any of the rocks may have porphyritic texture. Essentially, this means that a given rock has grains of two distinctly different sizes: conspicuously large phenocrysts embedded in a finer-grained groundmass. When the phenocrysts constitute less than 25 percent of the total, the adjective porphyritic is used to modify the rock name, as in porphyritic granite or porphyritic andesite. When the phenocrysts constitute more than 25 percent, the rock is called a porphyry (see Figure 3.32). The composition of a porphyry and the texture of its groundmass are indicated by using rock names as modifiers, as granite porphyry or andesite porphyry.

PEGMATITE

The solutions that develop late in the cooling of a magma are called **hydrothermal**. These crystallize into exceptionally granular igneous rock, called **pegmatite** (from the Greek *pēgmat-*, "fastened together"), which embodies the chief minerals to form from the hydrothermal solutions: potassium feldspar and quartz. So intimately intergrown are the grains of these minerals that they form what is essentially a single unit. The quartz is darker than the feldspar, and the overall pattern suggests the wedge-shaped figures of the writing of ancient Assyria, Babylonia, and Persia. As a result, this has become known as *graphic structure* (from the Greek *graphein*, "to write") (see Figure 3.33).

Pegmatite is found in dikes at the margins of batholiths and stocks. The dikes range in length from a few centimeters to a few hundred meters and contain crystals of very large size. In fact, some of the largest crystals known have been found in pegmatite. Crystals of spodumene (a lithium mineral) that measure 12 m in length have been found in the Black Hills of South Dakota; crystals of beryl (a silicate of beryllium and aluminum) that measure 1 by 5 m have been discovered in Albany, Maine. Great masses of potassium feldspar, weighing over 2,000 t yet showing the characteristics of a single crystal, have been mined from pegmatite in the Karelo-Finnish Soviet Socialist Republic.

Nearly 90 percent of all pegmatite is *simple* pegmatite of quartz, orthoclase, and unimportant percentages of micas. It is more generally called *granite pegmatite* because the composition is that of granite and

FIGURE 3.31

Columnar jointing, a pattern sometimes found in basalt, outlines a series of columns. Giant's Causeway, near Portrush, Antrim, Northern Ireland, is one of the best-known exposures of this feature. Large columns are approximately 1 m across. [L. Don Leet.]

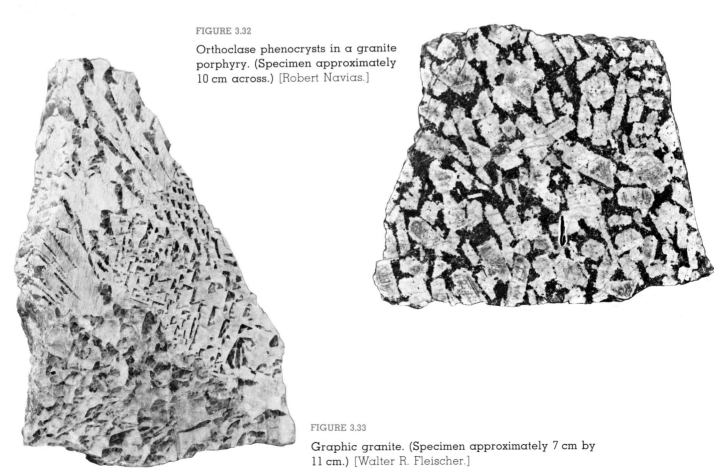

FIGURE 3.32

Orthoclase phenocrysts in a granite porphyry. (Specimen approximately 10 cm across.) [Robert Navias.]

FIGURE 3.33

Graphic granite. (Specimen approximately 7 cm by 11 cm.) [Walter R. Fleischer.]

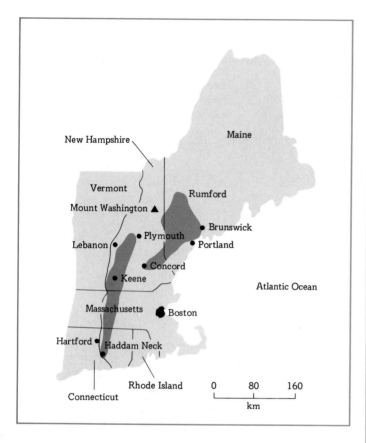

the texture that of pegmatite. The remaining 10 percent includes *complex* pegmatites. The major components of complex pegmatites are the same sialic minerals that we find in simple pegmatites, but in addition they contain a variety of rare minerals: lepidolite, tourmaline (best known as a semiprecious gem), topaz (also a gem), tantalite, and uraninite (*pitchblende*).

Simple pegmatites are common in some regions and complex pegmatites in others. In southwestern New Zealand, for example, pegmatites are uniformly simple, but throughout the Appalachian regions of North America complex pegmatites are more abundant. Figure 3.34 shows the areas in New England where pegmatites are found.

IGNEOUS ROCKS ON THE MOON

Rocks brought back from the moon by the *Apollo* missions are very much like terrestrial rocks except that they have no water in their compositions and almost no free oxygen. Most of the rocks from the moon are igneous in origin[12]—basalt lavas from the maria; gabbro, norite (a gabbro consisting of plagioclase and orthorhombic pyroxene), and anorthosite (composed essentially of plagioclase with more calcium than sodium) from the highlands (see also Chapter 17).

[12]A few rocks are made of angular particles formed by fragmentation of surface material by meteorite impact and subsequent compaction into a coherent rock (breccia) by shock compression during cratering by later meteorite impacts.

FIGURE 3.34

Map of regions where pegmatites are extensively developed, east of the Connecticut Valley in New England. Pegmatite areas are in dark color.

OUTLINE

Igneous activity consists of movements of molten rock inside and outside the earth and the variety of effects associated with these movements.

Volcanoes are surface piles of material that have accumulated around vents during successive eruptions.

    Evolution of volcanoes starts with basaltic lava flows, then andesitic lava flows, and then ejection of fragmental materials; the cycle ends when acidic lava congeals in the volcanic vent.

    A **volcanic eruption** is started and maintained by gases, of which steam is the most important.

    Composition and temperature of lavas have been measured from 750 to 1175°C.

    Volcanic gases are two-thirds steam but include carbon monoxide and dioxide, nitrogen, sulfur, hydrogen, fluorine, and chlorine.

    Pyroclastic debris consists of fragments blown out by explosive eruptions and includes ash, cinders, lapilli, blocks, bombs, and pumice.

        Fiery clouds of ash mixed with gases are heavier than air and flow down a volcano's side.

        Worldwide effects include volcanic dust in the stratosphere and reduction of solar heat reaching the earth's surface.

Formation of calderas may be by explosion, by collapse, or by a combination of both.

Classification of volcanoes according to the materials that have accumulated around their vents designates them as shield, composite, and cinder cones.

History of some volcanoes shows they may be dormant for long periods and again become active.

> **Vesuvius** was dormant for centuries and then in A.D. 79 started to erupt again.
>
> **Krakatoa** erupted in 1883 with one of the world's greatest explosions.
>
> **Disappearing islands** are submarine volcanoes that blow their heads off.
>
> **Parícutin** sprang into being on February 20, 1943, and was active for 9 years.
>
> **Hawaiian Islands** are peaks of volcanoes projecting above sea level and strung out along a line running 2,400 km to the northwest.

Distribution of active volcanoes locates many in the circum-Pacific belt and the Alpine-Himalayan belts and others in the Pacific Ocean, the Atlantic Ocean, the Indian Ocean, Africa, and the Antarctic, for a total of more than 500.

Basalt Plateaus are built by fissure eruptions.

Igneous activity and earthquakes seem to be closely related.

> **Depth of magma source** suggested by associated earthquakes may be 60 km at Kilauea and 77 km at Raoul Island.

The earth's heat is indicated by the thermal gradient to be more than adequate for igneous activity. **Speculations on the origin of the earth's heat** have included original heat and radioactivity.

> **Cause of igneous activity** seems to be the same internal force that elevates mountains, causes earthquakes, and causes metamorphism, but it is not surely known.

Igneous rocks are formed from the solidification of molten matter.

> **Masses of igneous rocks** are called **plutons,** which are classified according to size, shape, and relationships to surrounding rocks.
>
> > **Sills** are concordant tabular plutons.
> >
> > **Dikes** are discordant tabular plutons.
> >
> > **Lopoliths** are tabular concordant plutons shaped like a spoon.
> >
> > **Laccoliths** are massive concordant plutons with domed tops.
> >
> > **Batholiths** are massive discordant plutons 10 to 40 km thick.

Igneous rocks at the surface today were formed from magma.

> **Magma** solidifies through the process of crystallization.
>
> **Bowen's reaction series** are incorporated in a hypothesis accounting for all igneous rocks coming from an olivine basaltic magma.
>
> > **Limitations of Bowen's hypothesis** include failure to account for large undifferentiated masses of granite.
> >
> > **The rate of crystallization** is an important control over the rocks that form.

Texture of igneous rocks is the size, shape, and arrangement of their interlocking mineral grains.

> **Granular texture** includes large mineral grains from slow-cooling or low-viscosity magma.
>
> **Aphanitic texture** from rapid cooling consists of individual minerals so small that they cannot be identified without the aid of a microscope.
>
> **Glassy texture** results from ions disorganized as in a liquid but frozen in place by quick cooling.
>
> **Porphyritic texture** is a mixture of large mineral grains in an aphanitic or glassy groundmass.

Types of igneous rock are arbitrarily defined in terms of texture and composition.

Light-colored igneous rocks, sometimes called sialic, are dominated by granites and granodiorites.

Dark-colored igneous rocks (basalts and andesites) constitute 98 percent of rock formed from magma that has poured out onto the earth's surface.

Intermediate types of composition are given arbitrary names, such as andesite and diorite, because igneous rock compositions blend continuously from one to another from the light to the dark side of the classification chart.

Intermediate types of texture are also given arbitrary names such as granite, rhyolite, and obsidian, because rocks of a given composition grade continuously from granular to aphanitic to glassy texture.

Pegmatite is an exceptionally granular rock formed by hydrothermal solutions late in the cooling of a magma.

Igneous rocks on the moon are similar in composition to basalt, gabbro, norite, and anorthosite. Moon rocks have no water and almost no free oxygen.

SUPPLEMENTARY READINGS

Carmichael, I. S. E., F. J. Turner, and J. Verhoogen *Igneous Petrology,* McGraw-Hill Book Company, New York, 1974. *An advanced-level treatment of all aspects of igneous rocks.*

Green, J., and N. M. Short (eds.) *Volcanic Landforms and Surface Features: A Photographic Atlas and Glossary,* Springer-Verlag New York Inc., 1971. *A beautifully illustrated systematic treatment of volcanoes and volcanism.*

Kay, R., N. J. Hubbard, and P. W. Gast "Chemical Characteristics and Origin of Oceanic Ridge Volcanic Rocks," *J. Geophys. Res.,* vol. 75, pp. 1585–1613, 1970. *A discussion of the character and occurrence of igneous rocks commonly occurring as volcanic extrusions along trailing edges of lithospheric plates.*

Kruger, C. *Volcanoes,* G.P. Putnam's Sons, New York, 1971. *Finely illustrated survey of the history and science of volcanoes, from early mythology to recent concepts.*

Macdonald, G. A. *Volcanoes,* Prentice-Hall, Inc., Englewood Cliffs, N.J., 1972. *A relatively moderate-level introduction to volcanic activity, volcanic types, and characteristics of eruption.*

Pitcher, W. S., and A. R. Berger *The Geology of Donegal: A Study of Granite Emplacement and Unroofing,* John Wiley & Sons, Inc., New York, 1972. *A regionally specialized discussion of granitic plutons from emplacement through uplift, erosion, and resulting sedimentary residues.*

4

WEATHERING, SEDIMENTS, AND SEDIMENTARY ROCKS

The blurred inscription on a gravestone, the crumbling foundation of an ancient building, the broken rock exposed along a roadside—all tell us that rocks are subject to constant destruction. Marked changes of temperature, moisture soaking into the ground, the ceaseless activity of living things—all work to destroy rock material. This process of destruction we call **weathering** and we define it as the changes that take place in minerals and rocks at or near the surface of the earth in response to the atmosphere, to water, and to plant and animal life. Later on we shall extend this definition slightly, but for the time being it will serve our purpose.

Weathering leaves its mark everywhere about us. The process is so common, in fact, that we tend to overlook the way in which it functions and the significance of its results. It plays a vital role in the rock cycle; for by attacking the exposed material of the earth's crust—both solid rock and unconsolidated deposits—and converting it into different forms, it produces the raw materials for new rocks (see Chapter 1).

We have all dug our toes into a sandy beach, or picked our way over the gravels of a rushing stream, or perhaps slogged through the mud of a swamp. None of these—sand, gravel, or mud—immediately suggests

hard, solid rock. Yet deposits of this sort or materials very similar to them are the stuff from which the great bulk of the rocks exposed at the earth's surface were formed. When we look down into the mile-deep Grand Canyon of the Colorado River in Arizona (Figure 4.1), we can see there layer upon layer of rocks that were once unconsolidated deposits of sand, gravel, and mud. Over the course of time these loose sediments have been hardened into rocks that we call **sedimentary.**

The story of sedimentary rocks begins with weathering processes; for the products of chemical and mechanical weathering are the raw materials of sedimentary rocks. Streams, glaciers, wind, and ocean currents move the weathered materials to new localities and deposit them as sand, gravel, or mud. Sometimes, however, the products of weathering remain right where they are formed and are incorporated into the rock record. Certain ores, for example, such as those of aluminum, are actually old zones of weathering (see Chapter 15).

Some sediments, particularly sand and gravel, are consolidated into rock by a cementing process: Subsurface water trickling through the open spaces leaves behind a mineral deposit that serves to cement the

FIGURE 4.1

Terraces and templelike forms of the Grand Canyon have developed on alternating resistant and nonresistant sedimentary rocks. Differential weathering of rocks of varying resistance helps explain the topography in this region. The Colorado River flows in the steep-walled inner gorge cut in igneous and metamorphic rocks. [M. E. Kauffman.]

individual grains firmly together, giving the entire deposit the strength we associate with rock. Other sediments, such as fine deposits of mud, are transformed into rock by the weight of overlying deposits, which press, or compact, them into a smaller and smaller space. The sedimentary rock that results from either of these processes may eventually be exposed at the earth's surface. If the rock was formed beneath the bottom of the ocean, it may be exposed either by the slow withdrawal of the seas or by the upward motion of the sea floor, forming new areas of dry land.

Before we examine the processes of weathering and sedimentation, however, we pause a moment to look at the forces that are responsible for their action.

4.1 ENERGY SOURCES

Energy is a word that sounds familiar enough. We speak of energy-producing foods, of the energy needed to climb stairs, of energy radiated to us from the sun. Matter and energy are intimately related, but what is energy?

Energy is the capacity for producing motion. No form of matter is entirely devoid of motion. The motion may be in things we see, such as an automobile speeding down the highway, or in things we cannot see, such as atoms and molecules. All motion is produced by energy in one form or another. In fact, we can think of the entire universe as a great bundle of energy. Heat and light from the sun represent energy in one form; the revolution of the earth around the sun represents energy in another form; and the chemical transformation of food into heat and body activity represents energy in yet another form. So energy manifests itself in different forms, which have descriptive names: potential, kinetic, heat, chemical, electrical, atomic, and radiant.

In geological processes, through the motion of running water, energy sculptures the land; through the deformation of rocks, it builds mountains; and through the rupture of earth materials, it produces earthquakes. In fact, energy is involved in every geological process.

POTENTIAL ENERGY

Potential energy is stored energy waiting to be used. Uranium, petroleum, coal, and natural gas are eagerly sought after because the potential energy they contain can be effectively released and put to work. Water in clouds or in lakes and reservoirs at high altitudes has potential energy that is released when the water falls or runs downhill. A boulder poised on a hilltop has potential energy. Less obvious, but just as real, is the potential energy stored in the nucleus of every atom and in the molecules of every compound.

KINETIC ENERGY

The potential energy of a 10-t boulder resting on a hillside is transformed into the energy of movement when the boulder is dislodged and rolls downhill (see Figure 4.2). This energy of movement is called **kinetic**. *Every moving object possesses kinetic energy.*

The amount of kinetic energy possessed by an object depends on the mass of the object and on the speed with which it moves:

$$E_K = \frac{Mv^2}{2}$$

A 10-t boulder rolling down a hill has more energy at the bottom of the slope than does a pebble that has rolled down the same slope at the same speed. On the other hand, this same 10-t boulder moving a few meters per hour would have less energy than if it were hurtling along at 30 m/s.

Running water owes much of its effectiveness as a geological agent to kinetic energy; so do waves beating on the coast. Massive glaciers creeping down a mountain slope do geological work by means of kinetic energy.

ENERGY AND WEATHERING

The energy that drives the weathering process comes both from within the earth and from outside. From time to time motions originating inside the earth elevate some portions of its surface above other portions. These are the same motions that express themselves in earthquakes (Chapter 6) and mountain building (Chapter 7). Whatever their cause, they arrange the earth's surface materials so that gravity can be effective in the breakdown of rock material. Thus rock raised in mountain building has potential energy, which may be transformed to kinetic energy if gravity is strong enough to pull it downward to a lower level. The shattering of this rock as it falls is truly a form of weathering.

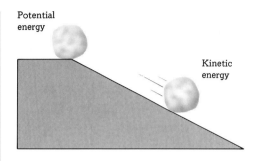

Potential energy

Kinetic energy

FIGURE 4.2

The potential energy of a 10-ton boulder resting on a hillside is transformed into kinetic energy, the energy of movement, when the boulder is dislodged and rolls downhill.

We already have seen that there is a measurable amount of heat that flows from the earth's interior to its surface and is there dissipated. Far more heat is received at the earth's surface from the sun. It is the distribution of this heat that causes the differential heating of the atmosphere and of the oceans. This differential heating brings about circulation of atmosphere and ocean, causes our weather and climate, and determines the pattern of organic activity. All these in turn help to modify the earth's surface materials—in short, to determine the process of weathering.

How much solar energy is available at the earth's surface? The sun radiates about 100,000 cal/cm²-min from its surface. A plane at the earth's outer surface perpendicular to the sun's rays would receive approximately 2 cal/cm²-min, which is 2×10^{-5} the amount of heat generated by an equivalent area on the sun's surface. This figure is the **solar constant.**

If our hypothetical plane perpendicular to the sun's rays at the edge of the outer atmosphere has the same diameter as the earth, geometry tells us that its area must be one-quarter that of the earth. Then we can express the solar constant for the earth's surface as 0.5 cal/cm²-min. This is an average for the entire globe. But of this energy approximately one-third is unavailable to heat either the atmosphere or the earth: It is reflected back into space from clouds or from the earth, or it is scattered into space from particles within the atmosphere. Therefore, approximately 0.3 cal/cm²-min is what is available to heat the atmosphere and the earth.

It is obvious that this energy is not evenly distributed over the earth's surface. Less is available toward either pole than is available toward the equator. We shall have occasion to return to this distribution later in this chapter as we consider the process of chemical weathering.

4.2 TYPES OF WEATHERING

There are two general types of weathering: **mechanical** and **chemical**. It is difficult to separate these two types in nature because they often go hand in hand, although in some environments one or the other predominates. Still, for our purposes here it is more convenient to discuss them separately.

MECHANICAL WEATHERING

Mechanical weathering, which is also referred to as **disintegration,** is the process by which rock is broken down into smaller and smaller fragments as the result of energy developed by physical forces. For example, when water freezes in a fractured rock, sufficient energy may develop from the pressure caused by expansion of the frozen water to split off pieces of the rock. Or a boulder moved by gravity down a rocky slope may be shattered into smaller fragments.

Expansion and contraction Changes in temperature, if they are rapid enough and great enough, may bring about the mechanical weathering of rock. In areas where bare rock is exposed at the surface and is unprotected by a cloak of soil, forest or brush fires can generate heat adequate to break up the rock. The rapid and violent heating of the exterior zone of the rock causes it to expand; and if the expansion is sufficiently great, flakes and larger fragments of the rock are split off. Lightning often starts such fires and, in rare instances, may even shatter exposed rock by means of a direct strike.

The debate continues concerning whether variations in temperature from day to night or from summer to winter are enough to cause mechanical weathering. Theoretically, such variations cause disintegration. For instance, we know that the different minerals forming a granite expand and contract at different rates as they react to rising and falling temperatures. We expect that even minor expansion and contraction of adjacent minerals would, over long periods of time, weaken the bonds between mineral grains and that it would thus be possible for disintegration to occur along these boundaries. In deserts we may find fragments of a single stone lying close beside one another. Obviously, the stone has split. But how? Many think the cause lies in expansion and contraction caused by heating and cooling.

But laboratory evidence to support these speculations is inconclusive. In one laboratory experiment coarse-grained granite was subjected to temperatures ranging from 14.5 to 135.5°C every 15 min. This alternate heating and cooling eventually simulated 244 years of daily heating and cooling; yet the granite showed no signs of disintegration. Perhaps experiments extended over longer periods of time would produce observable effects. In any event, we are still uncertain of the mechanical effect of daily or seasonal temperature changes; if these fluctuations bring about the disintegration of rock, they must do so very slowly.

Frost action Frost is much more effective than heat in producing mechanical weathering. When water trickles down into the cracks, crevices, and pores of a rock mass and then freezes, its volume increases about 9 percent. This expansion of water as it passes from the liquid to the solid state sets up pressures that are directed outward from the inside of the rock. These pressures can dislodge fragments from the rock's surface.

The dislodged fragments of mechanically weathered rock are angular, and their size depends largely on the nature of the bedrock from which they have been displaced. Usually the fragments are only a few centimeters in maximum dimension, but in some places—along the cliffs bordering Devil's Lake, Wisconsin, for instance—they reach sizes of up to 3 m.

A second type of mechanical weathering produced by freezing water is **frost heaving.** This action usually occurs in fine-grained, unconsolidated deposits rather than in solid rock. Much of the water that falls as rain or snow soaks into the ground, where it freezes during the winter months. If conditions are right, more and more ice accumulates in the zone of freezing as water is added from the atmosphere above and drawn upward from the unfrozen ground below, much as a blotter soaks up moisture. In time, lens-shaped masses of ice

FIGURE 4.3

Vigorous frost action has pried off this granitic rubble from the underlying bedrock along the crest of the Beartooth Mountains near the Montana-Wyoming state line. In addition, frost action around the edges of snow banks has combined with the water of melting snow to create irregularities called *nivation hollows* on the slopes. [Sheldon Judson.]

are built up, and the soil above them is heaved upward. Frost heaving of this sort is common on poorly constructed roads, and lawns and gardens are often soft and spongy in the springtime as a result of the soil's heaving up during the winter.

Certain conditions must exist before either type of frost action can take place: There must be an adequate supply of moisture; the moisture must be able to enter the rock or soil; and temperatures must move back and forth across the freezing line. As we might expect, frost action is more pronounced in high mountains and in moist regions where temperatures fluctuate across the freezing line, either daily or seasonally (see Figure 4.3).

Exfoliation Exfoliation is a mechanical weathering process in which curved plates of rock are stripped from a larger rock mass by the action of physical forces. This process produces two features that are fairly common in the landscape: large, domelike hills, called **exfoliation domes,** and rounded boulders, usually referred to as **spheroidally weathered boulders.** It seems likely that the forces that produce these two forms originate in different ways.

Let us look first at the manner in which exfoliation domes develop. Fractures, or parting planes, called **joints,** occur in many massive rocks. These joints are broadly curved and run more or less parallel to the rock surface. The distance between joints is only a few centimeters near the surface, but it increases to several meters as we move deeper in the rock (see Chapter 7). Under certain conditions one after another of the curved slabs between the joints is spalled, or sloughed, off the rock mass. Finally, a broadly curved hill of bedrock develops, as shown in Figure 4.4.

Just how these slabs of rock come into being in the first place is still a matter of dispute. Most observers believe that, as erosion strips away the surface cover, the downward pressure on the underlying rock is reduced. Then, as the rock mass begins to expand upward, lines of fracture develop, marking off the slabs

FIGURE 4.4

North Dome, Yosemite National Park, California, is an example of an exfoliation dome. The massive granite in this dome has developed a series of partings, or joints, more or less parallel to the surface. Rock slabs spall off the dome, giving it its rounded aspect. The jointing probably originated as the granite expanded after the erosion of the overlying material. [William C. Bradley.]

FIGURE 4.5

Near Boulder, Colorado, only isolated patches of partially weathered granite remain as most of the bedrock wears away, producing rounded masses, which tend to undergo further spheroidal weathering. [Sheldon Judson.]

FIGURE 4.6

Spheroidally weathered granite boulders almost completely isolated from the bedrock. [Sheldon Judson.]

that later fall away. Precise measurements made on granite blocks in New England quarries provide some support for this theory. Selected blocks were accurately measured and then removed from the quarry face, away from the confining pressures of the enclosing rock mass. When the free-standing blocks were measured again, it was found that they had increased in size by a small but measurable amount. Massive rock does expand, then, as confining pressures are reduced, and this slight degree of expansion may be enough to start the exfoliation process.

Well-known examples of exfoliation domes are Stone Mountain, Georgia, the domes of Yosemite Park, California, and Sugar Loaf in the harbor of Rio de Janeiro, Brazil.

Now let us look at smaller examples of exfoliation: spheroidally weathered boulders. These boulders have been rounded by the spalling off of a series of concentric shells of rock (Figures 4.5 to 4.7). But here the shells develop from pressures set up within the rock by chemical weathering rather than from the lessening of pressure from above by erosion. We shall see later that, when certain minerals are chemically weathered, the resulting products occupy a greater volume than does the original material, and it is this increase in volume that creates the pressures responsible for spheroidal weathering. Because most chemical weathering takes place in the portions of the rock most exposed to air and moisture, it is there that we find the most expansion and hence the greatest number of shells. Spheroidally weathered boulders are sometimes produced by the crumbling off of concentric shells. If the cohesive strength of the rock is low, individual grains are partially weathered and dissociated, and the rock simply crumbles away. The underlying process is the same in both cases, however.

Certain types of rocks are more vulnerable to spheroidal weathering than are others. Igneous rocks such as granite, diorite, and gabbro are particularly susceptible; for they contain large amounts of the mineral feldspar, which, when chemically weathered, produces new minerals of greater volume.

Other types of mechanical weathering Plants also play a role in mechanical weathering. The roots of trees and shrubs growing in rock crevices sometimes exert sufficient pressure to dislodge previously loosened fragments of rock, much as tree roots heave and crack sidewalk pavements (Figure 4.8).

More important, though, is the mechanical mixing of the soil by ants, worms, rodents, and other small animals. Constant activity of this sort makes the soil particles more susceptible to chemical weathering (see below) and may even assist in the mechanical breakdown of the particles.

Finally, agents such as running water, glacier ice, wind, and ocean waves all help to reduce rock material to smaller and smaller fragments. The role of these agents in mechanical weathering will be discussed in later chapters.

CHEMICAL WEATHERING

Chemical weathering, sometimes called **decomposition**, is a more complex process than mechanical weathering. As we have seen, mechanical weathering merely breaks rock material down into smaller and smaller particles, without changing its composition. Chemical weathering, however, actually transforms the original material into something different. The chemical weathering of the mineral feldspar, for example, produces the clay minerals, which have a different composition and different physical characteristics from those of the original feldspar. Sometimes the products of chemical weathering have no mineral form at all, as in the salty solution that results from the transformation of halite.

Particle size The size of the individual particles of rock is an extremely important factor in chemical weathering because substances can react chemically only where they come into contact with one another. The greater the surface area of a particle, the more vulnerable it is to chemical attack. If we were to take a pebble, for example, and grind it up into a fine powder, the total surface area exposed would be greatly increased. As a result, the materials that make up the pebble would undergo more rapid chemical weathering.

Figure 4.9 shows how the surface area of a 1-cm (or any other unit) cube increases as we cut it up into smaller and smaller cubes. The initial cube has a surface area of 6 cm² and a volume of 1 cm³. If we divide the cube into smaller cubes, each 0.5 cm on a side, the total surface area increases to 12 cm² although, of course, the total volume remains the same. Further subdivision into 0.25-cm cubes increases the surface to 24 cm². And if we divide the original cube into units 0.125 cm on a side, the surface area increases to 96 cm². As we have seen, this same process is performed by mechanical weathering: It reduces the size of the individual particles of rock, increases the surface area

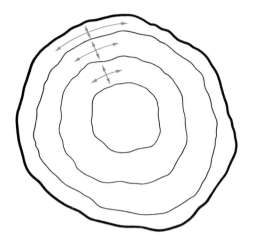

FIGURE 4.7

This cross section through a spheroidally weathered boulder suggests the stresses set up within the rock. The stress is thought to develop as a result of the change in volume as feldspar is converted to clay. (See also Figures 4.5 and 4.6.)

FIGURE 4.8

A white birch tree growing in a crevice pries a large block from a low rock cliff in Hermosa Park, Colorado. [U.S. Geological Survey.]

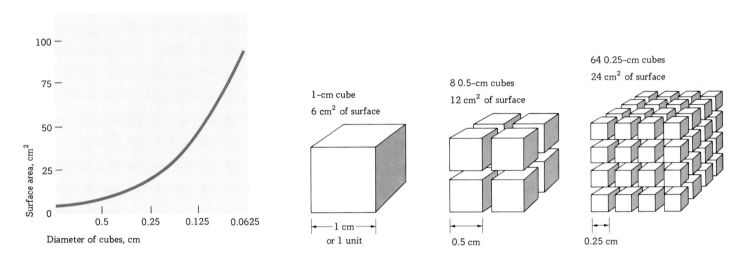

FIGURE 4.9

Relationships of volume, particle size, and surface area. In this illustration a cube
1 cm (or any other unit) on a side is divided into smaller and smaller units. The
volume remains unchanged, but as the particle size decreases, the surface area
increases. Because chemical weathering is confined to surfaces, the more finely a
given volume of material is divided, the greater is the surface area exposed to
chemical activity and the more rapid is the process of chemical weathering.

exposed, and thus promotes more rapid chemical
weathering.

Other factors The rate of chemical weathering is
affected by other factors as well—the composition of the
original mineral, for example. As we shall see later, a
mineral such as quartz, SiO_2, responds much more
slowly to chemical weathering than does a mineral
such as olivine, $(Fe, Mg)_2SiO_4$.

Climate also plays a key role in chemical weathering. Moisture, particularly when it is accompanied by
warmth, speeds up the rate of chemical weathering;
conversely, dryness slows it down. Finally, plants and
animals contribute directly or indirectly to chemical
weathering because their life processes produce oxygen, carbon dioxide, and certain acids that enter into
chemical reactions with earth materials.

The interrelation of some of these factors is shown
in Figure 4.10. This is a generalized section from the
pole to the equator and shows the fluctuation of precipitation, temperature, and amount of vegetation. At the
same time the figure shows the relative depth of weathering as these three factors vary. Thus weathering is
most pronounced in the equatorial zone, where the
factors of precipitation, temperature, and vegetation
reach a maximum. Weathering is least in the desert
and semidesert areas of the subtropics and in the far

north. A secondary zone of maximum weathering exists
in the zone of temperate climates. Here both the precipitation and the vegetation can reach secondary
maxima.

CHEMICAL WEATHERING
OF IGNEOUS ROCKS

In Chapter 3 we found that the most common minerals
in igneous rocks are silicates and that the most important silicates are quartz, the feldspars, and certain ferromagnesian minerals. Let us see how chemical
weathering acts on each of these three types.

Weathering of quartz Chemical weathering affects quartz very slowly, and for this reason we speak of
quartz as a relatively stable mineral. When a rock such
as granite, which contains a high percentage of quartz,
decomposes, a great deal of unaltered quartz is left
behind. The quartz grains (commonly called **sand
grains**) found in the weathered debris of granite are
the same as those that appeared in the unweathered
granite.

When these quartz grains are first set free from the
mother rock, they are sharp and angular; but because
even quartz slowly responds to chemical weathering,

the grains become more or less rounded as time passes. After many years of weathering, they look as though they had been abraded and worn by the action along a stream bed or a beach. And yet the change may have come about solely through chemical action. Indeed, the presence of silica in natural waters of lakes and rivers reminds us that the silicate minerals are soluble and that some of this may come from the chemical weathering of quartz.

Weathering of feldspars In the Bowen reaction series (Section 3.6) we saw that, when a magma cools to form an igneous rock such as granite, feldspars crystallize before quartz. When granite is exposed to weathering at the earth's surface, the feldspars are also the first minerals to be broken down. Mineralogists and soil scientists still do not understand the precise process by which feldspars weather, and some of the end products of this action—the clay minerals—offer many puzzles. But the general direction and results of the process seem fairly clear.

Aluminum silicate, derived from the chemical breakdown of the original feldspar, combines with water to form hydrous aluminum silicate, which is the basis for another group of silicate minerals, the clays. Let us examine the decomposition of orthoclase, a good example of the chemical weathering of the feldspar

group of silicates. In this instance a source of hydrogen ions is necessary to the weathering process, and two substances play an important role in producing them: carbon dioxide and water. The atmosphere contains small amounts of carbon dioxide, and the soil contains much greater amounts. Because carbon dioxide is extremely soluble in water, it unites with rainwater and water in the soil to form the weak acid H_2CO_3, *carbonic acid*. This ionizes to form hydrogen and bicarbonate ions as follows:

Water *plus* carbon *yields* carbonic *yields* hydrogen *plus* bicarbonate
       dioxide   acid   ion   ion

$$H_2O + CO_2 \rightarrow H_2CO_3 \rightarrow H^+ + (HCO_3)^-$$

And when orthoclase comes into contact with hydrogen ions, the following reaction takes place:

Orthoclase   *plus* hydrogen *plus* water *yields*
               ions

$$2K(AlSi_3O_8) + 2H^+ + H_2O \rightarrow$$

     clay   *plus* potassium *plus* silica
                  ions

$$Al_2Si_2O_5(OH)_4 + 2K^+ + 4SiO_2$$

FIGURE 4.10

Variation in temperature, precipitation, and organic matter from the poles to the tropics is related to the depth of chemical weathering.

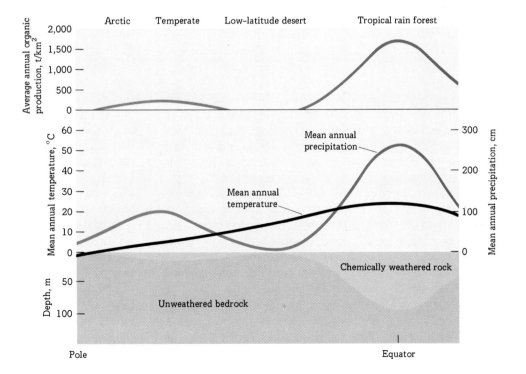

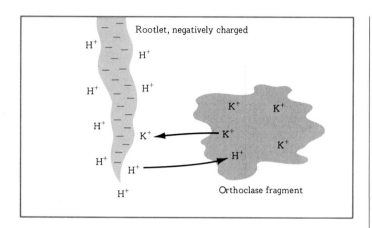

FIGURE 4.11

The conversion of orthoclase to a clay mineral by plant roots. In this diagram, a swarm of hydrogen ions (positive) are shown surrounding a negatively charged plant rootlet. The suggestion has been made that a hydrogen ion from the rootlet may replace a potassium ion in a nearby orthoclase fragment and there bond with the oxygen within the original mineral, to begin the conversion of the orthoclase to clay. The potassium ion thus ejected replaces the hydrogen ion along the negatively charged rootlet and is eventually utilized in plant growth. [Redrawn from W. D. Keller and A. F. Frederickson, "Role of Plants and Colloidal Acids in the Mechanism of Weathering," *Am. J. Sci.*, vol. 250, p. 603, 1952.]

In this reaction, the hydrogen ions from the water force the potassium out of the orthoclase, disrupting its crystal structure. The hydrogen ion combines with the aluminum silicate radical of the orthoclase to form the new clay minerals. (The process by which water combines chemically with other molecules is called **hydration**.) The disruption of the orthoclase crystal yields a second product, potassium ions. These may join with the bicarbonate ions formed by the ionization of the carbonic acid to form potassium bicarbonate. The third product, silica, is formed by the silicon and oxygen that are left after the potassium has combined with the aluminum silicate to form the clay mineral.

The action of living plants may also bring about the chemical breakdown of orthoclase. A plant root in

the soil is negatively charged and is surrounded by a swarm of hydrogen ions, H^+. If there happens to be a fragment of orthoclase lying nearby, these positive ions may change places with the potassium of the orthoclase and disrupt its crystal structure (see Figure 4.11). Once again, a clay mineral is formed, as in the equation above.

Now let us look more closely at each of the three products of the decomposition of orthoclase: first, the clay minerals. At the start, these minerals are very finely divided. In fact, they are sometimes of **colloidal** size, a size variously estimated as between 0.2 and 1 μm (micrometer). Immediately after it is formed, the aluminum silicate may possibly be amorphous; that is, its atoms are not arranged in any orderly pattern. It seems more probable, however, that even at this stage the atoms are arranged according to the definite pattern of a true crystal. In any event, as time passes, the small individual particles join together to form larger crystals, which, when analyzed by such means as X rays, exhibit the crystalline pattern of true minerals.

There are many different clay minerals and each has its own chemical behavior, physical structure, and evolution. Most of the clay minerals fall into three major groups: **kaolinite, montmorillonite,** and **illite.** Kaolinite is derived from the Chinese *Kao-ling*, "High Hill," the name of the mountain from which the first kaolinite was shipped to Europe for ceramic uses; the mineral montmorillonite was first described from samples collected near Montmorillon, a town in west-central France; and the name illite was selected by geologists of the Illinois Geological Survey in honor of their state. Like the micas shown in Figure 2.22, the clay minerals are built up of silicon-oxygen tetrahedra linked together in sheets. These sheets combine in different ways with sheets composed of aluminum atoms and hydroxyl molecules. (For this reason we refer to the clay minerals as hydrous aluminum silicates.) In addition, montmorillonite may contain magnesium and some sodium and calcium, and illite contains potassium, occasionally with some magnesium and iron.

We still do not understand exactly what factors determine which clay minerals will form when a feldspar is weathered. Climate is important, for we know that kaolinite tends to form as a result of the intense chemical weathering in warm, humid climates and that illite and montmorillonite seem to develop more commonly in cooler, drier climates. The history of the rock also seems to be influential. For example, when a soil forms from a sedimentary rock in which a clay has been incorporated, we often find that the soil contains the same type of clay as does the parent rock. The analysis of a number of sedimentary rocks and the soils developed on these rocks has shown that, when illite is present in the original rock, it is usually the dominant clay in the soil, regardless of climate. Clearly, then, both environment and inheritance seem to influence the type of clay that will develop from the chemical weathering of a feldspar.

Let us look back for a minute to the equation for the decomposition of orthoclase. Notice that the second product is potassium ions. We might expect that these would be carried off by water percolating through the ground and that all the potassium would eventually find its way to the rivers and finally to the sea. Yet analyses show that not nearly so much potassium is present in river and ocean water as we should expect. What happens to the rest of it? Some of it is used by growing plants before it can be carried away in solution, and some of it is absorbed by clay minerals or even taken into their crystal structure.

The third product resulting from the decomposition of orthoclase is silica, which appears either in solution (for even silica is slightly soluble in water) or as very finely divided quartz in the size range of the colloids. In the colloidal state silica may exhibit some of the properties of silica in solution.

So far we have been talking about the weathering of only orthoclase feldspar. But the products of the chemical weathering of plagioclase feldspars are very much the same. Instead of potassium carbonate, however, either sodium or calcium carbonate is produced, depending on whether the feldspar is the sodium albite or the calcium anorthite (see Table 4.1). As we found in Section 2.5 (see Silicates), the plagioclase feldspars almost invariably contain both sodium and calcium. The carbonates of sodium and calcium are soluble in water and may eventually reach the sea. We should note here, however, that calcium carbonate also forms the mineral calcite (see Carbonate and Sulfate Minerals). Calcite, in turn, forms the greater part of limestone (a sedimentary rock) and marble (a metamorphic rock). Both limestone and marble are discussed in subsequent chapters.

Weathering of ferromagnesians Now let us turn to the chemical weathering of the third group of common minerals in igneous rocks: the ferromagnesian silicates.

TABLE 4.1

Chemical-weathering products of common rock-forming silicate minerals

Mineral	Composition	Important decomposition products	
		Minerals	Others
Quartz	SiO_2	Quartz grains	Some silica in solution
Feldspars:			
Orthoclase	$K(AlSi_3O_8)$	Clay	Potassium carbonate (soluble)
		Silica	Some silica in solution
Albite (sodium plagioclase)	$Na(AlSi_3O_8)$	Clay	Some silica in solution
Anorthite (calcium plagioclase)	$Ca(Al_2Si_2O_8)$	Silica	Sodium and calcium carbonates (soluble)
		Calcite	
Ferromagnesians:			
Biotite	Fe, Mg, Ca silicates of Al	Clay	Calcium and magnesium carbonates (soluble)
Augite		Hematite	
Hornblende		Limonite	Some silica in solution
		Silica	
		Calcite	
Olivine	$(Fe, Mg)_2SiO_4$	Hematite	Iron and magnesium carbonates (soluble)
		Limonite	
		Silica	Some silica in solution

The chemical weathering of these minerals produces the same products as the weathering of the feldspars: clay, soluble salts, and finely divided silica. But the presence of iron and magnesium in the ferromagnesian minerals makes possible certain other products as well.

The iron may be incorporated into one of the clay minerals or into an iron carbonate mineral. Usually, however, it unites with oxygen to form hematite, Fe_2O_3, one of the most common of the iron oxides. Hematite commonly has a deep red color, and in powdered form it is always red; this characteristic gives it its name, from the Greek *haimatitēs*, "bloodlike." Sometimes the iron unites with oxygen and a hydroxyl ion to form *goethite*, $FeO(OH)$, generally brownish in color. (Goethite was named after the German poet Goethe, because of his lively scientific interests.) Chemical weathering of the ferromagnesian minerals often produces a substance called **limonite,** yellowish to brownish in color and referred to in everyday language as just plain "rust." Limonite is not a true mineral because its composition is not fixed within narrow limits, but the term is universally applied to the iron oxides of uncertain composition that contain a variable amount of water. Limonite and some of the other iron oxides are responsible for the characteristic colors of most soils.

What happens to the magnesium produced by the weathering of the ferromagnesian minerals? Some of it may be removed in solution as a carbonate, but most of it tends to stay behind in newly formed minerals, particularly in the illite and montmorillonite clays.

Summary of weathering products If we know the mineral composition of an igneous rock, we can determine in a general way the products that the chemical weathering of that rock will probably yield. The chemical-weathering products of the common rock-forming minerals are listed in Table 4.1. These products include the minerals that make up most of our sedimentary rocks, and we shall discuss them again later in this chapter.

4.3 RATES OF WEATHERING

Some rocks weather very rapidly and others only slowly. Rate of weathering is governed by the type of rock and a variety of other factors, from minerals and moisture, temperature and topography, to plant and animal activity.

RATE OF MINERAL WEATHERING

On the basis of field observations and laboratory experiments, the minerals commonly found in igneous rocks can be arranged according to the order in which they are chemically decomposed at the surface. We are not sure of all the details, and different investigators report different conclusions, but we can make the following general observations:

1 Quartz is highly resistant to chemical weathering.
2 The plagioclase feldspars weather more rapidly than orthoclase feldspar does.
3 Calcium plagioclase (anorthite) tends to weather more rapidly than sodium plagioclase (albite).
4 Olivine is less resistant than augite, and in many instances augite seems to weather more rapidly than hornblende.
5 Biotite mica weathers more slowly than do the other dark minerals and muscovite mica is more resistant than biotite.

Notice that these points suggest a pattern (Figure 4.12) similar to that of Bowen's reaction series for crystallization from magma, discussed in Section 3.6 (illustrated in Figure 3.28). But there is one important difference: In weathering, the successive minerals formed do not react with one another as they do in a continuous reaction series. The relative resistance of these minerals to decomposition may reflect the difference between the surface conditions under which they weather and the conditions that existed when they were formed. Olivine, for example, forms at high temperatures and pressures, early in the crystallization of a melt. Consequently, as we might expect, it is extremely unstable under the low temperatures and pressures that prevail at the surface, and it weathers quite rapidly. On the other hand, quartz forms late in the reaction series, under considerably lower temperatures and pressures. Because these conditions are more similar to those at the surface, quartz is relatively stable and is very resistant to weathering.

Now we can qualify slightly the definition of weathering given at the beginning of this chapter. We

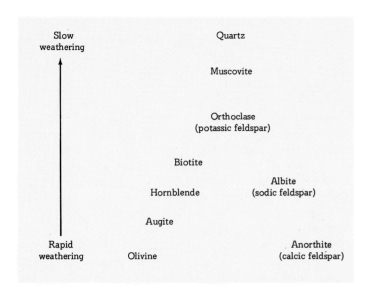

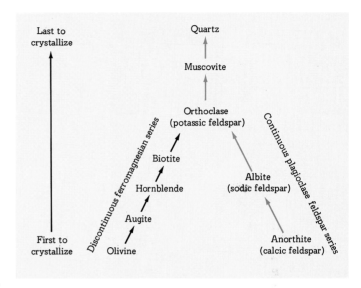

FIGURE 4.12

Relative rapidity of chemical weathering of the common igneous rock-forming minerals. The rate of weathering is most rapid at the bottom and decreases toward the top. Note that this table is in the same order as Bowen's reaction series (right). The discrepancy in the rate of chemical weathering between, for instance, olivine and quartz is explained by the fact that in the zone of weathering olivine is farther from its environment of formation than is quartz. It therefore reacts more rapidly than quartz does to its new environment and thus weathers more rapidly.

have found that weathering disrupts the equilibrium that existed while the minerals were still buried in the earth's crust and that this disruption converts them into new minerals. Following Parry Reiche,[1] we may revise our definition as follows: *Weathering is the response of materials that were once in equilibrium within the earth's crust to new conditions at or near contact with air, water, and living matter.*

DEPTH AND RAPIDITY OF WEATHERING

Most weathering takes place in the upper few meters or tens of meters of the earth's crust, where rock is in closest contact with air, moisture, and organic matter. But some factors operate well below the surface and permit weathering to penetrate to great depths. For instance, when erosion strips away great quantities of material from the surface, the underlying rocks are free

to expand. As a result, parting planes, or fractures—the joints that we spoke of earlier in the chapter—develop hundreds of meters below the surface.

Then, too, great quantities of water move through the soil and deep underground, transforming some of the materials there long before they are ever exposed at the surface. Rock salt that is located far below the surface in the form of a sedimentary rock often undergoes exactly this transformation. If a large quantity of underground water is present, the salt is dissolved and carried off long before erosion can expose it.

Weathering is sometimes so rapid that it can actually be recorded. The Krakatoa eruption of August, 1883, described in Chapter 3, threw great quantities of volcanic ash into the air and deposited it to a depth exceeding 30 m on the nearby island of Long-Eiland. By 1928, 45 years later, a soil nearly 35 cm deep had developed on top of this deposit, and laboratory analyses showed that a significant change had taken place in the original materials. Chemical weathering had removed a measurable amount of the original potassium and sodium. Furthermore, mechanical weathering

[1]Parry Reiche, "A Survey of Weathering Processes and Products," *Univ. New Mexico Publ. Geol.*, no. 3, p. 5, 1950.

FIGURE 4.13

Weathering of a marble headstone, burying ground of Christ Church, Cambridge, Massachusetts. The inscription, carved in 1818, is illegible. The monument illustrates the instability and rapid weathering of calcite (the predominant mineral in marble) in a humid climate. [Sheldon Judson, 1968.]

or chemical weathering or both had broken down the original particles so that they were generally smaller than the particles in the unweathered ash beneath.

In a more recent study, scientists from Ohio State University have demonstrated the nature and rate of soil development in unweathered material that has been exposed to the atmosphere with the retreat of Muir glacier in Glacier Bay National Park in southeastern Alaska. Geological and human records establish the successive positions of the retreating ice front since about 1700. Between this date and 1965 the front of the glacier retreated approximately 65 km. The age of soils at various points along this line of retreat can thus be demonstrated. Over a period of about 250 years a soil about 35 cm thick has developed. Of this, the upper half is represented chiefly by the accumulation of organic material. The lower half, however, shows changes in the materials left by the retreating ice. After 250 years, virtually all the calcite and dolomite had been removed to a depth of at least 15 cm, and the soil acidity had increased markedly. The amount of iron oxide in the form of hematite had increased measurably, particularly in the lowest 10 cm.

Graveyards provide many fine examples of weathering within historic time. Calcite in the headstone pictured in Figure 4.13 has weathered so rapidly that the inscription to the memory of Moses L. Gould, carved in 1818, is only partially legible after more than a century and a half. (Examination of other marble slabs in the burying ground indicates that the earliest legible date is 1811.) Undoubtedly the rate of weathering has increased with time, for two reasons. First, continued weathering roughens the marble surface, exposing more and more of it to chemical attack and quickening the rate of decomposition. Second, as the number of factories and dwellings in Cambridge and neighboring towns has increased, the amount of carbon dioxide in the atmosphere also has increased. Consequently, rainwater in the twentieth century carries more carbonic acid than it did in the nineteenth, and it attacks calcite more rapidly.

In contrast to the strongly weathered marble is a headstone of slate erected in 1699 (Figure 4.14). Two

FIGURE 4.14

Weathering of a slate headstone, burying ground of Christ Church, Cambridge, Massachusetts. The inscription date, 1699, testifies to the durability of the slate. [Sheldon Judson, 1968.]

hundred sixty-nine years later, the inscription was still plainly visible. Slate is usually a metamorphosed shale, which, in turn, is composed largely of clay minerals formed by the weathering of feldspars. The clay minerals in the headstone were originally formed in the zone of weathering. Slight metamorphism has since changed many of the clay minerals to muscovite, a white mica,

and the shale to slate (Section 5.3). The muscovite, however, is relatively stable in the zone of weathering, as indicated in Figure 4.12.

These examples show, then, that weathering often occurs rapidly enough to be measured during a lifetime. Let us now turn to the process that moves these products of weathering: erosion.

4.4 RATES OF EROSION

We found that chemical and mechanical weathering produces certain materials from the rocks of the earth. It is these materials that the agents of erosion move from one place to another until they finally reach the settling basins of the world's oceans. Several agents are involved in this movement, including gravity, water, ice, and wind. These agents are individually treated in Chapters 9 to 13, but here it will be instructive to pause and consider how much material is being removed from the continents and at what rate.

Sometimes ancient ruins provide an index to the rates of erosion. Thus Figure 4.15 is a photograph of the remains of a cistern built 60 km north of Rome, Italy, in the second century. The footings exposed at the base of the finished wall indicate the amount of erosion that has occurred here since the structure was built. The rate of erosion from then to the present averages 30 cm/1,000 years.

This, however, is only a spot measurement at a specific place. What method can we use to measure the rates over large areas? One way is to measure the amount of material carried by a stream each year from its drainage basin. The amount averaged over the area of the basin gives an average figure of erosion. Now, obviously, the rate of erosion is not the same at every place in the river basin. In some places the material will be removed more rapidly than at others. Furthermore, there will be places at which deposition takes place and the material temporarily halted on its way out of the basin. Nevertheless, this method gives us an average figure for a unit area within the drainage basin.

A stream carries material in solution as well as solid material in the form of sediments. Most of the solid material is buoyed up by the flow of water and is said to be carried in **suspension**. A relatively smaller amount is pushed and bounced along the stream bottom in what is known as **bed load** or **traction load**. Suspended and

dissolved matter can be measured without much difficulty. It is very difficult to measure traction load, but it is generally small and usually considered to be about 10 percent of the suspended load in the average stream.

With these facts in mind we may ask how much material is carried annually by streams out of the

FIGURE 4.15

Ruins of a cistern built in the second century for a villa about 60 km north of Rome. The exposed footings measure 1.3 m, as indicated by the tape. This is the amount of erosion since the cistern was built. [Sheldon Judson.]

TABLE 4.2

Rates of regional erosion in the United States[a]

Drainage region	Drainage area,[b] thousands of km²	Runoff, thousands of m³/s	Load, t/km²-yr			Erosion, cm/1,000 yr	Area sampled, %
			Dissolved	Solid	Total		
Colorado	629	0.6	23	417	440	17	56
Pacific Slopes, California	303	2.3	36	209	245	9	44
Western Gulf	829	1.6	41	101	142	5	9
Mississippi	3,238	17.5	39	94	133	5	99
South Atlantic and Eastern Gulf	736	9.2	61	48	109	4	19
North Atlantic	383	5.9	57	69	126	5	10
Columbia	679	9.8	57	44	101	4	39
Total (overall)	6,797	46.9	43	119	162	6	

[a] After Sheldon Judson and D. F. Ritter, "Rates of Regional Denudation in the United States," *J. Geophys. Res.*, vol. 69, p. 3399, 1964.
[b] Great Basin, Saint Lawrence, and Hudson Bay drainage not considered.

drainage basin in which is located the ancient Roman cistern in Figure 4.15. The major stream here is the Tiber River. If we measure the load of the Tiber at Rome, we find that it is removing sediments on the average of about 7.5 million t of solid material each year from the area upstream. This converts to an erosion rate of 17 cm/1,000 years over the entire basin. If we had data on bed load and dissolved load for the Tiber, the figures would be higher but not by much.

Turning to the United States, we can examine Table 4.2, which describes the erosion going on in the major regions of the country. Not included here is the drainage to Hudson Bay and the Saint Lawrence River area. Also, the table does not include the area in the western states known as the Great Basin, where topography and rainfall are such that streams do not reach the sea. We see that the rates vary from region to region but average approximately 6 cm/1,000 years.

Data from the Amazon River, the world's largest, indicate that it is removing material from its basin at the rate of 4.7 cm/1,000 years. Another large tropical river, the Congo, is carrying enough material out of its basin each year to reduce its drainage basin by approximately 2 cm/1,000 years.

A single large drainage basin integrates many factors affecting the rapidity of erosion. One we should consider carefully is the human influence. There is enough information from disciplines as diverse as archaeology and nuclear physics to indicate that when people occupy an area intensively and turn it to crop land, they increase the erosion rate 10 to 100 times over that of a naturally forested or grassed area.

How much material is transported each year to the oceans? We have already cited some figures for portions of the earth. These figures, however, cover only 10 percent of the land surface. Therefore, any estimate of the total amount of erosion of the earth's surface per unit time can be no better than that, a mere estimate. The information now at hand suggests, however, that something approaching 10 billion t/year was being delivered to the ocean before human beings became effective geological agents. This does not include windblown material although it is thought to be negligible. Glacier ice is not considered but probably would not appreciably affect our figure. The rate of 10 billion t/year corresponds to a continental reduction of approximately 2.5 cm/1,000 years. If we take into consideration the human effect on erosion, it is estimated that at present approximately 24 billion t are moved annually by the rivers to the oceans. This is 2.5 times the amount of material moved before our intervention.

DIFFERENTIAL EROSION

Differential erosion is the process by which different rock masses or different sections of the same rock mass erode at different rates. Almost all rock masses of any size weather in this manner. The results vary from the boldly sculptured forms of the Grand Canyon to the slightly uneven surface of a marble tombstone. Unequal rates of erosion are chiefly caused by variations in the composition of the rock. The more resistant zones stand out as ridges, ribs, or pinnacles above the more rapidly weathered rock on either side (Figure 4.16).

A second cause of differential erosion is simply that the intensity of weathering varies from one section to another in the same rock. Figure 4.17 shows a memorial that has undergone mechanical weathering in certain spots. The rock is a coarse, red, homogeneous sandstone made up of quartz, feldspar, and mica, with some red iron oxides. The inner sides of the pillars have disintegrated so badly that the original fluting has been entirely destroyed and the underside of the horizontal slab has scaled noticeably. Notice that these are the areas least accessible to the drying action of the sun. Consequently, moisture has tended to persist here, and the frost action has pried off flakes and loosened individual grains.

On a larger scale, differential erosion is shown in

FIGURE 4.16

Differential weathering and erosion of Ledger Dolomite bedrock of Cambrian age in Lancaster, Pennsylvania. Solution has dissolved out less resistant portions, leaving behind pinnacles of more resistant rock, exposed when the soil cover was removed. [M. E. Kauffman.]

FIGURE 4.17

Differential weathering in a sandstone monument, burying ground of Christ Church, Cambridge, Massachusetts. Mechanical weathering has been most rapid on the shaded portions (far right). Here moisture remains longer, and consequently frost action has been more effective than on the less shaded portions. [Sheldon Judson, 1968.]

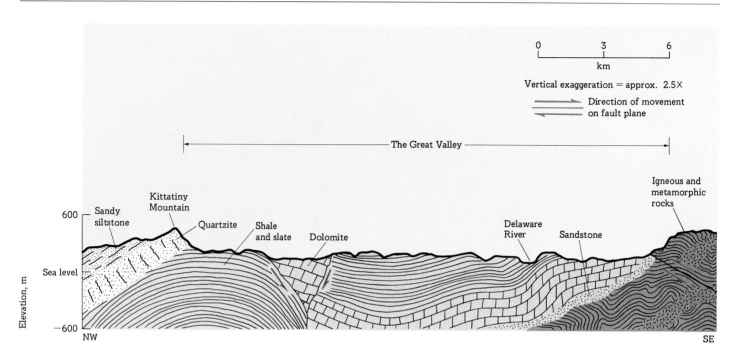

FIGURE 4.18

A generalized cross section in northeastern Pennsylvania and northwestern New Jersey (in the vicinity of the Delaware Water Gap) to show the relation of topography to the different rock types below the surface. See the text for discussion.

the cross section in Figure 4.18. The section is drawn across a portion of northeastern Pennsylvania near the Delaware Water Gap. Note that the highest ridge is underlain by a tough quartzite, which weathers very slowly. The major portion of the Great Valley is underlain by a shaley slate, which weathers easily. In this same valley are still lower spots underlain by a dolomitic limestone, which is susceptible to rapid chemical weathering in this humid climate. To the southeast lie the New Jersey Highlands, held up by resistant units of crystalline rock, both igneous and metamorphic.

4.5 SOILS

So far we have been discussing the ways in which weathering acts to break down existing rocks and to provide the material for new rocks, but weathering also plays a crucial role in the creation of the soils that cover the earth's surface and sustain all life. In fairly recent years the study of soils has developed into the science of **pedology** (from the Greek *pedon*, "soil," with *logos*, "reason," hence "soil science").[2]

SOIL CLASSIFICATION

In the early years of soil study in the United States researchers thought that the parent material almost wholly determined the type of soil that would result from it. Thus, they reasoned, granite would weather to one type of soil and limestone to another.

It is true that a soil reflects to some degree the

[2]In the United States, the term *pedology* is sometimes confused with words based on *ped-* and *pedi-*, combining forms meaning "foot," or with words based on *ped-* and *pedo-*, combining forms meaning

"boy," "child," as in *pediatrics*, the medical science that treats of the hygiene and diseases of children. Consequently, there is a tendency in this country to use **soil science** instead of *pedology*.

material from which it developed, and in some instances one can even map the distribution of rocks on the basis of the types of soil that lie above them. But as more and more information became available, it became apparent that the bedrock is not the only factor determining soil type. Russian soil scientists, following the pioneer work of V. V. Dokuchaev (1846–1903), have demonstrated that different soils develop over identical bedrock material in different areas when the climate varies from one area to another. The idea that climate exerts a major control over soil formation was introduced into this country in the 1920s by C. F. Marbut (1863–1935), for many years chief of the United States Soil Survey in the Department of Agriculture. Since that time, soil scientists have discovered that still other factors exercise important influences on soil development. For instance, the relief of the land surface plays a significant role. The soil on the crest of a hill is somewhat different from the soil on the slope, which in turn differs from the soil on the level ground at the foot of the hill; yet all three soils rest on identical bedrock. The passage of time is another factor: A soil that has only begun to form differs from one that has been developing for thousands of years although the climate, bedrock, and topography are the same in each instance. Finally, the vegetation in an area influences the type of soil that develops there: One type of soil will form beneath a pine forest, another beneath a forest of deciduous trees, and yet another on a grass-covered prairie.

Exactly what is a soil? It is a natural, surficial material that supports plant life. Since each soil exhibits certain properties that are determined by climate and living organisms operating over periods of time on earth materials and on landscape of varying relief and because all these factors are combined in various ways all over the land areas of the globe, the number of possible soil types is almost unlimited.

And yet certain valid generalizations can be made about soils. We know, for example, that the composition of a soil varies with depth. A natural or artificial exposure of a soil reveals a series of zones, each recognizably different from the one above. Each of these zones is called a **soil horizon,** or, more simply, a *horizon.* The three major zones or horizons in a typical soil, shown in Figure 4.19, may be described as follows from the bottom upward.

C horizon The *C* horizon is a zone of partially disintegrated and decomposed rock material. Some of the original bedrock minerals are still present, but others have been transformed into new materials. The *C*

horizon grades downward into the unweathered rock material.

B horizon The *B* horizon lies directly above the *C* horizon. Weathering here has proceeded still further than in the underlying zone, and only those minerals of the parent rock that are most resistant to decomposition (quartz, for example) are still recognizable. The others have been converted into new minerals or into soluble salts. In moist climates the *B* horizon contains an accumulation of clayey material and iron oxides delivered by water percolating downward from the surface. In dry climates we generally find, in addition to the clay and iron oxides, deposits of more soluble minerals, such as calcite. This mineral too may have been brought down from above, but some is brought into the *B* horizon from below, as soil water is drawn upward by high evaporation rates. Because material is deposited in the *B* horizon, it is known as the "zone of accumulation" (see Figure 4.19).

A horizon The *A* horizon is the uppermost zone—

FIGURE 4.19

The three major horizons of a soil. In many places it is possible to subdivide the zones themselves. Here the soil is shown as having developed from limestone.

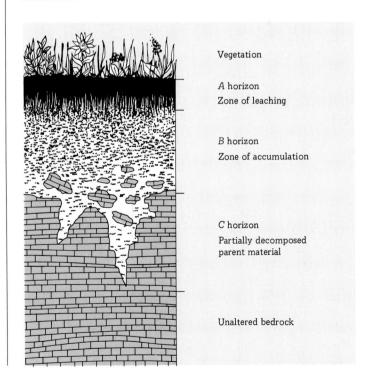

Vegetation

A horizon
Zone of leaching

B horizon
Zone of accumulation

C horizon
Partially decomposed
parent material

Unaltered bedrock

the one into which we sink a spade when we dig a garden. This is the zone from which the iron oxides have been carried to the B horizon, and in dry climates it is the source of some soluble material that may be deposited in the B horizon. The process by which these materials have been moved downward by soil water is called *leaching*, and the A horizon is sometimes called the "zone of leaching" (Figure 4.19). Varying amounts of organic material tend to give the A horizon a gray to black color.

The three soil horizons have all developed from the underlying parent material. When this material is first exposed at the surface, the upper portion is subjected to intense weathering, and decomposition proceeds rapidly. As the decomposed material builds up, downward-percolating water begins to leach out some of the minerals and to deposit them lower down. Gradually the A horizon and the B horizon build up, but weathering continues (at a slower rate now) on the underlying parent material, giving rise to the C horizon. With the passage of time the C horizon reaches deeper and deeper into the unweathered material below, the B horizon keeps moving downward, and the A horizon in turn encroaches on the upper portion of the B horizon. Finally a "mature" soil is built up.

The thickness of the soil that forms depends on many factors, but in the northern United States and southern Canada, the material that was first exposed to weathering after the retreat of the last ice sheet some 10,000 years ago is now topped by a soil 60 to 80 cm thick. Farther south, where the surface was uncovered by the ice at an earlier time, the soils are thicker, and in some places the processes of weathering have extended themselves from 5 to 10 m below the present surface.

SOME SOIL TYPES

We can understand the farmer's interest in soil, but why is it important for the geologist to understand soils and the processes by which they are formed? There are several reasons: First, soils provide clues to the environment in which they were originally formed. By analyzing an ancient soil buried in the rock record, we may be able to determine the climate and physical conditions that prevailed when it was formed. Second, some soils are sources of valuable mineral deposits (see Chapter 15), and the weathering process often enriches

otherwise low-grade mineral deposits, making them profitable to mine. An understanding of soils and soil-forming processes, therefore, can serve as a guide in the search for ores. Third, because a soil reflects to some degree the nature of the rock material from which it has developed, we can sometimes determine the nature of the underlying rock by performing an analysis of the soil.

But most important of all, soils are the source of many of the sediments that are eventually converted into sedimentary rocks. And these in turn may be transformed into metamorphic rocks or, following another path in the rock cycle, may be converted into new soils. If we understand the processes and results of soil formation, we are in a better position to interpret the origin and evolution of many rock types.

The following pages will discuss three major types of soil. Two of them, the pedalfers and the pedocals, are typical of the middle latitudes. The third group, referred to as laterites, is found in tropical climates.

Pedalfers A *pedalfer* is a soil in which iron oxides, clays, or both have accumulated in the B horizon. The name is derived from *pedon* and the symbols Al and Fe for aluminum and iron. In general, soluble materials such as calcium carbonate or magnesium carbonate do not occur in the pedalfers. Pedalfers are commonly found in temperate, humid climates, usually beneath a forest vegetation. In the United States most of the pedalfers lie east of a line that corresponds roughly to about 63 cm/year of rainfall. Northward in Canada, however, where the temperature is more important than the total rainfall in determining the distribution of pedalfers, the zone extends northwestward across Saskatchewan and Alberta coincident with a mean annual temperature of about 4.5°C or colder. Farther west, the pedalfers extend southward into the United States along the mountainous region of the Rockies, where the rainfall is somewhat higher and the temperatures lower than in the rest of western United States (see Figure 4.20).

In the formation of pedalfers certain soluble compounds, particularly those that contain sodium, calcium, and magnesium, are rapidly removed from the A horizon by waters seeping into the soil from the surface. These soluble compounds proceed downward through the B horizon and are carried off by ground water. The less soluble iron oxides and clay are deposited in the B horizon, giving that zone a clayey character with a brownish to reddish color.

FIGURE 4.20

Generalized distribution of pedalfer and pedocal soils in the United States and southern Canada. In the United States the pedalfers have developed in the more humid climates to the east of the line that marks approximately 63 cm/year of precipitation. To the west of this line, where precipitation is generally less than 63 cm/year, pedocal soils predominate. In Canada and in the northern Rocky Mountains of the United States temperature is more critical than precipitation in determining the distribution of the two soils. There pedocals occur in areas where the average annual temperature is 4.5°C or less.

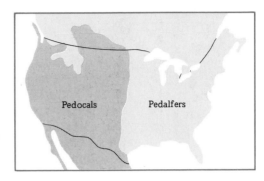

Using the information on the products of chemical weathering listed in Table 4.1, we can build up a picture of how a pedalfer develops from granite in temperate and humid areas. At some depth below the surface lies the unaltered granite. Directly above is the crumbly, partially disintegrated rock of the C horizon. Here we can still identify the minerals that made up the original granite although the feldspars have started to decompose and have become cloudy, and the iron-bearing minerals have been partially oxidized.

Moving upward to the B horizon, the zone of accumulation, we find that here the feldspars have been converted to clay and that the material has a compact, clayey texture. Iron oxides or limonite stain the soil a reddish or brownish color. And because the grains of quartz released from the granite have undergone little change, we find some sand in this otherwise clayey zone.

The A horizon, a few centimeters thick, has a grayish to ashen color; for the iron compounds have been leached from this zone and now color the B horizon below. Furthermore, the texture of this zone is sandier; for most of the finer materials have also been moved downward to the B horizon, and the soluble salts have been largely dissolved and removed by water. The very top of the A horizon is a thin zone of dark, humus material. The ultimately resulting pedalfer soil is shown in Figure 4.21.

Notice in Figure 4.22 how the original minerals of the granite are transformed as weathering progresses. On the left we have quartz, plagioclase, and orthoclase, which were released directly from the granite. Then, as weathering progresses, we find that the amount of kaolinite increases at the expense of the

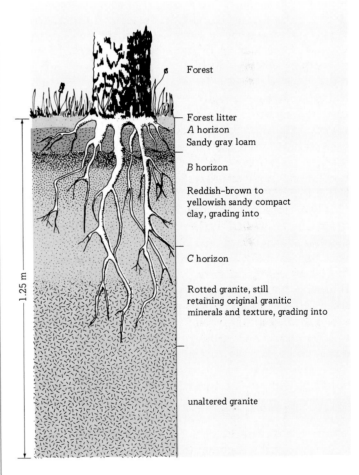

FIGURE 4.21

A pedalfer soil that has developed on a granite. Note the transition from unaltered granite, upward through partially decomposed granite of the C horizon, into the B horizon, where no trace of the original granite structure remains, and finally into the A horizon, just below the surface.

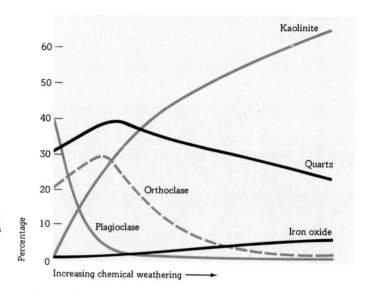

FIGURE 4.22

Change in mineral percentages as a granite is subjected to increased chemical weathering. As weathering progresses, the orthoclase and plagioclase of the granite decrease in abundance and give rise to an increasing amount of clay (kaolinite). At the same time the percentage of iron oxides increases at the expense of the original iron silicate minerals (not shown). [Redrawn from S. S. Goldich, "A Study of Rock Weathering," *J. Geol.,* vol. 46, p. 33, 1938.]

original minerals. The initial rise in the amount of quartz and orthoclase indicates simply that these minerals tend to accumulate in the soil because of their greater resistance to decomposition. Iron oxides also increase with weathering as the iron-bearing silicates decompose.

There are several varieties of soils in the pedalfer group, including the red and yellow soils of the southeastern states as well as **podsol** (from the Russian for "ashy gray soil") and the gray-brown podsolic soils of the northeastern quarter of the United States and of southern and eastern Canada (Figure 4.23). Prairie soils are transitional varieties between the pedalfers of the East and the pedocals of the West.

Pedocals *Pedocals* are soils that contain an accumulation of calcium carbonate. Their name is derived from a combination of *pedon* with an abbreviation for calcium. The soils of this major group are found in the temperate zones where the temperature is relatively high, the rainfall is low, and the vegetation is mostly grass or brush. In the United States the pedocals are found generally to the west of the pedalfers; and in Canada, to the southwest.

In the formation of pedocals calcium carbonate and, to a lesser extent, magnesium carbonate are deposited in the soil profile, particularly in the B horizon. This process occurs in areas where the temperature is high, the rainfall scant, and the upper level of the soil is hot and dry most of the time. Water evaporates before it can remove carbonates from the soil. Consequently

these compounds are precipitated as **caliche,** a whitish accumulation made up largely of calcium carbonate. *Caliche* is a Spanish word, a derivative of the Latin *calix,* "lime." The occasional rain may carry the soluble material down from the A horizon into the B horizon, there it is later precipitated as the water evaporates. Soluble material may also move up into the soil from below. In this case water beneath the soil or in its lower portion rises toward the surface through small, capillary openings. Then, as the water in the upper portions evaporates, the dissolved materials will be precipitated out.

The tendency for pedocals to develop under a growth of brush and grass also helps to concentrate the soluble carbonates: The plants intercept them before they can be moved downward in the soil. When the plants die, the carbonates are either added to the soil, where they are used by other plants, or simply precipitated in the soil by high evaporation rates.

Because rainfall is light in the climates where pedocals form, chemical weathering proceeds only slowly, and clay is produced less rapidly than in more humid climates. For this reason pedocals contain a lower percentage of clay minerals than do the pedalfer soils.

The pedocal group includes the black and chestnut-colored soils of southern Alberta and Saskatchewan in Canada and the northern plains of the United States, the reddish soils farther south, and the red and gray desert soils of the drier western states.

Tropical soil (laterites) The term *laterite* is applied to many tropical soils that are rich in hydrated aluminum and iron oxides. The name itself, from the Latin for "brick," suggests the characteristic color produced by the iron in these soils. The formation of laterites is not well understood. In fact, soil scientists are even not certain that the A, B, and C horizons characteristic of the pedalfers and the pedocals have their counterparts in the laterites even though these soils exhibit recognizable zones.

In the development of the laterites iron and aluminum accumulate in what is presumed to be the B horizon. The aluminum is in the form of $Al_2O_3 \cdot nH_2O$, which is generally called **bauxite**, an ore of aluminum. This ore appears to be developed when intense and prolonged weathering removes the silica from the clay minerals and leaves a residuum of hydrous aluminum oxide (that is, bauxite). In some laterites the concentration of iron oxides in the presumed B horizon is so great that it is profitable to mine them for iron, as is done in certain districts of Cuba.

The term *laterite* is most properly applied only to the zone in which iron and aluminum have accumulated. As we have seen, this is the zone that may be equivalent to the B horizon in more northerly soils. Overlying this zone, however, there is often a zone of crumbly loam, and below it is a light-colored, apparently leached zone that adjoins the parent material. Some soil scientists refer to these two zones as the A horizon and the C horizon, respectively.

PALEOSOLS

Not all soils occur at the surface. Some are buried beneath younger material and are known as **paleosols** (see Figure 4.24). Such buried soil zones are useful indicators of past periods of erosion and weathering.

FIGURE 4.24

Paleosol buried beneath younger material. The dark band in the center of this exposure is a buried soil and measures over 1 m in thickness. It is developed on water-laid pyroclastic material and is buried by beds of similar volcanic material. At the top of the younger beds the present surface soil has been almost completely destroyed by erosion processes accelerated by human activity. Crater of Baccano along the Via Cassia, north of Rome, Italy. [Sheldon Judson.]

FIGURE 4.23

This podsol, a member of the pedalfer group of soils, occurs on Cape Cod, Massachusetts. The dark layer at the very top just beneath the plant cover is the humic zone in the upper part of the A horizon. The light-gray, leached zone below makes up the bulk of the A horizon. The B horizon is thin and shows up dark in the photograph because of the iron oxides that have accumulated there. This zone grades down into the C horizon, the parent material, which is here an unconsolidated sandy deposit. [Paul MacClintock.]

4.6 AMOUNTS OF SEDIMENTS AND SEDIMENTARY ROCKS

It is extremely difficult to work out a concise, comprehensive definition of sedimentary rocks. The adjective, from the Latin *sedimentum,* means "settling." Therefore we might expect sedimentary rocks to be formed when individual particles settle out of a fluid, such as the water of a lake or an ocean. And as we mentioned at the beginning of this chapter, many sedimentary rocks are formed in just that way: Fragments or minerals derived from the breakdown of rocks are swept into bodies of water, where they settle out as unconsolidated sediments. Later, they are hardened into true rocks. But other rocks, such as rock salt, are made up of

FIGURE 4.25

Erosion has exposed and dissected these flat-lying beds of sedimentary rocks of mid-Tertiary age. The gentle, vegetated lower slopes are underlain by shales, and the steep-walled cliffs are held up by more resistant sandstones. The center butte is the Glenn L. Jepsen Butte in the Castle Butte district of South Dakota. [C. C. O'Hara.]

minerals left behind by the evaporation of large bodies of water, and these rocks are as truly sedimentary rocks as those formed from particles that have settled on an ocean floor. Still other sedimentary rocks are made up largely of the shells and hard parts of animals, particularly of invertebrate marine organisms.

Sedimentary rocks are often layered, or stratified. Unlike massive igneous rocks, such as granite, most sedimentary rocks are laid down in individual beds, one on another. The surface of each bed is essentially parallel to the horizon at the time of deposition, and a cross section exposes a series of layers like those of a giant cake. True, some igneous rocks, such as those formed from lava flows, are also layered. By and large, however, *stratification is the single most characteristic feature of sedimentary rocks* (see Figure 4.25).

About 75 percent of the rocks exposed at the earth's surface are sedimentary rocks or metamorphic rocks derived from them. Yet sedimentary rocks make up only about 5 percent by volume of the outer 15 km of the globe. The other 95 percent of the rocks in this zone are or once were igneous rocks (see Figure 4.26). The sedimentary cover is only as thick as a feather edge where it laps around the igneous rocks of the Adirondacks and the Rockies. In other places it is thousands of meters thick. In Oklahoma oil-drilling operations have cut into the crust nearly 9 km and have encountered nothing but sedimentary rocks. In the Ganges River basin of India the thickness of the sedimentary deposits has been estimated at between 13,500 and 18,000 m.

We have some estimates of the total mass of sedimentary rocks on the earth. Probably the best available estimate was made some time ago by Arie Poldervaart, who calculated their weight as $1,702 \times 10^{15}$ t. Of this total he estimated that 480×10^{15} t are presently on the continents and that the rest are in the oceans.

We have seen that sedimentary rocks are formed from the materials weathered from preexisting rocks. Eventually this material reaches the deep ocean basins. What estimate can we make about the amount of material delivered each year to the oceans? The figures presented in Table 4.3 are order-of-magnitude estimates. Therefore, a statement of the approximate amount of material delivered annually to the oceans— where most sedimentary rocks form—is approximately 10^{10} t. As Table 4.3 suggests and as we have implied in Section 4.4, the great bulk of this material is carried by rivers. That contributed by wind, by ice, or by extraterrestrial sources does not appreciably change the total amount of material deposited in the oceans.

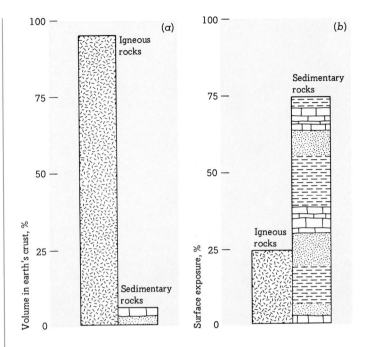

FIGURE 4.26

Graphs showing relative abundance of sedimentary rocks and igneous rocks. (a) The great bulk (95 percent) of the outer 15 km of the earth is made up of igneous rocks; only a small proportion (5 percent) is sedimentary. In contrast (b) the areal extent of sedimentary rocks at the earth's surface is three times that of igneous rocks. Metamorphic rocks are considered with either igneous or sedimentary rocks, depending on their origin.

TABLE 4.3

Order-of-magnitude estimates by source of materials delivered annually to the oceans[a]

Source	t/yr
Rivers	10 billion (10^{10})
Glaciers	100 million to 1 billion (10^8–10^9)
Wind	100 million (10^8)
Extraterrestrial	0.03 to 0.3 (3×10^{-2}–3×10^{-1})

[a]Estimated from various sources.

4.7 FORMATION OF SEDIMENTARY ROCKS

We found in Chapter 3 that igneous rocks harden from molten material that originates some place beneath the surface, under the high temperatures and pressures that prevail there. In contrast, sedimentary rocks form at the much lower temperatures and pressures that prevail at or near the earth's surface.

ORIGIN OF MATERIAL

The material from which sedimentary rocks are fashioned originates in two ways. First, the deposits may be accumulations of minerals and rocks derived either from the erosion of existing rock or from the weathered products of these rocks. Deposits of this type are called **detrital** (from the Latin for "worn down"), and sedimentary rocks formed from them are called *detrital sedimentary rocks.* Second, the deposits may be produced by chemical processes. We refer to these deposits as **chemical deposits** and to the rocks formed from them as *chemical sedimentary rocks.*

Gravel, sand, silt, and clay derived from the weathering and erosion of a land area are examples of detrital sediments. Let us take a specific example. The quartz grains freed by the weathering of a granite may be winnowed out by the running water of a stream and swept into the ocean. There they settle out as beds of sand, a detrital deposit. Later, when this deposit is cemented to form a hard rock, we have a sandstone, a detrital rock.

Chemically formed deposits are usually laid down by the precipitation of material dissolved in water. This process may take place either directly, through inorganic processes, or indirectly, through the intervention of plants or animals. The salt left behind after a salty body of water has evaporated is an example of a deposit laid down by inorganic chemical processes. On the other hand certain organisms, such as the corals, extract calcium carbonate from sea water and use it to build up skeletons of calcite. When the animals die, their skeletons collect as a biochemical (from the Greek for "life") deposit, and the rock that subsequently forms is called a *biochemical rock*—in this case limestone.

Although we distinguish between the two general groups of sedimentary rocks—detrital and chemical—most sedimentary rocks are mixtures of the two. We commonly find that a chemically formed rock contains a certain amount of detrital material. In similar fashion predominantly detrital rocks include some material that has been chemically deposited.

Geologists use various terms to describe the environment in which a sediment originally accumulated. For example, if a limestone contains fossils of an animal that is known to have lived only in the sea, the rock is known as a **marine** limestone. **Fluvial,** from the Latin for "river," is applied to rocks formed by deposits laid down by a river. **Eolian,** derived from Aeolus, the Greek wind god, describes rock made up of wind-deposited material. Rocks formed from lake deposits are termed **lacustrine,** from the Latin word for "lake."

Detrital and chemical, however, are the main divisions of sedimentary rocks based on the origin of material, and, as we shall see later, they form the two major divisions in the classification of sedimentary rocks.

SEDIMENTATION

The general process by which rock-forming material is laid down is called **sedimentation,** or deposition. The factors controlling sedimentation are easy to visualize: To have any deposition at all, there must obviously be something to deposit—another way of saying that a source of sediments must exist; we also need some process to transport this sediment; and finally there must be some place and some process for the deposition of the sedimentary material.

Methods of transportation Water—in streams and glaciers, underground and in ocean currents—is the principal means of transporting material from one place to another. Landslides and other movements induced by gravity also play a role, as does the wind, but we shall look more closely at these processes in Chapters 9 through 13.

Processes of sedimentation Detrital material is deposited when its agent of transportation no longer has sufficient energy to move it farther. For example, a stream flowing along at a certain velocity possesses energy to move particles up to a certain maximum size. If the stream loses velocity, it also loses energy, and it is no longer able to transport all the material that it has been carrying at the higher velocity. The solid particles, beginning with the heaviest, start to settle to the bottom. The effect is much the same when a wind that has been driving sand across a desert suddenly dies. A loss of energy accompanies the loss in velocity.

Material that has been carried in solution is deposited in a different way: by precipitation, a chemical process by which dissolved material is converted into a solid and separated from the liquid solvent. As already noted, precipitation may be either biochemical or inorganic.

Although at first glance the whole process of sedimentation seems quite simple, it is actually as complex as nature itself. Many factors are involved, and they can interact in a variety of ways. Consequently, the manner in which sedimentation takes place and the sediments that result from it differ greatly from one situation to another (see Figures 4.27 and 4.28). Think, for instance, of the different ways in which materials settle out of water. A swift, narrow mountain stream may deposit coarse sand and gravel along its bed, but farther downstream, as the valley widens, the same stream may overflow its banks and spread silt and mud over the surrounding country. A lake provides a different environment, varying from the delta of the inlet stream to the deep lake bottom and the shallow, sandy shore zones. In the oceans too environment and sedimentation differ from the brackish tidal lagoon to the zone of plunging surf and out to the broad, submerged shelves of the continents and to the ocean depths beyond.

MINERAL COMPOSITION
OF SEDIMENTARY ROCKS

Sedimentary rocks, like igneous and metamorphic rocks, are accumulations of minerals. In sedimentary rocks the three most common minerals are clay, quartz, and calcite although, as we shall see, a few others are important in certain localities.

Rarely is a sedimentary rock made up of a single mineral although one mineral may predominate. Limestone, for example, is composed mostly of calcite, but even the purest limestone contains small amounts of

FIGURE 4.27

The environment of this quiet pond near Lexington, Massachusetts, favors the deposition of fine-grained sediments, largely mud. Compare with Figure 4.28. [Sheldon Judson.]

FIGURE 4.28

Exposed to the direct attack of ocean surf, the environment of this cliffed California coastline favors the deposition of coarse sand and gravel. Compare with Figure 4.27. [Sheldon Judson.]

FIGURE 4.29

Quartz grains of varying sizes and shapes are cemented together by hematite. The large quartz fragment is about 1.2 cm long. [Willard Starks.]

quartz, as well as a number of much rarer forms, such as chert, flint, opal, and chalcedony.

The mechanical and chemical weathering of an igneous rock such as granite sets free individual grains of quartz that eventually may be incorporated into sediments. These quartz grains produce the detrital forms of silica and account for most of the volume of the sedimentary rock sandstone. But silica in solution or in particles of colloidal size is also produced by the weathering of an igneous rock. This silica may be precipitated or deposited in the form of quartz, particularly as a cementing agent in certain coarse-grained sedimentary rocks. Silica may also be precipitated in other forms such as *opal*, generally regarded as a hydrous silica, $SiO_2 \cdot nH_2O$. Opal is slightly softer than true quartz and has no true crystal structure.

Silica also occurs in sedimentary rocks in a form called **cryptocrystalline**. This term (from the Greek *kryptos*, "hidden," and *crystalline*) indicates crystalline structure so fine that it cannot be seen under most ordinary microscopes. The microscope does reveal, however, that some cryptocrystalline silica has a granular pattern and that some has a fibrous pattern. To the naked eye, the surface of the granular form is somewhat duller than that of the fibrous form. Among the dull-surfaced, or granular, varieties is *flint*, usually dark in color. Flint is commonly found in certain limestone beds—the chalk beds of southern England, for example. *Chert* is similar to flint but tends to be lighter in color; and *jasper* is a red variety of granular cryptocrystalline form. The general term *chalcedony* is often applied to the fibrous types of cryptocrystalline silica, which have a higher, more waxy luster than do granular varieties. Sometimes the term is also used to describe a specific variety of brown, translucent cryptocrystalline silica. *Agate* is a variegated form of silica, its bands of chalcedony alternating with bands of either opal or some variety of granular cryptocrystalline silica, such as jasper.

Calcite   The chief constituent of the sedimentary rock limestone, calcite, $CaCO_3$, is also the most common cementing material in the coarse-grained sedimentary rocks. The calcium is derived from igneous rocks that contain calcium-bearing minerals, such as calcium plagioclase and some of the ferromagnesian minerals. Calcium is carried from the zone of weathering as calcium bicarbonate, $Ca(HCO_3)_2$, and is eventually precipitated as $CaCO_3$ through the intervention of plants, animals, or inorganic processes. The carbon-

other minerals, such as clay or quartz. The grains of many sandstones are predominantly quartz, but the cementing material that holds these grains together may be calcite, dolomite, or iron oxide (see Figure 4.29). In general, we may say that most sedimentary rocks are mixtures of two or more minerals.

Clay   Earlier in this chapter we described how clay minerals develop from the weathering of the silicates, particularly the feldspars. These clays may be subsequently incorporated into sedimentary rocks; they may, for example, form an important constituent of mudstone and shale. Examination of recent and ancient marine deposits shows that the kaolinite and illite clays are the most common clays in sedimentary rocks and that the montmorillonite clays are relatively rare.

Quartz   Another important component of sedimentary rocks is silica, including the very common mineral

ate is ultimately derived from water and carbon dioxide.

Other materials in sedimentary rocks Accumulations of clay, quartz, and calcite, either alone or in combination, account for all but a very small percentage of the sedimentary rocks, but certain other materials occur in quantities large enough to form distinct strata. The mineral **dolomite**, $CaMg(CO_3)_2$, for example, is usually intimately associated with calcite although it is far less abundant. (It is named after an eighteenth-century French geologist, Déodat de Dolomieu.) When the mineral is present in large amounts in a rock, the rock itself is also known as dolomite. The mineral dolomite is easily confused with calcite; and because they often occur together, distinguishing them is important. Calcite effervesces freely in dilute hydrochloric acid; dolomite effervesces very slowly or not at all unless it is finely ground or powdered. The more rapid chemical activity results from the increase in surface area, an example of the general principle discussed in Section 4.2.

The feldspars and micas are abundant in some sedimentary rocks. We have found that chemical weathering converts these minerals into new minerals at a relatively rapid rate. Therefore, when we find mica and feldspar in a sedimentary rock, chances are that it was mechanical, rather than chemical, weathering that originally made them available for incorporation in the rock.

Iron produced by chemical weathering of the ferromagnesian minerals in igneous rocks may be caught up again in new minerals and incorporated into sedimentary deposits. The iron-bearing minerals that occur most frequently in sedimentary rocks are hematite, goethite, and limonite. In some deposits these minerals predominate, but more commonly they act simply as coloring matter or as a cementing material.

Halite, $NaCl$, and gypsum, $CaSO_4 \cdot 2H_2O$, are minerals precipitated from solution by evaporation of the water in which they were dissolved. The salinity of the water—that is, the proportion of the dissolved material to the water—determines the type of mineral that will precipitate out. The gypsum begins to separate from seawater when the salinity (at 30°C) reaches a little over three times its normal value. Then, when the salinity of the sea water has increased to about ten times its normal value, halite begins to precipitate.

Pyroclastic rocks, mentioned in Chapter 3, are sedimentary rocks composed mostly of fragments blown from volcanoes. The fragments may be large pieces that have fallen close to the volcano or extremely fine ash that has been carried by the wind and deposited hundreds of kilometers from the volcanic eruption.

Finally, organic matter may be present in sedimentary rocks. In the sedimentary rock known as coal, plant materials are almost the only components. More commonly, however, organic matter is very sparsely disseminated throughout sedimentary deposits and the resulting rocks.

TEXTURE

Texture, as we have explained, refers to the general physical appearance of a rock—to the size, shape, and arrangements of the particles that make it up. There are two major types of texture in sedimentary rocks: clastic and nonclastic.

Clastic texture The term **clastic** is derived from the Greek for "broken" or "fragmental," and rocks that have been formed from deposits of mineral and rock fragments are said to have clastic texture (see Figure 4.30). The size and shape of the original particles have a direct influence on the nature of the resulting texture. A rock formed from a bed of gravel and sand has a coarse, rubblelike texture that is very different from the sugary texture of a rock developed from a deposit of rounded, uniform sand grains. Furthermore, the process by which a sediment is deposited also affects the texture of the sedimentary rock that develops from it. Thus the debris dumped by a glacier is composed of a jumbled assortment of rock material ranging from particles of colloidal size to large boulders. A rock that develops from such a deposit has a very different texture from one that develops from a deposit of wind-blown sand, for instance, in which all the particles are approximately 0.15 to 0.30 mm in diameter.

Chemical sedimentary rocks may also show a clastic texture. A rock made up predominantly of shell fragments from a biochemical deposit has a clastic texture that is just as recognizable as the texture of a rock formed from sand deposits (see Figure 4.30).

One of the most useful factors in classifying sedimentary rocks is the size of the individual particles. In practice, we usually express the size of a particle in terms of its diameter rather than in terms of volume, weight, or surface area. When we speak of "diameter," we imply that the particle is a sphere; but it is very

FIGURE 4.30

Viewed under the microscope, some limestones are seen to have clastic texture, as does this example from an ancient reef deposit in the Austrian Alps. The individual particles include the shells of one-celled marine animals (Foraminifera), as well as the fragments of other marine organisms and pellets of calcite. The particles are cemented together by clear calcite (dark). The black lines are also calcite that fills cracks in the rock. The area shown is 2.8 cm wide. [E. C. Bierwagen.]

unlikely that any fragment in a sedimentary rock is a true sphere. In geological measurements the term simply means the diameter that an irregularly shaped particle *would* have if it were a sphere of equivalent volume. Obviously, it would be a time-consuming, if not impossible, task to determine the volume of each sand grain or pebble in a rock and then to convert these measurements into appropriate diameters. So the diameters we use for particles are only approximations of their actual sizes. They are accurate enough, however, for our needs.

Several scales have been proposed to describe particles ranging in size from large boulders to minerals of microscopic dimensions. The Wentworth scale, presented in Table 4.4, is used widely, though not universally, by American and Canadian geologists. Notice that, although the term *clay* is used in the table to designate all particles below $\frac{1}{256}$ (0.0039) mm in diameter, the same term is used to describe certain minerals. To avoid confusion, we must always refer specifically to either "clay size" or "clay mineral" unless the context makes the meaning clear.

Because determining the size of particles calls for the use of special equipment, the procedure is normally carried out only in the laboratory. In the examination of specimens in the field an educated guess based on careful observation usually suffices.

Nonclastic texture Some—but not all—sedimentary rocks formed by chemical processes have nonclastic texture, in which the grains are interlocked. These rocks have somewhat the same appearance as igneous rocks with crystalline texture. Actually, most of the sedimentary rocks with nonclastic texture have crystalline structure although a few of them, such as opal, do not exhibit this structure.

The mineral crystals that precipitate from an aqueous solution are usually small. Because the fluid in which they form has a very low density, they usually settle out rapidly and accumulate on the bottom as mud. Eventually, under the weight of additional sediments, the mud is compacted more and more. Now the size of the individual crystals may begin to increase. Their growth may be induced by added pressure's causing the favorably oriented grains to grow at the expense of less favorably oriented neighboring grains. Or crystals may grow as more and more mineral matter is added to them from the saturated solutions trapped in the original mud. In any event the resulting rock is made up of interlocking crystals and has a texture similar to that of crystalline igneous rocks. Depending on the size of the crystals, we refer to these nonclastic textures as fine-grained, medium-grained, or coarse-grained. A coarse-grained texture has grains larger than 5 mm in diameter, and a fine-grained texture has grains less than 1 mm in diameter.

LITHIFICATION

The process of **lithification** converts unconsolidated rock-forming materials into consolidated, coherent rock. The term is derived from the Greek and from the Latin "to make." In the following subsections, we shall discuss the various ways in which sedimentary deposits are lithified.

Cementation In cementation the spaces between the individual particles of an unconsolidated deposit are filled up by some binding agent. Of the many minerals that serve as cementing agents the most common are calcite, dolomite, and quartz. Others include iron oxide, opal, chalcedony, anhydrite, and pyrite. Apparently the cementing material is carried in solution by water that percolates through the open spaces between the particles of the deposit. Then some factor in the new environment causes the mineral to be deposited, and the former unconsolidated deposit is cemented into a sedimentary rock.

In coarse-grained deposits there are relatively large interconnecting spaces among the particles. As we should expect, these deposits are very susceptible to cementation because the percolating water can move through them with great ease. Deposits of sand and gravel are transformed by cementation into the sedimentary rocks sandstone and conglomerate.

Compaction and desiccation In a fine-grained clastic deposit of silt-sized and clay-sized particles the pore spaces are usually so small that water cannot freely circulate through them. Consequently, very little cementing material manages to find its way between the particles; but deposits of this sort are lithified by two other processes: compaction and desiccation.

In **compaction** the pore space between individual grains is gradually reduced by the pressure of overlying sediments or by pressures resulting from earth

TABLE 4.4

Wentworth scale of particle sizes for clastic sediments[a]

Wentworth scale		For next larger size multiply by
Size, mm	Fragment	
	Boulder	
256	Cobble	4
64	Pebble	16
4	Granule	2
2	Sand	32
$\frac{1}{16}$ (0.0625)	Silt[b]	16
$\frac{1}{256}$ (0.0039)	Clay[b]	

[a] Modified after C. K. Wentworth, "A Scale of Grade and Class Terms for Clastic Sediments," *J. Geol.*, vol. 30, p. 381, 1922.
[b] Dust.

movement. Coarse deposits of sand and gravel undergo some compaction, but fine-grained deposits of silt and clay respond much more readily. As the individual particles are pressed closer and closer together, the thickness of the deposit is reduced and its coherence is increased. It has been estimated that deposits of clay-sized particles buried to depths of 1,000 m have been compacted to about 60 percent of their original volume.

In **desiccation** the water that originally filled the pore spaces of water-laid clay and silt deposits is forced out. Sometimes this is the direct result of compaction, but desiccation also takes place when a deposit is simply exposed to the air and the water evaporates.

Crystallization The crystallization of certain chemical deposits is in itself a form of lithification. Crystallization also serves to harden deposits that have been laid down by mechanical processes of sedimentation. For example, new minerals may crystallize within a deposit, or the crystals of existing minerals may increase in size. New minerals are sometimes produced by chemical reactions among amorphous, colloidal materials in fine-grained muds. Exactly how and when these reactions occur is not yet generally understood, but that new crystals *have* formed after the deposit was initially laid down becomes increasingly apparent as we make more and more detailed studies of sedimentary rocks. Furthermore, it seems clear that this crystallization promotes the process of lithification, particularly in the finer sediments.

4.8 TYPES OF SEDIMENTARY ROCK

CLASSIFICATION

Having examined some of the factors involved in the formation of sedimentary rocks, we are in a better position to consider a classification for this rock family. The classification presented in Table 4.5 represents only one of many possible schemes, but it will serve our purposes very adequately. Notice that there are two main groups—detrital and chemical—based on the origin of the rocks and that the chemical category is further split into inorganic and biochemical. All the detrital rocks have clastic texture, whereas the chemical rocks have either clastic or nonclastic texture. We use particle size to subdivide the detrital rocks and composition to subdivide the chemical rocks.

DETRITAL SEDIMENTARY ROCKS

Conglomerate A *conglomerate* is a detrital rock made up of more or less rounded fragments an appreciable percentage of which are of granule size (2 to 4 mm in diameter) or larger. If the fragments are more angular than rounded, the rock is called a *breccia*. Another type of conglomerate is **tillite**, a rock formed from deposits laid down directly by glacier ice (see

Chapter 12). The large particles in a conglomerate are usually rock fragments, and the finer particles are usually minerals derived from preexisting rocks (see Figures 4.31 and 4.32).

Sandstone A *sandstone* is formed by the consolidation of grains of sand size, between $\frac{1}{16}$ (0.0625) and 2 mm in diameter. Sandstone is thus intermediate between coarse-grained conglomerate and fine-grained mudstone. Because the size of the grains varies from one sandstone to another, we speak of coarse-grained, medium-grained, and fine-grained sandstone.

Very often the grains of a sandstone are almost all quartz. When this is the case the rock is called an *orthoquartzite* (from the Greek here meaning "true," plus quartzite). The name *quartzose sandstone* is also used, as is *quartz arenite* (from the Latin for "sand"). If the minerals are predominately quartz and feldspar, the sandstone is called an *arkose*, a French word for the rock formed by the consolidation of debris derived from a mechanically weathered granite. Another variety of sandstone, called *graywacke*, or *lithic sandstone*, is characterized by its hardness and dark color and by angular grains of quartz, feldspar, and small fragments of rock set in a matrix of clay-sized particles.

Mudstone and shale Fine-grained detrital rocks composed of clay and silt-sized particles (less than

TABLE 4.5

Classification of sedimentary rocks

Origin	Texture	Particle size or composition	Rock name
Detrital	Clastic	Granule or larger	Conglomerate (round grains) breccia (angular grains)
		Sand	Sandstone
		Silt	Siltstone
		Clay	Mudstone and shale
Chemical: Inorganic	Clastic or nonclastic	Calcite, $CaCO_3$	Limestone
		Dolomite, $CaMg(CO_3)_2$	Dolomite
		Halite, $NaCl$	Salt
		Gypsum, $CaSO_4 \cdot 2H_2O$	Gypsum
Biochemical	Clastic or nonclastic	$CaCO_3$ shells	Limestone, chalk, coquina
		SiO_2 diatoms	Diatomite
		Plant remains	Coal

0.06 mm in diameter) are termed either *mudstone* or *shale*. Mudstones are fine-grained rocks with a massive or blocky aspect, whereas shales are fine-grained rocks that split into platy slabs more or less parallel to the bedding. The particles in these rocks are so small that it is difficult to determine the precise mineral composition of mudstone and shale. We do know, however, that they contain not only clay minerals but also clay-sized and silt-sized particles of quartz, feldspar, calcite, and dolomite, to mention but a few.

CHEMICAL SEDIMENTARY ROCKS

Limestone Limestone is a sedimentary rock that is made up chiefly of the mineral calcite that has been deposited by either inorganic or organic chemical processes. Most limestones have a clastic texture, but nonclastic, particularly crystalline, textures are common.

Inorganically formed limestone is made up of calcite that has been precipitated from solution by inorganic processes. Some calcite is precipitated from the fresh water of streams, springs, and caves although the total amount of rock formed in this way is negligible. When calcium-bearing rocks undergo chemical weathering, calcium bicarbonate, $Ca(HCO_3)_2$, is produced in solution. If enough of the water evaporates or if the temperature rises or if the pressure falls, calcite is precipitated from this solution. For example, most **dripstone**, or **travertine**, is formed in caves by the evaporation of water that is carrying calcium bicarbonate in solution. And **tufa** (from the Italian for "soft rock") is a spongy, porous limestone formed by the precipitation of calcite from the water of streams and springs.

Although geologists understand the inorganic processes by which limestone is formed by precipitation from fresh water, they are not quite sure how important these processes are in precipitation from seawater. Some observers have questioned whether they operate at all. On the floors of modern oceans and in rocks formed in ancient oceans, however, we find small spheroidal grains called **oölites**, the size of sand and often composed of calcite; these grains are thought to be formed by the inorganic precipitation of calcium carbonate from seawater. (The term comes from the Greek for "egg" because an accumulation of oölites resembles a cluster of fish roe.) Cross sections show that many oölites, though not all, have grown up around a mineral grain or around a small fragment of shell that

FIGURE 4.31

A conglomerate is made up of rounded pebbles, as shown in this deposit of partially consolidated material in Estes Park, Colorado. [Harvard University, Tozier Collection.]

FIGURE 4.32

Breccia is a lithified deposit containing many angular fragments. This 12-cm-wide specimen is from eastern Nevada. [Princeton University Museum of Natural History, Willard Starks.]

acts as a nucleus. Some limestones are made up largely of oölites. One, widely used for building, is the so-called Indiana or Spergen Limestone.

Dolomite In discussing the mineral dolomite, $CaMg(CO_3)_2$, we mentioned that, when it occurs in large concentrations, it forms a rock that is also called dolomite. The origins of extensive deposits of dolomite are still not well understood, but some uncertainties are beginning to disappear. There is no evidence that strata of dolomite develop by direct precipitation of the mineral in seawater. On the contrary, most dolomites appear to have been formed by replacement of preexisting deposits of calcite. Thus in some field situations it can be shown that igneous intrusions accompanied by solutions—presumably rich in magnesium—have altered limestone, some of the calcium in the calcite having been replaced by magnesium to form dolomite.

But dolomites caused by metamorphism are unimportant when compared with the bulk of dolomites in the geologic column. There is now increasing agreement that most dolomite is in some way related to the local increase of the amount of magnesium in solution. Previously deposited calcite is modified by the movement through it of these magnesium-rich solutions. Thus field and laboratory observations show that in shallow-water intertidal zones evaporation of seawater may cause precipitation of calcium-bearing deposits, both calcite and calcium sulfate, $CaSO_4$. Thus the waters may be increased by an order of magnitude in their content of magnesium relative to calcite. Such high-magnesium-content waters may then circulate through the underlying calcite deposits, replacing some of the calcium with magnesium and thus converting limestone to dolomite.

Evaporites An *evaporite* is a sedimentary rock composed of minerals that were precipitated from solution after the evaporation of the liquid in which they were dissolved. *Rock salt* (composed of the mineral halite) and *gypsum* are the most abundant evaporites. *Anhydrite* (from the Greek *anhydros*, "waterless") is an evaporite composed of the mineral of the same name, which is simply gypsum without its crystallization water, $CaSO_4$. Most evaporite deposits seem to have been precipitated from seawater according to a definite sequence. The less highly soluble minerals are the first to drop out of solution. Thus gypsum and anhydrite, both less soluble than halite, are deposited first. Then, as evaporation progresses, the more soluble halite is precipitated.

In the United States, the most extensive deposits of evaporites are found in Texas and New Mexico. Here gypsum, anhydrite, and rock salt make up over 90 percent of the Castile Formation, which has a maximum thickness of nearly 1,200 m. In central New York State there are thick deposits of rock salt, and in central Michigan there are layers of rock salt and gypsum. Some evaporite deposits are mined for their mineral content, and in certain areas, particularly in the Gulf Coast states, deposits of rock salt have pushed upward toward the surface to form salt domes containing commercially important reservoirs of petroleum (see Chapter 15).

ORGANIC ROCKS

A significant quantity of sedimentary rocks is the direct result of the accumulation of shells, plant fragments, and other remains of organisms. Together these are called **organic,** or **biochemical,** rocks. For example, diatoms are microscopic, single-celled plants that grow in marine or fresh water and secrete siliceous shells. When these shells accumulate in great numbers on the bottom of a basin, they may ultimately form a sedimentary rock called **diatomite.**

Biochemically formed limestones are created by the action of plants and animals that extract calcium carbonate from the water in which they live. The calcium carbonate may be either incorporated into the skeleton of the organism or precipitated directly. In any event, when the organism dies, it leaves behind a quantity of calcium carbonate, and over a long period of time thick deposits of this material may be built up. Reefs, ancient and modern, are well-known examples of such accumulations. The most important builders of modern reefs are algae, mollusks, corals, and one-celled animals—the same animals whose ancestors built up the reefs of ancient seas—the reefs, now old and deeply buried, that are often valuable reservoirs of petroleum.

Chalk is made up in part of biochemically derived calcite in the form of the skeletons or skeletal fragments of microscopic oceanic plants and animals. These organic remains are found mixed with very fine-grained calcite deposits of either biochemical or inorganic chemical origin. A much coarser type of limestone composed of organic remains is known as *coquina* (from the Spanish for "shellfish" or "cockle") and is characterized by the accumulation of many large fragments of shells.

Coal Coal is a rock composed of combustible matter derived from the partial decomposition of plants. We shall consider coal as a biochemically formed sedimentary rock although some geologists prefer to think of it as a metamorphic rock because it passes through various stages.

The process of coal formation begins with an accumulation of plant remains in a swamp. This accumulation is known as **peat,** a soft, spongy, brownish deposit in which plant structures are easily recognizable. Time, coupled with the pressure produced by deep burial and sometimes by earth movement, gradually transforms the organic matter into coal. During this process, the percentage of carbon increases as the volatile hydrocarbons and water are forced out of the deposit. Coals are ranked according to the percentage of carbon they contain. Peat, with the least amount of carbon, is the lowest ranking; then come lignite, or brown coal, bituminous, or soft, coal, and finally anthracite, or hard, coal, the highest of all the coals in its percentage of carbon.

RELATIVE ABUNDANCE OF SEDIMENTARY ROCKS

Sandstone, mudstone and shale, and limestone constitute about 99 percent of all sedimentary rocks. Of these, mudstone and shale are the most abundant. On the basis of extrapolation from measurements made in the field, the estimates of the percentages of mudstone and shale approximate 50 percent of all sedimentary rocks. Similar calculations for limestone and sandstone suggest that the limestone forms about 22 percent of these rock types and that sandstone accounts for the remaining 28 percent. These percentages, however, do not agree with theoretical determinations of relative abundances. These are based on the determination of the products to be expected from the weathering of an average igneous rock. If these weathering products are assigned to the three major sedimentary-rock types, then we find that shale should be considerably more important volumetrically than it appears to be on the basis of field measurements. On these theoretical grounds, mudstone and shale constitute approximately 75 percent of the three major sedimentary-rock types. Sandstone and limestone are approximately of equivalent volume and together constitute the other 25 percent. The discrepancy between the two estimates has not yet been resolved.

Estimates have been made of the average chemical composition of the world's sediments. One of these estimates is presented in Table 4.6.

TABLE 4.6

Average composition of all sediments (as oxides)[a]

Oxide	Wt. %
SiO_2	44.5
TiO_2	0.6
Al_2O_3	10.9
Fe_2O_3	4.0
FeO	0.9
MnO	0.3
MgO	2.6
CaO	19.7
Na_2O	1.1
K_2O	1.9
P_2O_5	0.1
CO_2	13.4
Total	100.0

[a]From Arie Poldervaart, "Chemistry of the Earth's Crust," *Geol. Soc. Am. Spec. Paper,* no. 62, p. 132, 1955.

4.9 FEATURES OF SEDIMENTARY ROCKS

We have mentioned that the stratification, or bedding, of sedimentary rocks is their single most characteristic feature. Now we shall look more closely at this feature, along with certain other characteristics of sedimentary rocks, including mud cracks and ripple marks, nodules, concretions, geodes, fossils, and color.

BEDDING

The beds, or layers, of sedimentary rocks are separated by **bedding planes,** along which the rocks tend to separate or break (see Figure 4.33). The varying thickness of the layers in a given sedimentary rock reflects the

FIGURE 4.33

Layered limestone of Ordovician age at Trenton Falls, New York. [Sheldon Judson.]

changing conditions that prevailed when each deposit was laid down. In general each bedding plane marks the termination of one deposit and the beginning of another. For example, let us imagine a bay of an ocean into which rivers normally carry fine silt from the nearby land. This silt settles out from the seawater to form a layer of mud. Now heavy rains or melting snows may cause the rivers suddenly to flood and thereby pick up coarser material, such as sand, from the river bed. This material will be carried along and dumped into the bay. There it settles to the bottom and blankets the silt that was deposited earlier. The plane of contact between the mud and the sand represents a bedding plane. If, later on, the silt and the sand are lithified into

shale and sandstone, the bedding plane persists in the sedimentary rock. In fact, it marks a plane of weakness along which the rock tends to break.

Bedding planes are usually horizontal, and, when they are, the bedding is called **parallel bedding.** Closely spaced horizontal bedding planes form **laminated bedding.** But some beds are laid down at an angle to the horizontal, and such bedding is variously called **cross bedding** or **inclined bedding** (see Figure 4.34). Such nonhorizontal bedding can occur in several situations. Thus the bedding in sand dunes may have high angles on the leeward side of the dune (see Chapter 13). The deposits laid down at the growing edge of a delta may be inclined from 5° to 30° and such

FIGURE 4.34

Inclined beds (cross bedding) in eolian deposits (consolidated sand dunes) of the Navajo Formation in Zion National Park, Utah. The large tree in the right foreground is about 6 m tall. [M. E. Kauffman.]

beds are usually given a special name, **foreset beds** (see Chapter 10). In flowing water scouring of the stream floor by turbulent water may create small depressions in the channel deposits, which later will be refilled by inclined beds. Alternatively, cross bedding can occur as flowing water passes to somewhat quieter conditions and drops its load in a way similar to that in which a stream builds a foreset bed in a delta.

When the particles in a sedimentary bed vary from coarse at the bottom to fine at the top, the bedding is said to be **graded.** Such bedding is characteristic of rapid deposition from a water turbid with sedimentary particles of differing sizes. The largest particles settle most rapidly, and the finest most slowly. Such deposits are called **turbidites,** a term from the Latin *turbidus* for "disturbed," referring to the stirring up of sediments and water (see also Chapter 13).

Bedding may be distorted before the material becomes consolidated into firm rock. Thus unconsolidated sediments may sometimes slump or flow, as suggested in Figure 4.35. In addition, animals may burrow or tunnel through unconsolidated layers to produce **disturbed,** or **mottled, bedding.**

RIPPLE MARKS, MUD CRACKS,
AND SOLE MARKS

Ripple marks are the little waves of sand that commonly develop on the surface of a sand dune, along a beach, or on the bottom of a stream. **Mud cracks** are familiar on the dried surface of mud left exposed by the subsiding waters of a river or pond. **Sole marks** are the casts, or filling, of a primary sedimentary structure such as a groove or track, found on the underside, or lower surface, of a siltstone or sandstone that had overlain such a structure in soft underlying sediment.

Ripple marks preserved in sedimentary rocks often furnish clues to the conditions that prevailed when a sediment was originally deposited. For instance, if the ripple marks are symmetric, with sharp or slightly rounded ridges separated by more gently rounded

FIGURE 4.35

Some sedimentary deposits are deformed before they have become lithified. Shown is a section of Precambrian sedimentary rocks west of Knob Lake, Quebec, Canada. The lower beds are still undisturbed, but the upper beds have been strongly contorted, an event that took place when the beds were still unconsolidated and was probably caused by slumping of the deposits along the original slope of deposition. [J. E. Howell.]

FIGURE 4.36

Ripple marks of oscillation on a 35-cm-long slab of sandstone. The symmetrical nature of the ripples indicates that the current moved back and forth rather than continuously in one direction. [Willard Starks.]

FIGURE 4.37

Polygonal pattern of mud cracks resulting from dessication of modern fine-grained sediments in a playa lake in Nevada. Scale shown by pocket knife. [William C. Bradley.]

troughs, we are fairly safe in assuming that they were formed by the back-and-forth movement of water such as we find along a sea coast outside the surf zone. These marks are called ripple marks of **oscillation** (see Figure 4.36). If, on the other hand, the ripple marks are asymmetric, we can assume that they were formed by air or water moving more or less continuously in one direction. These marks are called **current** ripple marks.

Mud cracks make their appearance when a deposit of silt or clay dries out and shrinks (see Figures 4.37 and 4.38). The cracks outline roughly polygonal areas, making the surface of the deposit look like a section cut through a large honeycomb. Eventually, another deposit may come along to bury the first. If the deposits are later lithified, the outlines of the cracks may be accurately preserved for millions of years. Then, when the rock is split along the bedding plane between the two deposits, the cracks will be found much as they appeared when they were first formed, providing evidence that the original deposit underwent alternate flooding and drying.

Sole marks develop when a groove or trail is formed on the surface of soft sediment by a stick or other object carried along in the current. Under ideal conditions this depression will be preserved and filled by silt or sand. When lithified, this sequence of sedimentary beds can be used to decipher current directions, tops and bottoms of strata, and conditions of the environment.

NODULES, CONCRETIONS, AND GEODES

Many sedimentary rocks contain structures that were formed only *after* the original sediment was deposited. Among these are nodules, concretions, and geodes.

A **nodule** is an irregular, knobby-surfaced body of mineral matter that differs in composition from the sedimentary rock in which it has formed. It usually lies parallel to the bedding planes of the enclosing rock, and sometimes adjoining nodules coalesce to form a continuous bed. Nodules average about 30 cm in maximum dimension. Silica in the form of chert or flint is the major component of these bodies. They are most commonly found in limestone or dolomite. Most nodules are thought to have formed when silica replaced some of the materials of the original deposit; some, however, may consist of silica that was deposited at the same time as the main beds were laid down.

A **concretion** is a local concentration of the cementing material that has lithified a deposit into a sedimentary rock. Concretions range in size from a fraction of a centimeter to a meter or more in maximum dimension. Most are shaped like simple spheres or disks although some have fantastic and complex forms. For some reason, when the cementing material entered the unconsolidated sediment, it tended to concentrate around a common center point or along a common center line. The particles of the resulting concretion are cemented together more firmly than the particles of the host rock that surrounds it. The cementing material usually consists of calcite, dolomite, iron oxide, or silica—in other words, the same cementing materials that we find in the sedimentary rocks themselves.

Geodes, more eye-catching than either concretions or nodules, are roughly spherical, hollow structures up to 30 cm or more in diameter (see Figure 4.39). An outer layer of chalcedony is lined with crystals that project inward toward the hollow center. The crystals, often perfectly formed, are usually quartz although crystals of calcite and dolomite have been found and, more rarely, crystals of other minerals. Geodes occur most commonly in limestone, but they also occur in shale.

How does a geode form? First, a water-filled pocket develops in a sedimentary deposit, probably as a result of the decay of some plant or animal that was buried in the sediments. As the deposit begins to consolidate into a sedimentary rock, a wall of silica with a jellylike consistency forms around the water, isolating it from the surrounding material. As time passes, fresh water may enter the sediments. The water inside the pocket has a higher salt concentration than does the water outside. To equalize the concentrations, there is a slow mixing of the two liquids through the silica wall or membrane that separates them. This process of mixing is called **osmosis**. As long as the osmotic action continues, pressure is exerted outward toward the surrounding rock. The original pocket expands bit by bit until the salt concentrations of the liquids inside and outside are equalized. At this point osmosis stops, the outward pressure ceases, and the pocket stops growing. Now the silica wall dries, crystallizes to form chalcedony, contracts, and cracks. If, at some later time, mineral-bearing water finds its way into the deposit, it may seep in through the cracks in the wall of chalcedony. There the minerals are precipitated, and crystals begin to grow inward, toward the center, from the interior walls. Finally, we have a crystal-lined geode imbedded in the surrounding rock. Notice that the crystals in a geode grow inward; in a concretion they grow outward.

FIGURE 4.38

Mud cracks preserved in limy shale at Morgantown, Pennsylvania, in the Cambrian Conococheague Formation. [M. E. Kauffman.]

FIGURE 4.39

Two geodes broken open to show their internal structure. The dark outer layer of chalcedony is lined with milky-to-clear quartz crystals that project inward toward a hollow center. These structures are most commonly found in limestone, where they apparently form by the modification and enlargement of an original void. The smaller geode is about 8 cm in diameter. [Specimen from Harvard University Geological Museum, Walter R. Fleischer.]

FIGURE 4.41

Casts of burrows made by worms that lived in Devonian muds are shown on the underside of this slab from a locality near Ithaca, New York. [Sheldon Judson.]

FIGURE 4.40

This fossil bat, *Icaronycteris index*, oldest known flying mammal, was found in the Green River Formation of the early Eocene age in southwestern Wyoming. It measures 12 cm in length, and its wing span (here restored) is 30 cm. Detailed study suggests that the bat was a young male that died on a summer evening while foraging for insects or small fish close to the surface of the lake in which the Green River Formation was laid down. Just to the right of the tail, a small flower has been preserved. [Princeton University Museum of Natural History, Willard Starks.]

FOSSILS

The word **fossil** (derived from the Latin *fodere*, "to dig up") originally referred to anything that was dug from the ground, particularly a mineral or some inexplicable form. It is still used in that sense occasionally, as in the term *fossil fuel* (see Chapter 14). But today fossil generally means any direct evidence of past life—for example, the bones of a dinosaur, the shell of an ancient clam, the footprints of a long-extinct animal, or the delicate impression of a leaf (see Figures 4.40 and 4.41).

Fossils are usually found in sedimentary rocks although they sometimes turn up in igneous and metamorphic rocks. They are most abundant in mudstone, shale, and limestone but are also found in sandstone, dolomite, and conglomerate. Fossils account for almost the entire volume of certain rocks, such as the coquina and limestones that have been formed from ancient reefs.

The remains of plants and animals are completely destroyed if they are left exposed on the earth's surface; but if they are somehow protected from destructive forces, they may become incorporated in a sedimentary deposit, where they will be preserved for millions of years. In the quiet water of the ocean, for example, the remains of starfish, snails, and fish may be buried by sediments as they settle slowly to the bottom. If these sediments are subsequently lithified, the remains are preserved as fossils that tell us about the sort of life that existed at the time when the sediments were laid down.

Fossils are also preserved in deposits that have settled out of fresh water. Countless remains of land animals, large and small, have been dug from the beds of such once-watery environments as extinct lakes, flood plains, and swamps.

The detailed story of the development of life as recorded by fossils is properly a part of historical geology, and although we do not have time to trace it here, in Chapter 18 we shall find that fossils are extremely useful in subdividing geologic time and constructing the geologic column.

COLOR OF SEDIMENTARY ROCKS

Throughout the western and southwestern areas of the United States, bare cliffs and steep-walled canyons provide a brilliant display of the great variety of colors exhibited by sedimentary rocks. The Grand Canyon of the Colorado River in Arizona cuts through rocks that vary in color from gray, through purple and red, to brown, buff, and green. Bryce Canyon in southern Utah is fashioned of rocks tinted a delicate pink, and the Painted Desert, farther south in Arizona, exhibits a wide range of colors, including red, gray, purple, and pink.

The most important sources of color in sedimentary rocks are the iron oxides. Hematite, Fe_2O_3, for example, gives rocks a red or pink color, and limonite or goethite produces tones of yellow and brown. Some of the green, purple, and black colors may be caused by iron but in exactly what form is not completely understood. Only a very small amount of iron oxide is needed to color a rock. In fact, few sedimentary rocks contain more than 6 percent of iron, and most contain very much less.

Organic matter, when present, may also contribute to the coloring of sedimentary rocks, usually making them gray to black. Generally, but not always, the higher the organic content, the darker the rock.

The size of the individual particles in a rock also influences the color or, at least, the intensity of the color. For example, fine-grained clastic rocks are usually somewhat darker than coarse-grained rocks of the same mineral composition.

SEDIMENTARY FACIES

If we examine the environments of deposition that exist at any one time over a wide area, we find that they differ from place to place. Thus the freshwater environment of a river changes to a brackish-water environment as the river nears the ocean. In the ocean itself

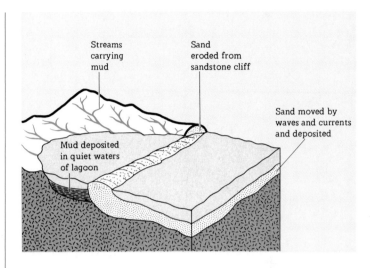

FIGURE 4.42

Diagram to illustrate a change in sedimentary facies. Here the fine-grained muds are deposited in a lagoon close to shore. A sandbar separates them from sand deposits farther away from shore. The sand in this instance has been derived from a sea cliff and transported by waves and currents.

marine conditions prevail. But even here the marine environment changes—from shallow water to deep water, for example. And as the environment changes, the nature of the sediments that are laid down also changes. The deposits in one environment show characteristics that are different from the characteristics of deposits laid down at the same time in another environment. This change in the "look" of the sediments is called a change in **sedimentary facies,** the latter word derived from the Latin word for the notion of "aspect" or "form."

We may define sedimentary facies as an accumulation of deposits that exhibits specific characteristics and grades laterally into other sedimentary accumulations formed at the same time but exhibiting different characteristics. The concept of facies is widely used in studying sedimentary rocks and the conditions that gave rise to metamorphic rocks (see Chapter 5). The concept is generally not used in referring to igneous rocks although there is no valid reason why it should not be used.

Let us consider a specific example of facies. Figure 4.42 shows a coastline where rivers from the land empty into a lagoon. The lagoon is separated from the open

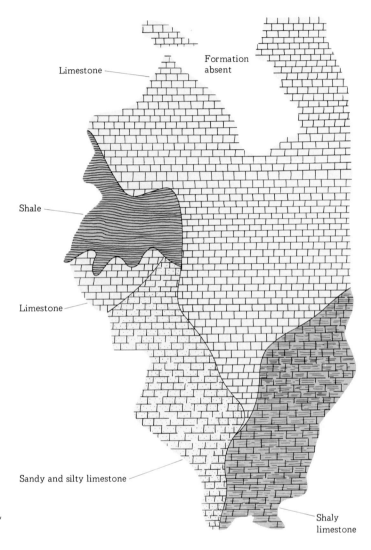

FIGURE 4.43

Variation in facies of ancient sediments comprising the Maquoketa Formation in Illinois. (After E. P. DuBois, "Subsurface Relations of the Maquoketa and 'Trenton' Formations in Illinois," *Ill. State Geol. Surv. Rep. Invest.*, 105, pp. 7–33, 1945.)

ocean by a sandbar. The fine silts and clays dumped into the quiet waters of the lagoon settle to the bottom as a layer of mud. At the same time, waves are eroding coarse sand from a nearby headland outside the lagoon. This sand is transported by currents and waves and deposited as a sandy layer seaward of the sandbar. Different environments exist inside and outside the lagoon; therefore different deposits are being laid down simultaneously. Notice that the mud and the sand grade into each other along the sandbar. Now imagine that these deposits have been consolidated into rock and then exposed to view at the earth's surface. We should find a shale layer grading into sandstone; that is, one sedimentary facies is grading into another.

But the picture is not always so simple. Recent sediments range from sand, through a sandy mud, to mud, and in a few areas there are limy deposits. Where the sea floor is rocky, little or no recent sedimentation has taken place. Should the soft sediments become lithified, a sedimentary rock ranging from sandstone, through sandy shale, to shale and limestone would result.

Ancient sedimentary deposits show exactly this kind of variation in facies. Figure 4.43 pictures the actual pattern of rock types that have been identified in the middle portion of the Maquoketa Formation in Illinois. Notice that the types include limestone, shale, limy shale, and a limy shale containing some sand.

4.10 PLATE TECTONICS AND SEDIMENTATION

As noted in Chapter 1, we are just beginning to understand the crustal motions of the colossal portions of the lithosphere known as plates. Areas exist on the earth's crust where sediments accumulate to great thicknesses. Many of these regions are ultimately pushed up into mountains, developing most commonly along the margins of the plates.

Thick accumulations of sediments may occur either where there is a deep hole into which the sediments are dumped or where the crust subsides as more and more sediments are deposited in relatively shallow water. We have evidence of both in the geologic record, but the overwhelming bulk of sedimentary rocks found in mountain ranges apparently formed originally as shallow-water sediments, as proved by indicators such as certain kinds of ripple marks, mud cracks, and shallow-water organisms.

GEOSYNCLINES

Following the work of James Hall, a nineteenth-century pioneer in American geology, Marshall Kay, and other twentieth-century proponents of the idea, the concept of geosynclines has developed into a dominant theme in the interpretation of many mountain belts. Folded rocks in mountain ranges of the world have apparently developed from thick deposits of sediments that have accumulated in these geosynclines—literally, "earth syncline," or large-scale downwarp in the sedimentary rocks found in the earth's crust. A geosyncline is thus a large sediment-filled, elongate basin. We shall examine the problem of deformation of these sedimentary rocks and the actual formation of mountains in Chapter 7.

Parallel adjacent belts characterize geosynclines. These have been called **miogeosyncline** (or simply **miogeocline** by Robert Dietz and others) where sediments are deposited on continental basement and where volcanic rocks and deeper-water sediments are lacking; and **eugeosyncline** (or **eugeocline**) where sediments are deposited on noncontinental basement and where there is an abundance of interstratified volcanic rocks and deeper-water sediments.

PLATE BOUNDARIES AND SEDIMENTATION

As discussed in Chapter 1, there are three basic types of plate boundaries: **convergent**, where old lithosphere is carried downward along subduction zones and magma rises to form volcanic regions, as the west coast of South America or the Aleutians; **divergent**, where old lithosphere pulls apart and new lithosphere forms along midoceanic spreading centers, as the Atlantic coasts; and **transform**, or **parallel**, where two plates slide past one another with no new lithosphere forming and no old lithosphere being destroyed, as along the San Andreas fault.

In relating sedimentation and geosynclinal concepts to those of plate-tectonic theory, we can interpret the miogeosynclines as representing thick sediment wedges forming on divergent continental margins and the eugeosynclines as representing sediments deposited farther offshore or in areas marked by volcanic island arcs, perhaps at a convergent plate boundary. Transform margins are marked by numerous basins that are later filled with thick sequences of sediments similar to those found along the San Andreas fault zone.

The relative motion of the plates on which or between which the sediments have accumulated will determine the ultimate structure of the mountain systems that later form from these sediments and sedimentary rocks (see Chapter 7).

OUTLINE

Energy, the capacity for producing motion, is involved in every geological process.
Potential energy is stored energy waiting to be used.
Kinetic energy, possessed by every moving object, depends on the mass of the object and on the speed with which it is moving.
The energy that drives the weathering process comes from within the earth in terms of uplift in mountainous areas and from outside the earth—such as solar energy.

Weathering is the response of surface or near-surface material to contact with water, air, and living matter.

Types of weathering are mechanical and chemical.

Mechanical weathering (disintegration) involves a reduction in the size of rock and mineral particles but no change in composition.

Chemical weathering (decomposition) involves a change in the composition of the material weathered. The rate of chemical weathering increases with the decrease in the size of particles and with increase in temperature and moisture.

Chemical weathering of igneous rocks, which include quartz, feldspar, and ferromagnesian minerals, gives clay, iron oxides, quartz, and soluble salts.

Rates of weathering vary with material weathered and the environment. For example, olivine chemically weathers more rapidly than does quartz, and limestone weathers very rapidly in a moist climate but very slowly in a dry climate.

Rates of erosion indicate that, before human beings began to use the landscape intensively, rivers annually transported about 10 billion t of material to the seas. Today it is two or three times that, chiefly because of human activities.

Differential erosion is the process by which different rock masses or different sections of the same rock mass erode at different rates. For example, limestone is more resistant to erosion in a dry climate than is mudstone.

Soil is a naturally occurring surface material that supports life and generally is the product of weathering.

Soil zones are the **A, B,** and **C** horizons from the surface downward.

Soil types include pedalfers in moist temperate climates, pedocals from dry temperate climates, and laterites from moist tropical climates.

Paleosols are buried soil zones indicative of past periods of erosion and weathering.

Sedimentary rocks cover about 75 percent of the earth's surface and make up about 5 percent by volume of the outer 15 km of the solid earth.

Formation of sedimentary rocks takes place at or near the earth's surface.

Detrital material worn from the landmasses and **chemical** deposits precipitated from solution are the two chief types of sediments.

Sedimentation is the process by which rock-forming materials are laid down, and the resulting deposits vary with the source of material, the methods of transportation, and the processes of deposition.

Clay, quartz, and **calcite** are the most common minerals in sedimentary rocks. Other minerals include dolomite, goethite, hematite, limonite, feldspar, mica, halite, and gypsum.

Texture depends on the size, shape, and arrangement of the particles. Texture may be clastic or nonclastic.

Lithification converts unconsolidated sediments to firm rock by cementation, compaction, desiccation, and crystallization.

Types of sedimentary rocks include detrital, chemical and organic forms.

Detrital rocks include conglomerate, sandstone, mudstone, and shale.

Chemical rocks include limestone, dolomite, and evaporites.

Organic rocks include chalk, coquina, diatomite, and coal. Most abundant are shale and mudstone, sandstone, and limestone, in that order. They form 99 percent of the sedimentary-rock family.

Features of sedimentary rocks include bedding, mud cracks, nodules, concretions, geodes, and fossils.

Color of sedimentary rocks is due largely to small amounts of the iron oxide minerals and less importantly to organic matter.

Sedimentary facies refers to an accumulation of deposits that exhibits specific characteristics and grades laterally into other accumulations formed at the same time but showing different characteristics.

Sedimentation is closely related to **plate boundaries,** with thick accumulations along some margins.

Geosynclines are large-scale downwarps in the earth's crust in which large accumulations of sediments occur.

Miogeoclines are thick accumulations of strata on margins of continental blocks.

Eugeoclines represent rocks developed farther offshore or in areas marked by volcanic island arcs.

SUPPLEMENTARY READINGS

Donahue, Roy L. *Soils: An Introduction to Soils and Plant Growth*, Prentice-Hall, Inc., Englewood Cliffs, N.J., 1965. *A basic introduction to the subject of soils and the interaction between soils and plants.*

Judson, Sheldon "Erosion of the Land—or What's Happening to our Continents?" *Am. Sci.*, vol. 56, pp. 356–379, 1968. *An overview of worldwide erosion.*

LaPorte, Leo F. *Ancient Environments*, Prentice-Hall, Inc., Englewood Cliffs, N.J., 1968. *An excellent treatment of the way sedimentary materials can be used to determine environments.*

Pettijohn, F. J. *Sedimentary Rocks*, 3rd ed., Harper & Row, Publishers, New York, 1975. *An excellent standard text on the whole field of sedimentary rocks.*

Reineck, H. E., and I. B. Singh *Depositional Sedimentary Environments*, Springer-Verlag New York Inc., 1973. *Profusely illustrated textbook on modern environmental studies, with reference to examples of ancient environments as found in the stratigraphic record.*

5

METAMORPHISM AND METAMORPHIC ROCKS

Many rocks exposed at the earth's surface today show evidence of change. At first glance some of these rocks resemble familiar igneous rocks, but then we discover that their mineral grains are arranged in a peculiar manner (see Figure 5.1). Other rocks have the same composition as limestone, but they seem to have developed larger mineral grains. Still others are strikingly different from both igneous rocks and sedimentary rocks. All these are metamorphic ("changed-form") rocks.

5.1 METAMORPHISM

Some sedimentary and igneous rocks have been changed while in the solid state, in response to pronounced changes in their environment. These changes may bring about modifications *within* the rocks themselves through the process called **metamorphism.** Metamorphism occurs within the earth's crust, below the zone of weathering and cementation and outside the zone of remelting. In this environment rocks undergo chemical and structural changes to adjust to conditions different from those under which they were originally formed.

Since metamorphism normally takes place in the crust, it is not readily available for direct study as are weathering, sedimentation, and igneous activity. Nevertheless, laboratory simulation of temperature and pressure conditions permits us to examine changes that occur during the processes of metamorphism.

AGENTS OF METAMORPHISM

Metamorphism is limited to changes that take place in the texture or composition of solid rocks because, after their melting point has been reached, a magma is formed and we are in the realm of igneous activity. (Rock is also considered to be solid when it is in the plastic state.) The agents of metamorphism are heat, pressure, and chemically active fluids.

Heat Heat, an essential agent of metamorphism, may even be *the* essential agent. Some geologists question whether pressure alone could produce changes in rocks without a simultaneous increase in temperature. In fact, they say that metamorphism appears as if it were invariably controlled by temperature.

Pressure Under the influence of pressure, changes take place to reduce the space occupied by the mineral components of a rock mass. Pressure may produce a closer atomic packing of the elements in a mineral, recrystallization of the mineral, or formation of new minerals.

When rocks are buried to depths of several kilometers, they gradually become plastic and responsive to the heat and deforming forces that are active in the earth's crust and mantle (see Figure 5.2). When plastic, they deform by intergranular motion, by the formation of minute shear planes within the rock, by changes in texture, by reorientation of grains, and by crystal growth. Francis J. Turner[1] says that depth of burial combined with a thermal gradient of from 10 to 15°C/km explains "burial metamorphism." The type of the original rock, again, has an important effect on the results achieved by burial and deformation.

Chemically active fluids Hydrothermal solutions released in the solidification of magma often percolate beyond the margins of the magma reservoir and react on surrounding rocks. Sometimes they remove ions and substitute others. Or they may add ions to the rocks' minerals to produce new minerals. When chemical reactions within the rock or the introduction of ions from an external source cause one mineral to grow or change into another of different composition, the process is called **metasomatism.** The term describes all ionic transfers, not just those that involve gases or solutions from a magma. Fine-grained rocks are more readily changed than are others because they expose greater areas of grain surface to chemically active fluids.

Some of the chemically active fluid of metamorphism is the liquid already present in the pores of a rock. It is believed that such pore liquid may often act as a catalyst; that is, it expedites changes without itself undergoing change. Sometimes the changes of metamorphism consist essentially of progressive dehydration.

[1] *Metamorphic Petrology*, McGraw-Hill Book Company, New York, 1968.

FIGURE 5.1

Contorted gneiss from Bedford, Westchester County, New York. Dark bands of ferromagnesian minerals trace out the pattern of distortion to which this rock was subjected by the processes of metamorphism. [Walter R. Fleischer.]

FIGURE 5.2

Metamorphosed conglomerate with stretched pebbles, north shore of Lake Superior. [M. E. Kauffman.]

5.2 TYPES OF METAMORPHISM

Several types of metamorphism occur, but we shall concern ourselves here with the two basic ones: contact metamorphism and regional metamorphism.

CONTACT METAMORPHISM

When magma is intruded into the earth's crust, it alters the surrounding rock. The alteration of rocks at or near their contact with a body of magma is called **contact metamorphism.** Minerals formed by this process are called *contact metamorphic minerals.* The type of reaction depends on the temperature, the composition of the intruding mass, and the properties of the intruded rock. At the actual surface of contact, all the elements of a rock may be changed or replaced by other elements introduced by the hydrothermal solutions escaping from the magma. Farther away, the replacement may be only partial.

Contact metamorphism occurs in zones called **aureoles** ("halos"), which seldom measure more than a few hundred meters in width and may be only a fraction of a centimeter wide. Aureoles are found bordering plutons (see Figure 5.3): sills, dikes, laccoliths, stocks, lopoliths, and batholiths. During contact metamorphism, temperatures may range from 300 to 800°C, and load pressures may range from 100 to 3,000 atm (atmospheres).[2]

Two kinds of contact metamorphic mineral are recognized: those produced by heating up the intruded rock and those produced by the hydrothermal solutions reacting with the intruded rock. Hydrothermal metamorphism develops at relatively shallow depths: It is only late in the cooling of a body of magma, when it nears the surface, that large quantities of hydrothermal solutions are released.

Contact metamorphic minerals Many new minerals are formed during contact metamorphism. For example, when an impure limestone is subjected to thermal contact metamorphism, its dolomite, clay, or quartz may be changed to new minerals. Calcite and quartz may combine to form wollastonite. Dolomite may react with the quartz to form diopside. Aluminum, in the clay, will react to form corundum, spinel, or garnet. If carbonaceous materials are present, they may be converted to graphite. Many more minerals are produced by hydrothermal contact metamorphism. Solutions given off by the magma react to produce new minerals contain-

[2]One atmosphere is the "unit of pressure equal to the pressure exerted per unit area at sea level by the column of air from sea level to the top of the earth's atmosphere."

FIGURE 5.3

(a) Aureole of Onawa, Maine, granodiorite pluton. (b) Inferred cross section at time of intrusion of pluton. [After S. S. Philbrick, *Am. J. Sci.*, vol. 31, pp. 1–40, 1936.]

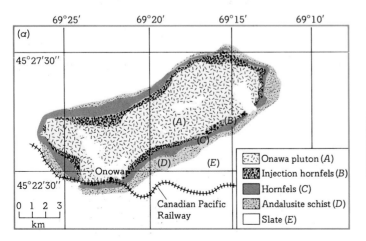

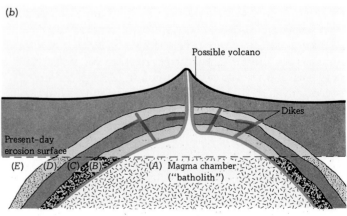

ing elements not present in the limestone. In this type of metamorphism, oxide and sulfide minerals are frequently formed and constitute ore deposits of economic importance (see Chapter 15).

REGIONAL METAMORPHISM

Regional metamorphism is developed over extensive areas, often involving thousands of square kilometers of rock thousands of meters thick. It is believed that regional metamorphism is related to the building of mountain ranges (see Chapter 7). This has not been proved. However, metamorphic rocks are found in the root regions of old mountains and in the Precambrian continental shields. Thousands of meters of rock have had to be eroded in order to expose these metamorphic rocks.

Regional metamorphic minerals During regional metamorphism, new minerals are developed as rocks respond to increases in temperature and pressure. These include some silicate minerals not found in igneous and sedimentary rocks, such as sillimanite, kyanite, andalusite, staurolite, almandite, brown biotite, epidote, and chlorite.

The first three of these new minerals are aluminosilicates and have the formula Al_2SiO_5. Their independent SiO_4 tetrahedra are bound together by positive ions of aluminum. *Sillimanite* develops in long, slender crystals that are brown, green, or white. *Kyanite* forms bladelike blue crystals. *Andalusite* forms coarse, nearly square prisms.

Staurolite is a nesosilicate composed of independent tetrahedra bound together by positive ions of iron and aluminum. It has a unique crystal habit, which is striking and easy to recognize: It develops six-sided prisms that intersect either at 90°, forming a cross, or at 60°, forming an ex (see Figure 5.4).

Garnets are a group of metamorphic nesosilicate minerals. All have the same atomic structure of independent SiO_4 tetrahedra, but a wide variety of chemical compositions is produced by the many positive ions that bind the tetrahedra together. These ions may be iron, magnesium, aluminum, calcium, manganese, or chromium. But whatever the chemical composition, garnets appear as distinctive 12-sided or 24-sided fully developed crystals. Actually, it is difficult to distinguish one kind of garnet from another without resorting to chemical analysis. A common deep-red garnet containing iron and aluminum is called *almandite*.

FIGURE 5.4

Crystal of staurolite, 2 cm wide, from Georgia. [Benjamin M. Shaub.]

Epidote is a complex silicate of calcium, aluminum, and iron, in which the tetrahedra are in pairs independent of each other. This mineral is pistachio green or yellowish to blackish green.

Chlorite is a phyllosilicate of calcium, magnesium, aluminum, and iron. The characteristic green color of chlorite was the basis for its name, from the Greek *chlōros*, "green." Chlorite exhibits a cleavage similar to that of mica, but the small scales produced by the cleavage are not elastic, like those of mica. Chlorite occurs either as aggregates of minute scales or as individual scales that will be scattered throughout a rock.

Regional metamorphic zones Regional metamorphism may be divided into zones: high grade, middle grade, and low grade. Each grade is related to the temperature and pressure reached during metamorphism. High-grade metamorphism occurs in rocks nearest the magma reservoir, outside the zone of contact metamorphism. Low-grade metamorphism is found farthest away from the reservoir and blends into unchanged sedimentary rock.

Metamorphic zones are identified by certain diagnostic metamorphic minerals called **index minerals.** Zones of regional metamorphism reflect the varied min-

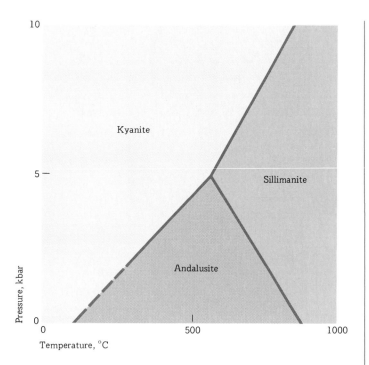

FIGURE 5.5

Temperatures and pressures at which Al$_2$SiO$_5$ forms different minerals. [After Sheldon Judson, Kenneth S. Deffeyes, and Robert B. Hargraves, *Physical Geology*, Prentice-Hall, Inc., Englewood Cliffs, N.J., 1976.]

eralogical response of chemically similar rocks to different physical conditions, and each index mineral gives an indication of the conditions at the time of its formation (see Figure 5.5). The first appearance of chlorite, for example, tells us that we are at the beginning of a low-grade metamorphic zone. The first appearance of almandite is evidence of the beginning of a middle-grade metamorphic zone. And the first appearance of sillimanite marks a high-grade zone. Other minerals sometimes occur in association with each of these index minerals (see Table 5.1), but they are usually of little help in determining the degree of metamorphism of a given zone.

By noting the appearance of the minerals that are characteristic of each metamorphic zone, it is possible to draw a map of the regional metamorphism of an entire area. Of course the rocks must have the proper chemical composition to allow these minerals to form.

Regional metamorphic facies   In Chapter 4 we defined facies as applied to sedimentary rocks. Actually, it

has a broader application: **A facies is an assemblage of mineral, rock, or fossil features reflecting the environment in which the rock was formed.**

A *metamorphic facies* is an assemblage of minerals that reached equilibrium during metamorphism under a specific set of conditions. Each facies is named after a common metamorphic rock that belongs to it, and every metamorphic rock is assigned to a facies according to the conditions that attended its formation, not according to its composition. Although it is not always possible to assign a rock to a particular metamorphic facies on the basis of a single hand specimen, it is possible to do so after examining various rocks in the region.

Widely recognized regional metamorphic facies are shown in Table 5.2. They are the greenschist-blueschist, amphibolite, granulite, and eclogite facies. Typical mineral assemblages are shown for metamorphism of a shale and of a basalt.

Table 5.3 shows the general relationships among zones, facies, and mineral assemblages in the progressive metamorphism of several different rock types.

LOW-GRADE REGIONAL METAMORPHISM IN PROGRESS

Studies on the cuttings and cores of three wells in sedimentary deposits of the Imperial Valley of California have indicated that low-grade metamorphism is in progress on a regional scale and at relatively shallow depth.[3] Two of the wells are located 400 m apart in a geothermal field at the southeastern end of the Salton Sea. They are part of a group of 10 wells that have been drilled since 1960 to extract the elements contained in solution in the hot brines of the region and to tap the heat energy. The third was a dry, oil test well drilled in 1963. It is located 35 km south-southeast of the geothermal field, in the Salton Trough (see Figure 5.6). Table 5.4 lists the wells and gives their depths and bottom-hole temperatures.

In this region the sedimentary deposits are geologically young. The oldest deposits in the Trough were deposited in the early Pliocene age, and those in the geothermal field during the Pleistocene age. They consist of poorly sorted sandstones and siltstones of the

[3]L. J. Patrick Muffler and Donald E. White, "Active Metamorphism of Upper Cenozoic Sediments in the Salton Sea Geothermal Field and the Salton Trough, Southeastern California," *Bull. Geol. Soc. Am.*, vol. 80, pp. 157–182, 1969.

TABLE 5.1
Regional metamorphic minerals

Zone	Grade of metamorphism	Minerals
Chlorite	Low	Chlorite, muscovite, quartz
Biotite	Low	Biotite, muscovite, chlorite, quartz
Garnet	Middle	Garnet (almandite), muscovite, biotite, quartz
Staurolite	Middle	Staurolite, garnet, biotite, muscovite, quartz
Kyanite	Middle	Kyanite, garnet, biotite, muscovite, quartz
Sillimanite	High	Sillimanite, quartz, garnet, muscovite, biotite, oligoclase, orthoclase

TABLE 5.2
Regional metamorphic facies and minerals characteristically produced in the metamorphism of a shale and a basalt[a]

| Metamorphic facies | Mineral assemblages developed from | |
	Shale	Basalt
Greenschist-blueschist	Muscovite, chlorite, quartz, albite	Albite, epidote, chlorite, actinolite
Amphibolite	Muscovite, biotite, garnet, quartz, plagioclase	Amphibole, plagioclase, garnet, epidote
Granulite	Garnet, sillimanite, plagioclase, quartz, biotite, pyroxene	Calcic pyroxene, plagioclase, garnet
Eclogite	Garnet, sodic pyroxene, quartz, kyanite	Sodic pyroxene, magnesian garnet

[a] After W. G. Ernst, *Earth Materials*, Prentice-Hall, Inc., Englewood Cliffs, N.J., 1969.

TABLE 5.3
Products of regional metamorphism[a]

| Original rock | Metamorphic rock | | | | |
	Chlorite zone	Biotite zone	Almandite zone	Staurolite zone	Sillimanite zone
Shale	Slate	Biotite phyllite	Biotite-garnet phyllite	Biotite-garnet-staurolite schist	Sillimanite schist or gneiss
Clayey sandstone	Micaceous sandstone	Quartz-mica schist	Quartz-mica-garnet schist or gneiss	Quartz-mica-garnet schist or gneiss	Quartz-mica-garnet schist or gneiss
Quartz sandstone	Quartzite	Quartzite	Quartzite	Quartzite	Quartzite
Limestone and dolomite	Limestone and dolomite	Marble	Marble	Marble	Marble
Basalt	Chlorite-epidote-albite schist (greenschists)	Chlorite-epidote-albite schist (greenschists)	Albite-epidote amphibolite	Amphibolite	Amphibolite
Granite	Granite	Granite gneiss	Granite gneiss	Granite gneiss	Granite gneiss
Rhyolite	Rhyolite	Biotite schist or gneiss	Biotite schist or gneiss	Biotite schist or gneiss	Biotite schist or gneiss

[a] After Marland P. Billings, *The Geology of New Hampshire. Part II: Bedrock Geology*, p. 139, Granite State Press, Inc., Manchester, 1956.

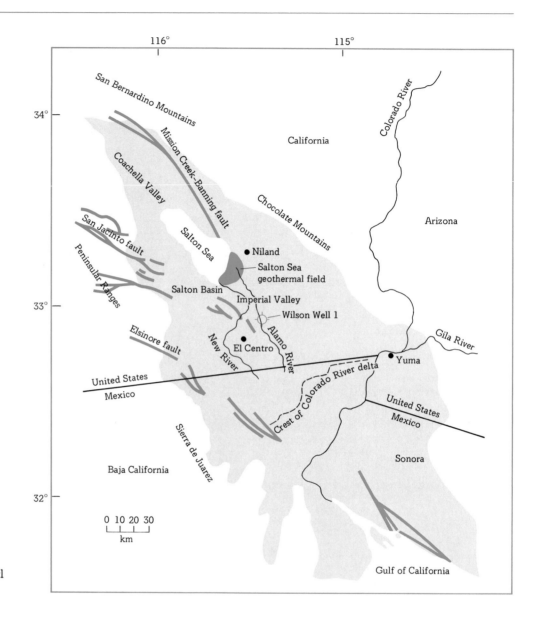

FIGURE 5.6

Location of Salton Sea geothermal field.

TABLE 5.4

Wells providing data on metamorphism

Location	Depth of well, m	Bottom-hole temperature, °C	Gradient, °/15 m
Sportsman 1 (geothermal field)	1,419	310	3
I.I.D. 1 (geothermal field)	1,570	328	3
Wilson 1 (Salton Trough)	4,033	260	1

Colorado River delta. The original dominant minerals were quartz, calcite, subordinate dolomite, plagioclase, potassium feldspar, montmorillonite, illite, and kaolinite.

Samples from the three wells show increasing hardness and a regular sequence of mineral changes with depth as the temperature increases. Some detrital minerals drop out, and other minerals are formed. The rocks look like ordinary sedimentary rocks, but all have undergone mineralogical change at depth since deposition. This change is brought about by progressive metamorphism.

FIGURE 5.7

Mineral occurrence as a function of temperature in the Salton Sea geothermal field. [Data from L. J. Patrick Muffler and Donald E. White, "Active Metamorphism of Upper Cenozoic Sediments in the Salton Sea Geothermal Field and the Salton Trough, Southeastern California," *Bull. Geol. Soc. Am.*, vol. 80, p. 165, 1969.]

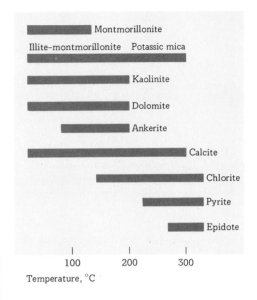

Geothermal field The wells in the geothermal field produce a brine containing over 250,000 ppm (parts per million) of dissolved solids, primarily Cl, Na, Ca, K, and Fe plus a host of minor elements. Mineral transformations are due to hydrothermal metamorphism in the high-temperature environment (see Figure 5.7). There is continuous transition from sediments through indurated sedimentary rocks to low-grade metamorphic rocks of the greenschist facies. The resultant rocks are more porous than are most metamorphic rocks, probably because of the low pressures and short duration of metamorphism. No schistosity is developed. The rocks are analogous to products of burial metamorphism.

Outside the geothermal field Wilson 1 was a dry, oil test well drilled in 1963. The temperature at the bottom of the well, at 4,033 m, was 260°C, which represents an average gradient of 60°C/km. Mineralogical transformations occur over the same temperature ranges in this well as in the geothermal field, but because the thermal gradient is less, new minerals form at greater depths and extend over a much wider depth range. The temperatures encountered at the bottom of the well were not high enough for the development of high-temperature mineral forms.

Because the same mineralogical changes occurred at the same temperature in all three of the wells, it seems clear that temperature is the controlling factor.

5.3 METAMORPHIC ROCKS

Contact and regional metamorphic rocks are found in mountain ranges, at roots of mountain ranges, and on continental shields. The regional metamorphic rocks are by far the most widespread. They vary greatly in appearance, texture, and composition.

Even when different regional metamorphic rocks have all been formed by changes of a single uniform rock type, such as a shale, they are sometimes so drastically changed that they are thought to be unrelated.

These and other characteristics of regionally metamorphosed shale are well illustrated in New Hampshire. Along the Connecticut River west of the White Mountains, rocks that were originally shales are found in a low-grade metamorphic zone as slate, with chlorite. Southeast of these rocks, the original shale is now found as phyllite, grading into schist. New metamorphic minerals appear one after the other toward the southeast: almandite, staurolite, and then sillimanite. These metamorphosed shales occur in a belt surrounding the White Mountain batholith (see Figure 5.8). The closer the area is to the batholith, the higher is its grade of regional metamorphism. This correlation is indicated by the presence of the index minerals found in the metamorphic rock.

TEXTURE OF METAMORPHIC ROCKS

In most rocks that have been subjected to heat and deforming pressures during regional metamorphism, the minerals tend to be arranged in parallel layers of flat or elongated grains. This arrangement gives the rocks a property called **foliation** (from the Latin *foliatus*, "leaved," or "leafy," hence consisting of thin sheets).

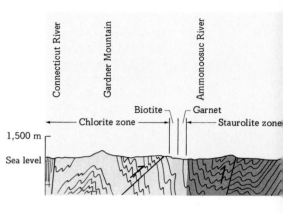

FIGURE 5.8

Cross section of New Hampshire, showing metamorphic zones around the White Mountain batholith of the Older Appalachians. The length of the section is approximately 80 km. [After Marland P. Billings, *The Geology of New Hampshire. Part II: Bedrock Geology*, p. 139, Granite State Press, Inc., Manchester, 1956.]

The textures most commonly used to classify metamorphic rocks are simply *unfoliated* (either aphanitic or granular) and *foliated*. Let us look first at the unfoliated textures. In rocks with aphanitic texture the individual grains cannot be distinguished by the unaided eye, and these rocks do not exhibit rock cleavage. You will remember that we have used the term cleavage to describe the relative ease with which a mineral breaks along parallel planes. But notice here that we are using the modifier "rock" to distinguish rock cleavage from mineral cleavage. In rocks with granular texture the individual grains are clearly visible, but again no rock cleavage is evident.

Rocks with foliated texture, however, invariably exhibit rock cleavage. There are four degrees of rock cleavage:

FIGURE 5.9

Vermont slate quarry, showing how rock cleavage controls breaking. [Harvard University, Gardner Collection.]

1 *Slaty* (from the old French *esclat*, "fragment," "splinter"), in which the cleavage occurs along planes separated by distances of microscopic dimensions.

2 *Phyllitic* (from the Greek *phyllon*, "leaf"), in which the cleavage produces flakes barely visible to the unaided eye. Phyllitic cleavage produces fragments thicker than those of slaty cleavage.

3 *Schistose* (from the Greek *schistos*, "divided" or "divisible"), in which the cleavage produces flakes that are clearly visible. Here cleavage surfaces are rougher than in slaty or phyllitic cleavage.

4 *Gneissic* (from the Greek *gneis*, "spark," for the luster of certain of the components), in which the surfaces of breaking are from a few millimeters to a centimeter or so apart.

TYPES OF METAMORPHIC ROCK

The many types of metamorphic rock stem from the great variety of original rocks and the varying kinds and degrees of metamorphism. Metamorphic rocks may be derived from any of the sedimentary or igneous rocks and other metamorphic rocks of lower grade.

Metamorphic rocks are usually named on the basis of texture. A few may be further classified by including the name of a mineral, such as chlorite schist, mica schist, and hornblende schist.

Slate   *Slate* is a metamorphic rock that has been produced from the low-grade metamorphism of shale or pyroclastic igneous rock (see Figure 5.9). It is aphanitic with a slaty cleavage caused by the alignment of platy minerals under the pressures of metamorphism. Some of the clay minerals in the original shale have been transformed by heat into chlorite and mica. In fact slate is composed predominantly of small colorless mica flakes and some chlorite. It occurs in a wide variety of colors. Dark-colored slate owes its color to the presence of carbonaceous material or iron sulfides.

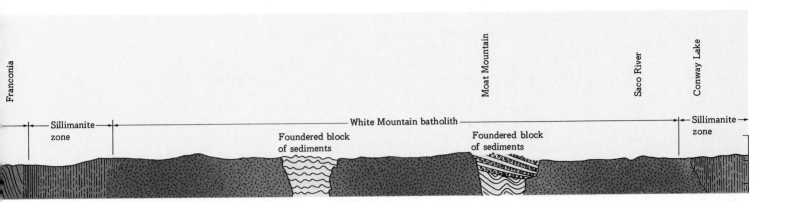

Phyllite *Phyllite* is a metamorphic rock with much the same composition as slate, but its minerals exist in larger units. Phyllite is actually slate that has undergone further metamorphism. When slate is subjected to heat greater than 250 to 300°C, the chlorite and mica minerals of which it is composed develop large flakes, giving the resulting rock its characteristic phyllitic cleavage and a silky sheen on freshly broken surfaces. The predominant minerals in phyllite are chlorite and muscovite. This rock usually contains the same impurities as slate, but sometimes a new metamorphic mineral, such as tourmaline or magnesium garnet, makes its appearance.

Schist Of the metamorphic rocks formed by regional metamorphism *schist* is the most abundant. There are many varieties of schist; for it can be derived from many igneous, sedimentary, or lower-grade metamorphic rocks; but all schists are dominated by clearly visible flakes of some platy mineral, such as mica, talc, chlorite, or hematite. Fibrous minerals are commonly present as well. Schist tends to break between the platy or fibrous minerals, giving the rock its characteristic schistose cleavage.

Schists often contain large quantities of quartz and feldspar as well as lesser amounts of minerals such as augite, hornblende, garnet, epidote, and magnetite. A green schistose rock produced by low-grade metamorphism, sometimes called a *greenschist*, owes its color to the presence of the minerals chlorite and epidote.

Table 5.5 lists some of the more common varieties of schist, together with the names of the rocks from which they were derived.

Amphibolite *Amphibolite* is composed mainly of hornblende and plagioclase. There is some foliation or lineation due to alignment of hornblende grains, but it is less conspicuous than in schists. Amphibolites may

be green, gray, or black and sometimes contain such minerals as epidote, green augite, biotite, and almandite. They are products of the medium-grade to high-grade regional metamorphism of ferromagnesian igneous rocks and of some impure calcareous sediments.

Gneiss A granular metamorphic rock, *gneiss* is most commonly formed during high-grade regional metamorphism (see Figure 5.10). A banded appearance makes it easy to recognize in the field. Although gneiss exhibits rock cleavage, it is far less pronounced than in the schists.

In gneiss derived from igneous rocks such as granite, gabbro, or diorite, the component minerals are arranged in parallel layers: The quartz and the feldspars alternate with the ferromagnesians. In gneiss formed from the metamorphism of clayey sedimentary rocks such as graywackes, bands of quartz or feldspar usually alternate with layers of platy or fibrous minerals such as chlorite, mica, graphite, hornblende, kyanite, staurolite, sillimanite, and wollastonite.

Marble A familiar metamorphic rock, *marble* is essentially composed of calcite or dolomite, is granular, and was derived during the contact or regional meta-

TABLE 5.5
Common schists

Variety	Rock from which derived
Chlorite schist	Shale
Mica schist	Shale
Hornblende schist	Basalt or gabbro
Biotite schist	Basalt or gabbro
Quartz schist	Impure sandstone
Calc-schist	Impure limestone

FIGURE 5.10

Metamorphic rock, showing alignment of previously unoriented minerals. The light-colored bands are mainly orthoclase and quartz; the dark streaks are biotite and other ferromagnesian minerals. The bulk composition is that of granite, but in contrast to the random mixing in granite the minerals here are distributed in relatively systematic patterns. [Robert Navias.]

FIGURE 5.11

Marble quarry, showing how blocks are sawed out for dimension stone to be used in buildings. [Benjamin M. Shaub.]

morphism of limestone or dolomite. It does not exhibit rock cleavage. Marble differs from the original rock in having larger mineral grains. In most marble the crystallographic direction of its calcite is nearly parallel; this is in response to the metamorphic pressures to which it was subjected. The rock shows no foliation, however, because the grains have the same color, and lineation does not show up.

Although the purest variety of marble is snow white, many marbles contain small percentages of other minerals that were formed during metamorphism from impurities in the original sedimentary rock. These impurities account for the wide variety of color in marble. Black marbles are colored by bituminous matter; green marbles, by diopside, hornblende, serpentine, or talc; red marbles, by the iron oxide hematite; and brown marbles, by the iron oxide limonite. Garnets have often been found in marble, as have rubies on rare occasions.

Marble occurs most commonly in areas of regional metamorphism (see Figure 5.11), where it is often found in layers between mica schists or phyllites.

Quartzite The metamorphism of quartz-rich sandstone forms the rock *quartzite*. The quartz in the original sandstone has become firmly bonded by the entry of silica into the pore spaces. Quartzite may also be formed by percolating water under the temperatures and pressures of ordinary sedimentary processes working near the surface of the earth. Many quartzites, however, are true metamorphic rocks and may have been formed by metamorphism of any grade.

Quartzite is unfoliated and is distinguishable from sandstone in two ways: There are no pore spaces in the quartzite, and the rock breaks right through the sand grains that make it up, rather than around them. The structure of quartzite cannot be recognized without a microscope; but when we cut it into thin sections, we can identify both the original rounded sand grains and the silica that has filled the old pore spaces.

Pure quartzite is white, but iron or other impurities sometimes give the rock a reddish or dark color. Among the minor minerals that often occur in quartzite are feldspar, muscovite, chlorite, zircon, tourmaline, garnet, biotite, epidote, hornblende, and sillimanite.

5.4 METAMORPHISM AND PLATE BOUNDARIES

Different types of metamorphism and metamorphic products can be expected to develop along the margins of lithospheric plates, depending upon the kind of boundary and its geologic setting.

CONVERGENT-BOUNDARY METAMORPHISM

Crustal plates are commonly subducted at trenches under convergent island-arc regions (such as the island arcs of the western Pacific) and at convergent ocean-continent boundaries (such as the Andean coast of South America). In these regions downgoing slabs of lithosphere interact with sediments that have accumulated on the sea floor. These sediments tend to consist of a *mélange* (heterogeneous mixture) of fine-grained deep-sea facies combined with exotic blocks of varying ages up to several kilometers in size. A good example of such an assemblage is the Franciscan Formation (Jurassic to Eocene in age) of California, which consists of a mélange of deep-sea sediments, oceanic-crust fragments, and various metamorphic assemblages (greenschist-blueschist facies).

A second convergent-boundary setting involves the downwarping of thick piles of sediments that have accumulated between an island arc and an adjacent continent or on the ocean edge of a continental block. Low-grade regional metamorphism commonly occurs at the bottom of such a wedge of sediments. Plutonic activity also develops so that contact metamorphism affects the rocks adjacent to these intrusions.

The third principal convergent boundary involves collision of two continents (such as the Alpine belt). The dominant metamorphic assemblage here is the blueschist facies involving extensive thrusting of rocks caught between the converging continental plates.

In all these settings descending slabs of lithosphere undergo metamorphism caused by increasing temperature and pressure. Such effects can be calculated for slabs of a given composition, for varying thicknesses and sizes of slabs, and for different rates of movement. These measurements become very complicated but are possible within certain limiting conditions for these parameters.

The lithospheric slabs apparently retain their mechanical competence and are relatively cool in con-

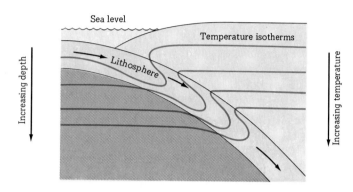

FIGURE 5.12

Relative temperature increase at convergent boundary. The lithospheric plate undergoing subduction retains its cooler temperature, deflecting the isotherms downward. [From various sources.]

trast to their surroundings until thermal equilibrium is reached (Figure 5.12). One analysis estimated that a slab would reach thermal equilibrium with the surrounding mantle at a depth of about 650 km for a lithospheric plate being subducted at a rate of 8 cm/year.[4]

Metamorphism is caused by heat generated along

[4]M. N. Toksoz, J. W. Minea, and B. R. Julian: "Temperature Field and Geophysical Effects of a Downgoing Slab," *J. Geophys. Res.* vol. 76, pp. 1114–1138, 1971.

this subduction zone from friction of the descending lithosphere's slipping past other rocks. At a given depth melting can occur, producing magmas that rise to supply surface volcanoes (Figure 5.13).

Different pressure-temperature conditions prevail at different parts of this descending plate: (1) A zone of low heat flow has been observed over what is considered to be the upper portion of this descending slab. Here low-temperature metamorphic minerals are created. (2) An intermediate zone of relatively higher temperature and pressure occurs along the slab at intermediate depths. (3) A zone of high heat flow occurs over the deepest part of the downward-moving lithosphere, where melting occurs and heat is transferred upward by rising magmas. High-temperature–high-pressure metamorphic minerals are created here (Figure 5.13).

DIVERGENT-BOUNDARY METAMORPHISM

Metamorphism (and attendant mineralization) occurs along divergent plate boundaries, notably at midocean spreading centers such as the Midatlantic ridge. Outflowings of basaltic lavas heat adjacent rocks to produce contact metamorphic minerals.

Seawater over midocean ridges forms hydrothermal solutions, penetrating fissures, dissolving minerals

FIGURE 5.13

Thermal and metamorphic zones associated with descending lithospheric plate under a converging island-arc region. The outer zone near the trench is a zone of relatively low heat flow. The zone associated with the deepest portion of the descending plate is a zone of high heat flow, with melting and the production of magma, contact metamorphism, and overlying volcanic island chain. Types of metamorphism are indicated for various parts of the system. [Modified from various sources.]

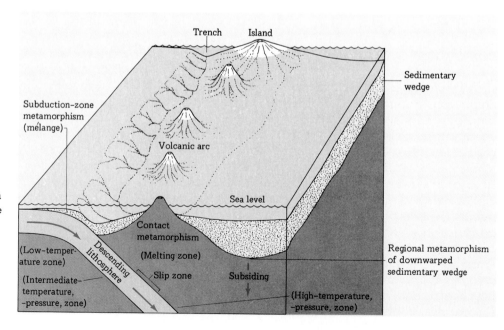

from rocks under the ridge crests and precipitating those minerals in highly concentrated deposits, often of economic importance (see Chapter 15 and Figure 5.14).

Metallic sulfides have been found in rocks dredged from midoceanic ridges at a number of places, including the Midatlantic and Indian Ridges.

The island of Cyprus is famous for a variety of rich mineral deposits, first mined by the Phoenicians. The origin of this island is closely related to sea-floor spreading from a midoceanic ridge that existed in the earliest development of the eastern Mediterranean.

The Red Sea is perhaps the very early product of a divergent plate boundary currently forming between the African and Eurasian plates. Very salty hot brines have been found associated with sediments containing significant quantities of sulfide minerals, including iron, zinc, and copper sulfides, plus some silver and gold. The brines are thought to be hydrothermal solutions from which these minerals precipitate or into which minerals are introduced from volcanic sources beneath the Red Sea or from sediments with high mineral content found adjacent to the basins.

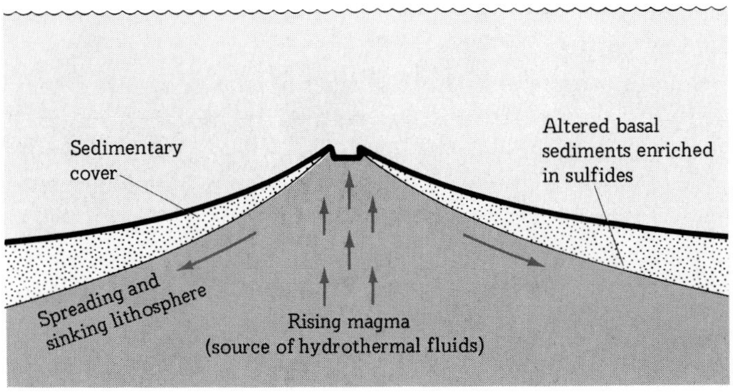

TRANSFORM- (PARALLEL-) BOUNDARY METAMORPHISM

Fractures occur along transform boundaries as plates slide past one another. These fractures and faults pro-

FIGURE 5.14

Cross section of divergent boundary at spreading center, showing hydrothermal fluids and their contact-metamorphosed sediments.

vide zones of weakness along which magma can rise. Volcanoes occur along such zones in oceanic regions. Traces of plates moving past one another can also be observed on land, for example, along the San Andreas fault system. Heat and pressure generated by these rock masses moving past one another produce metamorphic minerals. Mineralization of economic significance occurs in some important mining districts along major fault zones, especially in California as well as similar geologic settings around the world.

5.5 ORIGIN OF GRANITE

The eighteenth-century geologist James Hutton once stated that granite was produced by the crystallization of minerals from a molten mass, and ever since most geologists have accepted the magmatic origin of granite. However, several investigators have questioned this conclusion, suggesting instead that granite is a metamorphic rock produced from preexisting rocks by a process called **granitization.**

In discussing batholiths, we mentioned that one of the reasons for questioning the magmatic origin of granite was the mystery of what happened to the great mass of rock that must have been displaced by the intrusion of the granite batholiths. This so-called space problem has led some geologists to conclude that batholiths actually represent preexisting rocks transformed into granite by metasomatic processes. Certain

rock formations support this theory: These sedimentary rocks were originally formed in a continuous layer but now grade into schists and then into **migmatites** ("mixed rocks"), apparently formed when magma squeezed in between the layers of schist.

The concept of migmatite as a rock group was introduced by J. J. Sederholm in 1907.[5] It is a generic term that covers a large number of petrological combinations and is to be applied to rock formed from two principal ingredients: the host of early-formed material and the material introduced by permeation, metasomatism, or injection of liquids. The average composition of a migmatite is granitic. Its diagnostic character-

[5]J. J. Sederholm, "Om Granit och Gneiss," *Comm. Geol. Finland Bull.* 23, p. 110, 1907.

istics are structure, texture, and broad regional relationships.

The migmatites, in turn, grade into rocks that contain the large, abundant feldspars characteristic of granite but that also seem to show shadowy remnants of schistose structure. Finally, these rocks grade into pure granite. The proponents of the granitization theory say that the granite is the result of extreme metasomatism and that the schists, migmatites, and granite-like rocks with schistose structure are way stations in transforming sedimentary rocks into granite.

What mechanism could have brought about granitization? Perhaps ions migrated through the original solid rock, building up the elements characteristic of granite, such as sodium and potassium, and removing superfluous elements, such as calcium, iron, and magnesium. The limit to which the migrating ions are supposed to have deposited the sodium and potassium is called the **granitic front.** The limit to which the migrating ions are supposed to have carried the calcium, iron, and magnesium is called the **simatic front.**

In the middle of the twentieth century, geologists were carrying on an enthusiastic debate over the origin of granite, but they had reached agreement on one fundamental point: Various rocks with the composition and structure of granite may have different histories. In other words, some may be igneous and others metasomatic. So the debate between "magmatists" and "granitizationists" has been reduced to the question of what percentage of the world's granite is metasomatic and what percentage is magmatic. Those who favor magmatic origin admit that perhaps as much as a quarter of the granite exposed at the earth's surface is metasomatic. But the granitizationists reverse the percentages and insist that about three-quarters or more is metasomatic and only one-quarter or less is of magmatic origin.

Field relationships in some cases suggest that great granite bodies have formed as second-generation igneous rocks in the cores of mountain ranges. This would involve remelting of deep portions of geosynclines and would leave no space problem or need for widespread metasomatism.

OUTLINE

Metamorphism produces metamorphic rocks by changing igneous and sedimentary rocks while they are in the solid state.

> **Agents of metamorphism** are heat, pressure, and chemically active fluids.
>
>> **Heat** may be the essential agent.
>>
>> **Pressure** may be great enough to induce plastic deformation.
>>
>> **Chemically active fluids,** particularly those released late in the solidification of magma, react on surrounding rocks.

Types of metamorphism are contact and regional.

> **Contact metamorphism** occurs at or near an intrusive body of magma.
>
>> **Contact metamorphic minerals** include wollastonite, diopside, and some oxides and sulfides constituting ore minerals.
>
> **Regional metamorphism** is developed over extensive areas and is related to the formation of some mountain ranges.
>
>> **Regional metamorphic** minerals include sillimanite, kyanite, andalusite, staurolite, almandite, garnet, brown biotite, epidote, and chlorite.
>>
>> **Regional metamorphic zones** are identified by diagnostic index minerals.
>>
>> **Regional metamorphic facies** is an assemblage of minerals that reached equilibrium during metamorphism under a specific set of conditions.

Metamorphic rocks are found in mountain ranges, at mountain roots, and on continental shields.

> **Textures of metamorphic rocks** are unfoliated and foliated.
>
>> **Unfoliated rocks** do not exhibit rock cleavage.
>>
>> **Foliated rocks** exhibit rock cleavage classified as slaty, phyllitic, schistose, or gneissic.
>
> **Types of metamorphic rocks** are many because of the variety of original rocks and the varying kinds of metamorphism.

Metamorphism and plate boundaries are closely related.

Convergent boundaries show increased temperature and pressure effects on downgoing slabs of lithosphere with sediments intruded by batholiths, down-warped thick sedimentary sequences, and plates that have collided.

Divergent boundaries have high heat flow at spreading centers with mineralization including sulfides and evaporites.

Transform or parallel boundaries have less igneous activity but include some metamorphism and some economic mineralization.

Origin of granite may be igneous or metamorphic or by melting of sediments.

SUPPLEMENTARY READINGS

Read, H. H. *The Granite Controversy*, Thomas Murby and Company, London, 1957. *Eight addresses, delivered between 1939 and 1954, concerned with the origin of granitic and associated rocks.*

Turner, Francis J. *Metamorphic Petrology*, McGraw-Hill Book Company, New York, 1968. *A general text for advanced students. Embodies a survey of mineralogical aspects of metamorphism.*

Winkler, H. G. F. *Petrogenesis of Metamorphic Rocks*, 4th ed., Springer-Verlag New York Inc., 1976. *Stresses the mineralogical-chemical and physicochemical aspects of metamorphism, especially as related to specific rock groups of common composition.*

6

EARTHQUAKES AND EARTH'S INTERIOR

Earthquakes have profound effects on many peoples and on their works. They also cause changes in the earth's surface and give us controlling data on the structure and nature of the earth's interior. For example, one of the most significant measurements in physical geology is that some earthquakes have sources as deep as 700 km but that *no earthquakes are deeper.* Another is their distribution: Figures 6.1 and 6.2.

6.1 SEISMOLOGY

The scientific study of earthquakes is called **seismology,** from the Greek *seismos*, "earthquake," and *logos*, "reason" or "speech." At the turn of the twentieth century there were approximately a dozen scientists in the world who would have been classified as professional seismologists. The subject matured, however, until by midcentury there were close to 400 seismograph stations that had seismologists recording and studying earthquakes and other ground vibrations. Data from seismology have become an integral part of physical geology in its growth from a descriptive natural history to a science that includes geophysics, a category that cuts across the groupings of knowledge labeled "physics" and "geology."

FIGURE 6.1 (ABOVE)

Locations of earthquakes during a 7-year period, with focal depths of 0 to 100 km. [Data from ESSA, U.S. Coast and Geodetic Survey.]

FIGURE 6.2 (BELOW)

Locations of earthquakes during a 7-year period, with focal depths of 100 to 700 km. [Data from ESSA, U.S. Coast and Geodetic Survey.]

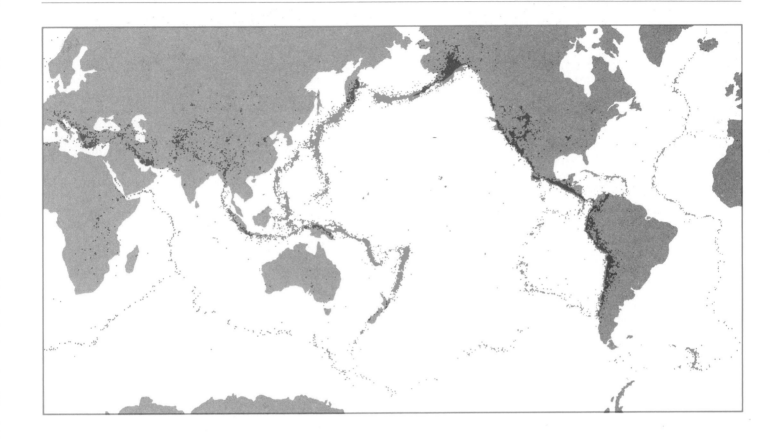

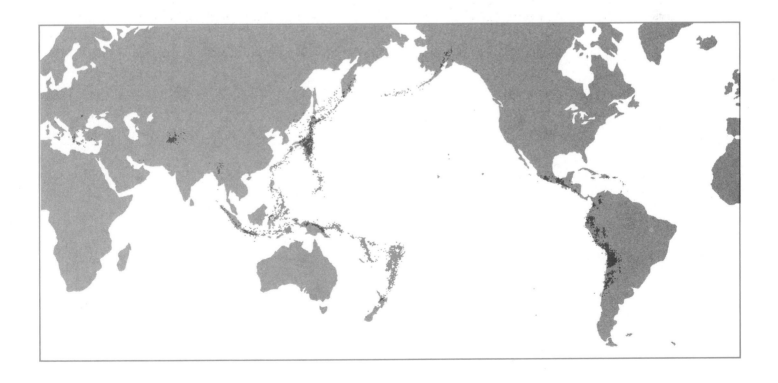

6.2 EFFECTS OF EARTHQUAKES

Earthquakes are interesting to most people because of their effects on human beings and structures. Their geological effects may also be profound. However, of all the earthquakes that occur every year, only one or two are likely to produce spectacular geological effects such as landslides or the elevation or depression of large landmasses. A hundred or so may be strong enough near their sources to destroy human life and property. The rest are too small to have serious effects. One of the best-known damaging earthquakes was that of 1923 in Japan.

At 11:58.5 A.M. on September 1, 1923, there occurred an earthquake whose center was under Sagami Bay, 80 km from Yokohama and 110 km from Tokyo. The vibrations spread outward with such energy that they caused serious destruction along the Japanese coast over an area 150 km long and 100 km wide. When the earthquake began, Professor Akitsune Imamura was sitting in his office at the Seismological Institute of Tokyo Imperial University, and his account is one of the few accurate eyewitness reports of an earthquake from within a zone of heavy damage, carefully documented by a knowledgeable observer.

At first, the movement was rather slow and feeble, so I did not take it to be the forerunner of so big a shock. As usual, I began to estimate the duration of the preliminary tremors, and determined, if possible, to ascertain the direction of the principal movements. Soon the vibration became large, and after 3 or 4 seconds from the commencement, I felt the shock to be very strong indeed. Seven or 8 seconds passed and the building was shaking to an extraordinary extent, but I considered these movements not yet to be the principal portion. At the 12th second from the start, according to my calculation, came a very big vibration, which I took at once to be the beginning of the principal portion. Now the motion, instead of becoming less and less as usual, went on increasing in intensity very quickly, and after 4 or 5 seconds I felt it to have reached its strongest. During this epoch the tiles were showering down from the roof making a loud noise, and I wondered whether the building could stand or not. I was able accurately to ascertain the directions of the principal movements and found them to have been about NW or SE. During the following 10 seconds the motion, though still violent, became somewhat less severe, and its character gradually changed, the vibrations becoming slower but bigger. For the next few minutes we felt an undulatory movement like that which we experience on a boat in windy weather, and we were now and then threatened by severe aftershocks. After 5 minutes from the beginning, I stood up and went over to see the instruments. . . . Soon after the first shock, fire broke out at two places in the University, and within one and a half hours our Institute was enveloped in raging smoke and heat; the shingles now exposed to the open air as the tiles had fallen down due to the shock, began to smoke and eventually took fire three times. . . . It was 10 o'clock at night before I found our Institute and Observatory quite safe. . . . We all, 10 in number, did our best, partly in continuing earthquake observations and partly in extinguishing the fire, taking no food or drink till midnight, while four of us who were residing in the lower part of the town lost our houses and property by fire.

FIRE

When an earthquake occurs near a modern city, fire can be a greater hazard than the shaking of the ground. In fact fire has caused an estimated 95 percent of the total loss caused by some earthquakes. This was dramatically demonstrated in 1923.

Within 30 min after the beginning of the just-described earthquake, fire had broken out in 136 places in Tokyo. In all, 252 fires were started, and only 40 were extinguished. Authorities estimated that at least 44 were started by chemicals. A 20-km/h wind from the south spread the flames rapidly. The wind shifted to the west in the evening, increased to 40 km/h, and then shifted to the north. These changes in wind direction added greatly to the extent of the area burned. Within 18 h, 64 percent of the houses in Tokyo had burned. The fires died away after 56 h, with 71 percent of the houses consumed—a total of 366,262. The spread of fire in Yokohama, a city of 500,000 population, was even more rapid. Within 12 h, 65 percent of the structures in the city had burned, and eventually the city was completely destroyed (see Figure 6.3).

Besides destroying buildings, fire also killed many people. A spectacular example occurred in Tokyo in an area of 1 million m² of open ground on the eastern bank of the Sumida River. People gathered there with their belongings, seeking refuge from the circling fires. By 4 P.M. of September 1, men, women, and children were so closely packed that movement was almost impossible. Meanwhile, fire had closed in from three sides, pinning the people against the river and blanketing them with sparks and suffocating fumes. Suddenly a fire whirlwind approached, the result of the rapid heating of the air superposed on unstable meteorological conditions. It had the characteristics of a true tornado. The central tube, with winds of incredible violence whirling around and upward and drawing smoke, flames, and debris from the surrounding fires, swept over the area. When it had passed, 38,000 were left dead, only about 2,000 persons, who had been close to the river on the southern part of the ground, survived. One report states that most of the dead were terribly burned but that many showed no effects of the heat on their skin or clothing. They were apparently suffocated as the tornado sucked up the breathable air and replaced it with smoke and fumes.

Final government statistics for the entire section of Japan devastated by this earthquake reported 99,333 killed, 43,476 missing, and 103,733 injured, with a total of 576,262 houses completely destroyed.

One reason for rapid spread of fire after an earthquake is that the vibrations often disrupt the water system in the area. In San Francisco, for example, some 23,000 service pipes were broken by the great earthquake of 1906. Water pressure throughout the city fell so sharply that, when the hoses were attached to fire hydrants, only a small stream of water trickled out. Since that time, a system of valves has been installed to isolate any affected area and keep water pressure high in unbroken pipes in the rest of the city.

DAMAGE TO STRUCTURES

Modern, well-designed buildings of steel-frame construction have withstood the shaking of even some of the most severe earthquakes. In the Japanese earthquake of 1923, for example, the famed Imperial Hotel in Tokyo escaped with no structural failure even though it was surrounded by many badly damaged structures. The 43-story Latino-Americano tower in Mexico City rode the waves of the earthquake of July 28, 1957, undamaged while nearby buildings suffered greatly.

FIGURE 6.3

Yokohama, Japan, showing devastation caused by earthquake and fire of September 1, 1923. [L. Don Leet.]

FIGURE 6.4

Contrast in building response to the Prince William Sound earthquake, Alaska, 1964.
[U.S. Coast and Geodetic Survey.]

Chimneys have often been particularly sensitive to earthquake vibrations because they tend to shake in one direction while the building shakes in another. Consequently chimneys often break off at the roof line. Two small earthquakes in New Hampshire in 1940 severed dozens of chimneys but caused no other damage. In contrast, tunnels and other underground structures are little affected by the vibrations of even the largest earthquakes because they move as a unit with the surrounding ground.

In some earthquakes the extent to which buildings are affected by vibrations depends in part on the type of ground on which they stand. For example, in the San Francisco earthquake of 1906, buildings set on water-soaked sand, gravel, or clay suffered up to 12 times as much damage as similar structures built on solid rock nearby. (See Liquefaction, page 139.)

It has been found that the duration as well as the intensity of ground motion is a factor in causing damage to buildings.[1] Continued motion may cause damage even if it does not do so at the start. Reinforced concrete buildings start to show hairline shear cracks at the beginning of damaging motion. With continuing motion the cracks become enlarged, and eventually disintegration results if shaking continues long enough. Repeated strong movements will bring destruction even to steel buildings.

[1]Fergus J. Wood (ed.), *The Prince William Sound, Alaska, Earthquake of 1964 and Aftershocks*, vol. II, Publication 10-3, U.S. Coast and Geodetic Survey, Washington, 1967.

In the Alaskan earthquake of March 27, 1964, the duration of damaging motion was approximately 3 min, or three times as long as the damaging shaking in San Francisco in 1906. Many buildings withstood the early vibrations only to collapse in the later stages. Had the buildings in Alaska been subjected to vibrations for only 1 min, many probably would not have collapsed. Some of the buildings in Alaska that suffered severe damage had been built to conform to the earthquake provisions of the Uniform Building Code recommended by the Structural Engineers Association of California (see Figure 6.4). These provisions were intended to safeguard against major structural failures during earthquakes. However, they were formulated on building damage resulting from California earthquakes, and no California earthquake has ever had damaging shaking exceeding 1 min in duration. It will therefore be necessary to establish new building codes that will take into account the stresses involved in long-lasting damaging shaking.

The duration of strong shaking is also blamed for the extensive landslides that destroyed many homes in Anchorage.

SEISMIC SEA WAVES

If you are ever fortunate enough to be basking in the sun on Waikiki Beach and the water suddenly pulls away from the shore and disappears over the horizon, do not start picking up seashells or digging clams—a seismic sea wave is coming, and the withdrawal of the water is the first warning of its approach.

Some submarine earthquakes abruptly elevate or lower portions of the sea bottom, setting up great sea waves in the water. The same effect may also be produced by submarine landslides at the time of a quake. These giant waves are called **seismic sea waves,** or **tsunami,** a Japanese term, which has the same form for both singular and plural. Seismic sea waves may be generated by volcanic eruptions, as in 1883, when Krakatoa erupted. These waves killed 36,500 in the East Indies. Seismic sea waves have devastated oceanic islands and continental coastlines from time to time throughout history. The Hawaiian Islands have been hit 37 times since their discovery by Captain Cook in 1778.

On April 1, 1946, a severe earthquake occurred at 53.5°N, 163°W, 130 km southeast of Unimak Island, Alaska, in the Aleutian Trench. Here the ocean is 4,000 m deep. Minutes after the earthquake occurred, waves more than 33 m high smashed the lighthouse at Scotch Cap, Unimak Island, killing five persons. About 4.5 h later, the first seismic sea wave from this quake reached Oahu, Hawaii, after traveling 3,600 km at 800 km/h. At the time, marine geologist Francis P. Shepard and his wife were living in a seashore cottage on northern Oahu and were awakened by a loud hissing noise. They dashed to the window just in time to see waters of the ocean boiling up over a high ridge and heading toward their house. Shepard grabbed his camera, but when he got to the door, much to his disappointment, he saw the water retreating rapidly oceanward. The sea's level quickly dropped 10 m. It was then that he realized he had seen a seismic sea wave. Several more times after this, the water surged up over high-tide levels and then was sucked back into its basin.

Starting in the Alaskan waters where the earthquake occurred, these seismic sea waves had spread out much as waves do when a rock is thrown into a pond. But these waves were tremendous in length. Their crests were about 160 km apart. They swept out into the Pacific Ocean, moving at terrific speeds of over 800 km/h. As they passed ships, however, these waves went unnoticed. They were about 1 m high in water 3,000 m deep, with 160 km between crests; so their effect was similar to the ground level's rising 1 m as you walk 160 km. But when they reached Oahu and other Pacific shores, the effect was dramatic. The energy that moved thousands of meters of water in the open ocean became concentrated on moving a few meters of water at a shallow shore. There the water curled into giant crests that increased in height until they washed up over shores meters above high tide: 12 m on Oahu and 18 m on Hawaii. These same seismic sea waves swept on down the Pacific. They reached Valparaiso, Chile, 18 h after their launching, when they still had enough energy to cause 2 m rises of the water after traveling 13,000 km. Some even returned to hit the other side of Hawaii 18 h later. In fact, tide gauges showed that seismic sea waves sloshed around the Pacific basin for days after the earthquake was over.

This 1946 seismic sea wave was one of the most destructive to hit Hawaii. This one came without warning and was rated the worst natural disaster in Hawaii's history, but it was the last destructive seismic sea wave to surprise the Hawaiian Islands.

Seismic-sea-wave warning system In 1948 the United States Coast and Geodetic Survey established a seismic-sea-wave warning system for Hawaii (see Fig-

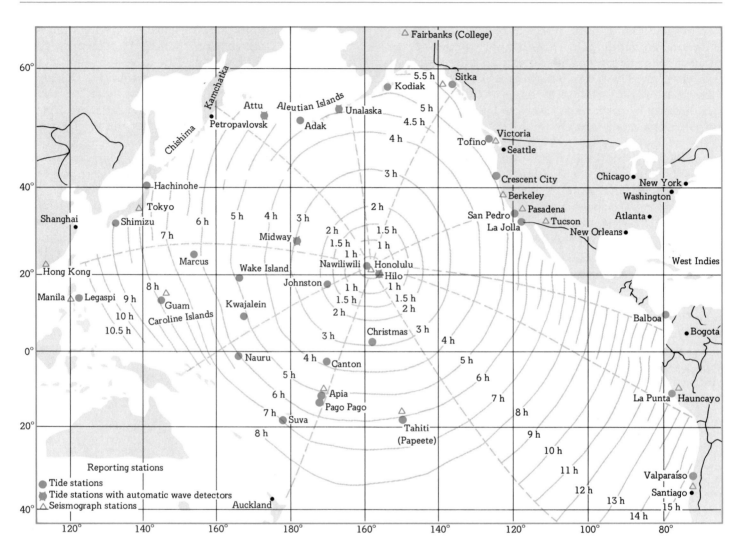

FIGURE 6.5

Seismic-sea-wave travel times to Honolulu.

ure 6.5). It operates continuously recording visible seismographs at its seismological observatory in Honolulu. These are equipped with an automatic alarm system that sounds whenever an unusually strong earthquake is recorded. Tide stations have also been set up to detect waves with characteristics of seismic sea waves. These gauges filter out the normal tides and the wind waves. When the alarm rings, requests are made for immediate readings from other seismograph stations around the Pacific. Within approximately 1 h the earthquake's location is determined. If it is found that the earthquake was in the Pacific Ocean or near its perim-

eter, tide gauges are then checked to see whether they show the existence of a seismic sea wave. If they do, an estimate of its arrival time is made, an alert is sounded, and people are warned to evacuate coastal areas.

The seismic-sea-wave warning system proved very effective on November 4, 1962, when there was an earthquake under the sea off Kamchatka Peninsula at 17 h 7 min Greenwich time. Within about 1 h, with the help of reports from seismograph stations in Alaska, Arizona, and California, Honolulu located the earthquake at 51°N, 158°E. Reports from tide stations at Attu and Dutch Harbor (Aleutian Islands) indicated that seismic sea waves had been started by the quake. Honolulu thereupon computed the time it would take for the first wave to reach Oahu and Hawaii, as well as the other Hawaiian Islands. It was due at Honolulu at 23 h 30 min Greenwich time, 6 h and 23 min after the quake occurred off Kamchatka. In a little over 3 h Midway reported that it was covered by 3 m of water as the first wave raced over it. At Honolulu and Hilo, the waves were not so large as those of 1946 although damage was great. But not a single life was lost, thanks in great measure to the warning system.

The Chilean seismic sea wave of May 22, 1960, was the most destructive in recent history, causing deaths and extensive damage in Chile, Hawaii, the Philippines, Okinawa, and Japan. As usual, Honolulu had computed the location and time of the earthquake; so when word was flashed that seismic sea waves were racing out over the Pacific, the Honolulu observatory issued warnings of the danger, urging evacuation of coastal areas and correctly predicting the arrival time of the first waves. These arrived on schedule 6 h after the warnings were broadcast, 15 h after they started off South America. Through failure of many people to heed the warnings, however, 61 lives were lost in Hilo. Two hundred and twenty-nine dwellings were destroyed or severely damaged (see Figure 6.6). In Japan, no general seismic-sea-wave warning was issued; for it was not known that a seismic sea wave of such distant origin could be so destructive. About 8 h after hitting Hawaii, more than 22 h after the earthquake that started them, the seismic sea waves roared up the coasts of Honshu and Hokkaido. There, more than 17,000 km from where they started, they brought death to 180 people and caused extensive damage. The waves also did considerable damage in New Zealand. All Chilean coastal towns between the thirty-sixth and forty-fourth parallels were destroyed or severely damaged.

Fortunately, only an extremely small fraction of all submarine or coastal earthquakes cause seismic sea waves.

LANDSLIDES

In regions where there are many hills with steep slopes or areas with special soil conditions sensitive to vibration, earthquakes are often accompanied by landslides (see Figure 6.7). These slides occur within a zone seldom exceeding 40 to 50 km in radius although the very

FIGURE 6.6

A mail truck lies half-buried in the shambles of a residential area in Hilo, Hawaii, after seismic sea waves from the Chilean earthquake of May 22, 1960, devastated the community. [Wide World.]

FIGURE 6.7

Turnagain Heights, Anchorage, Alaska, showing slide damage caused by the 1964 earthquake. [ESSA, U.S. Coast and Geodetic Survey.]

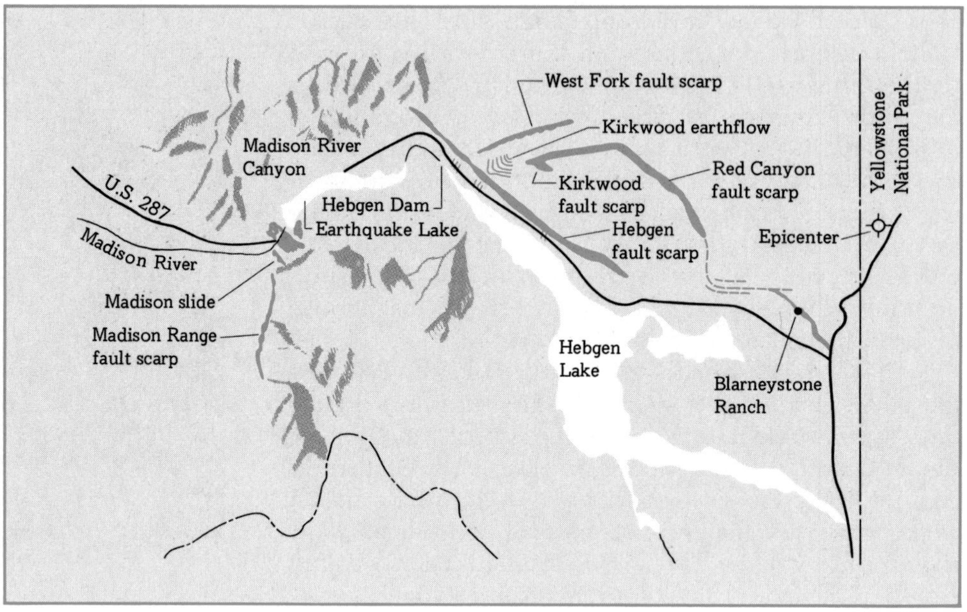

FIGURE 6.8

Region of Hebgen Lake, Montana: earthquake of August 17, 1959. [From John H. Hodgson, *Earthquakes and Earth Structure*, Prentice-Hall, Inc., Englewood Cliffs, N.J., 1964.]

largest earthquakes have affected areas as far away as 150 km. The Alaskan quake of 1964 did this.

One of the worst earthquake-caused landslides on record occurred on June 7, 1692. More than 20,000 lives were lost and much property destroyed in a large section of the then-bustling town of Port Royal, Jamaica. The whole waterfront was launched into the sea, together with several streets of two- and four-story brick houses. The houses and other buildings had been built on loose sands, gravel, and filled land. Shaken loose by the quake, the underlying gravel and sand gave way and slid into the sea; two-thirds of the town, consisting of the government buildings, wharf, streets, homes, and people, went with it.

In the province of Kansu, China, in deposits of loess (wind-deposited silt), an earthquake on December 16, 1920, caused some of the most spectacular landslides on record. The death toll was 100,000. Great masses of surface material moved nearly 2 km, and some of the blocks carried along undamaged roads, trees, and houses.

In the vicinity of the Japanese earthquake of 1923, one large slide moved down a valley as if it were a wall of water at a speed of 1.5 km/min, destroying a village and a railroad bridge at the mouth of the valley.

Several landslides occurred in the vicinity of Hebgen Reservoir, Montana, when the area was

shaken by an earthquake just before midnight on August 17, 1959. The largest slide dammed the Madison River to form Earthquake Lake, as shown in Figure 6.8.

CRACKS IN THE GROUND

One of the most persistent fears about earthquakes is that the earth is likely to open up and swallow everyone and everything in the vicinity. Such fears have been nourished by a good many tall tales, as well as by pictures like Figure 6.9, which shows shallow cracks in a pavement left unsupported by the slumping of a canal bank. One account of the Lisbon, Portugal, earthquake of November 1, 1755, claimed that about 40 km from Lisbon the earth opened up and swallowed a village's 10,000 inhabitants with all their cattle and belongings and then closed again. The story probably got its start when a landslide buried some village in the area.

In Japan widespread ideas about the earth's ability to swallow people may have sprung from an allusion in one of the Buddhist scriptures: When Devadatta, one of Sakya Muni's disciples, turned against his master and even made attempts on his life, Heaven punished him by consigning him to Hades, whereupon the ground opened and immediately swallowed him up.

It is true that landslides do bury people and buildings, and under special conditions they may even open small, shallow cracks. In California, in 1906, a cow did fall into such a crack and was buried with only her tail protruding. But there is no authenticated case in which solid rock has yawned open and swallowed anything.

LAND MOVEMENTS

Some earthquakes are accompanied by significant vertical and horizontal movement of the land surface. The surface sinks in some places, rises in others, and is often tilted, and portions can even be moved laterally great distances.

Fault creep is slow, periodic movement of the land on opposite sides of a fault (or break in rocks, along which movement has occurred; see Chapter 7). Measurable movement along the San Andreas is on the order of 1 or 2 cm/year. During the past 20 million years this has resulted in offsets of several hundred kilometers.

Vertical displacements in excess of 10 m have been recorded in a single earthquake, and areas of thousands of square kilometers may be affected. Details of some examples are discussed in Chapter 7.

LIQUEFACTION

Damage to structures as a result of earthquakes may be caused more by foundation failure and sliding due to **liquefaction** of saturated soil or loose sediments than by the actual shaking of the ground (see also Figure 6.9). Liquefaction often occurs several minutes to tens of minutes after a quake strikes a region, when underground sewers, storage tanks, pipes, and piles driven into the ground may be observed to float up to the surface. Nearby structures may settle several meters into the ground. Water and sand may be ejected a meter or more into the air for several minutes after the quake.

All these phenomena apparently result from turning what had been relatively stable deposits near the surface into liquefied material incapable of supporting structures. This most commonly occurs where the deposits are thoroughly saturated, such as along seacoasts, the shores of lakes or ponds, or anywhere that subsurface water has filled the fractures, voids, and pores in the ground. So it appears that soils and loose deposits that are free-draining are less likely to undergo liquefaction during earthquakes and are therefore better foundations for structures.

SOUND

When an earthquake occurs, the vibrations in the ground often disturb the air and produce sound waves that are within the range of the human ear. These are known as *earthquake sounds*. They have been variously described, usually as low, booming noises. Very near the source of an earthquake, sharp snaps are sometimes audible, suggesting the tearing apart of great blocks of rock. Farther away, the sounds have been likened to heavy vehicles passing rapidly over hard ground, the dragging of heavy boxes or furniture over the floor, a loud, distinct clap of thunder, an ex-

FIGURE 6.9

Cracks in pavement caused by the Japanese earthquake of 1923. Pictures of this kind are sometimes used to support the superstition that the earth yawns open during an earthquake. That is not what happened here: A hard-surface road lost its support when the bank of an adjacent canal underwent liquefaction and collapsed. [L. Don Leet.]

plosion, the boom of a distant cannon, or the fall of heavy bodies or great loads of stone. The true earthquake sound, of course, is quite distinct from the rumble and roar of shaking buildings, but in some cases the sounds are probably confused.

WORLDWIDE STANDARD SEISMOGRAPH NETWORK

In 1961 the United States Coast and Geodetic Survey began establishing a worldwide network of 125 standard seismograph stations (SSWWS—see Figure 6.10). Since then seismological knowledge has been ex-

panding rapidly because of the quantity and quality of earthquake measurements. Also, computers have helped to speed the processing of data for determining earthquake locations, depths of foci, and size. The United States Coast and Geodetic Survey now locates approximately 6,000 quakes per year, whereas, 10 years before, the number had been one-tenth that. This increase does not mean that the seismicity of the earth has increased. It only appears to have done so because the improved equipment in the increased number of recording stations gathers more, better data.

From this worldwide seismic network and from other seismograph stations come measurements that give us a new perspective on sequences of earthquake

FIGURE 6.10

Stations of the worldwide standard seismograph network.

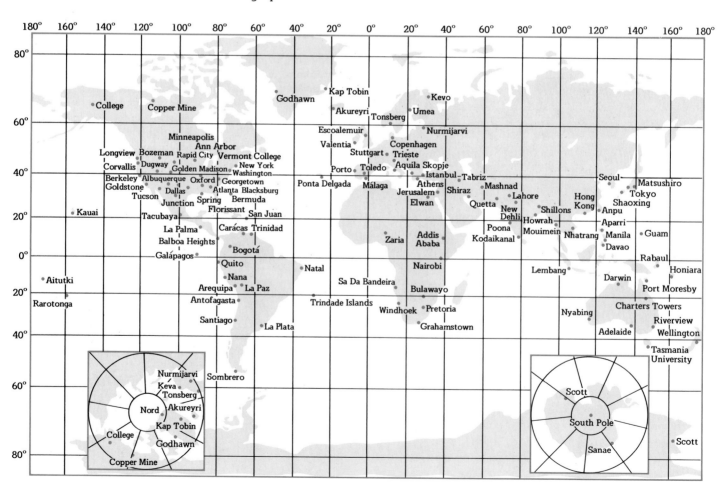

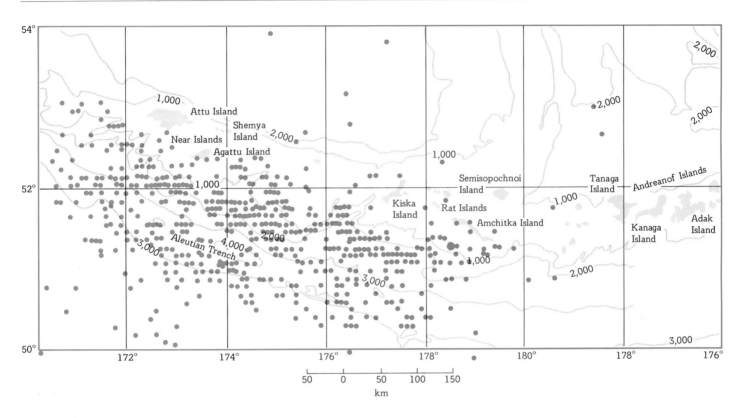

FIGURE 6.11

Aftershocks of the earthquake of February 4, 1965, in the Rat Islands of the Aleutian Islands.

activity associated with large shocks and information on the size of the area involved. For example, after the Rat Islands earthquake of February 4, 1965, 1,300 aftershocks were recorded during the first 45 days. These occurred over an area roughly 600 km long by 300 km wide (see Figure 6.11). It has been found that the largest earthquakes and their aftershocks involve areas up to 200,000 km².

6.3 INTERPRETING EARTHQUAKES

By examining the location and strength of earthquakes, we can learn much about how earthquakes do their damage.

EARTHQUAKE FOCUS

In seismology the term **focus** designates the source of a given set of earthquake waves. Just what is this source? As we know, the waves that constitute an earthquake are generated by the rupture of earth materials. When these waves are recorded by instruments at distant points, their pattern indicates that they originated within a limited region. Most sources have dimensions probably closer to 50 km in length and breadth than to 5 or 500 km. A few of the largest earthquakes may involve up to 1,000 km. But trying to fix these dimensions more accurately offers a real problem that has not yet been solved.

In any event, the focus of an earthquake is usually

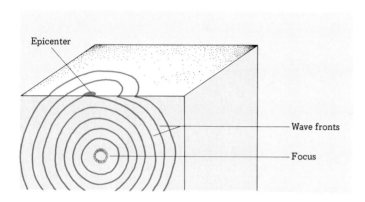

FIGURE 6.12

Diagram showing the positions of the focus and epicenter of an earthquake.

TABLE 6.1

Focal depths for a 7-year period[a]

Depth, km	Foci
< 75	27,650
100 ± 25	3,533
150 ± 25	1,942
200 ± 25	974
250 ± 25	447
300 ± 25	223
350 ± 25	196
400 ± 25	200
450 ± 25	205
500 ± 25	292
550 ± 25	488
600 ± 25	448
650 ± 25	154
>650 ± 25	8
Total	36,760

[a]Summarized from United States Coast and Geodetic Survey's preliminary determination of epicenter data (personal communication from Leonard M. Murphy, Chief, Seismology Division).

at some depth below the surface of the earth. An area on the surface vertically above the focus is called the *epicentral area*, or **epicenter**, from the Greek *epi*, "above," and *center* (Figure 6.12).

Foci have been located at all depths down to 700 km, a little more than one-tenth of the earth's radius (see Table 6.1). On some continental margins they have clustered along a plane dipping toward the interior of the continent, as in Figure 6.13. The deepest have been limited to the Tonga-Fiji area and the Andes, as shown in Figure 6.14.

EARTHQUAKE INTENSITY

How to specify the size of an earthquake has always posed a problem. Before the development of instrumental seismology, some of the early investigations of earthquakes led to various attempts to describe the intensity of the shaking at a specific place: For instance, a missionary in some remote region would keep a diary of earthquakes rated as weak, strong, or very strong. This was at best a personal scale.

In 1883 an intensity scale was developed that combined the efforts of M. S. de Rossi of Rome and F. A. Forel of Lausanne, Switzerland. For half a century the Rossi-Forel intensity scale was widely used throughout the world. According to this scale, earthquake effects were classified in terms of 10 degrees of intensity. It had definite limitations, however, from a scientific standpoint. For example, the definition of the sixth degree of intensity included "general awakening of those asleep; general ringing of bells; oscillation of chandeliers; stopping of clocks; some startled persons leave their dwellings." But an earthquake that produced those effects in Italy or Switzerland might not even wake the baby in Japan, or it might cause a general stampede in Boston.

Objections to the Rossi-Forel intensity scale led L. Mercalli of Italy to set up a new scale in 1902. This was modified in 1931 by H. O. Wood and Frank Neumann and is the scale currently in use by the United States Coast and Geodetic Survey to evaluate earthquake intensity. It has 12 degrees of intensity and takes into account varying types of construction:

I Not felt except by a very few under specially favorable circumstances. (I on the Rossi-Forel scale.)
II Felt only by a few persons at rest, especially on upper floors of buildings. Delicately suspended objects may swing. (I to II, Rossi-Forel scale.)

III Felt quite noticeably indoors, especially on upper floors of buildings, but many people do not recognize it as an earthquake. Standing motorcars may rock slightly. Vibration like passing of truck. Duration estimated. (III, Rossi-Forel scale.)

IV During the day, felt indoors by many, outdoors by few. At night, some awakened. Dishes, windows, doors disturbed; walls make creaking sound. Sensation like heavy truck striking building. Standing motorcars rocked noticeably. (IV to V, Rossi-Forel scale.)

V Felt by nearly everyone, many awakened. Some dishes, windows, and the like broken; a few instances of cracked plaster; unstable objects overturned. Disturbances of trees, poles, and other tall objects sometimes noticed. Pendulum clocks may stop. (V to VI, Rossi-Forel scale.)

VI Felt by all; many frightened and run outdoors. Some heavy furniture moved; a few instances of fallen plaster or damaged chimneys. Damage slight. (VI to VII, Rossi-Forel scale.)

VII Everybody runs outdoors. Damage negligible in buildings of good design and construction; slight to moderate in well-built ordinary structures; considerable in poorly built or badly designed structures; some chimneys broken. Noticed by persons driving motorcars. (VIII, Rossi-Forel scale.)

VIII Damage slight in specially designed structures; considerable in ordinary, substantial buildings, with partial collapse; great in poorly built structures. Panel walls thrown out of frame structures. Fall of chimneys, factory stacks, columns, monuments, walls. Heavy furniture overturned. Sand and mud ejected in small amounts. Changes in well water. Persons driving motorcars disturbed. (VIII+ to IX−, Rossi-Forel scale.)

IX Damage considerable in specially designed structures; well-designed frame structures thrown out of plumb; great in substantial buildings, with partial collapse. Buildings shifted off foundations. Ground cracked conspicuously. Underground pipes broken. (IX+, Rossi-Forel scale.)

X Some well-built wooden structures destroyed; most ma-

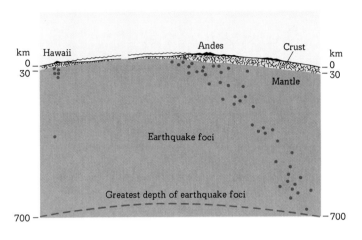

FIGURE 6.13

Earthquake foci under Hawaii and South America.

sonry and frame structures destroyed with their foundations; ground badly cracked. Rails bent. Landslides considerable from river banks and steep slopes. Shifted sand and mud. Water splashed (slopped) over banks. (X, Rossi-Forel scale.)

XI Few, if any, (masonry) structures remain standing. Bridges destroyed. Broad fissures in ground. Underground pipelines completely out of service. Earth slumps and land slips in soft ground. Rails bent greatly.

XII Damage total. Waves seen on ground surfaces. Line of sight and level distorted. Objects thrown upward into air.

By means of postcards, letters, and interviews, investigators make surveys of the effects caused by each earthquake. Then they determine what places had been shaken by about equal amounts. These are plot-

FIGURE 6.14

Earthquake foci between 600 and 700 km in depth.

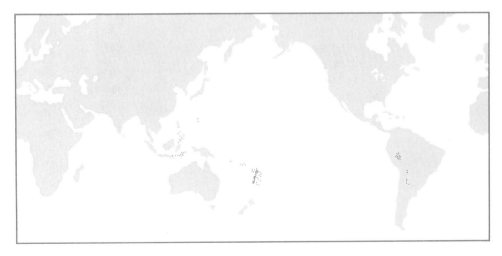

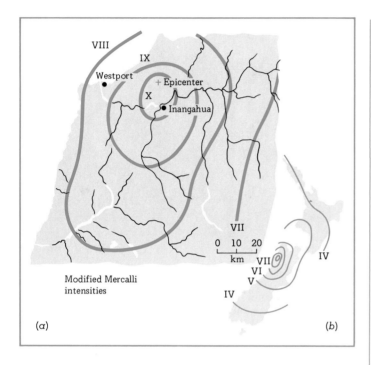

FIGURE 6.15

(a) Isoseismic map of epicentral region of Inangahua earthquake, New Zealand, May 24, 1968; (b) isoseismic map of New Zealand for the same earthquake.

ted on a map and the points connected by **isoseismic lines,** or lines of equal shaking. Figure 6.15 shows an isoseismic map for the New Zealand earthquake that occurred in May, 1968.

EARTHQUAKE MAGNITUDE AND ENERGY

Having to rely on the impressions of people is a very unsatisfactory way of compiling accurate information on the actual size of an earthquake. Therefore a scale was devised in 1935, based on instrumental records.[2] By computation based on the amount of motion in certain waves, it ascribed to each earthquake a number called the earthquake's **magnitude,** an index of the quake's energy at its source. This was refined in 1956 by B. Gutenberg and C. F. Richter and in 1967 by an International Committee on Magnitude. According to

[2]C. F. Richter, "An Instrumental Earthquake Magnitude Scale," *Bull. Seis. Soc. Am.,* vol. 25, pp. 1–32, 1935.

this scale, shallow earthquakes near inhabited areas can have the following effects: One of magnitude 2.5 would be just large enough to be felt nearby; one of magnitude 4.5 or over is capable of causing some very local damage; one of 6 or over is potentially destructive. A magnitude 7 or over represents a major earthquake.

The Richter scale is measured on a logarithmic basis: Each higher whole number on the scale represents an earthquake 10 times stronger than the next lower number. Therefore a quake of magnitude 6 on the Richter scale is 10 times as powerful as one of magnitude 5 and 100 times as powerful as one of magnitude 4. An earthquake of magnitude 5 releases approximately the same amount of energy as did the first atomic bomb when it was tested on the New Mexico desert, July 16, 1945. A megaton nuclear device releases the same amount of energy as an earthquake of magnitude 6. Of course, the energy is applied in quite different ways—highly concentrated in the device, widely dispersed in the earthquake—and the results are correspondingly different. The energy released by an earthquake of magnitude 8.6 is 3 million times as great as that of an earthquake of magnitude 5, or a million million (10^{12}) times the energy of the smallest earthquake.

Some earthquakes that are major catastrophes in a human and property sense are not of the greatest magnitude. Near the coast of northern Peru, at 9.2°S, 78.8°W, on May 31, 1970, a great quake killed an estimated 50,000 persons and left 800,000 homeless while causing $230 million in damage. Hualas Canyon towns

TABLE 6.2

Great earthquakes

Date	Location	Magnitude
September 10, 1899	Alaska, 60°N, 140°E	8.6
January 31, 1906	Colombia, 1°N, 82°W	8.6
August 17, 1906	Chile, 33°S, 72°W	8.4
January 3, 1911	Soviet Union-China, 44°N, 78°E	8.4
December 16, 1920	China, 36°N, 106°E	8.5
March 2, 1933	Japan, 39°N, 145°E	8.5
August 15, 1950	Pakistan–Tibet–Burma, 29°N, 97°E	8.6
May 22, 1960	Chile, 38°S, 73.5°W	8.4
March 27, 1964	Alaska, 61.1°N, 147.7°W	8.6

were flooded by burst dams and buried under land-slides and mudslides. The quake was felt along 1,000 km of Peru; yet its magnitude was 7.8, less than that of any of the quakes that have been listed in Table 6.2.

The largest magnitude assigned to an earthquake to date has been 8.6. As Table 6.2 shows, four of that size have occurred during the modern observational period: in Alaska, September 10, 1899; in Colombia, January 31, 1906; in the Tibet-Burma region, August 15, 1950; and in Alaska, March 27, 1964. The only earthquake in history that might have been larger—if we judge from the reported effects—was the Lisbon quake of 1755. Possibly its magnitude was between 8.7 and 9.0. An earthquake with a magnitude of over 10 should theoretically be perceptible in scattered areas over the entire earth, but such an occurrence has never been reported.

6.4 DISTRIBUTION OF EARTHQUAKES

The first extensive statistical study of earthquakes revealed that from year to year there were wide variations in the total energy released, as well as in the number of individual shocks. For example, 1906 showed 6 times the average energy released between 1904 and 1952 and 40 times the minimum; 1950 was another very large year, second only to 1906 (but see page 157).

Most of the energy released is concentrated in a relatively small number of very large earthquakes. A single earthquake of magnitude 8.4 releases just about as much energy as was released, on the average, each year during the first half of the twentieth century. It is not unusual for the energy of one great earthquake to exceed that of all the others in a given year or of several years put together. The nine great earthquakes listed in Table 6.2 represented nearly a quarter of the total energy released from 1899 through 1968.

The maximum energy released by earthquakes becomes progressively less as depth of focus increases. The nine earthquakes in the table were relatively shallow and had the magnitudes shown; the five largest intermediate-depth shocks over the same interval had magnitudes of 8.1, 8.2, 8.1, 7.9, and 8.0; and the three largest deep shocks had magnitudes of only 8.0, 7.75, and 7.75. This trend suggests that the force required to rupture earth materials decreases as depth becomes greater.

The average annual number of earthquakes from 1904 through 1946 was estimated by Gutenberg and Richter (Table 6.3), who placed the number of earth-

TABLE 6.3

Average annual number and magnitude of earthquakes[a]

	Magnitude	Av. number	Release of energy, approx. explosive equiv.
Actually observed:			
Great	7.7–8.6	2	50,000 1-megaton (hydrogen) bombs
Major	7.0–7.6	12	
Potentially destructive	6.0–6.9	108	
Estimates based on sampling special regions	5.0–5.9	800	1-megaton bomb
			1 small atom bomb (20,000 t TNT)
	4.0–4.9	6,200	
	3.0–3.9	49,000	
	2.5–2.9	100,000	
			0.5 kg TNT
	2.5	700,000	

[a] Modified from B. Gutenberg and C. F. Richter, *Seismicity of the Earth*, Princteon University Press, Princeton, 1949.

quakes large enough to be felt by someone nearby at more than 150,000 per year. Gutenberg and Richter estimated that the total number of earthquakes "may well be of the order of a million each year." Seismograph stations throughout the world are now supplying data leading to the location of close to 6,000 earthquakes per year although locations have been noted from seismographic records since 1899. Figures 6.1 and 6.2 show their distribution for 1961 through 1967, the first years of standardized networks and computer-programming techniques. These active zones were not significantly different in the past.

Earthquakes tend to occur in belts, or zones, also marked by active volcanoes. The earthquakes that occur in the most active zone, around the borders of the Pacific Ocean, account for a little over 80 percent of the total energy released throughout the world. The greatest activity is near Japan, western Mexico, Melanesia, and the Philippines. The loop of islands bordering the Pacific has a high proportion of great shocks at all focal depths. (See also the discussion of earthquakes resulting from plate movements, Section 6.6.)

Fifteen percent of the total energy released by all earthquakes is in a zone that extends from Burma through the Himalaya Range, into Baluchistan, across Iran, and westerly through the Alpine structures of Mediterranean Europe. This is sometimes called the "Mediterranean and trans-Asiatic" zone. Earthquakes in this zone have foci aligned along mountain chains. That leaves 5 percent of the energy to be released throughout the rest of the world. Narrow belts of activity are found to follow the oceanic ridge systems (see Figure 6.1).

With the widespread acceptance of the theory of plate tectonics (Chapters 1 and 8), it now becomes evident that most of these earthquakes occur along the margins of plates, where one plate comes into contact with another and stresses develop. The release of energy associated with these stress conditions is in the form of earthquakes.

Maps showing zones of earthquake expectancy in the United States have been issued by the United States government (Figures 6.16 and 6.17). The zones in Figure 6.16 were outlined after a 2-year study of 28,000 earthquakes and show where earthquakes may be expected in the next 100 years. Figure 6.17 gives a quick method for evaluating relative earthquake hazards: Levels of ground shaking are shown by contour lines that indicate the percentages of the force of gravity likely to occur at least once in a 50-year period. For example, a contour of 40 percent of gravity means one can be relatively certain that that region will not experience ground shaking more than 40 percent of the force of gravity. (All percentages are given with 90-percent relative certainty, or at the 90-percent probability level.)

FIGURE 6.16

Seismic risk map for the United States, issued January, 1969, shows earthquake damage zones of reasonable expectancy in the next hundred years [Developed by U.S. Coast and Geodetic Survey, *ESSA Rel.* ES-1, January 14, 1969.]

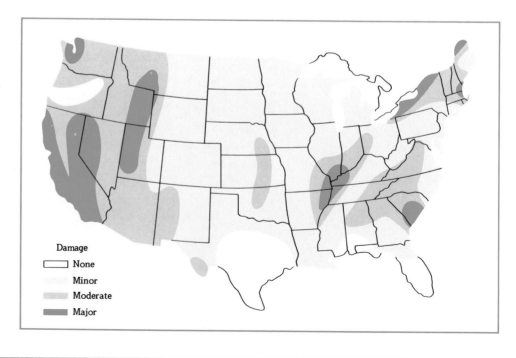

Damage
☐ None
  Minor
  Moderate
  Major

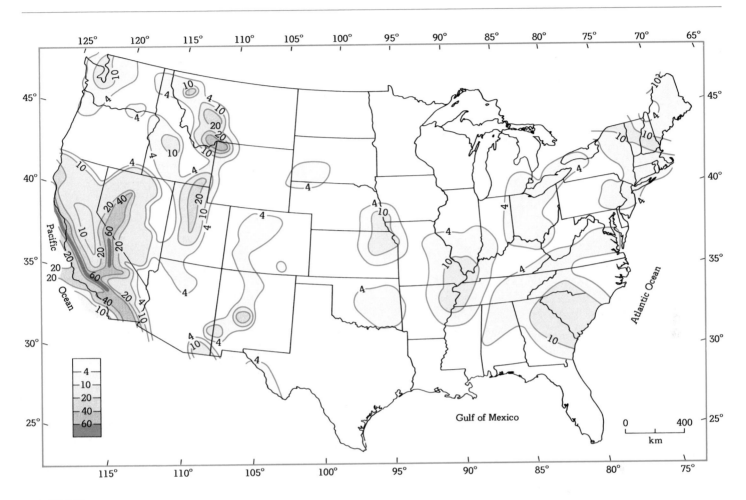

FIGURE 6.17

Map of expectable levels of earthquake shaking hazards. Levels of ground shaking for different regions are shown by contour lines, which express in percentages of the force of gravity the maximum amount of shaking likely to occur at least once in a 50-year period. [Modified from S. T. Algermissen, and D. M. Perkins, "A Probabilistic Estimate of Maximum Acceleration in the Contiguous United States," *U.S. Geol. Surv. Open-file Rep.* 76-416, July, 1976.]

SEISMICITY OF ALASKA AND ALEUTIAN ISLANDS

Before instruments were developed, reports of earthquakes were dependent on population distribution. Alaska was settled in 1783 by Russians on Kodiak Island, who first reported an earthquake occurring south of the Alaskan peninsula on July 27, 1788. From then more than 1,500 earthquakes have been felt by persons living in the area. Information on early earthquakes was obtained from historical records, newspapers, and ships' logs. From 1899 on there have been more than 300 instrumentally recorded earthquakes of which the magnitude has been 6 or greater (see Figure 6.18).

The Prince William Sound earthquake of 1964 On March 27, 1964, at 5:36 P.M. Alaskan standard time (or 3 h 36 min 13 s Greenwich mean time, March 28, 1964), an earthquake of magnitude 8.6 rocked the Prince William Sound area in south-central Alaska. It released twice as much energy as the great San Francisco earthquake of 1906 and was the most destructive earth-

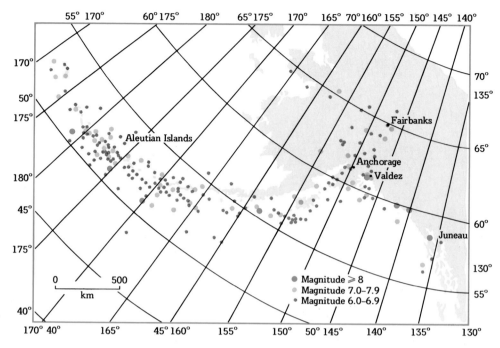

FIGURE 6.18

Earthquakes in Alaska with a magnitude of 6 or greater, 1899 to 1964.

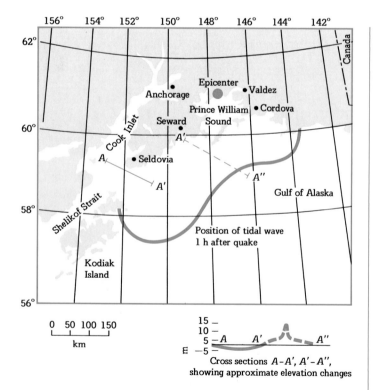

FIGURE 6.19

Epicenter of Alaskan earthquake of March 27, 1964. Elevation changes and the position of a tidal wave 1 h after the quake are also shown.

quake ever recorded on the North American continent—one of our greatest natural disasters.

Its center was located at 61.1°N, 147.7°W, at a depth of 33 km (see Figure 6.19). The earthquake was felt over 1.3 million km², and major damage to buildings, ports, and transportation occurred over an area of 130,000 km². Although 10 percent of the land area of Alaska was involved in the effects of the earthquake, this was occupied by 50 percent of Alaska's total population and included all its major seaports, highways, and railroads. Property damage was estimated at $86 million. In spite of all the damage, there were only nine lives lost in Anchorage and six elsewhere from building collapse and landslides. If it had been earlier in the day, the casualties could have been much higher. A seismic sea wave generated by the earthquake accounted for 98 deaths in Alaska; 11 in Crescent City, California; and 1 in Seaside, Oregon.

As soon as news of the disaster reached the United States Coast and Geodetic Survey in Washington, they took immediate action. They organized the largest project ever undertaken to obtain data about an earthquake. Investigations were conducted simultaneously from the land, sea, and air. Portable seismographs were sent into the area to record aftershocks. Survey teams determined the amounts and location of surface movements and alteration of the sea bottom. Engineers studied building damage. Much was learned from these investigations about the effects resulting from the unusually prolonged and violent shaking.

Many of the towns in the affected region had been built on unconsolidated deposits. As a result, landslides occurred, causing extensive damage to buildings and other structures. Displacements of the land and sea bottom modified shorelines and changed sea-floor topography. Because Alaska is dependent to a large

extent on marine commerce for its survival, the Coast and Geodetic Survey acted swiftly to determine which sea lanes were navigable and to chart new lanes where necessary.

Surveys showed that permanent vertical and horizontal displacements occurred across considerable distances (see Figure 6.20). At the entrance to Prince William Sound, 150 km from the epicenter, uplifts of the sea bottom exceeding 13 m were discovered (see Fig-

FIGURE 6.20

Uplift and subsidence associated with Alaskan quake of March 27, 1964.

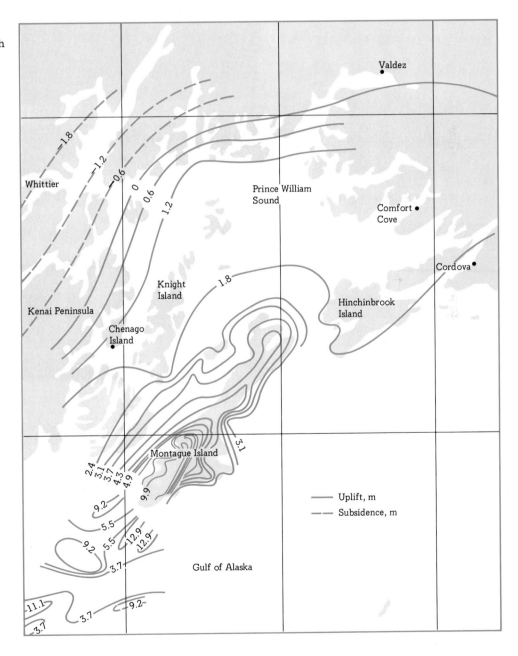

(a)

(b)

FIGURE 6.21

Hanning Bay Fault Cove, northeast of Macleod Harbor, Montague Island, Alaska. (a) Photograph taken by the United States Forest Service on June 8, 1959. The sharp break in the vegetation in the extreme lower right marks an old fault line. (b) Photograph taken on July 7, 1964, by the United States Coast and Geodetic Survey. Note the fault line extending across the harbor entrance and the amount of uplift. Comparison of the photographs shows the uplift and its effect in damming the outlet to the lake on the lower right. Note also the channels being cut through deltaic deposits of the two major streams.

ure 6.21). The entire land area from Anchorage to Seward, a distance of 150 km, subsided 1 to 2 m. Horizontal displacements resulted in changes of distances between points. These generally ranged from 1 to 2 m. The absolute amount and direction of movement could not be accurately determined, as reference points were also disturbed. Surveys across Montague Strait indicated that Montague and Latouche Islands were 5 to 7 m closer together after the quake than they had been in 1933. The general pattern of horizontal and vertical movements suggests turbulence of the earth's crust rather than simple displacement.

The earthquake was of sufficient magnitude to trigger the SSWWS at College, Alaska, and 8 min later the alarm sounded in the Honolulu Observatory. Within 1 h, Honolulu had determined a preliminary epicenter, and coastal cities were warned of an impending inundation.

The seismic sea wave was the first major one associated with an earthquake whose epicenter was on land. However, it was started when submarine landslides and vertical displacements disturbed a large area of the sea bottom. It was highly destructive around the Gulf of Alaska, causing extensive damage to many of Alaska's principal harbors: Anchorage, Cordova, Kodiak, Seward, Valdez, and Whittier. It also caused damage to the west coast of Canada and Crescent City, California. It accounted for the greatest loss of life.

(a)

(b)

FIGURE 6.22

Turnagain Heights, Anchorage, Alaska. (a) Photograph taken on August 12, 1961; (b) Photograph on July 26, 1964. Bootlegger Cove clay, which has little bearing strength when wet, underlies most of Anchorage. Earthquake-produced ground vibrations triggered multiple slides in the bluff area adjacent to Knik Arm. The slide covered an area about 2,400 m long and up to 400 m wide. The slide stopped when enough material collected at the toe to prevent further sliding. Clearing of debris and leveling of the ground has been started [lower left in (b)]. Note also the roadways through the slide area. [ESSA, U.S. Coast and Geodetic Survey.]

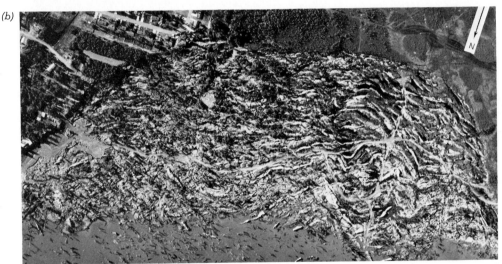

In Anchorage, 120 km from the center of the quake, a substantial portion of earthquake damage was attributable to landslides. The largest and most damaging slide occurred in the Turnagain Heights section (see Figure 6.22). This is a residential area built on a bluff overlooking an arm of Cook Inlet. The vibrations triggered multiple slumps in the bluff. Downward and outward movement occurred over a 2,400-m section of coastline extending from east to west. The slide extended inland approximately 200 m at the east end and 400 m at the west. The edge of the bluff slid more than 200 m into the inlet. Some blocks moved intact as much as 170 m. Eyewitness accounts indicated that the sliding retrogressed from the edge of the bluff, with move-

ments beginning approximately 2 min after the start of the earthquake. A total of 75 houses were destroyed in the Turnagain Heights section.

Valdez, 80 km from the epicenter, suffered extensive damage. The town was built on an alluvial fan that had grown out into the head of a deep, steep-sided fiord. The earthquake caused a major landslide, which brought total destruction to the entire waterfront. There were extensive shoreline changes, and nearly 6 months later the land was found to be still shifting. The town was relocated approximately 6.5 km northwest on more stable ground.

Seward, 150 km from the epicenter, suffered extensive damage. Waterfront for a length of 1,500 m slid into

FIGURE 6.23

Seward, Alaska. The tsunami of March 27, 1964, hit the Seward port section with sufficient force to hurl this fishing vessel across the road. Note also the displaced buoy and other debris. [ESSA, U.S. Coast and Geodetic Survey.]

(a)

(b)

FIGURE 6.24

Seward, Alaska: (a) in September, 1948; (b) in August, 1964. Extensive shoreline changes and railroad damage resulted from the 1964 earthquake (1,500 m of waterfront slumping into Resurrection Bay), tsunami, and fire. The steep and unstable newly formed shoreline was backed by many tension cracks. [ESSA, U.S. Coast and Geodetic Survey.]

Resurrection Bay. This slide, coupled with the destructive effect of an 8-m seismic sea wave, left no usable docks or piers. About 90 percent of Seward's industry was obliterated, and the rest was damaged. Investigations revealed that the northwest section of the bay was 27 to 30 m deeper than previously charted (see Figures 6.23 and 6.24).

Kodiak, on Kodiak Island, 450 km from the epicenter, had most of its business section and port facilities destroyed by the seismic sea wave.

Aftershocks A total of 7,500 shocks were instrumentally recorded in the months that followed. These occurred in a belt 300 km wide, stretching 800 km from the main shock southwestward to a region off the southwest coast of Kodiak Island. Their focal depths ranged between 20 and 60 km.

SEISMIC AND VOLCANIC ACTIVITY

Volcanoes' locations relative to major earthquake belts have been coming into focus as one of the results of increased data on number and quality of earthquakes.

At the scene of Chile's 1960 earthquake series there is a chain of 16 active volcanoes paralleling the coast. One of these, Puyehue, erupted the day after the largest earthquake of the series, and 10 were variously active during the following year. Figure 6.25 shows the affected region and the volcanoes. It has been estimated that approximately 6 percent of the world's large earthquakes occur in the Aleutian Islands and continental Alaska. The Aleutian Islands arc is more than 3,200 km in length and contains 76 volcanoes, 36 active since 1760. Two large earthquakes occurred off the adjacent coast after the 1912 eruption of Katmai.

It has been fashionable for decades to think of "volcanic earthquakes" as a special feature of explosive eruptions, small in energy and seldom damaging. But there is now emerging a realization that volcanoes and earthquakes may have a common ultimate cause in deep movements of mantle materials, and the coincidence of belts of major earthquake activity with belts that include active volcanoes supports the idea. The most obvious common cause of seismic and volcanic activity relates to plate interactions (see Chapters 1 and 8) where fracture zones afford access for volcanic materials welling up from the lower crust and mantle. These boundaries are also areas where earthquakes would naturally occur by interaction of the plates in zones of convergence, of divergence, or where two plates slide past along transform boundaries.

FIGURE 6.25

Location of volcanoes of Chile in the region affected by the earthquakes of 1960.

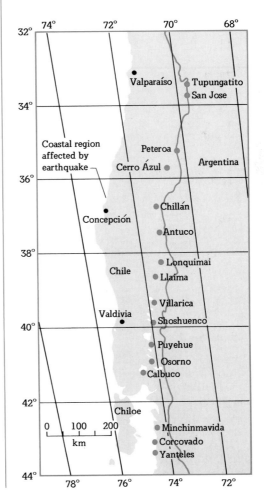

6.5 CAUSE OF EARTHQUAKES

Centuries ago (and in some places today), people believed that the mysterious shakings of the earth were caused by the restlessness of a monster that was supposed to be supporting the globe. In Japan it was first a great spider and then a giant catfish; in some parts of South America it was a whale; and some of the North American Indians decided that the earth rested on the back of a giant tortoise. The lamas of Mongolia had another idea. They assured their devout followers that, after God had made the earth, He placed it on the back of an immense frog; every time the frog shook his head or stretched one of his feet, an earthquake occurred immediately above the moving part (see Figure 6.26). This was a major advance in earthquake theory; for it attempted an explanation of the local character of earthquakes.

The Greek philosopher Aristotle (384–322 B.C.) held that all earthquakes were caused by air or gases struggling to escape from subterranean cavities. Because the wind must first have been forced into the cavities, he explained that just before an earthquake the atmosphere became close and stifling. As time passed, people began to refer to "earthquake weather," and to this day some insist that the air turns humid and stuffy when an earthquake is about to occur. A Japanese scientist checked on this belief by investigating the conditions that had preceded 18 catastrophic earthquakes in Japan between 1361 and 1891. He found that the weather was fair or clear on 12 occasions, cloudy on 2, rainy or snowy on 3, and rainy and windy once but that humid and sultry earthquake weather had never just preceded an earthquake.

Aristotle's idea that earthquakes were caused by imprisoned gases has long been abandoned as we have learned more about the behavior of gases, the structure of the earth's crust, and finally the depths at which earthquakes occur. What, then, causes earthquakes? Most are caused by deforming forces in the

FIGURE 6.26

Cause of earthquakes, according to the lamas of Mongolia. It was believed that, when the frog lifted a foot, there was an earthquake immediately above the part that had moved.

FIGURE 6.27

Reid's proposed representation of movements on the San Andreas fault associated with the 1906 earthquake, illustrating the elastic-rebound hypothesis.

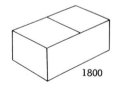

1800

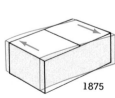

1875

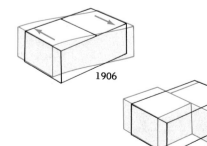

1906

1906

earth; and the immediate cause is the *sudden rupture* of earth materials distorted beyond the limit of their strength.

ELASTIC-REBOUND HYPOTHESIS

After the California earthquake of 1906, it was generally accepted that the immediate cause of earthquakes was faulting due to rupture of rocks of the earth's crust. The displacements along the San Andreas fault provided an unusual opportunity to study the mechanics of faulting and led to the formulation of Reid's **elastic-rebound hypothesis.** Surveys of part of the fault that broke in 1906 had been made during the years preceding the earthquake. H. F. Reid, of Johns Hopkins University, analyzed these measurements in three groups: 1851 to 1865, 1874 to 1892, and 1906 to 1907. The first two groups showed that the ground was twisting in the area of the fault. The third showed displacements that occurred at the time of the earthquake. From these Reid reconstructed a history of the movement.

Although there was no direct evidence, Reid assumed that the rocks in the earth's crust in the vicinity of the fault had been storing elastic energy at a uniform rate over the entire interval and that the region had started from an unstrained condition approximately a century before the earthquake. As the years passed, a line that in 1800 had cut straight across the fault at right angles was assumed to have become progressively more and more warped. When the relative movement on either side of the fault became as great as 6 m in places, the strength of the rock was exceeded, and rupture occurred. The blocks snapped back toward an unstrained position, driven by the stored elastic energy (see Figure 6.27).

The San Andreas fault runs roughly from northwest to southeast. Land on the western, or Pacific Ocean, side of the fault has moved northwest relative to land on the eastern side. In 1906 a section of the fault 450 km long broke. The strains and adjustments were greatest within a zone extending 10 km on each side of the fault. Imagine a straight line 20 km long crossing this zone at right angles to the fault, which was in the center of the zone. After the earthquake, this line was broken at the fault and was shifted into two curves. The broken ends were separated by 6 m at the break, but the other ends were unmoved. Actually, fences, roads, and rows of vegetation provided short sections of lines by which the fault displacement could be gauged (Figure 6.28).

FIGURE 6.28

Looking along a fence that crossed the San Andreas fault at approximately right angles. Before the earthquake of April 18, 1906, this fence was a continuous straight line. It is shown here as it appeared after the earthquake.
[J. C. Branner.]

The idea that earthquakes were caused by faulting as a result of elastic rebound was based on observed surface movements. It was also based on our knowledge of how rocks behave when they have been subjected to deforming forces. In the laboratory, rocks have been subjected to pressures that are equivalent to the pressures in the earth's crust. Under these pressures the rocks gradually change shape; but they resist more and more as the pressure builds up until they finally reach the breaking point. Then they tear apart and snap back into unstrained positions. This snapping back is called elastic rebound and was considered by Reid as the mechanism behind the generation of earthquakes.

EVIDENCE AGAINST ELASTIC-REBOUND HYPOTHESIS

Until recently, it was believed that all earthquakes were caused by faulting due to elastic rebound. However, elastic rebound assumes that earthquakes originate from distortion and rupture of elastic rocks capable of

FIGURE 6.29

The Parkfield-Cholame region.

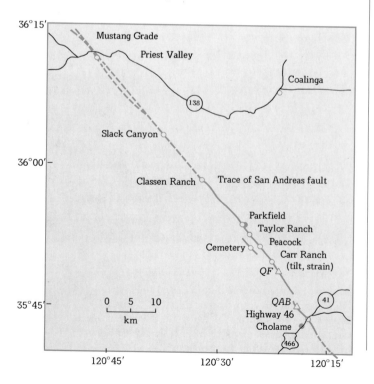

storing strain energy until it is released abruptly at the time of rupture. From laboratory measurements, it has been found that this is a reasonable assumption for rocks similar to those at the earth's surface. But with the increased pressures of burial under 20 or 30 km of overburden, all rocks deform plastically; and plastic rocks neither store strain elastically nor are able to rebound. Therefore the elastic-rebound hypothesis in its original form would apply only to earthquakes with surface foci or with foci that are at extremely shallow depths.

In the 1920s it became apparent that at least some foci were at depths where plastic deformation takes place. The first thoughts were that deep-focus earthquakes might represent a different mechanism of rupture than shallow-focus earthquakes do. However, there were so many features of earthquake records that indicated no really fundamental difference in the wave-generating mechanisms that an alternative suggestion was made involving the property of matter called **viscosity,** the ratio of deforming force to *rate* at which a substance changes shape in response to the force. A material may have such high viscosity that, although it deforms plastically, it does this slowly enough to have at the same time a buildup of force sufficient to allow rupture before complete adjustment has been accomplished through flow.

In the 1950s Percy Bridgman at Harvard University documented by laboratory measurements the mechanism of rupture of earth materials under pressures equivalent to depths of hundreds of kilometers. He described the deformation as taking place steadily by plastic flow but interrupted at times when the material ruptures, or in his words "lets go to get a fresh hold." Thus in an environment of continuous plastic deformation there would be sudden "jerks," or rupture, capable of causing earthquakes.

Since 1906 improvements in seismological recording and interpretation have accumulated an impressive number of measurements showing that an overwhelming number of earthquakes have their foci in the mantle and regions of the lower crust where earth materials are plastic. Therefore the elastic-rebound hypothesis would no longer seem adequate to explain most earthquakes.

Although a section of the San Andreas fault supplied data from which the elastic-rebound hypothesis was formulated, it has since become the scene where measurements are being made that show surface faulting's occurring *after* earthquakes rather than at the

FIGURE 6.30

Cumulative displacement across the San Andreas fault is shown in color. Creep rates are dotted. Felt earthquakes near the reporting station are shown by vertical lines. [After Stewart W. Smith and Max Wyss, "Displacement on the San Andreas Fault Subsequent to the 1966 Parkfield Earthquake," *Bull. Seis. Soc. Am.*, vol. 58, pp. 1955–1973, 1968.]

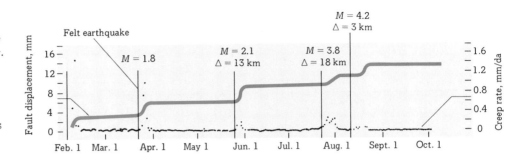

instant of an earthquake—which also goes to disprove Reid's hypothesis.

An earthquake occurred in the Parkfield-Cholame region on June 27, 1966.[3] Figure 6.29 shows the region. During the following year a special program of measurements was undertaken that produced some extremely significant data.

The earthquake occurred at night (21 h 26 min Pacific daylight time), and nothing is known of surface faulting at that time. There may have been none. When the first inspection was made 10 h later, a white line at the Highway 46 locality was offset 4.5 cm. By evening the displacement had increased to 6.4 cm, and by the following noon to 7.5 cm. During the following months the average rate of creep decreased from 10 mm/day 2 days after the earthquake to 0.17 mm/day 1 year later. The total average displacement during the year was about 20 cm over a 30-km section of the fault. Most of this occurred in 4- to 6-day intervals of rapid creep *following* aftershocks (see Figure 6.30).

Surface displacements are a result of earthquake-causing mechanisms and are not themselves the vibration-generating movements. In other words, faulting does not cause earthquakes.

6.6 EARTHQUAKES RESULTING FROM PLATE MOVEMENTS

As suggested in Chapter 1 and earlier in this chapter, most earthquakes occur in narrow belts that also mark the edges of crustal plates. It is as though these plates move about the earth like rafts over the plastic material of the asthenosphere. Earthquakes occur where these plates collide, separate, or simply slide past each other. Several examples of such earthquakes are given below from thousands that might have been used.

SEVERE EARTHQUAKES OF 1976

The year 1976 was a time of several rather severe earthquakes in widely separated parts of the earth, all occurring in active earthquake-prone regions.

Guatemala A severe quake jolted much of Central America on February 4, 1976, measuring 7.5 on the Richter scale. More than 23,000 persons died. This earthquake was apparently caused by movement of plates in the Caribbean region (Figure 6.31): The American plate moves westward relative to the Caribbean plate; the northern portion of the American plate carries Canada, the United States, and Mexico while the southern portion carries most of South America in this westerly direction. A small plate off the southwest coast of Central America, the Cocos plate, moves northeasterly and slides beneath Central America, forming a deep trench along a subduction zone just off the west coast.

Movement of these plates has averaged about 4 to 6 cm/year, building up enormous stress over thousands of years. There had been no major movement along these plates in Guatemala for at least 200 years. Suddenly the stresses could no longer be constrained, and rupture occurred. The American plate slipped westward as much as a meter along a zone extending more

[3]C. R. Allen and S. W. Smith, "Parkfield Earthquake of June 27–29, 1966, Pre-earthquake and Post-earthquake Surficial Displacement," *Bull. Seis. Soc. Am.*, vol. 56, pp. 955–967, 1966.

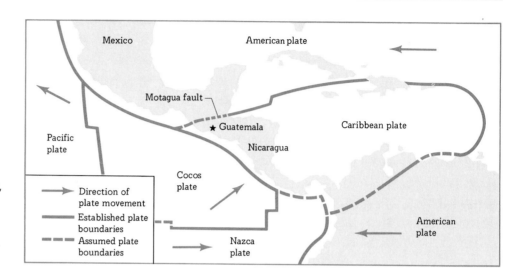

FIGURE 6.31

Guatemala earthquake of February, 1976. The quake was the result of movement along the Motagua fault zone at the boundary between the northern part of the American plate and the Caribbean plate.

than 150 km. This break occurred along the Motagua fault zone, where a number of faults had been mapped but where the exact location of the boundary between the American plate and the Caribbean plate had never before been confirmed.

China On July 27 a series of very severe earthquakes hit parts of northeast China in the province of Hopei, approximately 125 to 150 km east-southeast of Peking. This series of shocks reportedly devastated Tangshan, an industrial city of a million and a half, and caused casualties and damage in cities as far away as Peking. Estimates of resulting deaths reached staggering proportions, with fears of more than half a million. The initial shock registered 8.2 on the Richter scale. This was followed by many aftershocks, some registering as high as 7.9. Although there were no apparent precautionary measures taken prior to this particular series of quakes, accurate prediction of earlier quakes that had hit other provinces in China in February, 1975, and May, 1976, had permitted authorities to take measures to minimize casualties. China has been a leader in the study of the prediction of earthquakes (see Section 6.11).

Turkey A quake of magnitude 7.6 struck near the Turkish-Soviet border on November 24, 1976, destroying over 100 villages and killing almost 4,000 persons. This quake was the strongest in Turkey since tremors in 1939 that registered 7.9 and killed 30,000. The province hit by the 1976 quake lies along the Anatolia fault, stretching inland from Turkey's Aegean coast. It is part of the major Mediterranean boundary zone between the African and Eurasian plates.

6.7 EARLY INSTRUMENTAL OBSERVATIONS[4]

Information about the earth's interior comes from studying the records of earth waves. Earlier in this chapter we learned that some large earthquakes involve violent displacements of the earth's crust. These movements use up some of the energy released at the time of the earthquake. The rest is carried to great distances by earth waves. These may travel through the whole earth and around its entire surface if the earth-quake is large enough. At great distances sensitive instruments are needed to detect their passage. By studying the manner in which earth waves travel out from earthquakes, supplemented by the study of waves generated by dynamite and nuclear explosions, geologists have assembled a wealth of information about the structure of the globe from surface to center. But before we look at the results of these studies, let us review some of the methods used to obtain information.

The earliest instrument used to detect an earthquake was built around A.D. 136. Its design is credited

[4]James Dewey and Perry Byerly, "The Early History of Seismometry (to 1900)," *Bull. Seis. Soc. Am.*, vol. 59, pp. 183–227, 1969.

to a Chinese philosopher named Chang Hêng. His instrument was said to resemble a wine jar about 2 m in diameter. Evenly spaced around this were eight dragon heads under each of which was a toad with head back and mouth open. In each dragon's mouth was a ball (see Figure 6.32). When there was an earthquake, one of the balls was supposed to fall into a toad's mouth. There is no known record of what was inside the jar. Speculations over the centuries have assumed a pendulum of some sort, which would swing as the ground moved, knocking the ball out of a dragon's mouth in line with the swing. Chang Hêng had the notion that, if a frog on the south side caught a ball, the earthquake had happened to the north of the instrument. His instrument was intended to record the occurrence of an earthquake and show the direction of its origin from the observer. It was a seismoscope, as it had no provision for a written record of the motion.

In 1703 J. de la Haute Feuille had the same general idea. He proposed a seismoscope design that he felt would respond to the tilting of the earth's surface at the time of an earthquake. His instrument consisted of a bowl of mercury so placed that, when the earth's surface tilted, mercury would spill over the rim and fall into one of eight channels. Each channel represented a principal direction of the compass; the spilled mercury was to show the direction of the earthquake, and the amount of spillage the earthquake's intensity. However, there is no record that this seismoscope was ever built.

The first European to use a mechanical device for studying earthquakes was Nicholas Cirillo. He used his device to observe the motion of the ground from a series of earthquakes in Naples in 1731. His instrument was a simple pendulum, whose swing indicated the amplitude of ground motion. He made observations at different distances from earthquakes to see how the motion died out with distance. He found that the amplitude decreased with the inverse square of the distance.

In 1783 there occurred a series of earthquakes in Calabria that resulted in 50,000 deaths. These stimulated the development of other mechanical devices for studying such events. It also led to the appointment of the first earthquake commission. Shortly after the first large quake, a clockmaker and mechanic named D. Domenico Salsano had an instrument operating in Naples approximately 320 km from Calabria. It consisted of a long common pendulum with a brush attached that was supposed to trace out the motion with slow-drying ink on an ivory slab, the first attempt to preserve a written record. His instrument was also

FIGURE 6.32

Chang Hêng's seismoscope, as visualized by Wang Chen-To.

equipped with a bell that was to ring when motions were large enough. It has been reported that the bell rang on several occasions. A similar instrument was reported in use in Cincinnati around the time of the New Madrid, Missouri, earthquakes (1811 and 1812).

A series of small earthquakes near Comrie, Scotland, led to the development of an inverted-pendulum seismometer by James Forbes in 1844. His instrument consisted of a vertical metal rod with a movable mass on it. This was supported by a steel wire. The stiffness of the wire and the position of the mass could be adjusted to alter the free period of the pendulum. Forbes was trying to give his pendulum a long period so that it would remain stationary when the ground moved during an earthquake. At one end of the metal rod he attached a pencil that came into contact with a stationary, paper-lined dome above it, to preserve a written record of the movement (see Figure 6.33). By moving the pencil some distance from the mass, he was able to magnify the motion of the pendulum two or three times. This instrument, a crude seismograph, was not successful because friction between the writing pencil and the recording surface reduced the sensitivity of the instrument, ruining recording of felt local quakes.

The first record of a distant earthquake was obtained accidentally by Ernst von Rebeur-Paschwitz at Potsdam, Germany, on April 17, 1889.[5] The earthquake was felt in Tokyo about 1 h before it was recorded in

[5]E. von Rebeur-Paschwitz, "The Earthquake of Tokyo, April 18, 1889," *Nature*, vol. 40, pp. 294–295, 1889.

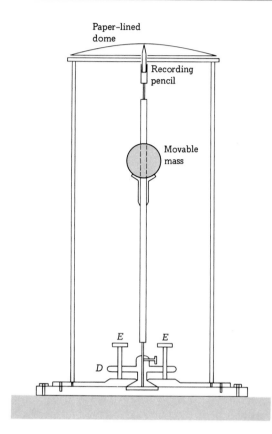

FIGURE 6.33

Forbes' seismometer. Screws, *E*, acting on a support, *D*, were to help set the pendulum in an upright position.

Germany. The motion was too small and was on a time scale that was too compressed to give much information. However, it showed that the energy from an earthquake could travel halfway around the world. And because the record gave the time of day, it indicated how long the waves took to travel that far, how long the motion lasted, and that the motion of the ground had a horizontal component.

6.8 MODERN SEISMOLOGY

With the discovery that earth waves could travel to such great distances from an earthquake, modern seismology was born. The first records of earthquake motion registered on a tilt instrument were too fuzzy to be analyzed accurately. They were too small in amplitude, and the events of 1 h were compressed into the space of 6 mm. To obtain meaningful data about earthquakes it was necessary to develop seismographs with increased recording speed and sensitivity so that the fuzzy lines would be spread out and enlarged.

A group of British professors teaching in Japan in the late nineteenth century initiated the activities that developed into all aspects of seismology as we find it today. Chief among these in his influence on the entire field was John Milne. He was the one mainly responsible for developing a seismograph capable of picking up and recording intelligible signals from distant earthquakes. Milne was also responsible for promoting the first worldwide network of seismograph stations for systematic, continuous registration and interpretation of their records. By 1900 Milne's seismographs were operating on all the inhabited continents (see Figure 6.34). Sixteen stations were regularly recording and sending their data to Milne.

MODERN INSTRUMENTS

A modern seismograph consists of three basic parts: an inertia member, a transducer, and a recorder.

The inertia member is a weight suspended by a wire, or spring, so that it acts like a pendulum but is so constructed that it can move in only one direction. The inertia member tends to remain at rest as earth waves pass by. It has to be damped so that the mass will not swing freely.

The transducer is a device that picks up the relative motion between the mass and the ground. It converts this into a recordable form. It may be a mechanical lever or an electrodynamic system. In one electrodynamic system a coil of wire moves back and forth in a magnetic field. This movement creates an electric current that passes through a galvanometer to be recorded on a sheet of paper.

To record motion in all directions, it is necessary to have three seismographs. One records vertical motion, and two record horizontal motion in directions at right angles to each other. A well-equipped seismograph station will have a set of three components: vertical, north-south, and east-west.

A mass on a spring will oscillate if displaced and then released. The time required to complete one oscillation is called the **period**. If the ground under this system oscillates with a shorter period, the mass hangs still in space, or nearly so. It then serves as a point of reference from which to measure the earth's motion. The inertia members are built to stand still during the passage of waves of a certain selected period range. Generally, two sets are used, one for short-period waves of 5 s or less and one for long-period waves of 5 to 60 s. Because the motion of the ground at distant stations is microscopic, the motion must be magnified to make a visible record. Therefore the transducer includes a system that will serve to magnify the ground motion.

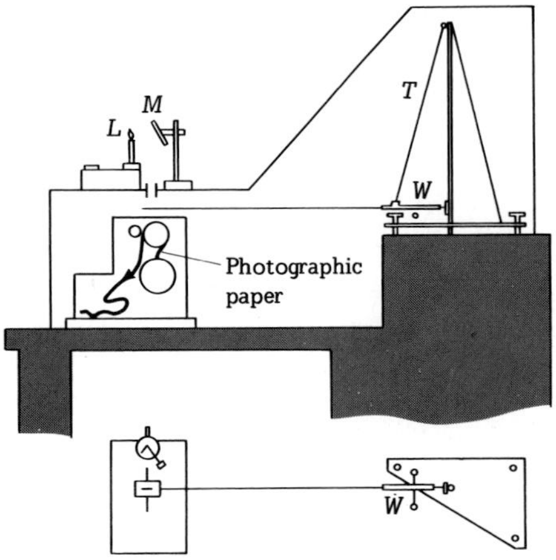

FIGURE 6.34

The Milne horizontal seismograph. Light from *L* was reflected by *M* onto photographic paper through the intersection of two crossed slits. The lower illustration is a top view of the instrument with its outer case removed. *T* is a flexible wire holding up the boom. The weight, *W*, was pivoted on the boom.

A standard recorder is a cylindrical drum on which is wrapped a sheet of recording paper. It rotates at a constant speed and by a helical drive moves sidewise as it turns. This produces a continuous record, a series of parallel lines. These lines appear straight when there is no ground motion; but if there is motion, the lines move in response. A clock-controlled device jogs the recording line briefly once per minute and for a longer time each hour. These jogs on the record identify the time of day. Each sheet of paper includes the capacity for 24 h of continuous recording (see Figure 6.35).

6.9 EARTH WAVES

When earth materials rupture and cause an earthquake, some of the energy released travels away by means of earth waves. The manner in which earth waves transmit energy can be illustrated by the behavior of waves on the surface of water: A pebble dropped into a quiet pool creates ripples that travel outward over the water's surface in concentric circles. These ripples carry away part of the energy that the pebble possessed as it struck the water. A listening device at some distant point beneath the surface can detect the

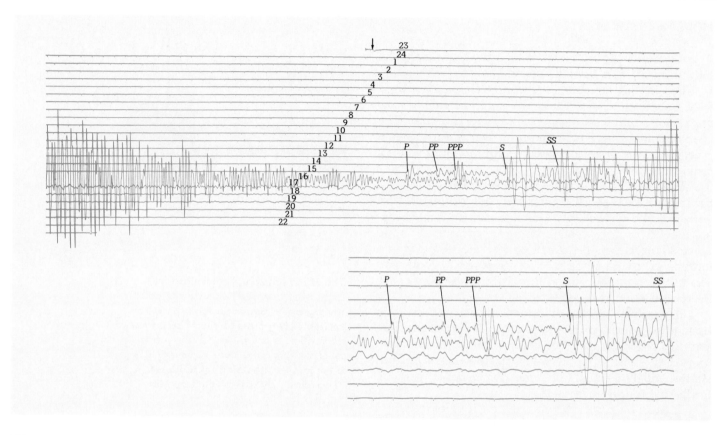

FIGURE 6.35

Long-period record from Alert, northern Canada, for May 18 and 19, 1962, part of it enlarged below. Hours down the middle of the record. For phases *PP*, *PPP*, and *SS* see Figure 6.39. [After John H. Hodgson, *Earthquakes and Earth Structure*, Prentice-Hall, Inc., Englewood Cliffs, N.J., 1964.]

noise of impact. The noise is transmitted through the body of the water by sound waves, far different from surface waves and not visible by ordinary means.

Just as with water-borne waves, there are two general classes of earth waves: **body waves,** which travel through the interior of the mass in which they are generated, and **surface waves,** which travel only along the surface.

BODY WAVES

Body waves are of two general types: **push-pull** and **shake.** Each is defined by its manner of moving particles as it travels along.

Push-pull waves, more commonly known as sound

waves, can travel through any material—solid, liquid, or gas. They move the particles forward and backward; consequently the materials in the path of these waves are alternately compressed and rarefied. For example, when we strike a tuning fork sharply, the prongs vibrate back and forth, first pushing and then pulling the molecules of air with which they come in contact. Each molecule bumps the next one, and a wave of pressure is set in motion through the air. If the molecules next to your eardrum are compressed at the rate of 440 times/s, you hear a tone that is called "middle A."

Shake waves can travel only through materials that resist a change in shape. These waves shake the particles in their path at right angles to the direction of their advance. Imagine that you are holding one end of a rope fastened to a wall. If you move your hand up

FIGURE 6.36

Motion produced by three earthquake wave types.

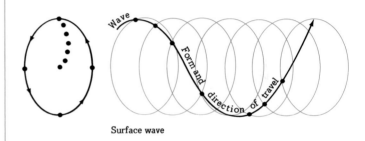

Motion of particle in wave's path
Push-pull (compressional) wave

and down regularly, a series of waves will travel along the rope to the wall. As each wave moves along, the particles in the rope move up and down, just as the particles in your hand did. In other words, the particles move at *right angles* to the direction of the wave's advance. The same is true when you move your hand from side to side instead of "vibrating" it up and down.

SURFACE WAVES

Surface waves can travel along the surface of any material. Let us look again at the manner in which waves transmit energy along the surface of water. If you stand on the shore and throw a pebble into a quiet pool, setting up surface waves, some of the water seems to be moving toward you. Actually, though, what is coming toward you is *energy* in the form of waves. The particles of water move in a definite pattern as each wave advances: up, forward, down, and back, in a small circle. We can observe this pattern by dropping a small cork on the surface into the path of the waves (see Figure 6.36).

When surface waves are generated in rock, one common type of particle motion is just the reverse of the water-particle motion—that is, forward, up, back, and down.

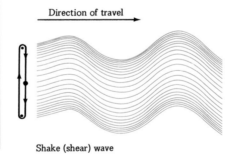

Surface wave

Shake (shear) wave

6.10 RECORDS OF EARTHQUAKE WAVES

As we have seen, the first records of earthquake motion were difficult to analyze accurately. But when the drum speed was increased to spread the events of 1 h over a space of 1 m, a sharp pattern emerged.

This pattern consisted of three sets of earth waves. The first waves to arrive at the recording station were named **primary;** the second to arrive were named **secondary;** and the last to arrive were named **large waves.** The symbols *P, S,* and *L* are commonly used for these three types. Closer study revealed that the *P* waves are push-pull waves and travel with speeds that vary with the density of the material. It also revealed that the *S* waves are shake waves and likewise travel at speeds that vary with density.

The *P* and *S* waves travel from the focus of an earthquake through the interior of the earth to the recording station. The *L* waves are surface waves that travel along the surface to the recording station from the area directly above the focus.

The *P* waves arrive at a station before the *S* waves because, although they follow the same general paths of travel, they go at different speeds. The push-pull

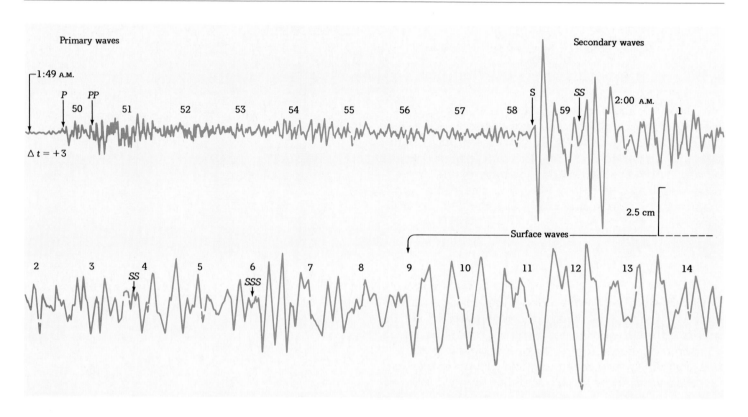

FIGURE 6.37

A record of an earthquake on a seismogram at Harvard, Massachusetts. All waves started in Rumania at the same instant. They arrived as indicated above because of different speeds and paths. The distance was 7,445 km.

mechanism by which P waves travel generates more rapid speed than does the shake mechanism of the S waves. The S waves travel at about two-fifths the speed of P waves in any given earth material. The L waves are last to arrive, because they travel at slower speeds and over longer routes (see Figure 6.37).

TIME-DISTANCE GRAPHS

As data accumulated, Milne began plotting the arrival time of the first P, S, and L waves from felt earthquakes. He set up time-distance graphs for each wave. He found that all three waves took longer and longer to travel to greater and greater distances. He also found that the time between the arrival of each wave type increased as the distance from the source increased.

Travel times If earth waves traveled through a material of uniform elasticity, their paths would be straight lines as they went at constant speed to greater and greater distances. However, if there is a progressive increase in elasticity with depth, they will gradually increase their speed, and their paths will become smooth curves. When there is a sharp increase in elasticity, the wave directions are changed, and this is reflected by sharp changes of direction in the line on a time-distance graph.

Distance is sometimes expressed as the number of kilometers for the length of a great-circle arc between two surface points or as the number of degrees for the angle at the earth's center subtended by that arc. For example, one-quarter of the way around the earth is 10,000 km, or 90°.

Less than 11,000 km From thousands of measurements the world over, it has been learned that P, S, and L waves have regular travel schedules for distances up to 11,000 km. From an earthquake in San Francisco, for example, we can predict that P will reach El Paso,

TABLE 6.4
Sample timetable for *P* and *S*

Distance from source, km	Travel time P		S		Interval between P and S (S − P)	
	min	s	min	s	min	s
2,000	4	06	7	25	3	19
4,000	6	58	12	36	5	38
6,000	9	21	16	56	7	35
8,000	11	23	20	45	9	22
10,000	12	57	23	56	10	59
11,000	13	39	25	18	11	39

1,600 km away, in 3 min 22 s and *S* in 6 min 3 s; *P* will reach Indianapolis, 3,220 km away, in 5 min 56 s and *S* in 10 min 48 s; *P* will reach Costa Rica, 4,800 km away, in 8 min 1 s and *S* in 14 min 28 s.

The travel schedules move along systematically out to a distance of 11,000 km, as shown in Table 6.4.

Beyond 11,000 km   Beyond 11,000 km, however, something happens to the schedule, and the *P* waves are delayed. By 16,000 km they are 3 min late. When we consider that up to 11,000 km we could predict their arrival by seconds, a 3-min delay is significant.

The fate of the *S* waves is even more spectacular: They disappear altogether, never to be heard from again.

When the strange case of the late *P* waves and the missing *S* waves was first recognized, seismologists became excited; for they realized that they were not just recording earthquakes but were developing a picture of the interior of the earth.

LOCATING EARTHQUAKES

We now have timetables for earth waves for all possible distances from an earthquake. These are represented in the graphs of Figures 6.38 and 6.39. Data of this sort are the essential tools of the seismologist.

When the records of a station give clear evidence of the *P*, *S*, and *L* waves from an earthquake, the observer first determines the intervals between them, as shown in Table 6.5. Next, he plots these times on the edge of a strip of paper, using the graph's time scale: (1) at 0, to represent *P*, (2) at 2 min 38 s, (3) at 4 min 16 s, (4) at 8 min 54 s, and (5) at 13 min 34 s. He then tries

FIGURE 6.38

Travel-time curves for earth waves. [After Harold Jeffreys and K. E. Bullen.]

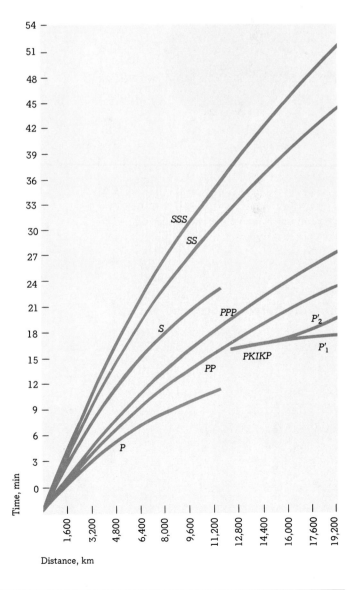

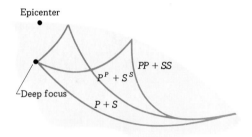

Deep-focus earthquake

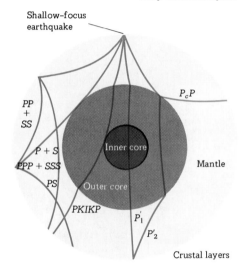

Crustal layers

FIGURE 6.39

The paths of some of the more common earthquake waves.

TABLE 6.5

Intervals between arriving waves

Time, beginning with P			Interval, time after P	
h	min	s	min	s
15	09	06		
15	11	44	2	38
15	13	22	4	16
15	18	00	8	54
15	22	40	13	34

these five marks on different parts of the graph, always keeping the first mark on the P line. At one place—and only one—the marks will each fall on a line of the graph. This will happen at the place that corresponds to the quake's distance, in this case 7,440 km, or 67°. The graph also shows that P travels 7,440 km in 10 min 52 s. The seismologist subtracts this travel time from the arrival time of P and has the time of the earthquake at its source:

P arrived	15 h 09 min 06 s
P had traveled	10 min 52 s
P started at	14 h 58 min 14 s

This process is carried out for several seismograph stations that have recorded the quake. Then an arc is drawn on a globe to represent the computed distance from each station. The point where all the arcs intersect indicates the center of the disturbance.

This example was based on a record taken at the station in Alert, Canada.[6] Palisades, New York, recorded it at 3,580 km, and Pasadena, California, at 2,675 km. The United States Coast and Geodetic Survey located it in south-central Mexico, where they reported extensive property damage, 3 deaths, and 16 injured.

Although this whole procedure is essentially very simple, some have found its accuracy hard to believe. On December 16, 1920, seismologists all over the world found the record of an exceptionally severe earthquake on their seismographs. Each computed the distance of the quake and sent the information along to the central bureaus, where the various reports were assembled. The next day, the location of the earthquake was announced to the press, unlike many lesser shocks that fail to make the news. The announcement stated that a very severe earthquake had occurred at 5 min 43 s after 12:00, Greenwich time, December 16, 1920, in the vicinity of 35.6°N, 105.7°E. That placed it in the Chinese province of Kansu, about 1,600 km inland from Shanghai, on the border of Tibet. This area is densely populated but quite isolated. No reports of damage came in, however, and the matter was soon forgotten by the general public. But it was not forgotten by the members of the press, who were sure that they had been misinformed. Then, 3 months later, a survivor staggered into the range of modern communications with a story of a catastrophe in Kansu on the day and

[6]John H. Hodgson, *Earthquakes and Earth Structure*, Prentice-Hall, Inc., Englewood Cliffs, N.J., 1964.

at the time announced, a catastrophe that had killed an estimated 100,000 persons and had created untold havoc by causing the great landslides described earlier in this chapter.

In contrast, however, on another occasion the news traveled faster than the waves themselves. On August 20, 1937, at 6:59 P.M., an earthquake occurred at Manila, in the Philippines, and the story was transmitted with unusual promptness. It was flashed to North America and found its way to the Boston office of a news agency. An operator there, 1 h after the earthquake happened, picked up a phone and called the Harvard Seismograph Station at Harvard, Massachusetts, 13,440 km from Manila, to inquire whether the disturbance had been recorded. The conversation took place 10 min before the earthquake's surface waves reached the station.

6.11 EARTHQUAKE PREDICTION AND CONTROL

From time to time observers have tried to relate the occurrence of earthquakes to sunspots, tides, positions of heavenly bodies, and other phenomena; but they have all ignored the facts in one way or another. Occasionally someone steps forward to predict earthquakes on the basis of some completely fancied correlation, usually keeping the supporting data secret. One such prophet stated that, on each of certain days during three months of one year, there would be "an earthquake in the southwest Pacific." Because he did not specify magnitude, time, or place in this highly active zone, statistically speaking, it was not possible for him to miss.

During the 1970s, however, great strides were made toward scientifically accurate measurement, interpretation, and potentially precise prediction of earthquakes. This has been an elusive goal of seismologists and astrologers alike for centuries. As a result of recent advances in the earth sciences, we appear to be on the verge of making this dream a practical reality.

EARTHQUAKE PRECURSORS

Many different kinds of both short-term and long-term effect have been observed to precede earthquakes. These **precursors** include crustal movements, unexpected changes in surface tilting, and changes in fluid pressure and in electrical and magnetic fields, as well as local small quakes.

It has long been alleged that dogs, cattle, and other animals behave strangely immediately prior to the occurrence of an earthquake. It now appears that these animals are not soothsayers but are merely sensing minute changes in the earth and its activities that tend to escape human senses.

DILATANCY MODEL

Observations for a number of earthquakes in Russia, Japan, and the Adirondack region of the United States have shown that prior to each earthquake there is a marked change in the velocity of seismic waves passing through the rocks near the quake area.

Laboratory experiments have long ago demonstrated that rocks undergo **dilatancy,** or increase in volume, before failure. This is apparently produced by the formation and gradual outward movement of cracks within the rock. This dilatancy begins to occur at stresses as low as half the breaking strength of the rock. Water is released as the cracks expand. The rocks become more fragile and ultimately are too weak to withstand the stress; rupture results.

A bowing up of the earth's crust commonly occurs in the region of the earthquake epicenter. This changes the tilt of the surface, which can often be measured with sensitive instruments.

All these changes are thought to be related to dilatancy. If such changes could be observed in sufficient time, it would be possible to predict the earthquake.

A formula has been developed by J. H. Whitcomb, J. D. Garmany, and D. L. Anderson of the California Institute of Technology, which relates the duration of the precursor events with the magnitude of the predicted earthquake. A quake with magnitude 5 on the Richter scale should have precursor events up to 4 months prior to the quake. One with magnitude 7 would be preceded by about 14 years of such precursor activity.

Up to the mid-1970s about 10 earthquakes had been successfully predicted prior to their actual occurrence. With continued close observation of earthquakes and their precursor events, earthquake prediction could soon be a reality. It has become apparent, however, that the dilatancy model is not sufficient to serve as the only explanation for precursor events. Indeed it has been established that earthquakes may be preceded by quite different physical events and changes in different geologic regions, and earthquake-prediction methods may thus vary according to different sets of criteria for different regions.

CONTROL OF EARTHQUAKES

A fortuitous discovery made some years ago has led to the distinct possibility of controlling or modifying at least some earthquakes.

Waste fluids had been injected into a deep well near Denver, Colorado, over a period of several years. There soon developed a series of minor earthquakes, which were subsequently demonstrated to coincide with periods of fluid injection. During lulls in the injections fewer or no quakes occurred. Since then, laboratory and field investigations have shown that movement along a fault zone can be initiated by reducing frictional resistance across the fault. This can be accomplished by injecting fluids along the fault. Conversely, faults can be locked when fluids are withdrawn. In the Rangeley oil fields in Colorado some United States Geological Survey workers demonstrated this ability to turn the seismic activity in an area on and off.

Although it is not likely that a major active fault zone will be treated in this manner in the near future, this method of control over fault movements may well be commonplace for future generations.

6.12 STRUCTURE OF THE EARTH'S INTERIOR

Studies of the travel habits of waves through the earth and of surface waves around the earth have given us information about its structure from surface to center. These studies have been made possible by our knowledge of the speed of these earth waves and of their behavior in different materials (for example, waves travel at greater speeds through denser materials than through less dense materials).

When earth waves move from one kind of material to another, they are refracted and reflected (see Figure 6.40). The waves have revealed several places within the earth where there are changes in physical properties. These could be due to changes in composition, atomic structure, or atomic state. The boundary where change takes place is called a **discontinuity.**

For body waves to reach greater and greater distances on the surface, they must penetrate deeper and deeper into the earth's interior. Thus in traveling from an earthquake in San Francisco to a station at Dallas, a surface distance of 2,400 km, the body waves penetrate to 480 km below the surface. This holds true for any other 2,400-km surface distance. To reach a station 11,200 km away, the body waves dip into the interior to a maximum depth of 2,900 km and bring out information from that depth.

On the basis of data assembled from studies of the travel habits of earth waves, the earth has been divided into three major zones: **crust, mantle,** and **core.**

CRUST

Information on the earth's crust comes primarily through observation of the velocities of P and S waves

FIGURE 6.40

Paths of refracted waves used to determine the thickness of a single layer. If velocity in V_2 material at depth d is greater than V_1, waves from O arrive before any others at stations S_1, S_2, S_3, and S_4 by way of surface paths. Waves from O through P to Q_5S_5, Q_6S_6, Q_7S_7, and Q_8S_8 take the lead after S_5 and are first to arrive. By plotting time-distance relationships here, seismologists measure V_1, V_2, and d.

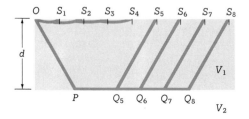

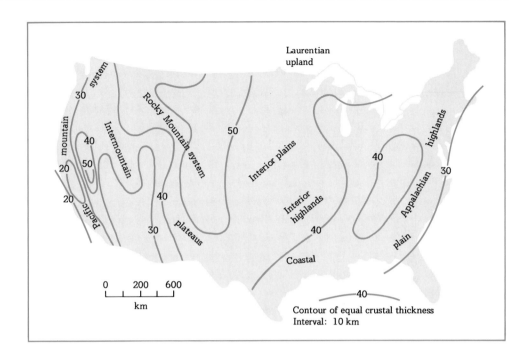

FIGURE 6.41

Crustal thickness determined from seismic data. [After L. C. Pakiser and I. Zietz.]

from local earthquakes (within 1,000 km) and from dynamite and nuclear blasts. One of the first things revealed is that the earth's crust is solid rock. Early in the history of crustal studies a seismologist in Yugoslavia, A. Mohorovičic, made a study of records of the earth waves from an earthquake on October 8, 1909, in the Kulpa Valley, Croatia. He observed two P waves and two S waves and concluded that the first P and S had encountered something that caused some of their energy to be reflected to the surface. He also concluded that velocities of P and S waves increased abruptly below a depth of about 50 km. This abrupt change in the speed of P and S waves indicated a change in material and became known as the **Mohorovičic discontinuity.** For convenience it is now referred to as the "Moho," or the **M discontinuity.** The Moho marks the base of the earth's crust, separating it from the mantle.

Crust of the continents The depth to the Moho varies in different parts of the continents. However, it is generally greater under mountain regions than it is under lowlands. Crust is also variable in composition.

In the United States the continental crust has an average thickness of 33 km but varies between 20 and 60 km (see Figure 6.41). Under the intermountain plateaus and the Great Basin Ranges it averages less than 30 km; under the Great Valley of California, 20 km; under the Sierra Nevada, more than 50 km. It is thickest

under the eastern front of the Rocky Mountain ranges. The crust under New England is about 36 km thick.

It has been difficult to get precise data on the earth's crust from the waves of earthquakes. Waves from dynamite blasts, however, with precisely known locations and times of detonation, have filled in some of the details. In 1941 the Harvard Seismograph Station determined the structure of the continental crust in New England. Analysis of many blast records revealed that in New England the continental crust has three layers, each one with different elastic properties, indicating different rock types. The layered structure of the crust may be representative of other sections of the continents. Table 6.6 summarizes the data.

The average velocity of waves traveling through the upper crustal layer is similar to the velocity expected for granite, granodiorite, or gneiss. Because these rocks are rich in silica and aluminum, we say that the material of the upper crust is **sialic** in composition; sial (Si for silicon and Al for aluminum) is generally used in speaking of this layer, which is found only on the continental areas of the earth. The second and third layers are more and more basic in composition. The third is believed to be basalt. These darker, heavier rocks are sometimes designated collectively as **sima,** a name coined from Si for silicon and ma for magnesium. A simatic layer encircles the earth, presumably under-

TABLE 6.6

Earth's crust under New England[a]

| Thickness, km (and zone) | Velocity, km/s | | Rock type |
	P	S	
16 (layer 1)	6.1	3.5	Sialic
13 (layer 2)	6.8	3.9	Intermediate
7 (layer 3)	7.2	4.3	Simatic
(Moho)			
(Top of mantle)	8.4	4.6	

[a]L. D. Leet, "Trial Travel Times for Northeastern America," *Bull. Seis. Soc. Am.*, vol. 31, pp. 325–334, 1941.

lying the sial of the continents, and is believed to be the outermost rock layer under deep permanent ocean basins.

Crust under oceans Our knowledge of the structure of the crust beneath the oceans is based on observations of rocks exposed on volcanic islands and on studies of the velocities of L waves from earthquakes, supplemented by dynamite-wave profiles and magnetic anomalies (Chapters 1 and 8).

The types of rock found on islands help to determine the edge of the Pacific basin. The **andesite line** (see Figure 6.42) has on its ocean side rocks composed primarily of basalt, whereas on the other side they are principally andesite. This has been viewed as the dividing line between the oceanic and the continental crusts.

On the basis of seismic-wave velocities it appears that the crust under the Pacific basin is not layered and is appreciably thinner than the crust of the continents. Its thickness also varies but averages about 5 km. It is thinner under the Gulf of California and in the northeast Pacific. It is apparently composed of simatic rocks.

Crust under Hawaii The crust under the island of Hawaii was surveyed by seismic-refraction methods in 1964.[7] It was concluded that it is 10 km thick under the summit of Kilauea and 12 to 15 km under other parts of the island.

The crust under Hawaii appears to be divided into two layers: The upper layer, 4 to 8 km thick, is probably an accumulation of lava flows that form the bulk of the island; the lower layer, also 4 to 8 km thick, is probably

[7]David P. Hill, "Crustal Structure of the Island of Hawaii from Seismic-refraction Measurements," *Bull. Seis. Soc. Am.*, vol. 59, pp. 101–130, 1969.

the original basaltic crust modified by a complex intrusive system associated with the central vents and rift zones of the island (see also Table 3.1).

MANTLE

Below the earth's crust is a second major zone, the *mantle*, which extends to a depth of approximately 2,900 km into the interior of the earth. Our knowledge of the mantle is based in part on evidence supplied by the behavior of P and S waves recorded between 1,100 and 11,000 km (Figure 6.43).

At the Moho the speeds of P and S waves increase sharply, an indication that the composition of the material suddenly changes. We have no direct evidence of the new material's nature, but the change in speed suggests that it may contain more ferromagnesian minerals than does the crust. Scientists have accepted the idea that the mantle is solid because it is capable of transmitting S waves. To explain mountain-building processes, observers have emphasized that the mantle material undergoes slow flow as it adapts to changing conditions on the surface.

There are also variations in the composition of the upper mantle. We know this from variations in the velocity of P_n, the wave that travels through the upper portion of the mantle. Speeds of P_n vary from 7.5 to 8.5 km/s. This difference may be due in part to variation in elastic properties of the mantle material.

LITHOSPHERE AND ASTHENOSPHERE

There are other discontinuities in the mantle. One occurs at the so-called low-velocity zone, which varies from about 70 to 80 km under oceans to about 100 km under continents (Figure 6.44). This apparently marks the boundary between the upper, colder portion of the mantle and the lower, hotter mantle. It also coincides, therefore, with the boundary between the **lithosphere** and the **asthenosphere**. When we speak of plates we are essentially referring to this lithosphere, which moves over the asthenosphere. It consists of crust *and* upper mantle although we often refer to it as crustal plates. This is the part of the earth that moves slowly but persistently. In essence the upper mantle is in motion, carrying the oceanic and continental crusts on its top in piggyback fashion.

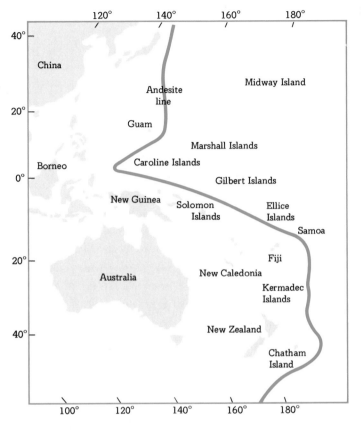

FIGURE 6.42

Position of the andesite line in the southwestern Pacific Ocean. This line marks the border of the Pacific basin in a geologic sense. On the Pacific side of the line young eruptive rocks are basaltic; on the other side they are principally andesitic. Islands east of the line are isolated or grouped volcanic peaks; west of the line they have the characteristic structure of folded continental mountain ranges. [After R. A. Daly.]

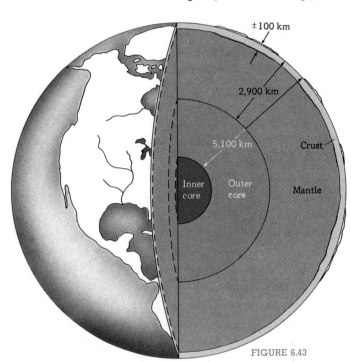

FIGURE 6.43

Structure of the mantle and core.

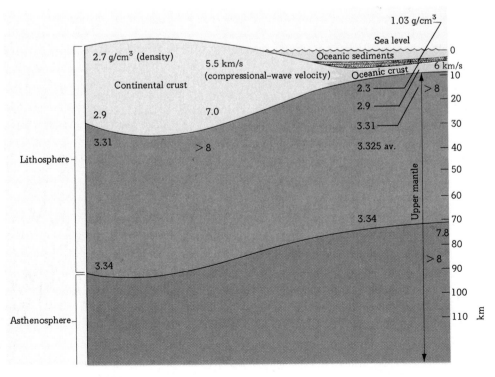

FIGURE 6.44

Crust, mantle, lithosphere, and asthenosphere.

CORE

We come now to the core, a zone that extends from the 2,900-km inner limit of the mantle to the center of the earth at a depth of 6,370 km. An analysis of seismographic records from earthquakes 11,200 km or more distant reveals that the core has two parts, an outer zone 2,200 km thick and an inner zone with a radius of 1,260 km.

In traveling between two points on the surface 11,200 km apart, P and S waves penetrate 2,900 km into the interior. But after they go deeper than that, they enter a material that delays P and eliminates S altogether. There have been various suggestions to explain these observations. The most widely favored postulates that the outer core is liquid. By definition, S waves are capable of traveling only through substances that possess rigidity; and since rigidity is a property of the solid state, it is assumed that the outer core, not transmitting S waves, is in a liquid state or at least behaving as a liquid.

P waves speed up again at a depth of about 5,150 km, indicating that the inner core is solid.

As the earth swings around the sun, it behaves like a sphere with a specific gravity of 5.5, but geologists have found that the average specific gravity of rocks exposed at the surface is less than 3.0. And even if rocks with this same specific gravity were squeezed under 2,900 km of similar rocks, their specific gravity would increase to only 5.7. Geophysicists have computed that the specific gravity of the core must be about 15.0 to give the whole globe an average of 5.5. To meet this requirement, it has been suggested that the core may be composed primarily of iron, possibly mixed with about 8 percent of nickel and some cobalt, in the same proportions found in metallic meteorites.

OUTLINE

Earthquakes are vibrations caused when earth materials have been distorted until they rupture.

Seismology is the scientific study of earthquakes.

Effects of earthquakes include fire, damage to structures, seismic sea waves, landslides, cracks in the ground, changes in land level, and sound.

> **Fire** has caused an estimated 95 percent of the total loss from some earthquakes.
>
> **Damage to structures** depends on intensity and duration of the shaking.
>
> **Seismic sea waves** are now being detected, and a **seismic-sea-wave warning system** operated by the United States Coast and Geodetic Survey alerts threatened communities.
>
> **Landslides** often accompany the largest earthquakes.
>
> **Cracks** in the ground, in the sense of yawning chasms in rock, have not happened.
>
> **Land movements** result in slow creep or rapid offsets both horizontally and vertically.
>
> **Liquefaction** of soil and loose sediments often follows earthquakes and commonly causes more damage than the actual shaking by the quake.
>
> **Sounds** associated with earthquakes vary from inaudible to booming and thunderous.
>
> **Worldwide standard seismograph network** is supervised by the Coast and Geodetic Survey.

Earthquakes are interpreted and evaluated from different facts and scales.

> **Earthquake focus** is a term used to designate the source of a given set of earthquake waves.
>
> **Earthquake-intensity** scales are used to estimate the amount of shaking at different places.
>
> **Earthquake magnitude** evaluates the amount of energy at the earthquake's source.

Distribution of earthquakes, in terms of energy released, is concentrated, with about 80 percent occurring around the borders of the Pacific Ocean.

Seismicity of Alaska and Aleutian Islands is estimated at 6 percent of the world's total.

The Prince William Sound earthquake of 1964 was the most destructive earthquake ever recorded on the North American continent.

Seismic and volcanic activity may have a common ultimate cause in deep movements of mantle materials, especially along crustal-plate margins.

Cause of earthquakes is the sudden rupture of earth materials distorted beyond the limit of their strength.

Elastic-rebound hypothesis proposed snapping back of distorted and ruptured rocks as the cause of earthquakes.

Evidence against elastic-rebound hypothesis includes observations that fault movements take place *after* earthquakes in some places.

Plate movements generate a large proportion of all earthquakes. **Rates of movement** average 4 to 6 cm/year.

Interior of the earth is known to us from studying the records of earth waves.

Early instrumental observations were mostly by seismoscopes; the first record of a distant earthquake was made in 1889.

Modern seismology had its beginnings with a group of British professors teaching in Japan in the late nineteenth century.

Modern instruments consist of an inertia member, a transducer, and a recorder.

Earth waves carry away some of the energy released when earth materials rupture.

Body waves travel through the interior of the earth.

Surface waves travel along the surface.

Records of earthquake waves are characterized by three sets of earth waves, **P, S,** and **L.**

Time-distance graphs are bases for analyzing the history of earth waves.

Travel times are governed by the materials through which waves have travelled.

Locating earthquakes requires data from several stations.

Earthquake prediction will be possible primarily by observing and interpreting **precursor** events, including **crustal movements, tilt of earth's surface,** and **changes in fluid pressure** and in **electrical and magnetic fields.**

The **dilatancy model** is based on volumetric increases in rocks immediately prior to their rupture.

Control of earthquakes may be achieved by injection and removal of fluids from potentially active fault zones.

Structure of the earth's interior is made up of crust, mantle, and core.

The earth's **crust** is separated from the mantle at a **discontinuity** called the Mohorovičic, or the **Moho,** or simply the **M discontinuity.**

The **crust of the continental** United States varies between 20 and 60 km in thickness.

The **crust under the oceans** averages about 5 km in thickness.

The **mantle** extends to 2,900 km in depth.

The **lithosphere** contains the crust and the upper, colder part of the mantle.

The **asthenosphere** extends downward from its border with the lithosphere.

The **core** is thought to have a liquid outer zone and a solid center; the outer zone behaves as a liquid in not allowing **S** waves to penetrate.

SUPPLEMENTARY
READINGS

Allen, C. R. (chm.) *Predicting Earthquakes, A Scientific and Technical Evaluation— With Implications for Society*, Panel on Earthquake Prediction of the Committee on Seismology, National Academy of Sciences, Washington, 1976. *Evaluates the status of earthquake prediction and previews future trends. Presents urgent scientific and technical recommendations for further study and implementation of earthquake-prediction capabilities.*

Clark, S. P. *Structure of the Earth*, Prentice-Hall, Inc., Englewood Cliffs, N.J., 1970. *An excellent treatment of the subject.*

Dewey, James, and Perry Byerly "The Early History of Seismometry (to 1900)," *Bull. Seis. Soc. Am.*, vol. 59, pp. 183–227, 1969. *Describes the first instruments tried for recording ground motion from earthquakes, the first successful record of a distant earthquake, and early research that was the forerunner of all aspects of modern seismology.*

Hodgson, John H. *Earthquakes and Earth Structure*, Prentice-Hall, Inc., Englewood Cliffs, N.J., 1964. *A good account written for the nonspecialists.*

Press, Frank "Earthquake Prediction," *Sci. Am.* vol. 232, no. 5, pp. 14–23, 1975. *A summary of recent technical advances, which have brought the United States and several other countries into a position of nearly achieving reliable long-term and short-term forecasts within a decade.*

Wood, Fergus J. (ed.) *The Prince William Sound, Alaska, Earthquake of 1964 and Aftershocks*, vols. I–III, Publication 10-3, U.S. Coast and Geodetic Survey, Washington, 1966, 1967, 1969. *The definitive reports on this greatest earthquake ever recorded on the North American Continent, making it also the most thoroughly studied.*

7

DEFORMATION

The crust of the earth is continuously changing. We have already discussed changes produced by weathering and erosion, but the crust is also being changed in other, more fundamental ways. It is being deformed by forces acting within it (Figure 7.1). These help to maintain the surface of the land above sea level and work to offset, or compensate for, the destructive effects of erosion.

7.1 DEFORMATION OF THE EARTH'S CRUST

Evidence of deformation can be seen everywhere. Sediments that were deposited on the bottom of the ocean are now found hardened into rocks in mountainous areas high above sea level. They contain fossils that attest to their sedimentary origin. Far below sea level, miners often find pieces of trees or other plants embedded in layers of coal. This shows that these beds, now deeply buried, were at one time lying on the earth's surface.

Deforming forces elevate and depress large landmasses, and they also work to distort the earth's crust. Rocks that are rigid at the surface become plastic at depth and respond to deforming forces by folding (see Figure 7.2). These shapes could not have been produced in a rigid rock.

ABRUPT MOVEMENTS

Crustal deformation has occurred during some large earthquakes. This has resulted in measurable displacements in historic times, surveys revealing that large portions of the earth's crust have been warped, tilted, or moved horizontally. Earthquakes have also altered the earth's surface by triggering landslides.

Three major earthquakes in 1811 and 1812 were centered near New Madrid, Missouri, on December 16, January 23, and February 7. These were accompanied by changes in topography over an area of 8,000 to 13,000 km². The crust sank 1.5 to 6 m in places, forming swamps and lakes and drowning forests. The largest lake to form was Reelfoot Lake in Tennessee.

FIGURE 7.1

Rocks tilted almost vertically: quartzite in the Great Smoky Mountains. [L. B. Gillett.]

FIGURE 7.2

Photograph of a deformed rock 50 cm across from the Auburn Mine near Virginia, Minnesota. The miniature forms seen here mimic many of the structural features produced in rock layers during the process of mountain building. [R. C. Gutschick.]

The Alaskan Prince William Sound earthquake of 1964 was accompanied by considerable vertical and horizontal displacements. These occurred over more than 120,000 km². The Coast and Geodetic Survey found that there was a general south-to-north rotation of a portion of the earth's crust centered at the northern coast of Prince William Sound. A maximum uplift of 13 m occurred off the southwestern tip of Montague Island. The largest subsidence of 2 m occurred 43 km north of Glennallen. Maximum horizontal movements were found between Montague and Latouche Islands, which, as we have said, were 4.5 to 6 m closer together after the earthquake than before. Surveys showed that the southwest side of Montague Island moved northwest with respect to Latouche Island. Figure 7.3 shows horizontal movements that occurred and the areas involved. One of the largest changes took place at Valdez when the earthquake triggered movement that deepened the harbor as much as 100 m in one place.

Vertical displacements Vertical displacements during one earthquake cannot account for the amount of elevation of the earth's crust associated with mountains or for the large depressions of sedimentary layers found in deep coal mines. However, if displacements occur often enough throughout geologic time, the total can become significant. For instance, on the shore of Sagami Bay, Japan, not far from Yokohama, a cliff reared up during an earthquake on September 1, 1923. The amount of movement was measured by using some marine bivalves called *Lithophaga* ("rock eaters") as a reference. These little animals scoop out small caves for their 5-cm shells at mean sea level and spend their lives waiting for the sea to bring food at each rise of the tide. After the 1923 earthquake, rows of *Lithophaga* were found starved to death 5 m above the waters that used to feed them. Other rows of *Lithophaga* holes in this same cliff were found and correlated with quakes in A.D. 1703, 818, and 33 (see Figure 7.4). The total elevation over that 2,000-year interval was 15 m. At this rate the elevation over a geologic time interval of 2 million years would be 15,000 m.

Horizontal displacements The San Andreas fault is a large scar in the earth's crust approximately 1,000 km long extending in a line from Cape Mendocino southeastward to the Gulf of California (see Figure 7.5). Displacements have occurred along portions of the fault at the times of some earthquakes but not over its entire length during any one quake (see Figure 7.6). At the time of the earthquake of April 18, 1906, there were horizontal displacements along 400 km of the northerly

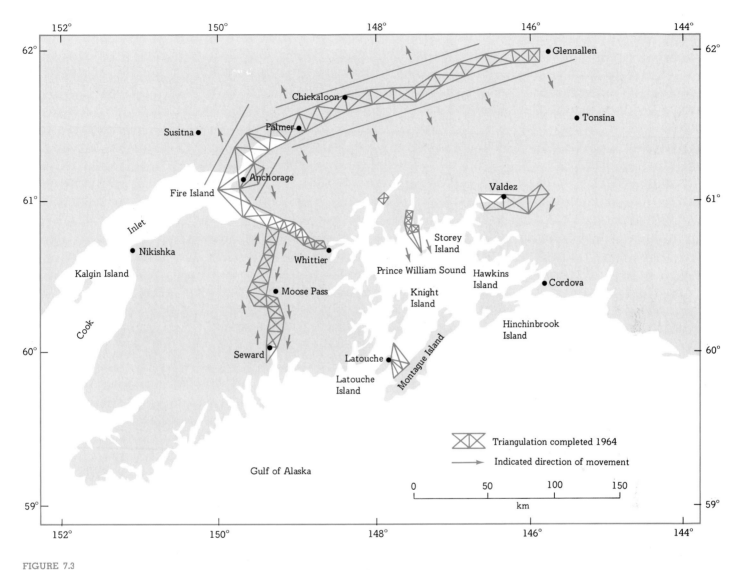

FIGURE 7.3

Relative horizontal movement of earth's crust near Prince William Sound as determined by 1964 surveys of United States Coast and Geodetic Survey.

FIGURE 7.4

Drawing showing elevated cliff on Sagami Bay, Japan. Inset, *Lithophaga* shells.

FIGURE 7.5

A section of the San Andreas fault in the Indio Hills, California; view looks northwest. An abrupt movement of 6 m along part of this fault occurred at the time of the San Francisco earthquake of 1906. [Spence Air Photos.]

end from Point Arena to San Juan Bautista. The maximum slippage was 7 m on the shore near Tomales Bay. This displacement died out in both directions.

Rocks on both sides of the San Andreas fault have been studied, and some geologists believe formations can be correlated that were joined across the fault as a single unit 150 million years ago. Estimates of the total displacement run in the scores of kilometers up to as many as 200.

SLOW MOVEMENTS

Not all crustal movements are accompanied by earthquakes. Slow movements of the crust are going on today. These are being measured along faults in Cali-

fornia by government geologists and university investigators in a broad, continuing program. They use cross-fault strain meters, which show that deformation occurs in zones up to 10 m across. Creep is in progress along the Hayward fault from Richmond to south of Fremont, along the Calaveras fault in the Hollister area, and along the San Andreas fault from San Juan Bautista to Cholame (See Figure 7.7). The time and amount of movement are shown in Figure 7.8. It is interesting to note that at San Juan Bautista and at the Cienega Winery the creep occurs as distinct events, whereas the pattern of movement is different at Stone Canyon Observatory. Creep does not occur simultaneously at San Juan Bautista and Cienega Winery but may be several days apart. The difference in time at which creep starts suggests that it propagates from San

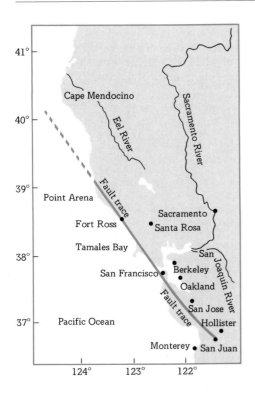

FIGURE 7.6

Section of the San Andreas fault that slipped at the time of the earthquake in 1906. [After John H. Hodgson, *Earthquakes and Earth Structure*, p. 15, Prentice-Hall, Inc., Englewood Cliffs, N.J., 1964.]

FIGURE 7.7

Sections of the San Andreas, Hayward, and Calaveras faults, showing places of creep measurement. Heavier lines indicate sections of faults undergoing active movement.

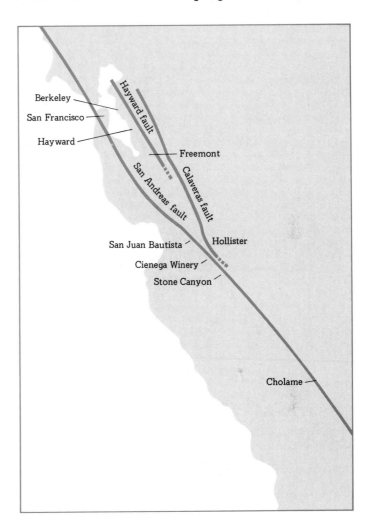

FIGURE 7.8

Fault movement at Cienega Winery, San Juan Bautista, and Stone Canyon (18 km southeast of Cienega Winery).

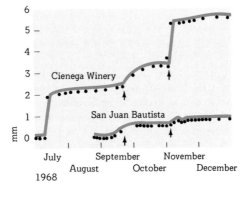

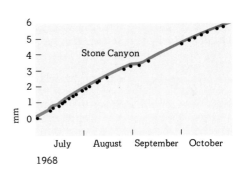

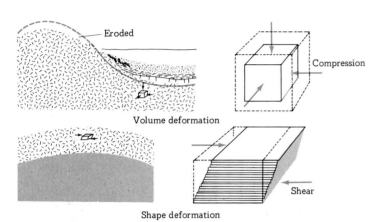

Eroded

Volume deformation

Shape deformation

Compression

Shear

FIGURE 7.9

Deformation may produce change in volume without change in shape or change in shape without change in volume or a combination of the two.

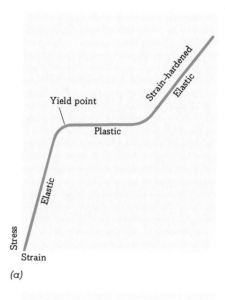

Yield point

Strain-hardened

Elastic

Plastic

Elastic

Stress

Strain

(a)

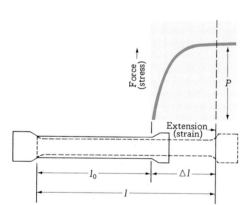

FIGURE 7.10

Deformation of a cylindrical specimen of rock by extension. See the text for discussion.

Force (stress)

P

Extension (strain)

l_0

Δl

l

FIGURE 7.11

Behavior of material under stress. (a) As stress increases, for a while strain is proportional to stress. This is the range of elastic deformation. Beyond the yield point deformation becomes plastic for a while until the material again resists, and stress must be increased to produce more strain.

(b) After deformation to point B, if the stress is removed, the elastic part of the deformation is recovered, but the rest is not. A retardation of the recovery occurs. If stress is again applied, deformation picks up where it left off at B and continues as though there had been no interruption.

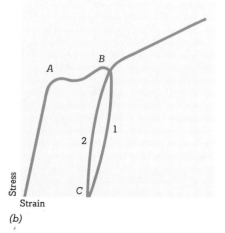

A

B

1

2

C

Stress

Strain

(b)

Juan Bautista to the Cienega Winery at a rate of approximately 10 km/day.[1] The average movement along a 100-km section of the San Andreas in the vicinity of Parkfield is 5 cm/year, with the Pacific side of the fault moving northwestward.[2]

Another example is that of Mount Suribachi on Iwo Jima, in 1945 scene of a famous flag raising by the United States Marines: According to geologist Takeyo Kosaka, pressure from underlying magma caused the surface to rise 7 m in 23 years from that date.

7.2 STRESS, STRAIN, AND STRENGTH OF ROCKS

Deformation of the earth's crust can be described in terms of change of volume, change of shape, or a combination of both. The kinds of deformation are illustrated in Figure 7.9. If a material changes shape without a change of volume, the deformation is called **shear.** An analogy for pure shear is supplied by staggering a deck of cards. If the bottom card is held fixed and the others are slid forward by amounts proportional to their distances from the bottom, the shape of the space occupied by the deck is altered, but its volume (the total amount of cardboard) remains the same. A force applied to material that tends to change that material's dimensions is called **stress.** This is commonly *unit stress,* defined as the total force divided by the area over which it is applied.

STRESS AND STRAIN

The effects of some deforming forces on earth materials have been studied in the laboratory by measuring the strain produced by application of stress. For example, if a cylindrical specimen of rock is stretched, the amount of its extension can be plotted against the force producing it. If its original length is l_0 and the force used to stretch it is P, the increase in length Δl leaves it at a new length l. The **strain,** sometimes called *unit strain,* is the change in length per unit length, or $\Delta l/l_0$. The stress, or unit stress, is the total force P divided by the area of the cross section of the specimen, or P/A (see Figure 7.10). Graphs of stress versus strain are usually set up with unit stress and unit strain as the axes.

In most materials, as stress increases, for a time strain is proportional to stress. If stress is removed,

strain goes back to zero. This is the range of elastic deformation. Then, when the stress reaches the yield point, the strength of the material has been overcome. Deformation becomes plastic; that is, with no additional stress, strain continues to increase. When stress is removed, the plastic portion is not recovered (Figure 7.11).

Elastic deformation So elastic deformation of rock is recoverable. A solid is deformed elastically if it can recover its size and shape after the deforming force has been removed. An elastic solid resists deformation up to a certain point. Its resistance to a change in volume without a change of shape is defined as the ratio of stress to strain:

$$\frac{\text{Increase in force per unit area}}{\text{Change in volume per unit volume}}$$

A material's resistance to elastic shear is called **rigidity.**

Retardation of recovery Rocks are said to be deformed elastically if the deformation disappears when the stress is removed. However, they do not ordinarily regain their former shape the instant the stress is removed. There is a time lag over which recovery is made.

It has been suggested that a demonstration of this retardation is provided on a grand scale by basining of the crust under loads of glacial ice and subsequent recovery upon melting of the ice. The meltings and upwarpings from the last ice advance began about 15,000 years ago, but recovery is still progressing in areas where the ice had completely disappeared as much as 10,000 years ago. Table 7.1 lists some areas where upwarping has occurred and the amounts recovered so far.

Plastic deformation Plastic deformation is permanent. It involves a property of rock called *viscosity.* A material that is deforming plastically does so by flow along an indefinitely large number of shear planes. If the rate of the flow is proportional to the stress causing it, the material is said to be viscous; and if we plot the rate of flow against the stress, the slope of the graph is

[1]Robert D. Nason, "Preliminary Instrumental Measurements of Fault Creep Slippage on the San Andreas Fault, California," *Earthquake Notes,* vol. 40, pp. 6–10, 1969.
[2]Stewart W. Smith and Max Wyss, "Displacement on the San Andreas Fault Subsequent to the 1966 Parkfield Earthquake," *Bull. Seis. Soc. Am.,* vol. 58, pp. 1955–1973, 1968.

TABLE 7.1

Upwarping in regions of complete or partial deglaciation[a]

Area	Maximum observed uplift, m
Scotland	30
Iceland	120
Greenland	146 +
Novaya Zemlya	100
New Zealand (South Island)	100 ±
Antarctica	100
Norway and Sweden	275 +
Eastern Canada–Labrador	270 +
Newfoundland	137

[a]R. A. Daly, *Strength and Structure of the Earth*, Prentice-Hall, Inc., Englewood Cliffs, N.J., 1940.

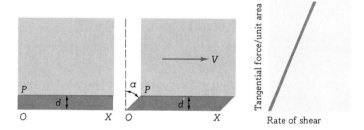

FIGURE 7.12

Demonstration of viscosity. If an oil film of thickness d rests on a surface OX under a block P and P is displaced sidewise with a velocity V, the change in the angle α that OP makes with the vertical can be used to measure the shear (change in shape) of the oil. If the rate of shear is proportional to the force causing it, the deformation is said to be viscous.

the viscosity (Figure 7.12). Viscosity equals stress per rate of flow. Viscosity is an important property in some geological processes. It is the property that governs the ability of magma to flow during igneous activity and is the property that enables the mantle to adjust to crustal loads. It is measured experimentally by observing rate of flow, the force needed to turn a paddle wheel at constant speed, or the rate of sinking of standard spheres (see Figure 7.13).

We do not know the absolute viscosity of materials in the earth's interior. This is difficult to determine experimentally, and its effects are difficult to visualize. However, the effect of viscosity may be illustrated by considering a granite magma with a viscosity of hard pitch, 10^{13} times that of water. A sphere of gneiss 2 m in diameter would sink in this magma about 10 cm/day. A sphere of 4 m in diameter would sink four times as fast.

Time factor   Within the earth, materials show a duality of response. They transmit earthquake waves elastically but move plastically during the formation of mountains and other surface features. Here an important factor that cannot be reduced to a graph is time. Material may be elastic for short-time stresses in transmitting earthquake waves but plastic over a longer time.

On a short time scale, a tuning fork formed from pitch can ring as though it were purest steel. Yet, given time, it will flow into a formless blob from its own weight. A steel bar supported at its ends and loaded in the middle will appear to have strength entirely adequate to support the load but will in time slowly bend. And some metamorphic rocks testify to the deformation of rocks under prolonged periods of stress and increased temperature. On geologic time scales the rocks of the crust have shown themselves to be weak.

Types of stress   There are different types of stress: **tension, compression,** and shear. Tension is a stretching stress and can increase the volume of a material. Compression tends to decrease the volume. A shear stress, as described, produces changes in shape. A stress beyond a material's strength can cause rupture.

STRENGTH OF ROCKS

Rocks possess strength. **Strength** is the stress at which a material begins to be permanently deformed. A special kind of permanent deformation is rupture.

Rocks possess several types of strength, as they respond differently to different stresses. Laboratory test results for rocks generally give compression strength, shear strength, and tension strength (Figure 7.14). These have shown that a rock's strength in tension is smaller than its strength in compression. When the difference between these two is large, we say the material is brittle. When it is small, we say the material is ductile. As the confining pressure increases, a brittle rock tends to become ductile. It does not rupture in the ordinary sense of disintegration but flows along an indefinite number of shear planes. Rupture in tension may be important near the earth's surface, but it will not occur at great depth. The strength in compression for brittle rocks is about 30 times larger than the strength in tension.

How a rock mass responds to stress depends on temperature and depth of burial. At depth, rocks will respond to stress by flowing before they rupture, whereas at the earth's surface they will break. The stresses at the earth's surface are quite different from those at depth, where temperatures and pressures increase considerably. Whether a rock breaks or flows under a given stress depends on the prevailing temperatures. Rocks are generally weaker at high temperatures than they are at low ones. If a rock is surrounded by equal pressure on all sides, it tends to increase in strength. Such pressure strengthens a rock and increased temperature weakens it.

Different rocks have different compressive strengths. Hard, brittle, competent rocks such as sandstone and limestone have a different strength than does clay or shale.

ISOSTASY AND EQUILIBRIUM

A mass standing high above its surroundings should exert a gravitational attraction that could be computed and measured. One device for measuring this attraction makes use of a plumb bob suspended on a plumb line. Like every other object on the globe, the suspended bob is pulled by gravity. On the surface of a perfect sphere with uniform density, the bob would be pulled straight down, and the plumb line would point directly toward the center of the sphere. But if there is any variation from these ideal conditions—that is, if there are surface irregularities on the sphere—the plumb line will be deflected as the bob is attracted by their concentrations of mass.

In 1749 Pierre Bouger found that plumb bobs were deflected by the Chimborazo Mountain in the Andes,

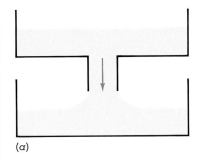

(a)

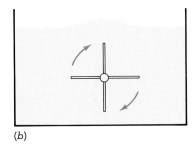

(b)

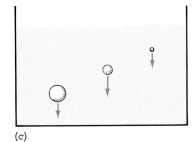

(c)

FIGURE 7.13

Methods of measuring viscosity: (a) rate of flow, (b) force necessary to turn paddle wheel at known speed, and (c) rate at which spheres of the same material but different radii fall through the fluid. For c, $x = 0.22gr^2(d - d')/v$, where x is the velocity of the sphere when the motion is steady (terminal velocity), g is the acceleration of gravity, d is the density of the sphere, d' is the density of the fluid, r is the radius of the sphere, and v is the viscosity of the fluid. In a granite magma with the viscosity of hard pitch (10^{13} times that of water, or 10^{10} times that of glycerine) a sphere of gneiss with a diameter of 2 m would sink about 10 cm/day. A 4-m sphere would sink four times as fast.

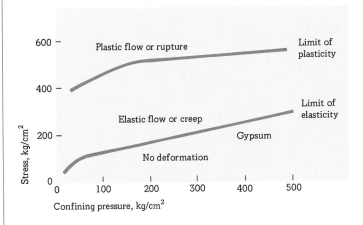

FIGURE 7.14

Diagram of stress versus confining pressure for gypsum. [After J. Goguel, *Mém. carte géol. France*, p. 530, 1943.]

but by amounts much less than calculated.[3] In 1849, in the Pyrenees, F. Petit actually found that the plumb bob appeared to be deflected away from the mountains. Similar discrepancies between calculated and measured values of gravitational attraction of mountains were also observed in the middle of the nineteenth century by British surveyors in India. They were using the plumb line to sight stars, in an attempt to fix the latitude of Kaliana, near Delhi in northern India, and of Kalianpur, about 600 km due south. They observed that the difference in latitude between these two stations was 5°23'37.06". Then they checked the difference directly by standard surveying methods. The difference computed from these measurements was 5°23'42.29". There was a discrepancy of 5.23", or about 150 m, between measurements. That may not seem very much over a distance of 600 km, but it was too large to be explained by errors of observation. Scientists then concluded that the plumb line at Kaliana had been deflected more by the attraction of the Himalaya Range than it had been at Kalianpur farther south. Actually,

[3]Pierre Bouger, *La Figure de la Terre*, p. 391, Paris, 1749.

FIGURE 7.15

Two explanations for why mountains stand high.

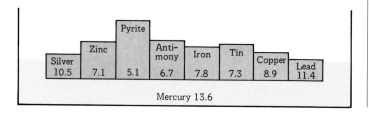

Airy hypothesis

however, the discrepancy should have been three times as large on the assumption that the mountains were of the same average density as the surrounding terrain and that they were resting as a dead load on the earth's crust.

Two quite different explanations were proposed to explain the discrepancies between computed and observed values of gravity. The first was made by G. B. Airy in 1855. He regarded the earth's crust as having the same density everywhere and suggested that differences in elevation resulted from differences in the thickness of the outer layer. Hydrostatic equilibrium was achieved by lighter material's floating in a denser substratum. The depth of compensation is variable. Continents, mountains, and other topographic features are in equilibrium because they have roots like icebergs and are floating in a denser material.

J. H. Pratt, on the other hand, proposed that all portions of the earth's crust have the same total mass above a certain uniform level, called the level of compensation. Any section with an elevation higher than its surroundings would have a proportionately lower density. These ideas of Airy and Pratt are illustrated in Figure 7.15.

In 1889 C. E. Dutton[4] suggested that different portions of the earth's crust should balance out, depending on differences in their volume and specific gravity (see Figure 7.16). He called this **isostasy.** In his words:

If the earth were composed of homogeneous matter its normal figure of equilibrium without strain would be a true spheroid of revolution; but if heterogeneous, if some parts were denser or lighter than others, its normal figure would no longer be spheroidal. Where the lighter matter was accumulated there would be a tendency to bulge, and where the denser matter existed there would be a tendency to flatten or depress the surface. For this condition of equilibrium of figure, to which gravitation tends to reduce a planetary body, irrespective of whether it be homogeneous or not, I propose the name *isostasy* (from the Greek *isostasios*, meaning "in equipoise with"; compare *isos*, equal, and *statikos*, stable). I would have preferred the word *isobary*, but it is preoccupied. We may also use the corresponding adjective *isostatic.* An isostatic earth, composed of homogeneous matter and without rotation, would be truly spherical. . . .

[4]"On Some of the Greater Problems of Physical Geology," *Bull. Phil. Soc. Wash.* vol. 11, p. 51, 1889. Reprinted in *J. Wash. Acad. Sci.*, vol. 15, p. 359, 1925; also in *Bull. Natl. Res. Council (U.S.)*, vol. 78, p. 203, 1931.

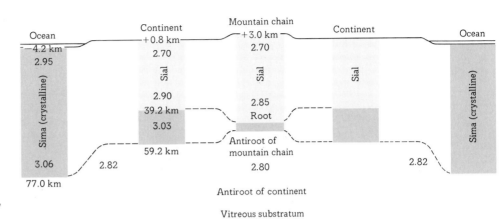

FIGURE 7.16

Hypothesis explaining isostasy. [After R. C. Daly, *Strength and Structure of the Earth*, p. 61, Prentice-Hall, Inc., Englewood Cliffs, N.J., 1940.]

GRAVITY

Gravity is a universal force. On earth it is the force that makes things fall to earth. In fact, it was long thought that this force did not apply to the heavenly bodies because they did not fall to earth. In 1666, when he was 24 years old, Isaac Newton proposed that gravity is a universal force: Every aggregation of matter attracts every other aggregation of matter by a force that is directly proportional to their respective masses and inversely proportional to the square of the distance between them. Newton reasoned that the moon does not fall to earth because its speed in orbit just balances the pull of the earth's and its own gravity. The universal law of gravity can be expressed as

$$F = \frac{mM}{r^2} \, G$$

where F is force, m and M are the masses, r is the distance between them, and G is the universal constant of gravitation. Newton's law provides an excellent summary of the observed motions in our solar system.

On the earth, masses can best be compared by a process called weighing. This actually compares the earth's attraction for one mass with its attraction for another. The **weight** of a body on earth is the force that gravity exerts upon it. This force is proportional to its mass, that is, to the quantity of matter it contains. This can be written

$$\frac{F}{m} = \frac{GM}{R^2}$$

where R is the earth's radius and M is its mass. G is again the universal gravitational constant. The quantities R, G, and M do not change.

Acceleration of gravity   Gravity governs the fall of a body toward the earth and causes it to fall faster during each second. This is known as the acceleration due to gravity, represented by g. Falling freely in a vacuum, a body increases its speed by 980 cm/s during each second of fall. This is written

$$g = 980 \text{ cm/s}^2$$

which is read "g equals 980 centimeters per second per second."

A body falling in the earth's atmosphere does not increase in speed without limit, however. Friction of the atmosphere causes it to reach a terminal velocity and go no faster after that. For a person falling freely this is about 200 km/h.

The force we call weight is mass times the acceleration of gravity:

$$W = mg$$

On the moon the acceleration of gravity is one-sixth of what it is on the earth; so a person would weigh one-sixth as much. But because the moon has no atmosphere, a body falling toward it would not have a terminal velocity as it does on earth.

Gravity anomalies—evidence for inequilibrium   If the earth were a smooth, homogeneous, nonrotating sphere, gravity on its surface would be the same everywhere. It is not the same, however, because of differ-

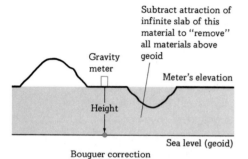

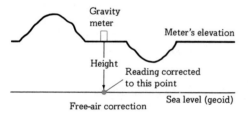

FIGURE 7.17

Correcting gravity-meter readings to a common level of reference at sea level. Because the hill rises above the meter, its attraction is upward and the meter would show a greater reading if the hill were not there. So the hill is "removed" by adding a figure to represent its attraction. In effect the topographic correction reduces the surface to a plane. The Bouguer correction leaves a reading that the meter should show if it were suspended with nothing but free air between it and sea level. The free-air correction leaves the reading at a value that it would have if the meter were at sea level vertically below the point at which it actually stands.

ences in elevation, which change the distance from the earth's center, effects of rotation, which are different at different latitudes, and differences in the density of materials in the earth's crust.

Because $W = mg$ and m for a given body remains the same, changes in g, the acceleration due to gravity, will be reflected in changes in the weight of a body. An instrument for making such measurements is called a *gravity meter*, or *gravimeter*.

An acceleration of 1 cm/s² has been defined as a unit called the *Gal*, after the renaissance physicist Galileo. The acceleration of gravity at the earth's surface is around 980 Gal, but changes as small as a ten-millionth of this may be significant. Consequently, the common, or practical, unit in which gravity differences are expressed is the *milligal* (mGal), or 0.001 Gal.

Measurements of the force of gravity have to be corrected for differences in altitude, latitude, and geological structure. Altitude of the point of measurement is taken into account by subtracting 0.3086 mGal for each meter of height above sea level or adding the same for each meter below sea level. Adjustment of an observed value to what it would be at sea level is known as the *free-air correction*.

The sea-level value of g varies from 978.049 Gal at the equator to 985.221 Gal at the poles because the centrifugal force of the earth's rotation tends to counteract the pull of gravity toward the center. Two kilometers away from the equator, the gravity is reduced by 1 mGal.

Corrections also need to be made for hills and valleys in the vicinity of the measuring point, as indicated in Figure 7.17.

After these factors have been evaluated, a number is obtained that represents the expected value of gravity at the point. If the measured value is different, this difference is called a **gravity anomaly**.

A sphere of salt 600 m in radius, with its center 1,200 m below the surface and surrounded by sedimentary formations about 10 percent more dense, causes gravity to be about 1 mGal less directly above it than at surface points beyond its range of influence. A buried granitic ridge under Kansas, with 1,500 m of relief, has a gravity anomaly of only 2 mGal. The effect of a volcanic pipe 3 km in diameter and of considerable depth may be 30 mGal.

There are significant negative gravity anomalies over ocean deeps, such as the Mindanao Deep off the Philippines, the Nero Deep off Guam, and the deep fronting Java and Sumatra in the East Indies. These have been cited as possibly caused by down-plunging currents involved in ocean-floor spreading (see Chapter 8).

WHY DO WE STILL HAVE MOUNTAINS?

If gravity were the only force acting on the earth's surface, all masses of surface rocks would be standing today at heights governed by their thickness and the ratio of their specific gravity to that of the rocks supporting them. Mountains, however, are great masses of rocks once deeply buried but now standing above adjacent rocks. One of the problems of geology is to explain why such masses stand high after millions and even billions of years of weathering and erosion, and three possible explanations for this phenomenon have been suggested.

One is that the crust underlying mountains is strong enough to support them as dead loads. Laboratory experiments, however, have shown that no rocks are sufficiently strong to support the weight of even comparatively low hills. Consequently we must conclude that the crust beneath mountains is not by itself capable of supporting their dead weight.

Another possible explanation is that mountains retain their elevations because the forces that originally formed them are still active. Certainly in areas where mountain-building forces still make their presence known through earthquakes and changes of level it is quite reasonable to assume that they are contributing to the continued elevation of mountains. In fact it would be difficult to point with assurance to any region on the globe that lacks at least some kind of active internal force. Plate boundaries are regions of especially active mountain building (see further particulars in Chapter 8).

A third explanation for the continued elevation of mountains is that they are isostatic in relation to surrounding portions of the crust. That is, mountains are merely the tops of great masses of rock floating in a substratum, as icebergs float in water (see Figure 7.18a). Such a situation requires a substratum of rock, at not too great a depth, that will flow to adjust itself to an excess load. This rock, however, need not be a liquid in the ordinary sense of the word. It could be rock in a state not unlike that of silicon putty, which can be shaped into a ball that bounces but which under a very slight load—even under its own weight—gradually loses its shape entirely. Hence this rock would be rigid

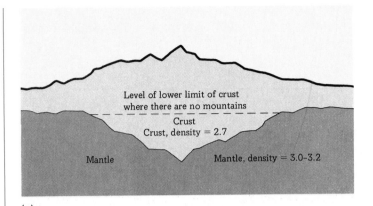

(a)

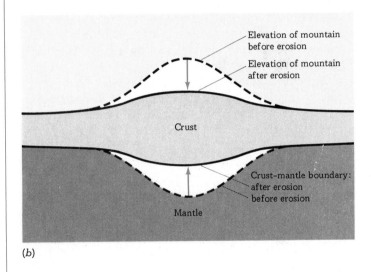

(b)

FIGURE 7.18

Schematic representation of (a) mountain roots and (b) isostasy.

enough for stresses of short duration but would have very little real strength over a long period of time. A small piece of erasing gum nine-tenths as dense as silicone putty would, in time, sink into the putty until it was floating with only one-tenth of its volume above the surface. Likewise, a mountain range with an average specific gravity of 2.7 (that of granite) can sink into a layer of plastic rock with a specific gravity of 3.0 to 3.2 (like the mantle) until the range is floating with a root of about nine-tenths and a mountain of one-tenth its total volume. When some of the top of the mountain is removed by weathering and erosion, the mass will float upward to a new position of equilibrium (Figure 7.18b), where the root will no longer be quite so deep but the same ratio of mass (roughly one-tenth) will form the mountain and nine-tenths of this new mass will serve as the root.

In this way a mountain range can continue to exist by floating isostatic equilibrium after millions of years of erosion.

7.3 STRUCTURAL FEATURES

Structural geology deals primarily with deformed masses of rock, their shapes, and stresses that caused the deformation. In describing structural features, the concept of **relief** is sometimes helpful. This is familiar in connection with the ground surface, where it denotes the difference in elevation between the highest and lowest points in a specified area. For example the topographical relief of California is 4,400 m. This is the difference between the elevation of Mount Whitney and that of Death Valley.

Structural relief is the difference in elevation of parts of a deformed stratigraphic horizon and is a measure of the extent of deformation. Sometimes the shape of a rock mass is determined by drilling and sometimes by hypothetical reconstruction of eroded portions.

Sedimentary rocks and some igneous rocks form in approximately horizontal layers. When these are found tilted, folded, or broken, they provide evidence of deformation of the earth's crust. In some areas erosion has stripped away as much as thousands of meters of uplifted portions of the crust, revealing structures which were once buried deeply. Deformation features include folds, joints, faults, and unconformities.

In describing the attitudes of structural features, geologists have found it convenient to use two special terms: dip and strike. These are more easily described with reference to layered rocks. If a rock layer is not

FIGURE 7.19

Measuring dip with a Brunton compass.

horizontal, the amount of its slope is called its **dip** (see Figure 7.19), measured by specifying the acute angle that the layer makes with the horizontal. The dip is measured in the direction of the greatest amount of inclination. Its **strike** is the course, or bearing, of the outcrop of an inclined bed or structure on a level surface. If the rock layer is tilted so that it disappears below the surface but protrudes somewhat because of resistance to weathering, as in Figure 7.20, strike is the direction in which the resulting ridge runs. A bed that dips either east or west has a north-south strike, usually designated simply as north. A bed that dips either to the north or to the south has an east-west strike.

FOLDS

Folds are a common feature of rock deformation. They are produced by compressive stress (see Figure 7.21). They range in size from microscopic wrinkles in a piece of metamorphic rock to huge folds involving thousands of meters of thickness for distances of hundreds of kilometers. Folds are seldom isolated structures but generally occur in closely related groups. Sometimes they are very broad and sometimes tight and narrow. Sometimes they are tilted to one side and sometimes to one end.

In some areas sedimentary layers are only slightly bent, whereas in others, usually associated with mountain structures, they are intensely deformed. The difference in the kind of folding is mainly dependent on the amount of deforming stress in relation to the strength of the rocks.

Mechanism of folding Folding can be classified as **concentric** or **flow**. *Concentric folding* is basically an elastic bending of an originally horizontal sheet with all internal movements parallel to a *basal plane*, which is the lower boundary of the fold. It occurs in surface layers of the crust. The size of a concentric fold is determined by the thickness of the beds involved and by

(a)

FIGURE 7.20

Dip and strike. (a) Photograph showing outcropping edges of tilted beds in southwestern Colorado, a few kilometers east of Durango. (b) Sketch illustrating terms used to describe the attitude of these beds. The beds strike north and dip 30° east. [U.S. Department of Agriculture, Soil Conservation Service.]

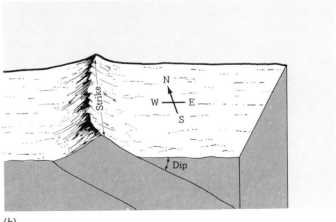

(b)

FIGURE 7.21

Syncline and adjacent anticline in sandstone beds of the Moccasin Formation along the river road to Goodwin's Ferry, Giles County, Virginia. [T. M. Gathright II.]

their elastic properties. Beds are shortened during concentric folding, but the thickness of the beds and the volume remain the same. Large-amplitude folds form from thick strata and small-amplitude folds from thin strata. The distance between crests is also controlled by the thickness of the beds.

Flow is a type of deformation that occurs in rocks when they are in the plastic state. It involves internal movement that may take any direction. It occurs in weak rocks near the surface and in strong rocks at depth. Flow is due to conditions of confining pressures and high temperatures. It is the only true plastic deformation of rocks.

Monoclines Monoclines are relatively simple examples of deformation, involving an elastic bending of sedimentary layers. A **monocline** is a double flexure connecting strata at one level with the same strata at another level. Extensive horizontal layers are bent down and pass beneath younger horizontal strata. The flexed layers did not accumulate with the bend in them. As bending progressed, erosion worked to strip away the higher portion.

There are many monoclines in the Colorado Pla-

teau. Some involve displacements as great as 4,000 m and lengths of 250 km. The Waterpocket fold in southern Utah is one of the better known. It involves displacements of close to 2,000 m. Estimates of the rate of its deformation range from a few centimeters to possibly a few meters per century.

Anticlines and synclines Local arching of layered rocks is the most common form of a structure called an **anticline** (Figures 7.22 to 7.24). The two sides of the fold, called its **limbs**, dip away from each other. The **axis** of an anticline is the direction of an imaginary line drawn on the surface of a single layer parallel to the length of the fold. The axes of most geological folds are inclined. The angle of dip of its axis is the **plunge** of the fold. Parts of a fold are illustrated in Figure 7.25.

Anticlines can be spectacular when their anatomy has been exposed by erosion (see Figure 7.26). The oldest rocks are at the core and the youngest on the outer flanks.

A **syncline** is a downward fold, the opposite of an anticline (Figure 7.27). The limbs dip toward the central axis. Synclines are best seen in mountainous areas where there has been uplift along with the folding. The

FIGURE 7.22

An anticline in the Appenines east of Assisi, Italy. [Sheldon Judson.]

FIGURE 7.23

Broad anticlinal fold in Devonian rocks along the southwestern coast of England (opening approximately 1 m high and 4 m wide). [M. E. Kauffman.]

FIGURE 7.24

Tight anticline in Devonian rocks along the southwestern coast of England. [M. E. Kauffman.]

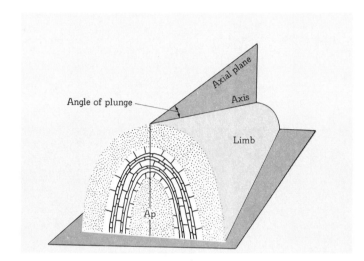

FIGURE 7.25

The parts of a fold, shown on a plunging anticline but applicable to a syncline also. The axis is the line joining the places of sharpest folding. The axial plane (Ap) includes the axis and divides the fold as symmetrically as possible. If the axis is not horizontal, the fold is plunging, and the angle between the axis and the horizontal is the angle of plunge. The sides of a fold are the limbs.

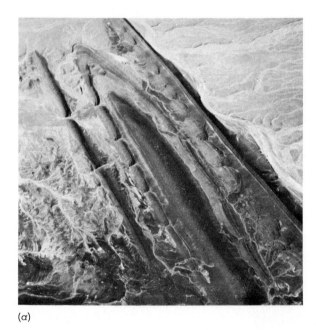

(a)

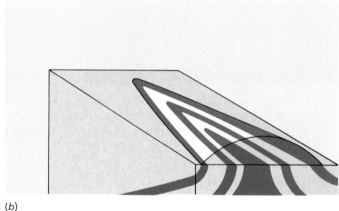

(b)

FIGURE 7.26

(a) Aerial photograph showing resistant beds of an eroded plunging anticline. (b) Sketch illustrating the relationship of outcrops to fold. Western Sahara, Africa. [U.S. Army Air Corps.]

FIGURE 7.27

Tight syncline in Devonian rocks along the south-western coast of England. Thickest bed is 0.5 m thick. [M. E. Kauffman.]

youngest rocks are in the center of the fold, with successively older rocks farther out (Figure 7.28).

Anticlines and synclines commonly occur together in elongate groups like belts of wrinkles in the earth's outer skin. Anticlines and synclines are said to be *symmetrical* if their opposite sides have approximately equal dips, and *asymmetrical* if one limb is steeper than the other. Types of fold are listed in Table 7.2 and illustrated in Figures 7.29 and 7.30.

Figure 7.31 shows synclines, anticlines, and faults underlying a section of the Berkeley Hills and adjacent ranges of the middle coast ranges of California near San Francisco. The Wildcat fault marks the edge of the hills, and Wildcat Creek appears to follow the axis of a syncline. Elsewhere, there is less marked connection between surface forms and subsurface structure except for a general trend of the principal streams along the strike of the structures.

Structural domes Structural domes come in many sizes and may be circular or elongated. Circular domes are formed when the active force is vertical, with horizontal stress equal and constant. Elongated domes are

FIGURE 7.28

View of plunging syncline from top of cliff overlooking tightly folded Devonian rocks along the coast of Wales. See figures on beach for scale. [M. E. Kauffman.]

TABLE 7.2

Types of fold

Name	Description
Anticline	Upfold, or arch
Syncline	Downfold, or trough
Monocline	Local steepening of an otherwise uniform dip
Dome	Anticline roughly as wide as it is long, with dips in all directions from a high point
Basin	Doubly plunging syncline
Asymmetrical	Strata of one fold limb dip more steeply than those of the other
Overturned	Limbs tilted beyond the vertical; both dip in the same direction although perhaps not the same amount
Recumbent anticline	Beds on the lower limb upside down
Recumbent syncline	Upper limb inverted
Isoclinal	Beds on both limbs nearly parallel, whether fold upright, overturned, or recumbent

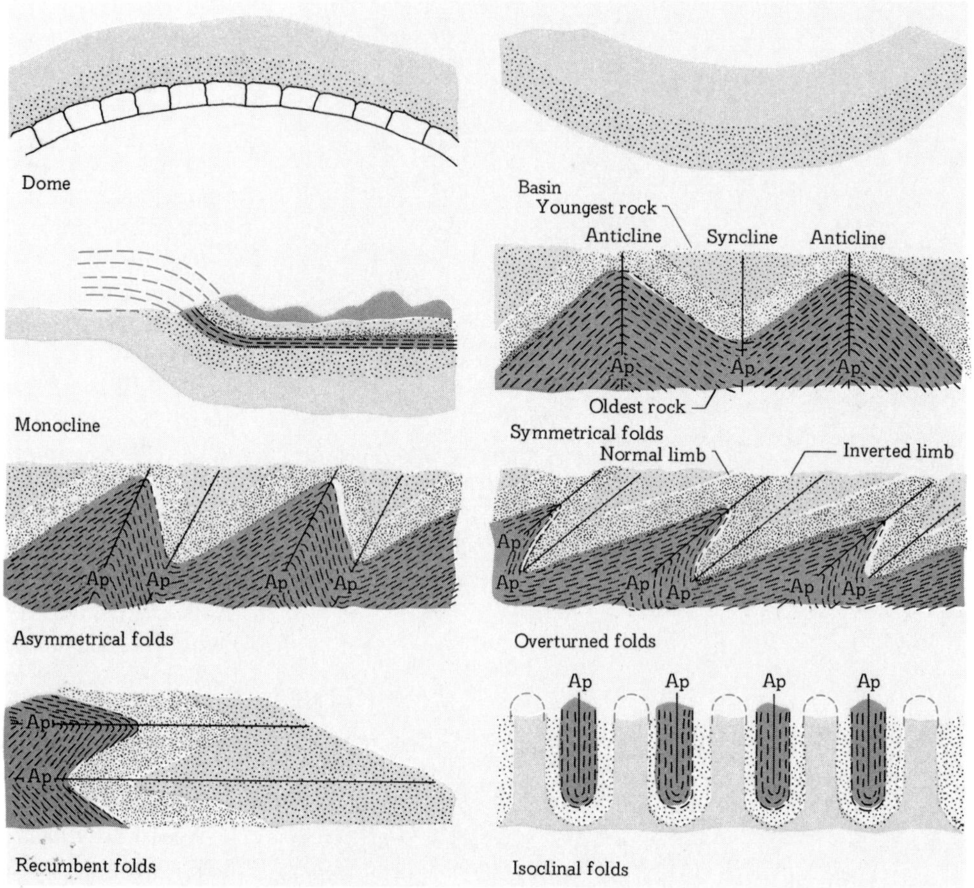

Dome

Basin

Youngest rock

Anticline   Syncline   Anticline

Ap   Ap   Ap

Oldest rock

Monocline

Symmetrical folds

Normal limb   Inverted limb

Ap

Ap   Ap

Ap   Ap   Ap   Ap

Asymmetrical folds

Ap   Ap   Ap   Ap

Overturned folds

Ap   Ap

Ap   Ap   Ap   Ap

Recumbent folds

Isoclinal folds

FIGURE 7.29

Types of fold. (Ap = axial plane.)

FIGURE 7.30

Recumbent fold in Rheems Quarry, Pennsylvania. Rocks are limestones and dolomites of the Beekmantown Formation of Ordovician age.
[M. E. Kauffman.]

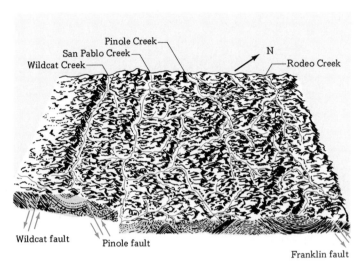

Pinole Creek
San Pablo Creek
Wildcat Creek
N
Rodeo Creek

Wildcat fault   Pinole fault

Franklin fault

FIGURE 7.31

Synclines, anticlines, and faults underlying a section of the Berkeley Hills and adjacent ranges, near San Francisco, of the middle coast ranges of California. The location of the corner of block near Franklin Fault is 122°07′W, 37°57′N.

formed when the horizontal stress is not uniform.

Structural domes are the result of pressure's acting upward from below to produce an uplifted portion of the crust with beds dipping outward on all sides. Where erosion has removed the top of a dome, concentric ridges are formed by one or more relatively resistant beds. The troughs are formed by erosion of weaker rocks. It is possible to project the dip and strike of the beds upward over the dome and reconstruct its whole shape before erosion removed parts.

Some structural domes are formed by the upwelling of plastic material such as salt or magma. It is impossible to know whether slight folding started the upwelling or whether hydrostatic adjustment of loads on the crust was the primary cause. Where there is a local thickening or thinning of the crust, the plastic material starts to rise and continues as a result of the unequal static load. Most domes show marked upward bending of the surrounding beds against the stock, often accompanied by faulting. The beds above are domed and stretched by the push from below and often

exhibit an intricate pattern of normal tension faults. A structural dome produced by an igneous intrusion is illustrated in Figure 7.32.

JOINTS

Masses of rock may be broken, with or without movement along the cracks. If there has been no slippage along the fracture surface, the crack is called a **joint.** (If there has been displacement, the structural feature is called a *fault.*)

Joints are found in all kinds of rock and are the most common feature of rocks exposed at the surface. They usually occur in sets, the spacing between them ranging from just a few centimeters to a few meters. As a rule, the joints in any given set are almost parallel to one another, but the whole set may run in any direction—vertically, horizontally, or at some angle (see Figure 7.33). Most rock masses are traversed by more than one set of joints, often with two sets intersecting at approximately right angles. Such a combination of intersecting joint sets is called a *joint system.* A regional pattern of joint systems may occur over areas of hundreds of square kilometers in a given type of rock exposed at the surface. As a rule, the kinds of rock have a marked influence on what joint systems develop. A massive sandstone and a sandy shale, for example, each show characteristic jointing directions and spacing.

FIGURE 7.32

Section across the Black Hills in South Dakota and Wyoming, illustrating a structural dome produced by the intrusion of an igneous mass after deposition of sediments. [After N. H. Darton.]

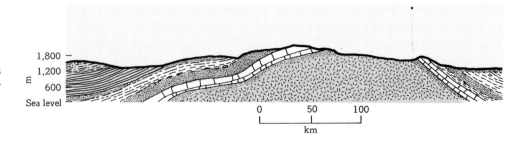

1,800
1,200
600
Sea level

0   50   100

km

FIGURE 7.33

The sawtooth shape of the American Falls of Niagara Falls results from the prevailing joint pattern. (Photographed during unwatering of June, 1960, for Niagara power project.) [State of New York Power Authority.]

FIGURE 7.34

Basalt sill and columnar jointing on the north shore of Snake River, Washington. Note the abrupt changes in column orientation. [U.S. Corps of Engineers, courtesy of Harlan E. Moore.]

FIGURE 7.34

Basalt sill and columnar jointing on the north shore of Snake River, Washington. Note the abrupt changes in column orientation. [U.S. Corps of Engineers, courtesy of Harlan E. Moore.]

It is rarely possible to determine a joint set's age or origin. It is generally accepted, however, that different sets in a joint system were probably made at different times and under different conditions. In sedimentary rocks one set could develop while the rock was being compressed by the weight of overlying rocks during burial and consolidation. Another set could be produced when pressure is released during unloading by erosion. In other cases sets may be developed during deformation by tension. Some joint sets are known to be formed during the cooling of a mass of igneous rock, and others during movements of magma.

Columnar jointing Columnar jointing is a special pattern of jointing found in some masses of basalt. It consists of sets of cracks produced by the mechanism of cooling. In certain lava flows and shallow intrusive bodies joints are formed as the rock shrinks in cooling. Dikes or sills cool by fairly uniform loss of heat over the entire surface that is in contact with cold wall rock. This allows shrinkage about many equally spaced centers, not unlike that which produces mud cracks. Such contraction means tension between the centers, which results in a system of short straight cracks separating each center of contraction from its neighbors. These form a polygonal pattern, ideally hexagonal but sometimes four- or five-sided because of irregularities in cooling. As cooling progresses into the body, contraction advances with it, and each column may grow until it reaches the opposite side.

Columnar jointing is not clearly developed in all masses of basalt. It seems to be most characteristic of tabular masses, where columns form across the narrow dimension of the mass. The columns are often broken into sections of varying lengths by transverse joints (see Figure 7.34).

One of the best-known areas in the world where columnar jointing can be seen is the Giant's Causeway, near Portrush, Antrim, Northern Ireland. It is also well developed in Devil's Post Pile National Monument, Sierra Nevada, California, along the Snake River in Washington, and in Devil's Tower, Wyoming.

Sheeting A pattern of essentially horizontal joints

FIGURE 7.35

Sheeting shown in granite. [N. Dale.]

is called **sheeting** (see Figure 7.35). Here the joints occur fairly close together near the surface, but less and less frequently the deeper we follow them down until they seem to disappear altogether a few tens of meters below the surface. But even at considerable depth the rock shows a tendency to break along surfaces parallel to the surfaces of sheeting above. This type of jointing is especially common in masses of granite, and engineers often put it to good use in planning blasting operations.

Joints formed during movements of magma   Some joint systems are formed during movements of magma within the crust. Moving magma exerts pressure on adjacent rocks, developing cracks. Magma is then injected into the cracks as they open, enlarging them. This produces such structural features as dikes and sills, which are thought to form before the magma

reaches the surface and is extruded. After extrusion begins, pressure is reduced. Because extrusion takes less pressure than intrusion, the magma wells out instead of intruding into adjacent rocks.

Dikes are exposed at the earth's surface after softer rocks have been eroded away. They vary in width from a few centimeters to kilometers across and may be many kilometers in length. The largest dike known is in Rhodesia. It is 480 km long and up to 8 km wide.

Dikes occur in different patterns: ring dikes, cone sheets, and swarms. **Ring dikes** and **cone sheets** consist of a concentric arrangement of dikes formed by a definite center of stress. They occupy fractures around a center where magma has pushed up a section of the earth's crust. Dikes are filled by upwelling magma (see Figure 7.36). Ring dikes have diameters ranging from 2 to 25 km, with individual dikes sometimes as much as

FIGURE 7.36

Ring dikes and cone sheets. [After E. M. Anderson, *Proc. Roy. Soc. Edinburgh,* vol. 56, pp. 128–157, 1936.]

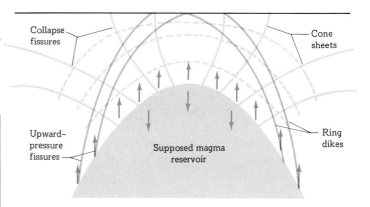

500 to 1,000 m wide. Ring dikes dip away from the magmatic body that cracked the crust to provide avenues of weakness along which they moved into place. Cone sheets outline inverted cones with sides dipping toward the magmatic body. From their angle of dip, it is possible to estimate the depth below the surface to the body of magma which supplied them.

In some places, dikes occur in roughly parallel groups called **dike swarms.** Some dikes in a swarm are up to 400 km in length. Dike swarms differ from ring dikes in their pattern which suggests that they originate in a different kind of stress field.

Some dike swarms are caused by an elastic release of pressure. This has been documented in several places in Scotland. The Caledonian orogenic belt contains dike swarms running parallel to the main folding axis, which trends northeast-southwest (see Figure 7.37). The trend of the dikes suggests that they were formed by an elastic release after compression. Dike swarms are also associated with strike-slip faults in Scotland. These include the Great Glen fault, the Highland Boundary fault, and the Southern Upland fault, which trend northeast-southwest. Their direction suggests that they were formed by a north-south compression. The dike swarms trend east-west, perpendicular to the compressive stress. This again suggests that the cracks in the earth's crust resulted from an elastic release of compression.

Some dike swarms are associated with the folding process but do not involve an elastic release of compression. They are the result of tension. Such a case has been reported in the bend of a monocline on the east coast of Greenland.[5] The monocline consists of a sheet of basalt blanketing metamorphic rocks stretching many hundreds of kilometers and varying in dip. The dike swarm is located on the convex side of the flexure and is absent in the concave part (see Figure 7.38). Where the flexure dips 55°, there are more than 70 dikes per kilometer across the structure. Where the dip is 12°, there are only about 15 dikes per kilometer; and where the dip is 7°, there are only a few dikes.

[5]L. R. Wager and W. A. Deer, "A Dike Swarm and Crustal Flexure in East Greenland," *Geol. Mag.,* vol. 75, pp. 39–46, 1938.

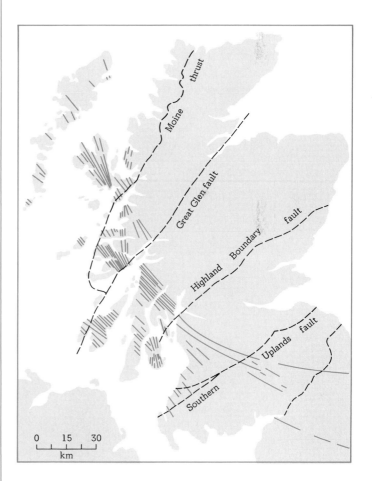

FIGURE 7.37

Tertiary dike swarms in Scotland. [After J. E. Richey, *Trans. Geol. Soc. Edinburgh,* vol. 13, p. 393, 1939.]

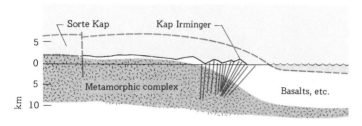

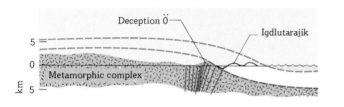

FIGURE 7.38

Dike swarms on the eastern Greenland coast.
[After L. R. Wager, *Med. om Grønland*, vol. 134, no. 5, pp. 1–62, 1947.]

The dikes dip at about right angles to the basalt sheet, and their position indicates that they fill fissures developed by tension as the crust bent. The monocline has a maximum relief of at least 8 km.

FAULTS

Faults are deformation by rupture in which the sections on each side of the break move relative to each other. Theoretically they may be of any size, but here we shall discuss only the types that are large shearing surfaces in the earth's crust.

A useful basis for the classification of faults is the nature of the relative movement of the rock masses on opposite sides of the fault. If displacement is in the direction of dip, the fault is a **dip-slip fault**. The sections separated by the fault are named as they were by miners who encountered faults underground. The block that is overhead is called the **hanging wall,** and the one beneath is the **foot wall.** Dip-slip faults are classified according to the relative movement of these blocks. If the hanging wall seems to have moved downward in relation to the foot wall, the fault is called **normal.**

Normal faults are a result of tensional stresses due to crustal movements that cause the surface to stretch and rupture. A **graben** is an elongated, trenchlike structural form bounded by parallel normal faults created when the block that forms the trench floor has moved downward relative to the blocks that form the sides. The term comes from the German for "trough" or "ditch" and is the same in both singular and plural. A **horst** is an elongated block bounded by parallel normal faults in such a way that it stands above the blocks on both sides. The term comes from the German and is used figuratively for a crag or height. The type of movement involved in formation of a graben and a horst is illustrated in Figure 7.39.

A dip-slip fault in which the hanging wall appears to have moved upward in relation to the foot wall is a **thrust fault** (or **reverse fault**). It is the result of largely horizontal compressive stresses.

A fault along which the movement has been predominantly horizontal is called a **strike-slip fault** be-

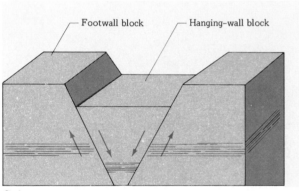

Graben

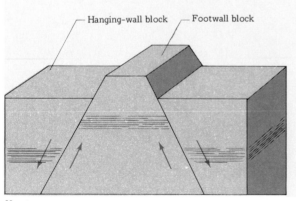

Horst

FIGURE 7.39

A graben and a horst.

cause the slipping has been parallel to the strike of the fault. Such faults have also been called **transcurrent faults.** Strike-slip faults are further designated as *right-*
lateral or *left-lateral*, depending on how the ground opposite you appears to have moved when you stand facing the fault. Types of fault are listed in Table 7.3 and are shown in Figure 7.40.

The directions of movement involved in faulting are entirely relative. It is convenient to indicate one block as having moved upward or downward or as moving left or right. However, the absolute direction of movement usually cannot be determined, and the best that can be done is to indicate relative movements. During the upward warping of a large region, for example, all blocks could have moved upward, but some may have moved less than others. These lagging blocks could be said to have dropped relative to their neighbors even though all parts at the finish were at higher elevations than when they started.

Normal faults In the earth's crust there are large zones that have repeatedly been disturbed by large-

TABLE 7.3
Types of fault

Type	Predominant relative movement
Dip-slip	Parallel to the dip
Normal	Hanging wall down
Thrust	Hanging wall up
Strike-slip (transcurrent)	Parallel to the strike
Right-lateral	Offset to the right
Left-lateral	Offset to the left
Oblique-slip	Components along both strike and dip
Hinge	Displacement dies out perceptibly along strike and ends at definite point

FIGURE 7.40

Types of fault.

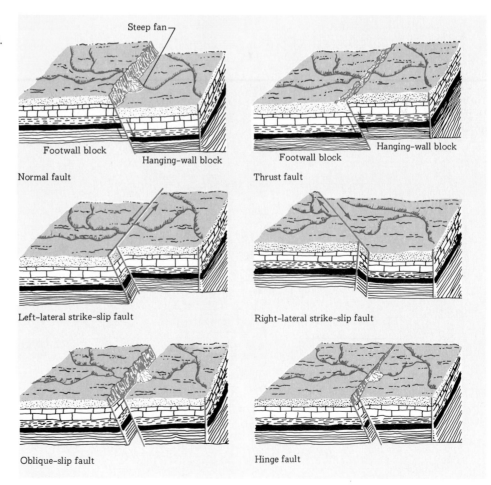

Normal fault

Thrust fault

Left-lateral strike–slip fault

Right-lateral strike–slip fault

Oblique–slip fault

Hinge fault

scale normal faulting. The faults form horsts and rift valleys. The most famous, the African rift zone, extends over 6,000 km in a north-south direction. In some areas volcanoes are found along the zone where magma has worked its way up along fault surfaces. In Europe there are several zones of normal faults, including the rift zone of the Upper Rhine valley. In the western part of the United States we find an extensive zone of normal faulting in the Basin and Range province, Innumerable smaller normal fault zones can be found all over the world.

The Basin and Range province consists of tilted fault blocks. Its eastern limit is the Wasatch fault. The main faults in the vicinity of Wasatch Mountain dip 50° to 55°. Net slip on major fault planes in the Wasatch Range is from 2,200 to 2,600 m. Farther west are the faulted Oquirrh Range, the Stansbury Range, and the Cedar Range. Each is a tilted fault block some 30 km in breadth. The western limit is the great normal fault that limits the tilted Sierra Nevada block on the east.

Thrust faults Thrust faults are usually closely associated with the folding process. Big, steep thrust faults are characteristic of the marginal stress of mountain chains and in their uplifted central blocks of crystalline rocks. Rocks above the fault plane seem to have been pushed up and over those below. Along the fault plane rocks are considerably crushed and sheared. Thrust faulting is more clearly seen in eroded regions where metamorphic rocks that were once deeply buried now overlie sedimentary rocks that were formed near the surface. Old rocks are generally thrust over younger ones.

The total movement along a thrust fault can be only estimated in most cases, but some involve total vertical displacements as much as 1,000 m, combined with horizontal movements of tens of kilometers. The dips of thrust faults vary from less than 10° up to 60°. *Overthrusts* are far-travelled, low-angle thrusts.

Some well-known regions of thrust faulting are in the Alps, the southern Appalachians, the central and northern Rocky Mountains, and southern Nevada. Besides thrust faults of considerable displacement, every deformed region contains numerous small thrust faults (Figure 7.41). These may have displacements as small

FIGURE 7.41

Thrust fault on the Pan American Highway 40 km northwest of San Salvador. [Thomas F. Thompson.]

Hanging Wall

Footwall

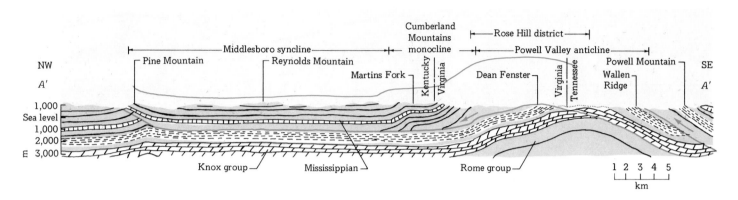

FIGURE 7.42

Cumberland overthrust, Valley and Ridge province of the Appalachians. [After
R. L. Miller and J. O. Fuller, "Geology and Oil Resources of the Rose Hill District,"
Bull. Va. Geol. Surv., vol. 71, p. 383, 1955.]

as 2 m but do not usually appear on geologic maps unless they occur in a mining area.

Thrust faults are often associated with asymmetric anticlines where rupture occurs in the highest part of the fold. A thrust fault of this type is a superficial structure because it extends down only to the basal plane and occurs in brittle rocks near the earth's surface. Some are found in the Valley and Ridge province of the Appalachians, with the best known being the Cumberland thrust in Virginia (Figure 7.42).

Strike-slip (transcurrent) faults Strike-slip faults are found all over the world. A well-known example of left-lateral strike-slip faulting is the Great Glen fault, which intersects Scotland from coast to coast. There is a string of lakes, including Loch Ness, along its eroded trace, which is marked by a belt of crushed, sheared rock up to 1.5 km in width. Horizontal displacements along the fault of as much as 100 km have been suggested on the basis of correlating geological structures on both sides. It dates from the Late Paleozoic.

Several left-lateral strike-slip faults rip across folds of the Jura Mountains, not far from Lake Geneva and Lake Neuchâtel. They clearly originated during the folding process.

The San Andreas fault in California is a right-lateral strike-slip fault that has been traced for nearly 1,000 km. Terrace deposits cut by the fault show offsets of as much as 10 km, and stream channels show shifts of 25 km within relatively short geological intervals. The total strike slip has been debated. Evidence has been interpreted as showing a displacement of 16 km since the Pleistocene, of 370 km since late Eocene time, and 580 km since pre-Cretaceous time. The Garlock fault separates two clusters of dike swarms that seem pretty clearly to have been intruded as a single event. They are now separated by 80 km in a sense that makes the Garlock a left-lateral strike-slip fault.

Transform faults Another basis for the classification of faults is their relationship to other structural features.[6] It has been suggested that deformation of the earth's crust produces major structural forms: mountains, deep sea trenches, midocean ridges, and strike-slip faults with large horizontal displacements. J. Tuzo Wilson proposed that these features are interrelated, that they do not come to dead ends but link in continuous networks girdling the globe, and that any major feature at its apparent termination may change into one of the other types. A junction where this takes place is called a *transform*. It has been proposed that the name **transform fault** be applied to a class of faults connecting other major structural features. Figure 7.43 suggests the relationships that exist along a transform fault connecting two segments of a fragmented oceanic ridge and compares them with the relationships along a transversely faulted ridge. Analysis of earthquakes along the faults connecting segments of ridges indicates the motion to be as shown in Figure 7.43*a*. This is

[6]J. Tuzo Wilson, "A New Class of Faults and Their Bearing on Continental Drift," *Nature*, vol. 207, pp. 343–347, 1965.

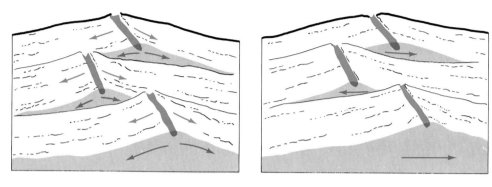

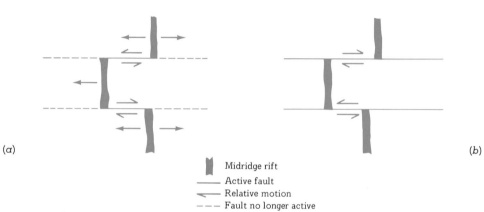

FIGURE 7.43

Comparison between transform and transcurrent faults. The perspective sketch and plan in (a) show a ridge-to-ridge transform fault. New crustal material is postulated as being continuously added at the crest and transferred laterally away from the crest. There is no net displacement of the forms of the ridge segments with time. Transcurrent faults are shown in perspective and plan in (b). The ridge fragments are increasingly offset with time.

(a)

(b)

◢ Midridge rift
── Active fault
← Relative motion
-- - Fault no longer active

FIGURE 7.44

Equatorial section of the Midatlantic Ridge showing ridge crest and fracture zones. [After L. R. Sykes, "Seismological Evidence for Transform Faults, Sea Floor Spreading and Continental Drift," in R. A. Phinney (ed.), *The History of the Earth's Crust*, Princeton University Press, Princeton, 1968. © 1968 by Princeton University Press. Reprinted by permission.]

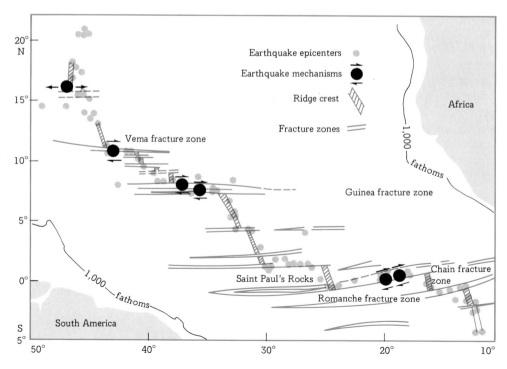

exactly opposite to that which would result from a simple offsetting of the ridge by transcurrent faults, as shown in Figure 7.43b. The original offset of the ridges along transform faults may well have been due to transcurrent faulting. But with the establishment of transform motion the ridge crests do not change position with time. In the case of continued transcurrent faulting the offset increases with time. Many of the faults across the ocean ridges have been interpreted as transform faults: for example, those of the equatorial Midatlantic Ridge shown in Figure 7.44.

UNCONFORMITIES

There is no known place on earth where sedimentation has been continuous throughout geologic time. During the formation of a continent large sections of the crust have been raised out of the shallow seas in which its rocks were formed, subjected to erosion, and then lowered again to levels where deposition of sediments is renewed. Activity of this sort produces a buried erosion surface with younger rocks overlying older rocks. Some surfaces of this kind can be seen today because of another cycle of uplift. A buried erosion surface separating two rock masses of which the older was exposed to erosion for a long interval of time before deposition of the younger is called an **unconformity.**

The time represented by an unconformity is important geologic evidence of the history of a region, marking an interval when the surface was above the sea and sediments were not being deposited. Some unconformities represent gaps of a few thousand years; others, as many as 400 million years.

There are three principal types of unconformity: angular unconformity, disconformity, and nonconformity.

Angular unconformity An unconformity in which the older strata dip at an angle different from that of the younger strata is called an **angular unconformity.** On a wall of Box Canyon, near Ouray, Colorado, can be seen some layers of Precambrian sedimentary rocks dipping at nearly 90°. Above them are some nearly horizontal Devonian sedimentary rocks. The older rocks were deposited under the waters of the ocean and were then folded and uplifted above the water. While exposed at the surface, the tilted beds were beveled by erosion. Then, as time passed, these tilted and eroded rocks sank again beneath the ocean, where they were covered by new layers of sediments. Both were later elevated and exposed to view (see Figure 7.45). On the

FIGURE 7.45

A striking unconformity between Precambrian sedimentary rocks that were twisted from their original horizontal position before deposition of overlying Devonian beds. In Box Canyon, near Ouray, Colorado. [Kirtley F. Mather.]

basis of fossil evidence we know that the second sedimentation began during the Devonian, indicating that more than 180 million years elapsed between the two periods of sedimentation. The Precambrian rocks meet the younger Devonian rocks at an angle; so this unconformity is an angular unconformity.

Disconformity　An unconformity with parallel strata on opposite sides is called a **disconformity.** It is formed when layered rocks are elevated, exposed to erosion, and then lowered to undergo further deposition without being folded. Careful study and long experience are required to recognize a disconformity because the younger beds are parallel to the older ones. Geologists rely heavily on fossils to correlate the times of deposition of beds above and below a disconformity. Fossils are a useful tool because certain ones lived during a definite geologic time range.

Nonconformity　An unconformity between profoundly different rocks is called a **nonconformity.** Nonconformities are formed where intrusive igneous rocks or metamorphic rocks are exposed to erosion and then downwarped to be covered by sedimentary rocks. A structure of this sort is illustrated in Figure 7.46, which shows a sandstone deposit lying on top of an eroded surface of granite. Field studies have revealed that pieces of weathered granite occur in the bottom layers of the sandstone and that cracks in the granite are filled with sandstone. This evidence supports the view that the granite did not come into its present position as an intrusion after the sandstone was formed.

Unconformities in the Grand Canyon　The Grand Canyon of the Colorado exposes several unconformities (see Figure 7.47). This region has undergone three major sequences of uplift and erosion, subsidence and deposition, and crustal deformation and crustal stabil-

ity. These have produced three major rock units: Vishnu Schist, Grand Canyon Series, and Paleozoic Series.

The Vishnu Schist is the base rock of the canyon. It is composed of highly metamorphosed sedimentary rock and some volcanics. It has yielded no fossils and may be 2 billion years old.

Resting unconformably on the Vishnu Schist in some places are tilted beds of the Precambrian Grand Canyon Series. The beds consist of 75 m of Bass Limestone, 180 to 240 m of sandy shale, and 300 m of relatively resistant beds of Shinumo Sandstone, topped off by 600 m of shaly sandstone. It has been estimated that an additional 3,000 m of Grand Canyon sediments were deposited on the beds before the region was elevated, deformed, and then eroded to the surface on which Paleozoic sediments were deposited. An uplift of at least 3,600 m had to take place to bring the lowest Grand Canyon layers above sea level. Erosion then removed all but the lower sections of some tilted blocks. It is not possible at the present time to say how long it took to complete the downwarping of the Vishnu Schist, deposition, sedimentation, deformation, elevation, and erosion.

When the Grand Canyon Series was eroded to a surface of low relief and then submerged, Paleozoic sediments were deposited. These sediments are found today in horizontal layers along the upper part of the canyon walls, indicating uplifting of the region without deformation. The Paleozoic Series contain at least four discontinuities. The longest gap is between the Muav limestones and the Temple Butte Formation. The Muav limestones were deposited in Cambrian time. The Temple Butte Formation immediately overlying these and essentially parallel to them contains fossils of primitive armored fish that occur only in Devonian rocks. Ap-

FIGURE 7.46

A nonconformity between light-colored granite and dark overlying Table Mountain sandstone on the Cape of Good Hope, South Africa. The cave, about 6 m above sea level, was cut by waves when sea level was higher than it is now. [R. A. Daly.]

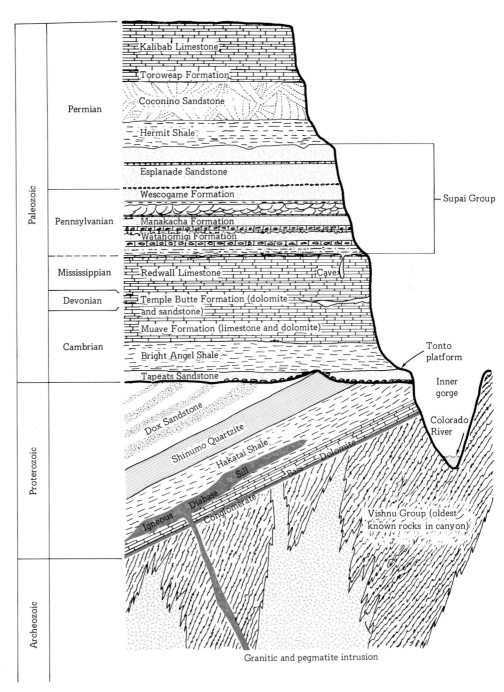

FIGURE 7.47

Diagrammatic representation of one wall of the Grand Canyon along the Kaibab Trail, showing unconformities. [After W. J. Breed and E. Roat (eds.), *Geology of the Grand Canyon*, Museum of Northern Arizona-Grand Canyon Natural History Association, Flagstaff and Grand Canyon, Ariz., 1976.]

parently, there was no deposition during Ordovician or Silurian time, a lapse of some 80 million years.

In areas where all the Grand Canyon Series had been eroded away before deposition of Paleozoic sediments, we find the "great unconformity." Its time gap is estimated as several hundred million years.

BROAD SURFACE FEATURES

Every continent has a nucleus of Precambrian rocks. These are called **shields**, and the continents have apparently grown around them. The oldest rocks were formed about 3 billion years ago along chains of volca-

FIGURE 7.48

Broad surface features of North America.

noes. By about 2 billion years ago platforms of the earliest rocks had developed on which sediments accumulated. The full range of mountain-building cycles has been operating since then. Deformation has produced the continental land features we find today—the plains, the plateaus, and the mountains.

North America's shield, the Canadian shield (Figure 7.48), is exposed over an appreciable part of the northeastern portion of the continent. It covers more than 5 million km² and includes all of Labrador, much of Quebec, Ontario, and the Northwest Territories, the northeastern parts of Manitoba and Saskatchewan, and part of the Arctic islands. Exposure at the surface could not have taken place without elevation—in this case brought about by a broad warping of the basement complex.

Bordering the shield are the interior lowlands, extending to eastern Ohio on the east, the Ouachita Mountains on the south, and the Arctic Ocean on the

north. They are composed of sedimentary rocks, the oldest having been deposited in seas lapping up onto the shield during Cambrian time. On 14 different occasions portions of this continental lowland sank beneath the ocean and later rose above the water. So it was a restless area during this building of the continent, and the rocks bear the traces of unsettled times. Gently tilted in places, folded into small domes and basins in others, their thickness varies to more than 3,000 m in areas of exceptional subsidence and deposition; yet they lie relatively low and flat and form plains.

The sedimentary rocks surrounding the interior lowland plains have been broadly upwarped into plateaus although the rocks still lie nearly flat. Still farther from the shield the plateaus merge into mountain ranges, which represent deformation at its greatest. As much as 15,000 m of sediments accumulated here through the formation of basining in the basement complex; they then consolidated into sedimentary rock.

Intense folding, rupture, and elevation of the entire region resulted in mountain ranges such as the Appalachians and the Rockies.

Plains are regions of low relief composed of nearly horizontal sedimentary rocks. Plateaus are higher regions of nearly horizontal strata with a relief generally over 150 m. Plateaus may originate by uplift or by deposition of lava flows. Mountains encompass a wide range of geological structures and relief, including fault-block mountains, thrust mountains, and mountain ranges. Youthful mountains stand today thousands of meters high, whereas the older mountain chains that have been attacked by many more years of erosion have low relief.

Broad warps Some large sections of the continents are composed of horizontal strata that have been gently bent upward or downward. In North America these strata are found in the lowland between the Rocky Mountains and the Appalachians. They are distinguished only by mapping the elevation of certain rock layers. Data for the mapping were obtained from many field studies or from the borings obtained from thousands of drill holes. Regional downwarping forms basins; upwarping forms arches or plateaus.

The Great Basin, 470,000 km^2 between the mountains, is centered in Nevada and includes adjacent parts of California, Oregon, and Utah. Its relief is close to 3,900 m.

OUTLINE

Evidence of deformation of the earth's crust is best preserved in sedimentary and metamorphic rocks.

    Abrupt movements of the earth's crust have occurred at the times of some large earthquakes.

        Vertical displacements have been such as to amount to 15,000 m if continued at the same rate for 2 million years.

        Horizontal displacements have been estimated as great as 200 km in 150 million years.

    Slow movements are being measured instrumentally in several places today, and on the San Andreas fault near Parkfield they average 5 cm/year.

Kinds of deformation are described in terms of change of volume, change of shape, or a combination of both.

    Stress-strain relationships are used to define a material's response to stress.

        Elastic deformation is recoverable; plastic deformation is not.

        Plastic deformation is permanent and involves a property of rock called **viscosity.**

        Time factor sometimes results in a material's being elastic for a short time but plastic over a longer time.

        Types of stress are tension, compression, and shear.

    Strength of rocks is the stress at which they begin to be permanently deformed.

    Gravitational attraction of mountains can be computed and measured; **isostasy** is the principle that different portions of the earth's crust should be in balance.

    Gravity is a universal force by which every aggregation of matter attracts every other aggregation of matter.

        Acceleration of gravity is the amount by which a freely falling body increases its speed of fall each second.

        Gravity anomalies are differences between expected and measured values of gravity.

    Mountain elevation may be maintained because the forces that originally formed mountains are still active or because they are isostatic.

Structural features are the shapes of deformed masses of rock, including folds, joints, faults, and unconformities.

    Folds are a common feature of rock deformation and are produced by stresses.

        Mechanism of folding is elastic bending or plastic flow.

        Monoclines are double flexures.

Anticlines and **synclines** occur together in elongate groups like belts of wrinkles in the earth's outer skin.

Structural domes result from pressures acting upward.

Joints usually occur in sets.

Columnar jointing is sets of cracks produced by the mechanism of cooling certain tabular plutons to outline columns.

Sheeting is a pattern of essentially horizontal joints.

Joints formed during movements of magma include dikes and sills.

Faults are deformation by rupture in which the sections on each side move relative to each other.

Normal faults have the hanging wall apparently dropped relative to the foot wall.

Thrust faults have the hanging wall apparently moving higher than the foot wall.

Strike-slip faults have movement predominantly horizontal, parallel to the strike of the fault.

Transform faults are strike-slip faults that terminate in a mountain range, midocean ridge, or deep sea trench.

Unconformities are buried erosion surfaces separating rock masses where the older was exposed to erosion for a long interval of time before deposition of the younger.

Angular unconformity is one in which the older strata dip at an angle different from that of the younger.

Disconformity is an unconformity with parallel strata on opposite sides.

Nonconformity is an unconformity between profoundly different rocks.

Unconformities in the Grand Canyon show three major sequences of uplift and erosion, subsidence and deposition, and crustal deformation and crustal stability.

Broad surface features are basins, plains, plateaus, mountains, and Precambrian shields with continents built around them.

Broad warps have produced basins and arches or plateaus.

SUPPLEMENTARY READINGS

Billings, Marland P. *Structural Geology,* 3rd ed., Prentice-Hall, Inc., Englewood Cliffs, N.J., 1972. *A textbook designed to follow an introductory course in physical geology.*

Dennis, J. G. *Structural Geology,* The Ronald Press Company, New York, 1972. *A very understandably written introduction to the subject.*

De Sitter, L. U. *Structural Geology,* 2nd ed., McGraw-Hill Book Company, New York, 1964. *A textbook presupposing a certain familiarity with the elements of structural geology and its terminology.*

Hobbs, Bruce E., W. D. Means, and P. F. Williams *An Outline of Structural Geology,* John Wiley & Sons, Inc., New York, 1976. *A textbook also assuming some background in basic concepts of structural geology; covers most modern concepts in the field.*

Umbgrove, J. H. F. *The Pulse of the Earth,* 2nd ed., Martinus Nijhoff, The Hague, 1947. *A treatise that must be rated a classic for its impact on the broad philosophical aspects of structural geology.*

8

CONTINENTS, OCEANS, PLATES, AND DRIFT

Recent observations from drill-hole depths of several kilometers inside the earth and measurements from satellites hundreds to thousands of kilometers above the earth permit us to gain a unique view of this planet and its crust. This picture was only indirectly observable prior to the last decade. A major result of the data obtained has been the development of a central theme called the "new global geology," relating paleomagnetism, sea-floor spreading, plate tectonics, and continental drift.

Building of mountains, erosion of continents, and shifting shorelines of the seas are changes of the earth that can be documented and visualized without overtaxing our credulity. But now we must consider change on a different scale, challenging not only our sense of direction but our concept of geographic permanency

as well. We wish to investigate whether the north and south poles have shifted position through time—and if so, why? We shall examine the question of **continental drift**: whether or not our landmasses have been rent from their moorings and have wandered over the globe's surface. We shall investigate possible mechanisms for continental drift via the opening up of ocean basins through the process of **sea-floor spreading**. Some attempts will be made to determine the movement rates, beginnings of most recent movements, and the implications of such global tectonics. This will lead to a consideration of **plate tectonics**, which pictures the earth's crust as a jigsaw puzzle of rigid plates jostling one another, growing in some places and decaying in others.

8.1 ORIGIN AND PERMANENCE OF CONTINENTS AND OCEAN BASINS

Questions of primary importance to geology are when and how the continents and ocean basins formed. Were they always distinct and separate entities, or have they only recently developed their characteristic identities?

The oldest rocks on earth have been dated by rubidium-strontium methods as 3.8 billion years old (see Chapter 18). These ages came from highly metamorphosed rocks from the Baltic Shield, Greenland,

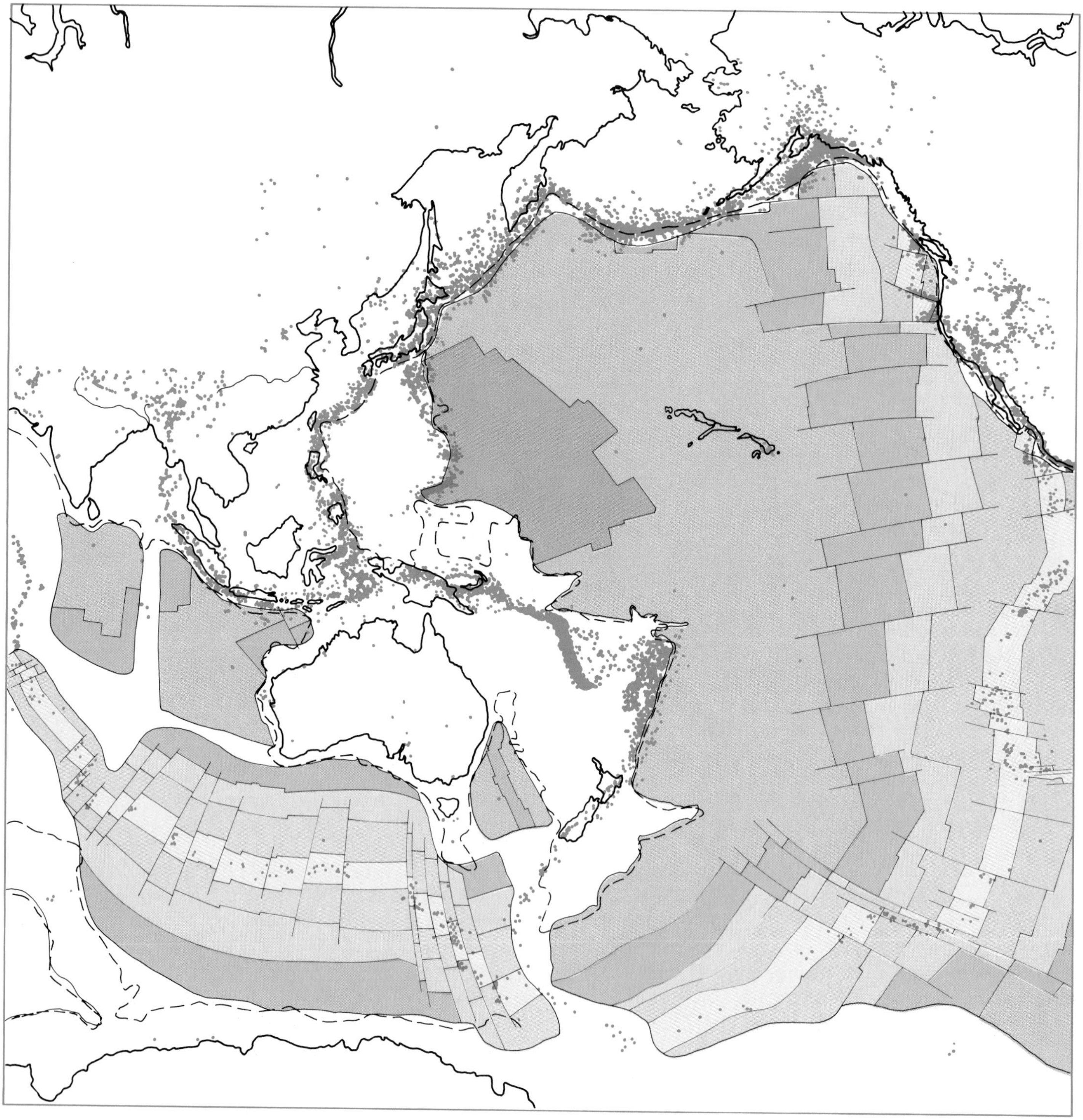

FIGURE 8.1

The age of the ocean floor is shown as bands of different colors, with the youngest
ocean floor near midoceanic spreading centers and the oldest farthest away. The
edge of the deep ocean floor is marked by the dashed line. [Modified from Sheldon

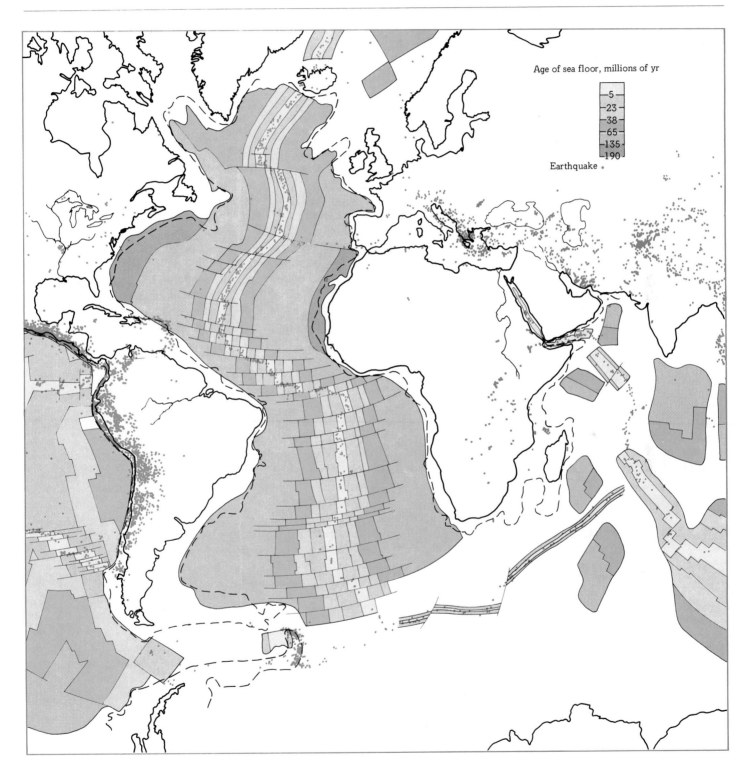

Age of sea floor, millions of yr

−5
−23
−38
−65
−135
−190

Earthquake

Judson, Kenneth S. Deffeyes, and Robert B. Hargraves, *Physical Geology*, Prentice-Hall, Inc., Englewood Cliffs, N.J., 1976; and from the map by W. C. Pitmann III, R. L. Larson, and E. M. Herron, "Age of the Ocean Basins Determined from Magnetic Anomaly Lineations," The Geological Society of America, Inc., Boulder, 1974.]

and Minnesota. Interpretation of the relationships from rocks dated by uranium-lead methods in different parts of the earth suggest that at least some crust was in existence about 4.0 billion years ago—which compares favorably with recent age dating of lunar crustal samples in the 4.0- to 4.1-billion-year range.

True crustal material can be distinguished from mantle material by its magmatic origin. The earliest crust must have crystallized from magma formed by the melting of a portion of the mantle; but how the crust differentiated into continental and oceanic portions is still unsettled. The various processes that are candidates for explaining differentiation need not concern us here. The theories about them, however, group into those assuming that differentiation occurred very early in the earth's history in a single, catastrophic event and those that posit a continuing phenomenon over a long period of geologic time, perhaps ending at a given stage in the earth's development.

CONSTANT FREEBOARD OF CONTINENTS

It is possible to conclude from the rocks in the earth's geologic column whether they developed under continental or oceanic (marine) conditions. Maps of such development were examined by D. U. Wise,[1] of the University of Massachusetts, who showed that the **freeboard of continents** (the relative elevation of continents with respect to sea level) has remained fairly constant throughout much of the last 0.5 billion years of the earth's history. Wise argues that the earth arrived fairly early at essentially its present volumes of continents

and oceans. Since then, a steady-state relationship has been maintained even though continental margins have been locally growing or retreating. The net effect is apparently to rework any continental-margin sedimentary material back into these continental rafts of nearly constant thickness, area, and volume. Isostatic balance exists between continental and oceanic crust and mantle (see Chapter 7). This equilibrium does not depend on the shape or number of continents or ocean basins, but it depends merely on the total volume of each of them.

The only change in this picture of equilibrium seems to have happened at the time of the breakup of the earth's supercontinents, about 200 million years ago (see Section 8.4). This change in the freeboard may have been a response to the magnitude of the initial breakup of this large landmass, an occurrence unique in most of the earth's history and one not duplicated at least since Precambrian time.

AGE OF OCEAN BASINS

As we show later in this chapter, the present ocean basins are geologically young. The oldest features found in the sedimentary cover on the ocean floor are of Jurassic age (see Figure 8.1), about 150 million years old. This is not to say that there were no earlier ocean basins with sediments. Any older sediments that had once existed were lost as these former ocean floors were consumed along subduction zones (see Chapter 1 and Section 8.3); all evidence of earlier ocean basins was thus destroyed.

8.2 MAGNETISM AND PALEOMAGNETISM

In Chapter 1 we referred briefly to the earth's magnetic field and defined some of the terms useful in describing it. We begin this section with a more extensive discussion of magnetism and particularly of magnetism of the geological past, **paleomagnetism.** The data derived from paleomagnetic studies not only gave the subject of continental drift a new push in the 1940s and early 1950s but also provided proponents of continental drift

with some of their strongest arguments. Furthermore, they have helped demonstrate the youthfulness of the ocean basins as compared with the continents and have been instrumental in the development of the idea of plate tectonics.

EARTH'S MAGNETISM

We are all familiar with the fact that the earth behaves as if it were a magnet and that consequently the compass needle seeks the north magnetic pole. We can

[1] D. U. Wise, "Continental Margins, Freeboard and the Volumes of Continents and Oceans through Time," in C. A. Burk and C. C. Drake (eds.), *The Geology of Continental Margins,* Springer-Verlag New York Inc., 1974.

FIGURE 8.2

The earth's magnetic field can be pictured as a series of lines of force. The arrows indicate positions that would be taken by a magnetic needle free to move in space and located at various places in the earth's field. The magnetic, or dip, poles do not coincide with the geographic poles, nor are the north and south magnetic poles directly opposite each other.

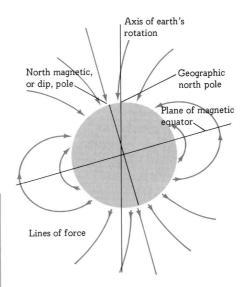

picture the earth's magnetic field as a series of lines of force. A magnet free to move in space will align itself parallel to one of these lines. At a point north of the Prince of Wales Island, at about 75°N and 100°W, the north-seeking end of the magnetic needle will dip vertically downward. This is the **north magnetic, or dip, pole.** Near the coast of Antarctica, at about 67°S and 143°E, the same end of our needle points directly skyward at the **south magnetic, or dip, pole.** Between these dip poles the magnetic needle assumes positions of intermediate tilt. Halfway between the dip poles the magnetic needle is horizontal and lies on the **magnetic equator.** Here the intensity of the earth's field is least, and it increases toward the dip poles, where the field is approximately twice as strong as it is at the magnetic equator (Figure 8.2).

The angle that the magnetic needle makes with the surface of the earth is called the **magnetic inclination,** or dip. The north and south dip poles do not correspond with the true north and south geographic poles as defined by the earth's rotation. Because of this, the direction of the magnetized needle will in most instances diverge from the true geographic poles. The angle of this divergence between a geographic meridian and the magnetic meridian is called the **magnetic declination,** and it is measured in degrees east and west of geographic north (see Figure 8.3).

SECULAR VARIATION OF THE MAGNETIC FIELD

As long ago as the midseventeenth century it was known that the magnetic declination changed with time. Since then we have been able to demonstrate not only slow changes in declination but also changes in inclination and intensity. These changes in magnetism take place over periods measured in hundreds of years. Because they are detectable only with long his-

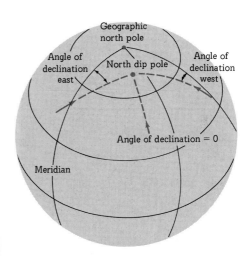

FIGURE 8.3

Because the dip poles and geographic poles do not coincide, the compass needle does not point to true north. The angle of divergence of the compass from the geographic pole is the declination and is measured east and west of true north.

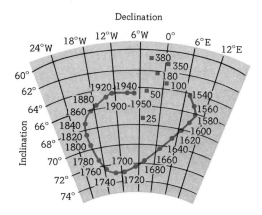

FIGURE 8.4

Magnetic inclination and declination vary with time. At Paris we have a continuous record of these changes since 1540. The data for inclination and declination for Gallo-Roman times (colored squares) are based on magnetic measurements of archaeological materials. [Adapted from Emile Thellier, "Recherches sur le champ magnétique terrestre," *L'Astronomie*, vol. 71, p. 182, 1957.]

torical records, these changes are called *secular changes*, from the Latin *saeculum*, meaning "age" or "generation," implying a long period of time.

Figure 8.4 shows changes in the declination and inclination records at Paris during Gallo-Roman times and from 1540 to 1950. Another long-range record, from 1573 to the present, is available from London and shows a pattern very similar to that of Paris from the sixteenth century on. Such records tempt us to suspect that these might represent worldwide, periodic variations in the earth's magnetic field. But when records are compiled for the entire earth, it becomes apparent that the changes in the magnetic field are regional rather than global. The centers of greatest change wax and wane and also move generally westward at a rate averaging approximately 0.2°/year.

THE GEOMAGNETIC POLE

The magnetic field at the earth's surface can be considered as composed of three separate components. There is, first, a small component that seems to result from activity above the earth's surface and is sometimes

referred to as the external field. Second, there is a quantitatively more important component, best described as if it were caused by a dipole—such as a simple bar magnet—passing through the center of the earth and inclined to the earth's axis of rotation. Finally, there is what we refer to as the nondipole field, that portion of the earth's field remaining after the dipole field and the external field are removed.

The dipole best approximating the earth's observed field is one inclined 11.5° from the axis of rotation. The points at which the ends of this *imaginary* magnetic axis intersect the earth's surface are known as the **geomagnetic poles** and should not be confused with the magnetic, or dip, poles. The north geomagnetic pole is about 78.5°N, 69°W, and the south geomagnetic pole is exactly antipodal to it at 78.5°S, 111°E (see Figure 8.5).

MAGNETOSPHERE

The earth's magnetic field performs a service for terrestrial life by warding off and trapping powerful radiation from space. This is in the form of electrons and very energetic ionized nuclei of atoms, mostly hydrogen. Some of the particles are believed to have been produced by cosmic-ray collisions in the atmosphere, but most are attributed to violent sprays of particles from the sun in solar flares.

The particles are diverted into a great doughnut-shaped racetrack around the earth, with the earth's axis under the doughnut's hole. They leak off and are turned back into space fairly rapidly. This region of trapped ions is called the **magnetosphere** (see Figure 8.6). It begins about 1,000 km above the earth and extends out to about 60,000 km. A similar band of charged particles has been reported encircling Jupiter.

CAUSE OF THE EARTH'S MAGNETISM

The cause of the earth's magnetism has remained one of the most vexing problems of earth study. A completely satisfactory answer to the question is still forthcoming.

As we have indicated, the earth's magnetic field is composed both of internal and external components. The external portion of the field is due largely to the activity of the sun. This activity affects the ionosphere and appears to explain magnetic storms and the north-

FIGURE 8.5

The geomagnetic poles are defined by an imaginary magnetic axis passing through the earth's center and inclined 11.5° from the axis of rotation. This magnetic axis is determined by a hypothetical, earth-centered bar magnet positioned to best approximate the earth's known magnetic field.

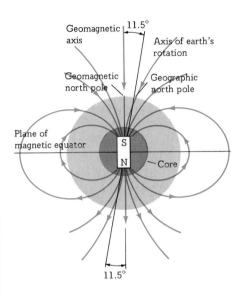

ern lights. The changes and effects of the external field may be rapid and dramatic, but they have little effect on the internal field of the earth, which is of greatest concern to us.

William Gilbert, who first showed that the earth behaves as if it were a magnet, suggested that the earth's magnetic field results from a large mass of permanently magnetized material beneath the surface. This notion might at first seem attractive not only because large quantities of magnetic minerals have been found in the earth's crust but also because geologists think that the earth's core is made largely of iron. However, close examination reveals that the average intensity of the earth's magnetization is greater than that of the observable crustal rocks. We must therefore look deeper for the source of magnetism.

The first difficulty we face is that materials normally magnetic at the earth's surface lose their magnetism above a certain temperature. This temperature is called the **Curie temperature** and varies with each material. The Curie temperature for pure iron is about 760°C; for hematite, 680°C; for magnetite, 580°C; and for nickel, 350°C. The temperature gradient for the earth's crust is estimated to average about 30°C/km, and at this rate of increase the temperature should approximate the Curie temperature of iron about 25 km below the surface and exceed the Curie temperatures for most normally magnetic materials. Therefore, below 25 km we should not expect earth materials to be magnetic, and permanent magnetism can exist only above this level.

On the other hand, if all the earth's magnetism were concentrated in the crustal rocks, then the intensity of magnetism of these rocks would have to be some 80 times that of the earth as a whole. And yet, as we have mentioned a few lines back, we know that the magnetic intensity of the surface rocks is less than that of the earth's average intensity.

From these observations we must conclude that the earth's magnetic field is *not* due to permanently magnetized masses either at depth or near the surface.

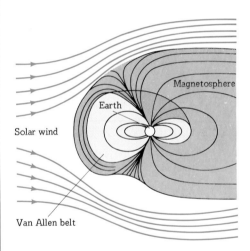

FIGURE 8.6

The magnetosphere: a doughnut-shaped region where the earth's magnetic field traps and wards off powerful radiation from space. The inner portion was first detected by James A. Van Allen in interpreting measurements from early United States satellites in 1958. It is sometimes referred to as the Van Allen belt. Later satellite data defined the region from 1,000 km out to about 60,000 km now known as the magnetosphere.

Some physicists have suggested that the rotation of the earth accounts for the earth's magnetic field. This explanation has also met with great difficulties.

The theory of earth magnetism most widely entertained at present is that the earth's core acts as a self-exciting dynamo. One model of the earth's core pictures its outer portion as a fluid composed largely of iron. This core therefore not only is an excellent conductor of electrical currents but also exists in a physical state in which motions can easily occur. Electromagnetic currents are pictured as generated and then amplified by motions within the current-conducting liquid. The energy to drive the fluid is thought to come from convection, which in turn results from temperature differences. The dynamo theory further requires that the random convective motions and their accompanying electromagnetic fields be ordered to produce a single, united magnetic field. It is thought that the rotation of the earth can impose such an order. The dynamo theory of earth magnetism still remains a theory, but so far it has proved the most satisfactory explanation of the earth's magnetism.

PALEOMAGNETISM

Some rocks, such as iron ores of hematite or magnetite, are strongly magnetic. Most rocks, however, are only weakly so. Actually the magnetism of a rock is located in its individual minerals, and we should be more correct to speak of the magnetism of minerals rather than of the rock. By convention, however, we refer to rock magnetism. This magnetism is referred to as the rock's **natural remanent magnetism (NRM)**. This remanent magnetism may or may not agree with the present orientation of the earth's field and may have been acquired by the rock in many different ways. Identifying, measuring, and interpreting the different components of a rock's NRM form the basis of paleomagnetism, the study of the earth's magnetic field in the geological past.

Let us examine some of the ways an igneous rock acquires its magnetism. As a melt cools, minerals begin to crystallize. Those which are magnetically susceptible acquire a permanent magnetism as they cool below their Curie temperatures. This magnetism has the orientation of the earth's field at the time of crystallization. It is called **thermoremanent magnetism (TRM)**. This remanent magnetism remains with the minerals—and hence with the rock—unless the rock is reheated past the Curie temperatures of the minerals involved. This new heating destroys the original magnetism; and when the temperature again drops below the Curie temperatures of the magnetic minerals, the rock acquires a new TRM.

The NRM of our igneous rock may include other components. One of these is an **induced magnetism** arising from the present magnetic field of the earth. This induced magnetism is parallel to the earth's present field, but it is weak when compared with the TRM of the rock.

Sedimentary rocks acquire remanent magnetism in a different way than do igneous rocks. Magnetic particles such as hematite tend to orient themselves in the earth's magnetic field as they are deposited. This orientation is retained as the soft sediments are lithified. This magnetism, known as **depositional remanent magnetism (DRM)**, records the earth's field at the time the rock particles were deposited. Of course, the sedimentary rock, like the igneous rock, may also acquire an induced magnetism reflecting the current magnetic field.

Virtual geomagnetic poles   Paleomagnetic studies define the earth's magnetic field at various localities at different moments in geologic time. Instead of expressing the data on declination and inclination for a given locality, we usually express them in terms of equivalent pole positions. We refer to these poles as **virtual geo-**

FIGURE 8.7

Intensity variation of the earth's magnetic field, from Czech archaeological samples. Each circle represents a sample or samples of the same age. The vertical bars represent standard deviation from the mean ratio. [After V. Bucha, *Archaeometry*, vol. 10, p. 20, 1967.]

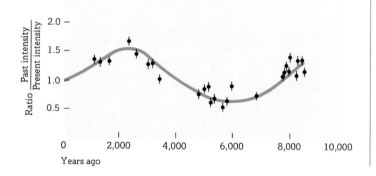

magnetic poles. These are different from the dip poles and the geomagnetic poles we have already discussed. The pole consistent with the magnetic field as measured at any one locality is the virtual geomagnetic pole of that locality. It differs from the geomagnetic pole because it refers to the field direction of a single observational station, whereas the geomagnetic pole is the best fit of a geocentric dipole for the entire earth's field. Inasmuch as it is impossible to describe the entire earth's field at various times in the past, the virtual geomagnetic pole is commonly used in expressing paleogeomagnetic data.

Geomagnetic intensity, variations, and reversals Earlier we stated that there has been a secular variation in the position of the geomagnetic pole over the last several hundred years. In addition to this change in pole position, we also know that there has been a variation in intensity of the earth's magnetic field. Thus by 1965 the intensity had decreased nearly 6 percent from the time it was first successfully analyzed in 1835 by K. F. Gauss, a German mathematician.[2]

We have been able to extend our knowledge of the variation in the field's intensity by techniques first successfully used by the French physicist Emile Thellier. Samples that can be precisely dated by historical or radiocarbon techniques show that the intensity of the field has been decreasing since about the time of Christ, when it was about 1.6 times its present intensity. Previously it had risen from a low value of about half the present intensity around 3500 B.C., this having been preceded by a steady decline from an intensity of about 1.5 times that of the present around 6500 B.C. (Figure 8.7).

[2]The measured decline was from 8.5×10^{25} to 8.0×10^{25} G (gauss), where "gauss" (or "oersted") is the unit of measure for magnetic-field intensity.

FIGURE 8.8

A time scale for geomagnetic reversal during the last 4.5 million years has been based on extrusive igneous rocks. Each horizontal line in the two left-hand columns represents a rock sample, whether its polarity is normal or reversed. In the normal-field column normal-polarity intervals are shown in color; in reversed-field reversed-polarity intervals are shown in color. (See also Figure 18.11.) [Modified from Allan Cox, "Geomagnetic Reversals," *Science*, vol. 168, pp. 237–245, 1969.]

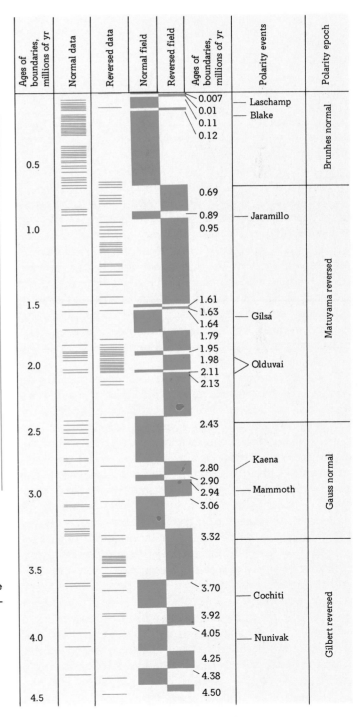

TABLE 8.1

Intervals for which the earth's magnetic field was of normal polarity during the last 4.5 million years[a]

Millions of yr ago

0.00–0.007	2.11–2.13
0.01–0.110	2.43–2.80
0.12–0.69	2.90–2.94
0.89–0.95	3.06–3.32
1.61–1.63	3.70–3.92
1.64–1.79	4.05–4.25
1.95–1.98	4.38–4.50

[a]Modified from Allan Cox, "Geomagnetic Reversals," *Science*, vol. 168, pp. 237–245, 1969.

If the decay of magnetic intensity were to continue at its present rate, it has been estimated by Allan Cox, Stanford University geophysicist, that the intensity would reach zero about 2,000 years from now. Thereafter the direction of the magnetic field would reverse itself. There is no assurance that this will happen, and it may just as well fluctuate as it has for the last several thousand years. But if we turn farther back into earth history, we find that the poles have been reversed, not once but many times. So we assume that such a change is possible in the future.

We now have a great deal of information about past reversals of the magnetic field. That is to say, if we were able to take our ordinary magnetic compass back into time, we should find some periods in which the north-seeking end of the needle would point toward the north pole as it does today. But there would be many other times during which the needle would point to the south pole instead. The present polarity is called "normal." A polarity at 180° to it is called "reversed." A succession of normal and reverse fields can be pieced together for the last several 10 million years. We call the longer intervals of dominance by a particular polarity a "polarity epoch," and we have named them after distinguished students of earth magnetism. During each polarity epoch are shorter periods in which the polarity is in the opposite direction, and these time intervals are referred to as "polarity events" (see Figure 8.8 and Table 8.1). We shall come back shortly to the use of this information when we take up the study of continental drift.

RESULTS OF PALEOMAGNETIC STUDIES

If we can measure the TRM or the DRM of a rock and relate it to the earth's present field, we can determine to what extent the orientation of the earth's magnetic field has varied at that spot through time.

Studies of ancient pole positions assume that the earth's field has been dipolar and, further, that this dipole has approximated the earth's axis of rotation. A consequence of these assumptions is that the earth's geographic poles must have coincided in the past with the earth's geomagnetic poles. Clearly this is not so at present. Why should we think that it was true in the past? A partial answer to the question lies in the dynamo theory of earth magnetism. If the dynamo theory is correct, then theoretical considerations suggest that the rotation of the earth should orient the axis of the magnetic field parallel to the axis of rotation.

Observational data support the theoretical considerations. When we plot the virtual geomagnetic poles of changing fields recorded over long periods at magnetic observatories around the world, we find a tendency for these pole positions to group around the geographic poles. More convincing are the paleomagnetic poles calculated on the basis of magnetic measurements of rocks of Pleistocene and Recent age. These materials, including lava flows and varves, reveal pole positions clustered around the present geographic pole rather than the present geomagnetic pole (see Figure 8.9).

As a result of theoretical and observational data, therefore, most authorities feel that the apparent, present-day discrepancy between magnetic and rotational poles would disappear if measurements were averaged out over a span of approximately 2,000 years. The same principle would apply for any 2,000-year period throughout geologic time. Thus, when we speak of a paleomagnetic pole, we have some confidence that it had essentially the same location as the true geographic pole of the time.

Paleomagnetic data derived from rocks of Tertiary age indicate, fairly conclusively, no significant shift of the geomagnetic poles from the Oligocene to the present. The farther backward we go in time beyond the Oligocene, however, the more convincing becomes the case for a changing magnetic pole and, on the basis of the above discussion, for a changing geographic pole. The magnitude of this change in suggested in Figure 8.10, which shows that the paleomagnetic poles for

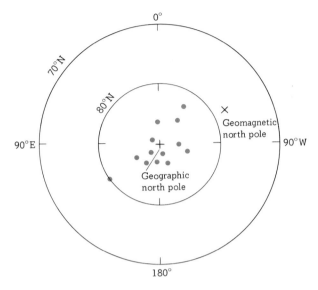

FIGURE 8.9

The virtual geomagnetic poles determined by magnetic measurements of earth materials of Pleistocene and Recent age cluster around the modern geographic pole rather than around the present geomagnetic pole. [Redrawn from Allan Cox and R. R. Doell, "Review of Paleomagnetism," *Geol. Soc. Am. Bull.*, vol. 71, p. 734, 1960.]

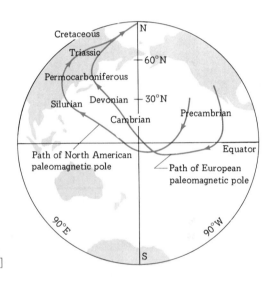

FIGURE 8.10

Paleomagnetic measurements of rocks from North America and Europe show the paths followed by the magnetic poles of these two continents from Precambrian times to the present. [Redrawn from Allan Cox and R. R. Doell, "Review of Paleomagnetism," *Geol. Soc. Am. Bull.*, vol. 71, p. 758, 1960.]

Europe and North America were in the eastern Pacific during Precambrian time. Thereafter they moved southwestward and crossed the equator into the southern hemisphere before moving northwestward toward Asia and eventually to the position on the present globe.

The extensive migration of the magnetic poles (and, by extension, the geographic poles), combined with the observation that the paths of polar migration of different continents fail to coincide, raises tantalizing questions. We find ourselves, in fact, faced with the entire concept of continental drift.

8.3 CONTINENTAL DRIFT

In considering the theory of continental drift, we should bear in mind that two general types of movement may be involved: movement of individual continents in relation to one another and movement of the earth's poles. Actually, if the latter movement took place, most geologists feel that a slippage of the earth's crust and upper mantle has produced an apparent motion of the poles. In other words, the magnetic and rotational poles remain fixed within the earth, but different points on the earth's crust would at different times be located at the polar positions as the crust moved. Such motion might involve the crust as a single unit or as fragments of it.

The first coherent theory that our continents have moved as individual blocks was presented by Alfred Wegener (1880–1930), a German meteorologist and a student of the earth. Wegener, as many before and after him, was intrigued with the apparent relationship in form between the opposing coasts of South America and Africa. Was it possible that these two landmasses were once part of the same general continent but have since drifted apart? Wegener's emphatic assent and his extensive documentation of the lateral motion of continents started a spirited discussion that has continued to the present.

EVIDENCE FOR CONTINENTAL DRIFT

Wegener's theory In 1912 Wegener published in detail his theory of continental drift. He pictured the dry land of the earth as included in a single, vast continent that he named Pangaea (from the Greek for "all" and "earth"). This primeval continent, he argued, began to split asunder toward the end of the Mesozoic. The fragments began a slow drift across the earth's face, and by the Pleistocene they had taken up their positions as our modern continents.

The idea of continental drift has been argued now for more than 60 years. The early evidence for the theory was drawn almost entirely from the geologic record. In the midtwentieth century the development of geophysical techniques brought new data into the discussion of continental drift. Let us look at some of the evidence.

Shape of continents Anyone observing the map of Africa and South America is struck by the jigsaw-puzzle match of the two continents. If we fit the eastern nose of South America into the large western bight of Africa, the two continents have a near perfect match. One distinguished geologist, while admitting to skepticism of continental drift, found this fit so credible that he opined, "If the fit between South America and Africa is not genetic, surely it is a device of Satan for our frustration."[3]

If we examine the outlines of the other continents, we can perhaps be persuaded that these landmasses too can be reassembled into a single large landmass, as suggested by Wegener. If the continents are fitted at

[3]Chester Longwell, "My Estimate of the Continental Drift Concept," in S. W. Carey (ed.), *Continental Drift, a Symposium*, p. 10, Tasmania University Press, Hobart, 1958.

the margin of the continental slope (a level of about 900 m below sea level), there is a very impressive match indeed, as shown in Figure 8.11. (Wegener himself first suggested using the edge of the continental shelf, *not* the edge of the continent at present sea level.)

Evidence from paleomagnetism We pointedly began this section with a consideration of the earth's magnetism; for it has been the increasing body of information on rock magnetism and paleomagnetism that has kindled new interest in the old theory of continental drift. We have seen in Figure 8.10 how the paleomagnetic poles have shifted during the last 500 million years. For the two continents represented, North America and Europe, there is a divergence between the ancient poles—that is, the virtual paleomagnetic poles—of these two landmasses as we go back in time. The case of North America and Europe is not unique. The pattern of polar migration, as sketched from paleomagnetic information, is different for each continent.

The divergence of the paths of polar migration among continents can be explained by the shifting of the landmasses in relation to each other. The course of polar migration suggests that the continental movement consists of a general drift of these continents away from each other, and in some instances there is an additional continental rotation.

Magnetic measurements at sea have revealed a characteristic pattern of anomalous magnetic intensities. These magnetic anomalies are most often associated with ocean ridges and are arranged in stripes of alternating intensity parallel to the ridge. An example is provided by the magnetics of the Reykjanes Ridge, a portion of the Midatlantic Ridge southwest of Iceland. Figure 8.12 shows the stripes parallel with the ridge and shows also a bilateral symmetry, the axis for which is the rift valley along the crest of the ridge.

Two geophysicists, Fred Vine and D. H. Matthews, suggested that the alternating stripes of high- and low-intensity magnetism were in reality zones of normal and reverse polarity in the rock of the sea floor. From this suggestion, now widely accepted, have flowed some intriguing concepts about the sea floor, the drift of continents, and the building of mountains.

Sea-floor spreading The Vine-Matthews hypothesis leads us to a consideration of sea-floor spreading. This is a mechanism that involves the active spreading of the sea floor outward, away from the crests of the main ocean ridges. As material moves from the ridge, new material replaces it along the ridge crest by welling upward from the mantle. As this mantle material

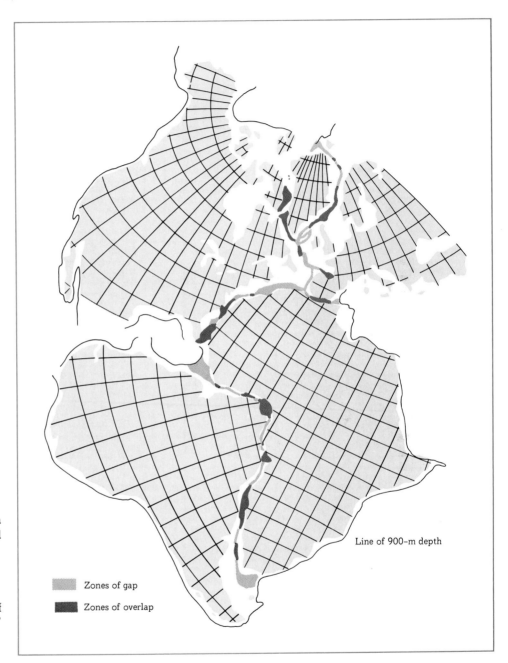

FIGURE 8.11

A statistically determined fit of North America, South America, Africa, and Europe at the 900-m depth in the ocean. Zones of overlap are shown in the dark tone; zones of gap are shown in the lighter tone. [After Edward Bullard and others, "The Fit of the Continents around the Atlantic," *Phil. Trans. Roy. Soc. London* 1088, pp. 41–51, 1965.]

Line of 900-m depth

Zones of gap

Zones of overlap

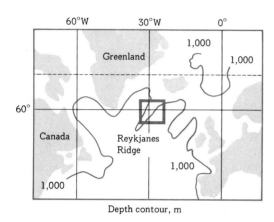

Depth contour, m

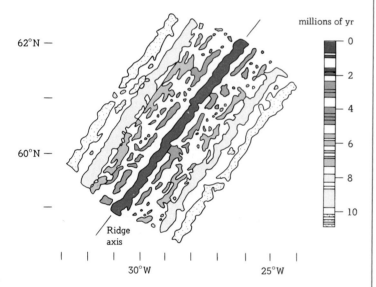

FIGURE 8.12

Magnetic anomalies over the Reykjanes Ridge
southwest of Iceland. The patterned areas display
normal polarity and are separated by areas of reversed polarity. Note that the belts of normal and
reversed polarity are generally symmetric to the
axis of the Midatlantic Ridge. Ages of the various
polarity belts are based on a time scale derived
from a study of magnetic anomalies on land and
on the sea floor. [After F. J. Vine, "Magnetic
Anomalies associated with Ocean Ridges," in
R. A. Phinney (ed.), *The History of the Earth's
Crust*, Princeton University Press, Princeton, 1968.
ⓒ 1968 by Princeton University Press. Reprinted
by permission.]

cools below the Curie temperature, it will take on the
magnetization of the earth's field at that time. Room for
the new material is made along the ridge by the continued pulling apart of the crust and its movement
laterally away from the ridge. At times of change in the
earth's polarity, then, newly added material along the
crest will record this change. Spreading laterally, it
carries this record with it, to be followed at a later time
by the record of the next polarity change. We can
visualize this as a gigantic magnetic tape preserving a
history of the earth's changing magnetic polarity. A
diagram to suggest the mechanism is shown in Figure
8.13.

The rates at which new sea floor forms at spreading centers is seen to vary. In the North Atlantic it is
about 1 cm/year, and in the South Atlantic about
2.3 cm/year. Along the Juan de Fuca Ridge off California the rate is 3 cm/year, and on the east Pacific rise it
is calculated to be 4.6 cm/year, reaching a rate of
9 cm/year in some sections. (These are all "half-
spreading rates": Each side of the ocean floor is moving at these rates; so the continents on either side of the
ocean basins would be moving apart at *twice* these
rates.)

Figure 8.14 shows the effects of magnetic reversals
on deep-sea sediments and associated oceanic lithosphere. If there were no sea-floor spreading, the layers
of sediments would simply show alternating polarity as
the earth's magnetic field changed. With sea-floor
spreading, each sedimentary layer begins at the ridge
with the magnetization of that particular time in the
earth's history. As the lithosphere migrates away from
this spreading center, the sediments are likewise carried farther and farther away.

Sea-floor spreading indicates that the ocean floor
is a dynamic system, and in terms of continental drift it
suggests that expansion of ocean basins may also
affect the location and position of continents. Furthermore, it raises many questions: What drives the ocean
floor, and where do the moving ocean floors go? The
proposed answers to these questions bear on continental drift, but we shall defer their consideration until we
examine the subject of plate tectonics (Section 8.4). In
the meantime let us take up additional geologic data
bearing on continental drift.

Evidence from ancient climates Much geologic
evidence cited in support of continental drift and polar
wandering is based on the reconstruction of climates of
the past. Their use is justified by the fact that modern
climatic belts are arranged in roughly parallel zones

FIGURE 8.13

A diagram to suggest the development of magnetic anomalies related to ocean ridges. New oceanic crust, cooling below the Curie point, forms along the axis of the ridge and assumes the magnetic characteristics prevailing at the time of its formation. As the crust moves laterally away from the ridge crest, changes in the magnetic field of the earth are retained in strips of the oceanic crust. Serpentinite, a rock made up chiefly of the mineral serpentine, is thought by many to form the lower portion of the oceanic crust. [After F. J. Vine, "Sea-floor Spreading—New Evidence," *J. Geol. Educ.*, vol. 17, pp. 6–16, 1969.]

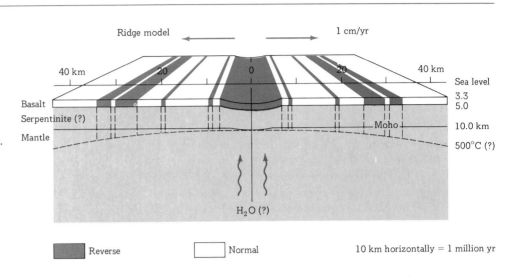

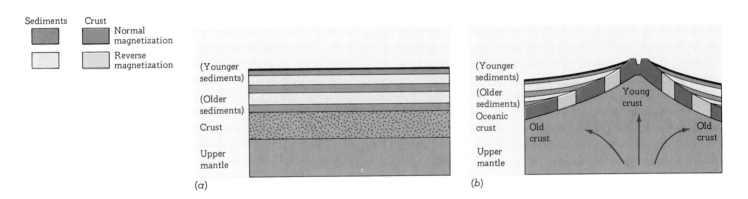

FIGURE 8.14

Magnetization of marine sediments: (a) If there were no sea-floor spreading, the layers would show alternating polarity in continuous sheets; (b) with sea-floor spreading, the alternate normal and reversed stripes in the crust have correspondingly magnetized sediments, beginning adjacent to that part of the crust formed at the ridge crest of the spreading center. [Modified from P. J. Wyllie, *The Way the Earth Works*, John Wiley & Sons, Inc., New York, 1976.]

whose boundaries are east-west and which range from the tropical equatorial climates to the polar ice climates, as suggested in Figure 8.15. Although climates at various times in earth history have been both colder and warmer than they are at present, we assume that basic climatic controls have remained the same and, therefore, that the climatic belts of the past have always paralleled the equator and been concentric outward from the poles.

Evidence from the geologic record allows us to map the distribution of ancient climates in a very general way. The pattern of some ancient climatic zones suggests that they were related to poles and an equator with locations different from those of today and that, therefore, the landmasses and the poles have varied relative to each other since those climates existed.

Late Paleozoic glaciation   On the Indian peninsula lies a sequence of rocks known as the Gondwana System, reaching in age from the late Paleozoic to the early Cretaceous. Beds of similar nature and age are recorded in South Africa, Malagasy, South America, the Falkland Islands, Australia, and Antarctica. In these other localities they are known by other names, but we can still refer to them as belonging to the Gondwana System.

Geologists who have worked on the Gondwana formations have discovered many similarities among the rocks of the various continents despite their wide geographic separation. Some of these similarities are so striking that many accept only one interpretation: The various southern lands must once have been part

of a single landmass, a great southern continent, early called *Gondwanaland* by these geologists.

The distribution of ancient glacial deposits is one of the most convincing lines of evidence for continental drift in these southern lands. During late Paleozoic time continental ice sheets covered sections of what are now South America, Africa, the Falkland Islands, India, Antarctica, and Australia. In southwestern Africa deposits related to these ancient glaciers are as much as 600 m thick. In many places the now-lithified deposits (tillites) rest on older rocks striated and polished by these vanished glaciers. If we plot the distribution of these deposits and the direction of ice flow on a map (see Figure 8.16), we can make two immediate observations: First, these traces of Paleozoic ice sheets occur in areas where no ice sheets (except for Antarctica) exist now or have existed during the glacial epochs of the Pleistocene Ice Age. Only an occasional towering peak may bear the scars of modern or Pleistocene ice. Continental ice sheets cannot exist in these latitudes today. Second, the direction of glacier flow is such that we can imagine the ice of Africa and South America to have been part of the same ice sheet when the two continents were one.

These observations have led most students of the earth to two conclusions: First, to account for glaciers in present tropical and subtropical areas, they locate the south pole at this time somewhere in or near southern Africa. Second, to account for, among other things, the apparent continuity of glacier flow, these same students have postulated a single southern continent. This continent later split into several sections that drifted apart to form the modern landmasses.

Ancient evaporite deposits and coral reefs   Evaporites, sedimentary rocks composed of minerals that have been precipitated from solutions concentrated by the evaporation of the solvents, are generally accepted as evidence of an arid climate. The ancient evaporite deposits represent the great arid belts of the past. The present hot-arid belts are located in the zones of subtropical high pressure, centered at about 30° north and south of the equator. In the northern hemisphere an evaporite belt has shifted through time from a near-polar location in the Ordovician and Silurian to its present position in the modern desert belts. This again suggests a relative motion of pole and landmasses of the past as compared with present-day conditions.

Turning to another line of evidence, we find that corals mark a climate shifting geographically through time. Today true coral reefs are restricted to warm, clear

FIGURE 8.15

Present-day climatic boundaries are arranged concentrically around the poles and thus are approximately parallel to lines of latitude.

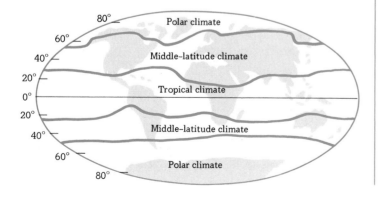

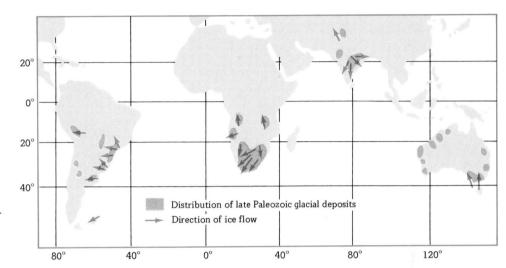

FIGURE 8.16

Direction of movement of late Paleozoic ice sheet and distribution of known late Paleozoic tillites (exclusive of Antarctica).

marine waters between 30° north and south of the equator. If we assume that ancient reef-forming corals had similar restrictions, then plotting their distribution in the past will show the distribution of tropical waters of the past and the location of the past equator. Doing this, we find that reef-forming corals did not approximate their present distribution until halfway through the Mesozoic. Prior to that time they lay well north of the present equator.

Plants, reptiles, and continental drift   Shortly after the disappearance of the southern hemisphere's late Paleozoic ice sheets, an assemblage of primitive land plants became widespread. This group of plants, known as *Glossopteris* flora, named for the tonguelike leaves of the seed fern *Glossopteris* (Figure 8.17), has been found in South America, South Africa, Australia, India, and within 480 km of the South Pole in Antarctica. The *Glossopteris* flora is very uniform in its composition and differs markedly from the more varied contemporary flora of the northern hemisphere. Many geologists have argued that the uniformity of the *Glossopteris* flora could not have been achieved across the wide expanses of water now separating the different collecting localities. In other words, in one way or another, there must have been continuous or near-continuous land connections between now-separate continents. To many this suggests that a single continent with a single uniform flora has been split apart into smaller continents that have since migrated to their present position. As we shall see, this conclusion has not gone unchallenged.

Among the vertebrate fossils of the late Paleozoic

FIGURE 8.17

These leaves of the fossil plant *Glossopteris* come from strata of Permian age in Australia. The *Glossopteris* flora is found also in South America, Africa, India, and Antarctica. The widespread occurrence of this very uniform flora has been used as evidence both by opponents and proponents of continental drift. Diameter of detail is 2.5 cm. [Specimen from Princeton University Paleobotanical Collections, Willard Starks.]

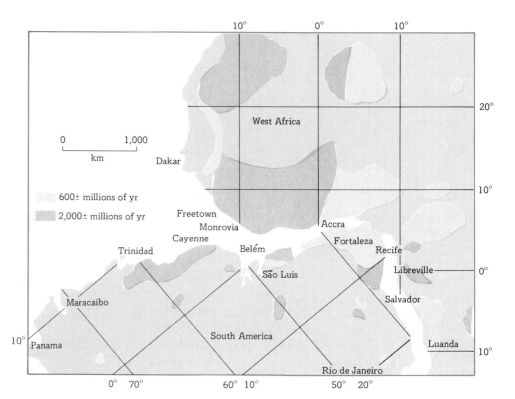

FIGURE 8.18

Rocks of two different ages mark the opposing shield areas of South America and Africa. If the two continents are fitted together, the zones of the same age match across the line of fit. [After P. M. Hurley and J. R. Rand, "Review of Age Data in West Africa and South America," in R. A. Phinney (ed.), *The History of the Earth's Crust*, Princeton University Press, Princeton, 1968. © 1968 by Princeton University Press. Redrawn by permission.]

and earliest Mesozoic we find a great number of different reptilian types. Two of them provide us with arguments for continental drift. *Mesosaurus*, a toothed early reptile of the late Permian, lived in the water and thus far is known only from Brazil, South Africa, and Antarctica. Although it was aquatic, most paleontologists do not believe that it could have made the trip across the South Atlantic. If this is so, then *Mesosaurus* may offer evidence for a closer proximity of South America and Africa and thus for continental drift. A second reptile, *Lystrosaurus*, dates from the Triassic period. *Lystrosaurus*, which is about 1 m in length, was adapted to an aquatic but nonmarine environment, and its remains are reported from South Africa and from Antarctica's Alesandra Range. It is argued that *Lystrosaurus* could not have made the journey across the Antarctic Ocean separating South Africa from Antarctica and that this implies the former joining of the two continents.

Other evidence Some anciently formed mountain chains now terminate abruptly at the continental margins. Join the continents together, and some of these geological structures match up between the two landmasses. Thus the Cape Mountains of South Africa are thought to be the broken extension of the Sierra de la

Ventana of Argentina in one direction and of the Great Dividing Range in eastern Australia in the other direction. The entire stretch is cited by some adherents of continental drift as a once-continuous chain of mountains now segmented and separated. Similarly, the Appalachian Mountain system ends in the sea on the northern shore of Newfoundland. Is its extension to be found in the orogenic belts of the British Isles and western Europe? Many geologists think so.

P. M. Hurley and his colleagues at the Massachusetts Institute of Technology have carried on extensive geochronologic studies of Precambrian igneous and metamorphic rocks in West Africa and the eastern bulge of South America. Radiometric-age determinations of many samples show distinct belts of roughly similar age on the two continents. If the continents are shifted back together, the several provinces of Precambrian rocks of differing ages on the rejoined landmasses match fairly well (Figure 8.18). We are tempted to believe that rocks of similar age in Africa were once continuous with rocks of the same age in South America. Sometime after they were formed, they have been separated so that we see them on two widely separated continents.

Planetary winds probably always existed. It would be interesting to see whether ancient eolian deposits might give some indication of such major wind belts as the northeast and southeast trades of the past. Sand beds accumulated on the face of a sand dune dip in the direction that the wind blows. We can measure such dip in ancient wind-deposited sediments. If enough of these directions are available, the direction of the dune-forming wind can be determined statistically. Preliminary studies on presumed eolian deposits of late Paleozoic age in Wyoming, Utah, Arizona, and England indicate that these areas fell within the northeast-trade-wind belt when the earth's pole (as suggested by the paleomagnetic evidence) was located on the East China coast.

Reliability of evidence The reliability of evidence cited in support of continental drift has been questioned by some workers. The paleowind directions discussed above, for example, are thought by some to be not only inconclusive but actually based on deposits some of which were not truly eolian.

Some of the evidence from the fossil record is attacked as too fragmentary and open to too many interpretations to be diagnostic. The use of *Mesosaurus* as representing evidence for continental drift is one such example.

Paleomagnetic evidence is currently regarded as one of the most conclusive arguments in favor of continental drift. But some argue that the evidence may not be so strong as it first appears, and they point out that this evidence may be overextrapolated. For instance, we said earlier that discussions of paleomagnetism assume a dipolar magnetic field for the earth. Yet the more ancient fields could have had another form—four poles instead of two, perhaps. If so, the entire paleomagnetic argument in support of continental drift would have to be reevaluated.

SUMMARY OF THE PROBLEM

Prior to the 1940s most of the supporters of the theory of continental drift lived in the southern hemisphere, particularly in South Africa. Geologists and geophysicists of the northern hemisphere, particularly in the United States and western Europe, felt that there was little or no support for the motion of continents or the wandering of the poles. In fact, it was hardly respectable to entertain the possibility of such movement.

After the 1940s the techniques of geophysics and particularly those of paleomagnetic measurement began to bring new data into the discussion. This new evidence reopened the entire question of continental drift for American and European geologists. By the 1970s a vast amount of data bearing on continental drift had accumulated. Most workers in the field now feel that the available evidence favors the reality of continental drift. Many, having accepted that continental drift has occurred, have turned to how it might have happened—and to what implications drift might have for our understanding of earth history and earth mechanics. In the following section we take up this thread of thought and consider the concept of plate tectonics.

8.4 PLATE TECTONICS

Geologists in recent years have pointed out that the earth's outer surface can be divided into large units, or plates. We have suggested at the beginning of this chapter and elsewhere that the origin, movement, and interrelation of these plates can be coordinated with sea-floor spreading, earthquake belts, crustal movement, and continental drift.

THE WORLD SYSTEM OF PLATES

At least six major plates and a host of minor plates are recognized, as shown in Figure 8.19. The major plates include the Indian, Pacific, Antarctic, American, African, and Eurasian plates. The boundaries of these plates are loci of present-day earthquakes and volcanic activity. Consider for example the Pacific plate. On its southern and southeastern side it is bounded by an oceanic rise that undergoes active spreading. On its northern and northwestern borders it is marked by island arcs, deep-sea trenches, and tectonic and volcanic activity. Its northeastern side along western United States is defined by the San Andreas fault, where the Pacific plate moves generally northwestward and jostles the American plate in the process.

The relationship of plates to the distribution of earthquakes is striking when one compares Figure 8.19

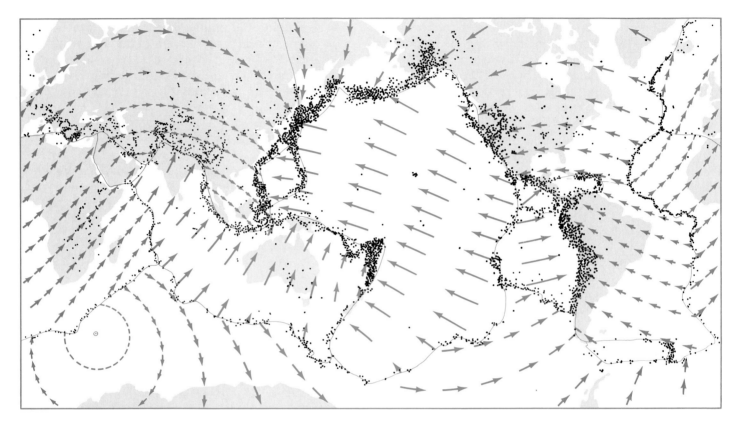

FIGURE 8.19

Crustal plates are outlined by zones of intensive earthquake activity. The directions of plate motion over the underlying asthenosphere are shown by arrows. Each plate moves the distance equivalent to the length of the arrow in 20 million years. [From Sheldon Judson, Kenneth S. Deffeyes, and Robert B. Hargraves, *Physical Geology*, Prentice-Hall, Inc., Englewood Cliffs, N.J., 1976.]

with Figures 6.1 and 6.2. In general the shallow quakes mark the ocean ridges; deep earthquakes are missing in these zones. Both shallow and deep quakes are found along the arcs and trenches and along young mountains, as the Himalayas (see Table 8.2).

The plates may include both continental and oceanic areas, as for example the American plate, which embraces North and South America and Greenland as well as the western Atlantic Ocean. On the other hand the Pacific plate is restricted to oceanic area.

MAKEUP OF PLATES

The rigid lithospheric plates consist of several distinct layers. The lowest portion comprises the cold, solid upper part of the mantle. The Moho discontinuity marks the upper contact of this layer with the crust.

Above this lowermost layer is the principal portion of the crust. As we have explained, under the oceans this crustal layer may be only a few kilometers thick and consist of coarse gabbrolike simatic rock (rock rich in silica plus iron and magnesium). It is capped by lava flows—solidified magma extruded along the midocean ridges, rises, and other volcanic centers. Under continents this layer may be 30 km or more thick. The lower crust may include a continuation of the oceanic gabbroic layer. The upper crust is sialic (rich in silica and alumina) and includes the crystalline granitic masses of the roots of mountains covered by a relatively thin layer of sedimentary and metamorphic rocks at the continental surfaces.

TABLE 8.2

Plate boundaries and their interactions with other plates under tension, compression, or neither[a]

Boundary	Divergent (tensional)	Convergent (compressional)	Transform (neither tensional nor compressional)
Ocean-ocean	Ridge crest; S[b] earthquakes in narrow belt; submarine lavas	Ocean trench and volcanic island arc; S, I, and D quakes over wide belt; volcanoes common	Ridge and valley fracture zone; S quakes in narrow belt only between offset ridges; volcanoes absent
Continent-continent	Rift valley; S quakes over wide zone; volcanoes present	Young mountain ranges; S and I quakes (or just S) over wide zone; no volcanoes	Fault zone; S quakes over broad zone; no volcanoes
Ocean-continent		Ocean trench and young mountain range; S and I quakes over wide belt; D quakes occasionally; volcanoes common	

[a] Modified from P. J. Wyllie, *The Way the Earth Works*, John Wiley & Sons, Inc., New York, 1976.
[b] S, shallow-focus earthquakes; I, intermediate focus; D, deep focus.

Topping this layer-cake structure in the oceanic regions are the sediments of the sea floor, consisting of fine inorganic debris plus the remains of organisms that lived in the upper portion of the ocean, died, and fell to the bottom, where they accumulate. This sedimentary layer is thin or nonexistent adjacent to the spreading centers and thickens progressively away from the ridge as a result of the outward movement of the crust away from the spreading center, the oldest crust receiving sediments for periods of time longer than those during which the younger crust receives them (refer to Figure 8.14 again).

CLASSIFICATION OF PLATE BOUNDARIES

Lithospheric plates originate at midocean ridges and spread laterally toward subduction zones. The motion of these plates with respect to other adjacent plates results in three principal kinds of boundary (Figures 8.20 and 8.21): Where plates move away from one another, **divergent** boundaries result; where they move toward each other, **convergent** boundaries result; and where one plate moves past another plate in a nearly parallel fashion, **transform** boundaries result.

The plates may be entirely dense oceanic lithosphere or may include the lighter rafts of continental lithosphere. Movement of these plates may involve ocean-ocean, ocean-continent, or continent-continent interactions.

Because divergent boundaries represent conditions of *tension*, shallow-focus earthquakes and volcanic emanations occur along the ridge crests of ocean-ocean regions and along the rift valleys of continent-continent zones (Table 8.2).

Compression characterizes convergent boundaries and results in trenches and island-arc volcanoes, with shallow, intermediate, and deep-focus quakes along ocean-ocean contacts. Trenches and young mountain ranges occur with volcanoes and shallow- and intermediate-focus quakes (with or without deep-focus quakes) at ocean-continent boundaries.

Where continental plates meet other continental plates, young mountain ranges result from the piling up, or doubling, of the lighter continental masses. Volcanoes are rare or nonexistent under these conditions, but shallow-focus quakes are common. Intermediate-focus quakes are rare, and deep-focus quakes entirely absent.

Neither tension nor compression occurs along transform boundaries. Continent-continent interactions are characterized by fault zones and shallow-focus earthquakes over a rather broad zone. Fracture zones with shallow-focus earthquakes in narrow belts mark ocean-ocean transform boundaries. Little volcanic activity is likely along any transform boundaries.

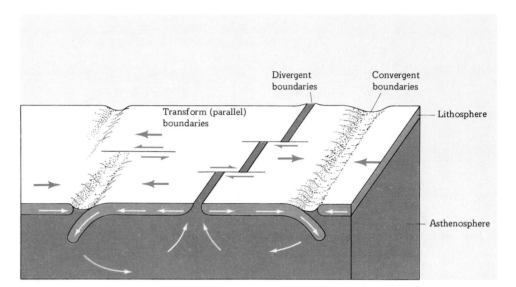

FIGURE 8.20

Major features of sea-floor spreading with the three principal types of plate boundary: convergent, divergent, and transform (or parallel). [After B. Isacks, J. Oliver, and L. R. Sykes, "Seismology and the New Global Tectonics," *J. Geophys. Res.*, vol. 73, pp. 5855–5899, 1968.]

MOVEMENT OF PLATES

We have seen in Figure 8.19 that some of the sutures between adjacent plates are zones of active sea-floor spreading, a process described in Section 8.3. Along such zones, therefore, crustal material is considered as continuously forming, and to make room for it the sea floor is pictured as moving laterally away from the ridge crest. Transform faults, described in Chapter 7 and illustrated in Figure 7.43, cut across the ridge and produce its offset. An analysis of the first movement of the crust during earthquakes on the ridge shows two types of movement. Along the rifts that mark the axes of the oceanic ridges motion during earthquakes indicates tensional stress, normal faulting, and dropping down of the blocks associated with the grabens along the ridges. In contrast is the movement along the fractures offsetting the ridges: Along these structures analysis of earthquake data shows that the movement is of a strike-slip nature on a plane at right angles to the ridge; and movement is away from the ridge crests, as in Figure 7.44 of the Midatlantic Ridge near the equator.

If plates are actually moving away from each other through a process of sea-floor spreading, then a number of questions arise: What drives the ocean floor? Where does the moving ocean floor go?

Convection cells Most students of the problem appeal to convection currents (described in Chapter 7) to provide a mechanism for moving the ocean floor. They envision the ocean ridges as lying over a rising con-

vection current. That seismic and volcanic activity are generally high on the ridges and that heat flow from the interior is higher at the ridge crest than it is at most other places add appeal to the suggestion. The current is thought to lie in the upper mantle, well below the Moho, and to represent ascending limbs of adjacent convection cells. At some point below the Moho the rising current is pictured as splitting into two currents flowing laterally away from the midoceanic ridge.

Plume theory Another possible driving mechanism for the movement of plates has been suggested by Jason Morgan (of Princeton University), who earlier had pieced together much of the evidence in support of plate tectonics. He infers the existence of hot spots, or plumes, of hot rising solid mantle in zones a few hundred kilometers in diameter. This is an extension of an earlier hypothesis by J. Tuzo Wilson of Toronto, who suggested similar structures to account for volcanic island chains unrelated to plate boundaries (see Figure 3.12). (Over 120 hot spots have been active during the past 10 million years according to a study by Kevin Burke and Wilson.[4]) By this hypothesis hot mantle rock rises and spreads out laterally in the asthenosphere in about 20 major thermal plumes around the world (Figure 8.22), creating hot spots of volcanic activity. The rest of the mantle undergoes very slow downward movement to balance the upward flow in the plumes. The

[4]Kevin C. Burke and J. Tuzo Wilson, "Hot Spots on the Earth's Surface," *Sci. Am.*, vol. 235, no. 2, pp. 46–57, 1976.

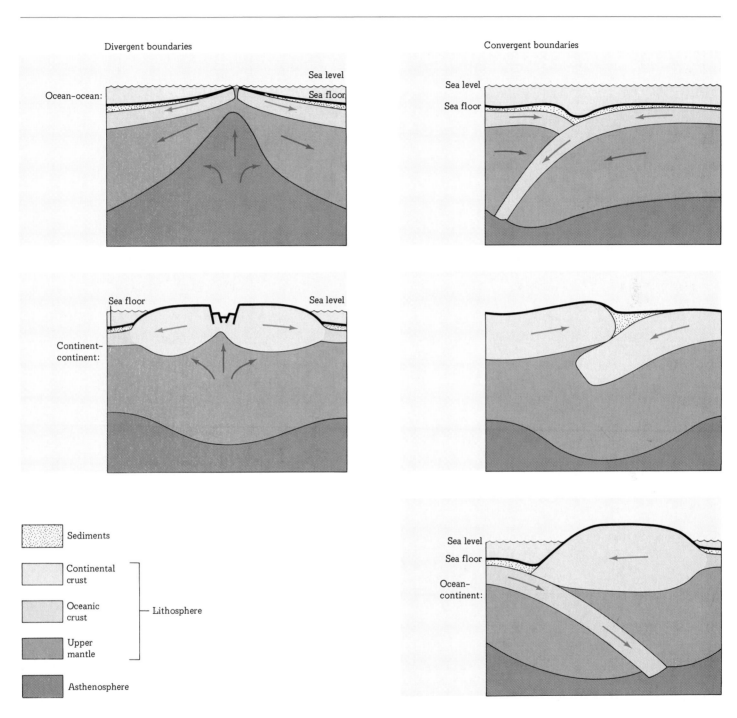

FIGURE 8.21

Illustrations of plate boundaries and their interactions with other plates. [From various sources.]

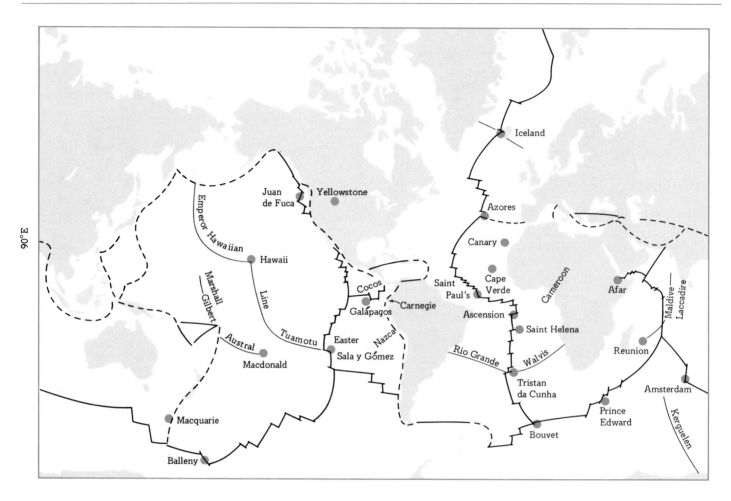

FIGURE 8.22

About 20 major mantle plumes. Thin black lines represent volcanic trails, or linear volcanic chains, tracing the hot spots. [After W. J. Morgan, "Plate Motions and Deep Mantle Convection," in R. Shagam (ed.), *Studies in Earth and Space Sciences*, pp. 7–22, Geological Society of America Memoir 132, Boulder, 1972; and K. C. Condie, *Plate Tectonics and Crustal Evolution*, Pergamon Press, Inc., Elmsford, N.Y., 1976.–

movement of any plate is caused by the combination of forces exerted by the upward flow in the plume, the outward radial flow away from the plume, and the return downward flow of the mantle.

Recycling the sea floor What happens to the old sea-floor crust as it is displaced from the ridge crest by younger and younger material? Those who explain the motion by convection see the crust and a part of the upper mantle carried laterally by convection currents toward the margins of the ocean basins. This is supported not only by the paleomagnetic measurements

already discussed but also by studies of oceanic sediments. These studies indicate not only that the sedimentary deposits are thinner in the zones of the mid-oceanic ridge but also that, as they become thicker away from the ridge, the basal layers of the sedimentary accumulations become older, as we discussed earlier in this section. This all suggests that the oldest sea floor lies farthest from the oceanic ridges.

If we follow this line of reasoning, we should expect to find that belts of varying width along the active ridges mark the most recently formed oceanic crust. If we accept the reconstruction shown in Figure 8.23, then an area approaching 50 percent of the ocean floor has been formed during the Cenozoic—in approximately the last 65 million years. In this regard it is interesting to note that no rocks older than 150 million years have as yet been taken from the deep ocean basins. Moreover, present-day sedimentation rates in the oceans can account for the sedimentary accumulations there within the last 100 to 200 million years. All this leads to the suggestion that the ocean basins, as we know them, are young. It would then follow that, although ocean basins have existed through most of geologic time, they have

been continuously reconstituted and recycled. Furthermore, their shapes and their geographic locations have been shifting constantly (Figure 8.24).

DATING THE MOVEMENTS

Interpretation of earthquake data tells us lithospheric plates have been carried as distinct entities downward along subduction zones to depths of 700 km or more. Although we can only tentatively decipher the present position of plate borders and their most recent direc-

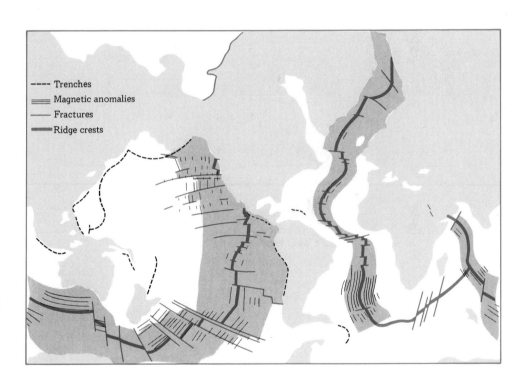

FIGURE 8.23

The shaded portion of the basins is thought to represent the area of oceanic crust formed within the last 65 million years. [From F. J. Vine, "Sea-floor Spreading—New Evidence," *J. Geol. Educ.*, vol. 17, pp. 6–16, 1969.]

FIGURE 8.24

Convection cell in upper mantle providing the energy to drive the lithospheric plates. The ocean opens up by spreading from the ocean ridge and is in turn consumed at the subduction zones. The leading edge of a drifting continent becomes an active margin, and the trailing edge an inactive margin. [From various sources.]

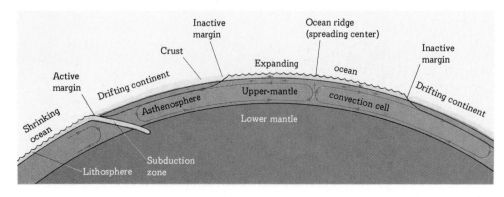

TABLE 8.3

Major events in continental drift and sea-floor spreading[a]

Era or period	Age, millions of yr	Southern hemisphere (Gondwanaland)	Northern hemisphere (Laurasia)
Cenozoic	0–		
	–		Gulf of California opens.
	–	Gulf of Aden opens.	Spreading direction in eastern Pacific changes.
	–		
	–	Red Sea opens.	Iceland begins to form.
	50–	India collides with Eurasia.	Arctic Basin opens.
Cretaceous	–		Greenland begins to separate from Norway.
	–		
	–	Australia begins to separate from Antarctica.	
	–		
	100–		
	–		
	–		
	–	Madagascar begins to separate from Africa, and South America from Africa.	North America separates from Eurasia.
Jurassic	–		
	–		
	150–		
	–		
	–	Africa begins to separate from India and Australia-Antarctica.	
	–		North America begins to separate from Africa.
Triassic	–		
	200–		

[a]From various sources.

tions and rates of movement, we know that these movements have occurred at least long enough for the generation of 700 km of new lithosphere at the spreading centers (otherwise a gap would exist in the ocean where plates have moved away from the midoceanic-ridge spreading centers).

Rates of sea-floor spreading have been determined from analyzing magnetic anomalies (Section 8.2). By relating these rates and distances of movement, we can determine the ages for various portions of the sea floor. The initial opening of several parts of the ocean basins has thus been determined for the northern-hemisphere landmass (Laurasia) and the southern-hemisphere Gondwanaland. Opening of the ocean basins coincides with, and results in, the drifting of the associated continents.

Physical and biological events continuing on a given landmass might be rather abruptly terminated by the rifting apart of that landmass. Dating of these events would permit the determination of a minimum age for the initial breaking apart of such a continent. By a combination of such methods and by the dating of the ocean-floor materials, it is possible to establish the principal sequence of events in the breakup of the supercontinents, as listed in Table 8.3.

SUMMARY SPECULATIONS

The idea of plate tectonics can be used to integrate some of the major features of the ocean basins. Thus the midoceanic ridges and rises, with their central rift valleys and a sedimentary cover that is young and thin (or lacking), reflect the upwelling of mantle material along the ascending currents of adjacent convection cells. The large fracture systems that offset the midoceanic ridges are transform faults. These faults come into being as the rising convection currents finally diverge laterally, carrying with them the overlying crustal plates and upper mantle. As the currents cool and turn downward, they carry with them the oceanic crust, with its overlying sediments, as well as the underlying upper mantle. To many students of the problem the deep, arcuate trenches of the oceans are the surface manifestations of the downturning of the convection currents along subduction zones. Great crustal plates, then, are seen as moving within a worldwide framework. The leading edges of the plates are consumed in the trench–island arc systems of the Pacific basin and along the Mediterranean-Himalayan belt of Eurasia.

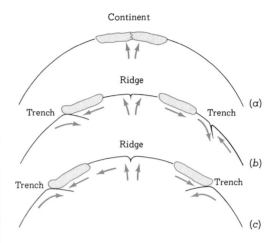

FIGURE 8.25

This sequence illustrates what might be expected if mantle material upwells beneath a continent. In (a) a continent is shown as rifting begins. In (b) the sea floor has opened between the two fragments of the original landmass. The leading edge of one is shown overriding a trench, and that of the other is still separated from an oceanic trench. Diagram (c) shows a still further stage of sea-floor spreading and continental drift. Both landmasses are shown overriding trenches. [After F. J. Vine, "Sea-floor Spreading—New Evidence," *J. Geol. Educ.*, vol. 17, pp. 6–16, 1969.]

Along the trailing edges, new crust is generated in the active ridges. The drifting of continents, then, becomes a consequence of the movement of crustal plates and the appearance and disappearance of oceanic crust.

If the mantle wells up beneath a continent we can expect that the continent will be rifted and that the two fragments will drift apart as a new ocean basin forms. Such a possibility is suggested in Figure 8.25. As the ocean basins enlarge and the continents ("tied" to the underlying but moving upper mantle) drift away from each other, their leading edges may be fronted by trenches. The continents, being lighter, override the trenches.

A modern example presents itself in the western margin of South America. Here Pacific oceanic crust appears to spread eastward toward the continent. South America's margin is marked by a deep ocean trench, a narrow continental shelf, and the towering

Andes. Earthquakes occur, becoming deeper in origin inland, the deepest being about 700 km below the surface. The earthquake foci lie along the Benioff zone that dips beneath the South American continent.[5]

The elevation of the Andean chain is interpreted in part as the result of underflow of oceanic crust. Seismic and igneous activity are seen as related to the motion between the Pacific and American plates along the Benioff zone. Farther north, the Mesozoic history of the western United States has been interpreted in terms of the underflow of 2,000 km of Pacific material beneath the American plate. The early Paleozoic mountain-building episodes in Newfoundland and Scotland have been ascribed by some workers to the interplay between two plates, one underthrusting another.

Thus mountain ranges, ocean basins, and most major features of the earth's surface owe their origin to the movement and interaction of lithospheric plates.

[5]A seismic zone dipping beneath a continent or continental margin and having a deep-sea trench as its surface expression is called a **Benioff zone** after Hugo Benioff, an American seismologist, who first described the feature. It is equivalent to our subduction zone.

OUTLINE

Oldest continental crust has been dated at 3.8 billion years from highly metamorphosed rocks from the Baltic Shield, Greenland, and Minnesota.

Constant freeboard of continents is assumed from study of the relative elevation of continents with respect to sea level through at least the latter part of geologic time.

Age of ocean basins is surprisingly young because older ocean floor is destroyed at subduction zones as new floor develops at spreading centers.

The earth's magnetism results from its behaving as if it were a magnet; its changes as recorded in rocks is closely tied to the case in favor of continental drift.

Secular variation of the magnetic field has been known through long-term changes in magnetic declination since the midseventeenth century.

The geomagnetic pole at the earth's surface does not coincide with the geographic pole.

The magnetosphere is a region above the earth that traps powerful radiation from space and wards it off.

Cause of magnetism is both internal and external in relation to the earth. A small part is external, caused by the sun. The largest part of the earth's field originates within the earth and is best explained as caused by circulation within the earth's liquid core, which acts as if it were a giant dynamo. The **direction** and **intensity** of the magnetic field varies with time. If the intensity decreases far enough, it may eventually pass a zero point and the earth's field *may* become polarized in a sense the reverse of that of the present.

Paleomagnetism is the study of the earth's field in past time. **Thermal remanent magnetism** (TRM) occurs in igneous rock-forming minerals as they cool below the Curie temperature. **Depositional remanent magnetism** (DRM) forms as magnetic particles align parallel to the earth's field at time of deposition. DRM (and TRM) reflects magnetic field existing at time of rock formation.

The **virtual geomagnetic pole** is the equivalent pole position representing the magnetic field as recorded at one observation point. The discrepancy between geomagnetic poles and geographic poles is thought to average out to zero in time. The ancient poles are thought to have been, like the modern poles, dipolar; the ancient poles have migrated thousands of miles since the Precambrian, and the path of polar migration is different for each continent. Direction of the earth's magnetic field reversed itself many times in the geological past.

Continental drift is supported by the data provided by paleomagnetism, the shape of continents, sea-floor spreading, ancient climatic patterns, and disrupted geological structures.

Plate tectonics is a hypothesis that postulates that the earth's crust is broken into large blocks being renewed in some zones and destroyed in others.

Lithospheric plate boundaries may be **convergent, divergent,** or **transform,** depending upon whether the driving forces are compressional, tensional, or parallel.

Movement of plates may be by means of convection cells or by interaction of mantle material over **hot spots,** or **plumes.**

Dating the movement of plates can be done by analyzing magnetic anomalies and rates of sea-floor spreading, as well as dating structural trends and stratigraphic units that developed on continents prior to their rifting.

Plate-tectonic theory can integrate notions of continental drift, sea-floor spreading, seismic zones, continents, and ocean basins.

SUPPLEMENTARY READINGS

Clague, D. A., and R. D. Jarrard "Tertiary Pacific Plate Motion Deduced from the Hawaiian-Emperor Chain," *Bull. Geol. Soc. Am.,* vol. 84, pp. 1135–1154, 1973. *Rates of rotation of the Pacific plate relative to the hot spot presently located near Hawaii are determined from radiometric age dates obtained from the Hawaiian-Emperor and other Pacific chains of volcanoes and seamounts.*

Condie, K. C. *Plate Tectonics and Crustal Evolution,* Pergamon Press, Inc., Elmsford, N.Y., 1976. *A good recent summation of the development of oceanic and continental crust in light of plate-tectonic theory; well illustrated.*

Dewey, J. F., W. C. Pitman, III, W. B. F. Ryan, and J. Bonnin "Plate Tectonics and the Evolution of the Alpine System," *Bull. Geol. Soc. Am.,* vol. 84, pp. 3137–3180, 1973. *A detailed and comprehensive summary of the history and development of the Alpine system in terms of stepwise location and motion of Africa relative to Europe and the opening of the Atlantic Ocean.*

Wyllie, P. J. *The Way the Earth Works,* John Wiley & Sons, Inc., New York, 1976. *A highly readable and well-illustrated paperback on recent developments, using as a central theme the new global geology with good coverage of plate tectonics and continental drift.*

9

MASS MOVEMENT OF SLOPE MATERIAL

The earth's surface is a collection of slopes: some gentle, some steep; some long, some short; some cloaked with vegetation, some bare of plants; some veneered with soil, some with naked rock. They lead from hill crests downward to streams and rivers. Taken together they make up our landscape. Down these slopes move rainwater (Figure 9.1) and the water from melting snow to collect in rivulets, brooks, and rivers flowing to the ocean. Down them also travel soil and rock, which, via the rivers, finally get to the sea, there to form sedimentary rocks. But because this material derives from the slopes themselves, it follows that they must change.

We have already learned that, when rocks are exposed at the earth's surface, weathering immediately sets to work to establish an equilibrium between the rock and its new environment. Among the other factors joining forces with the processes of weathering is the direct pull of gravity. It acts to move the products of weathering and even unweathered bedrock to lower and lower levels. The movement of surface material caused by gravity is known as **mass movement**. Sometimes it takes place suddenly in the form of great landslides and rock falls from precipitous cliffs, but often it occurs almost imperceptibly as the slow creep of soil across gently sloping fields.

Water also plays a direct role in the movement of slope material. The impact of a raindrop, the flow of water in thin sheets or tiny rills, and the solution of solid material as water seeps into the ground—all contribute to the erosion of hillslopes.

9.1 FACTORS OF MASS MOVEMENT

Gravity provides the energy for the downslope movement of surface debris and bedrock, but several other factors, particularly water, augment gravity and ease its work (see Figure 9.2).

Immediately after a heavy rainstorm you may have witnessed a landslide on a steep hillside or on the bank of a river. In many unconsolidated deposits the pore spaces between individual grains are filled partly with moisture and partly with air, and so long as this condition persists, the surface tension of the moisture gives a certain cohesion to the soil. When, however, a heavy rain forces all the air out of the pore spaces, this surface tension is completely destroyed. The cohesion of the soil is reduced, and the whole mass becomes more susceptible to downslope movement. The presence of water also adds weight to the soil on a slope although this added weight is probably not a very important factor in promoting mass movement.

Water that soaks into the ground and completely fills the pore spaces in the slope material contributes to instability in another way: The water in the pores is under pressure, which tends to push apart individual grains or even whole rock units and to decrease the internal friction or resistance of the material to movement. Here again water assists in mass movement.

Another factor, as we shall see, is air trapped beneath rapidly moving masses of rock debris as they hurtle down steep slopes. This air, confined by the falling material, acts as a cushion to reduce the friction of the debris with the ground and to make possible high-velocity movement of rock slides.

Gravity can move material only when it is able to overcome the material's internal resistance to being set into motion. Clearly, then, any factor that reduces this resistance to the point where gravity can take over contributes to mass movement. The erosive action of a stream, an ocean, or a glacier may so steepen a slope that the earth material can no longer resist the pull of gravity and is forced to give in to mass movement. In regions of cold climate alternate freezing and thawing of earth materials may suffice to set them in motion. The impetus needed to initiate movement may also be furnished by earthquakes, excavating or blasting operations, sonic booms of aircraft, or even the gentle activities of burrowing animals and growing plants.

BEHAVIOR OF MATERIAL

In Chapter 7, in discussing the deformation of earth material, we pointed out that it may behave as an **elastic solid**, a **plastic substance**, or a **fluid**. Material moving down a slope may behave in the same manner. We could actually study any type of mass movement and classify it on the basis of these three behavior types; but we should have to assemble an excessive amount of technical data and find the picture complicated by the fact that material often behaves in different ways during any one movement. Therefore we shall simply classify mass movement as either **rapid** or **slow**.

FIGURE 9.1

These badland slopes in Badlands National Monument, South Dakota, are almost devoid of vegetation. Gravity and rain wash are the major agents in slope maintenance. [National Park Service.]

FIGURE 9.2

A large slab of massive Wingate Sandstone has fallen and shattered at the foot of this precipitous cliff, where the San Juan River now begins to move away the fragments. [William C. Bradley.]

9.2 RAPID MOVEMENTS

Catastrophic and destructive movements of rock and soil, the most spectacular and readily identified examples of mass movement, are popularly known as **landslides,** but the geologist recognizes slump, rock slides, debris slides, mudflows, and earthflows.

LANDSLIDES

Landslides include a wide range of movements, from the slipping of a stream bank to the sudden, devastating release of a whole mountainside. Some landslides involve only the unconsolidated debris lying on bedrock; others involve movement of the bedrock itself.

Slump Sometimes called **slope failure,** *slump* is the downward and outward movement of rock or unconsolidated material traveling as a unit or as a series of units. Slump usually occurs where the original slope has been sharply steepened, either artificially or naturally. The material reacts to the pull of gravity as if it were an elastic solid, and large blocks of the slope move downward and outward along curved planes. The upper surface of each block is tilted backward as it moves.

Figure 9.3 shows a slump's beginning at Gay Head, Massachusetts. The action of the sea has cut away the unconsolidated material at the base of the slope, steepening it to a point where the earth mass can no longer support itself. Now the large block has begun to move along a single curving plane, as suggested in Figure 9.4.

After a slump has been started, it is often helped along by rainwater's collecting in basins between the tilted blocks and the original slope. The water drains down along the plane on which the block is sliding and promotes further movement.

A slump may involve a single block, as shown in Figures 9.3 and 9.4, but very often it consists of a series of slump blocks (Figure 9.5).

Rock slides The most catastrophic of all mass movements are **rock slides**—sudden, rapid slides of bedrock along planes of weakness.

A great rock slide occurred in 1925 on the flanks of Sheep Mountain, along the Gros Ventre River in northwestern Wyoming, not far from Yellowstone National Park (see Figure 9.6). An estimated 36.5 million m³ of rock and debris plunged down the valley wall and swept across the valley floor. The nose of the slide rushed some 110 m up the opposite wall and then settled back, like liquid being sloshed in a great basin. The debris formed a dam between 68 and 75 m high across the valley, the dammed-up river creating a lake almost 8 km long. The spring floods of 1927 raised the water level to the lip of the dam, and in mid-May the water flooded over the top. So rapid was the downcutting of the dam that the lake level was lowered about 15 m in 5 h. During the flood that followed, several lives were lost in the town of Kelly, in the valley below.

FIGURE 9.3

The beginning of slump, or slope failure, along the sea cliffs at Gay Head, Massachusetts. Note that the slump block is tilted back slightly, away from the ocean. This slump block eventually moved downward and outward toward the shore along a curving plane, a portion of which is represented by the face of the low scarp in the foreground. [Harvard University, Gardner Collection.]

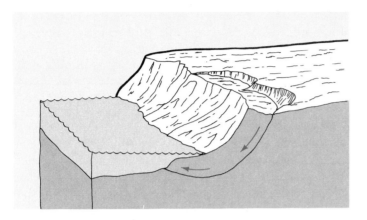

FIGURE 9.4

This diagram shows the type of movement found in a slump similar to that pictured in Figure 9.3. A block of earth material along the steepened cliff has begun to move downward along a plane that curves toward the ocean.

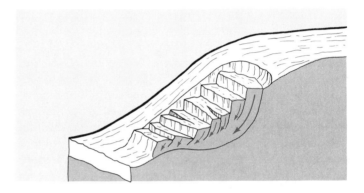

FIGURE 9.5

Many slump movements contain several discrete units, as suggested here.

FIGURE 9.6

Aerial photograph of the Gros Ventre rock slide in northwestern Wyoming. The lake in the lower left of the picture has been dammed by a landslide that moved down into the valley of the Gros Ventre River. The area from which the material slid is about 2.5 km long and is well marked by the white scar down the center of the photograph. When this photograph was taken, the vegetative cover of trees and bushes on the adjoining slopes had not yet reestablished itself in the slide area. [University of Washington, Austin Post.]

The Gros Ventre slide was a long time in the making, and there was probably nothing that could have been done to prevent it. Conditions immediately before the slide are shown in Figure 9.7a. In this part of Wyoming the Gros Ventre valley cuts through sedimentary beds inclined between 15° and 21° to the north. The slide occurred on the south side of the valley wall, where the beds dip into the valley. Notice that the sandstone bed is separated from the limestone strata by a thin layer of clay. Before the rock slide occurred, the sandstone bed near the bottom of the valley had been worn thin by erosion. The melting of winter snows and the heavy rains that fell during the spring of 1925 furnished an abundant supply of water, which seeped down to the thin layer of clay, soaking it and reducing the adhesion between it and the overlying sandstone. When the sandstone was no longer able to hold its position on the clay bed, the rock slide roared down the slope. Figure 9.7b suggests the amount of material that was moved from the spur of Sheep Mountain to its resting place on the valley floor.

Another rock slide, in 1903, killed 70 people in the coal-mining town of Frank, Alberta, when some 36.8 million m³ of rock crashed down from the crest of Turtle Mountain, which rises over 900 m above the valley.

Mining activities may have triggered this movement, but natural causes were basically responsible for it. As Figure 9.8 shows, Turtle Mountain has been sculptured from a series of limestone, sandstone, and shale beds, which have been tilted, folded, and faulted. The diagram shows that the greater part of the mountain is made of limestone and that the valley below is underlain by less-resistant beds of sandstone, siltstone, shale, and coal. It also shows four factors that contributed to the slide: the steepness of the mountain, the bedding planes in the limestone dipping parallel to the mountain face, the thrust plane that breaks the strata partway down the mountain, and the weak shale, siltstone, and coal beds in the valley.

The steep valley wall enhanced the effectiveness of gravity, and the bedding planes along with the fault planes served as potential planes of movement. The weak shale beds at the base of the mountain probably underwent slow plastic deformation under the weight of the overlying limestone, and as the shale was deformed, the limestone settled lower and lower. The set-

FIGURE 9.7

Diagrams to show the nature of the Gros Ventre slide: (a) Conditions existing before the slide took place; (b) the area of the slide and the location of the debris in the valley bottom. Note that the sedimentary beds dip into the valley from the south. The large section of sandstone slid downward along the clay bed. [Redrawn from William C. Alden, "Landslide and Flood at Gros Ventre, Wyoming," *Trans. AIME*, vol. 76, p. 348, 1928.]

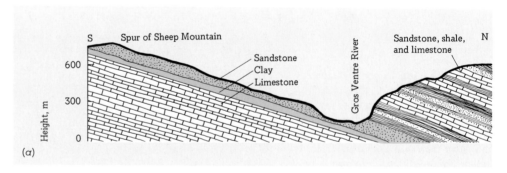

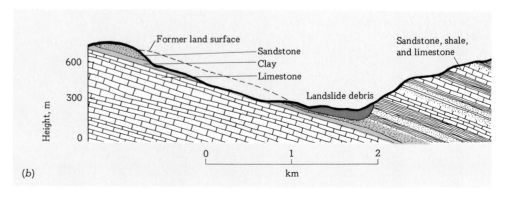

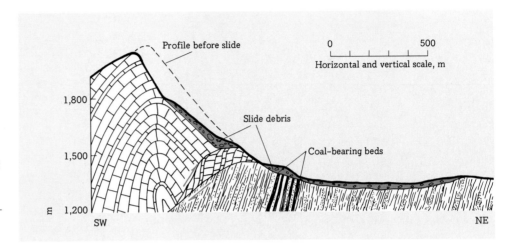

FIGURE 9.8

A cross section to show the conditions at Turtle Mountain that brought about the Frank, Alberta, landslide. [Modified from D. M. Cruden and J. Krahn, "A Reexamination of the Geology of the Frank Slide," *Canad. Geotech. J.*, vol. 10, no. 4, 1973.]

tling action may have been helped along by the coal-mining operations in the valley as well as by frost action, rain, melting snows, and earthquake tremors that had shaken the area 2 years before.

In any event stress finally reached the point where the limestone beds gave way along their bedding planes and pushed outward toward the valley along the uppermost fault; the great mass of rock hurtled down into the valley. The rock material behaved in three different ways. The shales underwent plastic deformation, producing a condition of extreme instability on the mountain slope. When the strata that still held the limestone mass on the slope sheared in the manner of an elastic solid, the slide actually began. Once under way, the rock debris bounced, ricocheted, slid, and rolled down the mountain slope until it was literally "launched" into the air by a ledge of rock. Thereafter it arched outward and downward to the valley below. Once on the valley floor, it moved at high speed up the hills on the far side of the river. In this phase it moved like a viscous fluid with a series of waves spreading out along its front, and there is some evidence that air was caught and compressed beneath the debris, drastically reducing its friction with the ground and allowing speeds of at least 100 km/h.

A more recent slide occurred in southwestern Montana on August 17, 1959. An earthquake whose focus was located just north of West Yellowstone, Montana, triggered a rock slide in the mouth of the Madison River Canyon, about 32 km to the west. A mass of rock estimated to be of over 32 million m³ fell from the south wall of the canyon. It climbed over 100 m up the opposite valley wall and dammed a lake 8 km long and over 30 m deep. More than a score of people lost their lives in the Madison River campground area below the slide (see Figures 9.9 and 9.10). Survivors reported extremely powerful winds along the margins of the slide. In fact some persons were blown away, never again to be seen, and a 2-t automobile was observed to have been carried by the wind over 10 m before smashing against a row of trees. It seems clear that, when the mass of rock fell to the valley bottom, it trapped and compressed large quantities of air that served to provide a near frictionless base on which it spread across the valley. The high winds represented the extrusion of this mass of air from beneath the moving debris.

The Alaska earthquake of March 27, 1964 (see Section 6.4) had many effects on the topography and shoreline of a large part of the southern portion of the state. Landslides were common and in inhabited places caused much damage. In the glaciated mountains of the Alaska Range rock slides were particularly spectacular. Austin Post of the University of Washington mapped some 78 earthquake-induced slides that dumped material down glacially steepened mountain slopes onto the surfaces of the glaciers. The debris of the largest of these slides covered over 11 km², and the total debris on all glaciers amounted to more than 120 km². One large slide on the Sherman glacier near Cordova contributed 28 million m³ of rock debris to the surface of the glacier. Assuming the density of the material to be about 2.3, the mass of rock material weighed 57.5 million t.

In addition to the modern landslides described above, there are vestiges of prehistoric slides of great magnitude. One, at Saidmarreh, Iran, involved the

FIGURE 9.9

The Madison River Canyon before the landslide of August, 1959. The slide began on the forested valley wall on the far side of the river. The debris clogged the river channel to form a lake. The spot where this fisherman once stood is now under 30 m of water. Compare with Figure 9.10. [Montana Power Co.]

FIGURE 9.10

An aerial view of the Madison Canyon after the landslide. An outlet channel for the newly formed lake has been cut across the landslide dam. [William C. Bradley.]

movement of 4,245 million m³ of rock for a distance of over 12 km. Data on this and some other landslides are given in Table 9.1.

Debris slides A *debris slide* is a small, rapid movement of largely unconsolidated material that slides or rolls downward and produces a surface of low hummocks with small, intervening depressions. Movements of this sort are common on grassy slopes, particularly after heavy rains, and in unconsolidated material along the steep slopes of stream banks and those of shorelines.

Warnings of landslides We usually think of a landslide as breaking loose without warning, but it is more accurate to say that people in the area simply fail to detect the warning.

For example, a disastrous rock slide at Goldau, Switzerland, in 1806, wiped out a whole village, killing 457 people. The few who lived to tell the tale reported that they themselves had no warning of the coming slide but that animals and insects in the region may have been more observant or more sensitive. For several hours before the slide, horses and cattle seemed to be extremely nervous, and even the bees abandoned their hives. Some slight preliminary movement probably took place before the rock mass broke loose.

During the spring of 1935 slides took place in clay deposits along a German superhighway that was being built between Munich and Salzburg. The slides came as a complete surprise to the engineers, but for a full week the workmen had been murmuring, *Der Abhang wird lebendig* ("The slope's becoming alive").

Landslides like the one on Turtle Mountain are often preceded by slowly widening fissures in the rock near the upward limit of the future movement.

There is some evidence that landslides may recur periodically in certain areas. In southeastern England, not far from Dover, extensive landslides have been occurring every 19 to 20 years. Some observers feel that there may be a correlation between such periodic mass

TABLE 9.1
Data on some large landslides[a]

Location and date	Debris volume, millions of m^3	Distance moved, km	Minimum speed, km/h	Height climbed, m
Elm, Switzerland, September 11, 1881	11.3	1.4	160	100
Sherman, Cordova, Alaska, March 27, 1964	28.3	4.3	185	145
Madison Canyon, Montana, August 17, 1959	32.6	0.8	130	110
Gros Ventre River, Wyoming, June 23, 1925	36.5	1.3	150	105
Frank, Alberta, Canada, April 29, 1903	36.8	1.7	175	120
Silver Reef, California, prehistoric	226	5.3	105	50
Blackhawk, California, prehistoric	283	8.0	120	60
Saidmarreh, Iran, prehistoric	4,245	12.8	340	460

[a] Compiled from various sources and arranged in order of volume of material moved.

movement and times of excessive rainfall. On steep slopes in very moist tropical or semitropical climates, for instance, landslides seem to follow a cyclic pattern: First, a landslide strips the soil and vegetation from a hill; in time new soil and vegetation develop, the old scar heals, and when the cover reaches a certain stage, the landsliding begins again. Although landslides may occur in cycles, our data are as yet far too scanty to support firm conclusions.

MUDFLOWS

A **mudflow** is a well-mixed mass of rock, earth, and water that flows down valley slopes with the consistency of newly mixed concrete. In mountainous, desert, and semiarid areas mudflows manage to transport great masses of material.

The typical mudflow originates in a small, steep-sided gulch or canyon where the slopes and floor are covered by unconsolidated or unstable material. A sudden flood of water, from cloudbursts in semiarid country or from spring thaws in mountainous regions, flushes the earth and rocks from the slopes and carries them to the stream channel. Here the debris blocks the channel until the growing pressure of the water be-

comes great enough to break through. Then the water and debris begin their down-valley course, mixing together with a rolling motion along the forward edge of the flow. The advance of the flow is intermittent; for sometimes it is slowed or halted by a narrowing of the stream channel; at other times it surges forward, pushing obstacles aside or carrying them along with it.

Eventually the mudflow spills out of the canyon mouth and spreads across the gentle slopes below. No longer confined by the valley walls or the stream channel, it splays out in a great tongue, spreading a layer of mud and boulders that ranges from a few centimeters to 1 m or more in thickness. Mudflows can move even large boulders weighing 80 t or more for hundreds of meters across slopes as gentle as 5°.

EARTHFLOWS

Earthflows involve the plastic movement of unconsolidated material lying on solid bedrock and are usually helped along by excessive moisture. They move slowly but perceptibly and may involve from a few to several million cubic meters of earth material. The line at which the material pulls away from the slope is marked by an abrupt scarp or cliff, as shown in Figure 9.11.

FIGURE 9.11

An earthflow in a roadcut near Dallas, Texas, shows a sharp scar high on the slope and, farther down, the area of soil movement by flow. [C. W. Brown.]

FIGURE 9.12

A series of convergent talus cones have formed along the steep slopes of the Elk Mountains, Colorado. [U.S. Geological Survey.]

TALUS

Strictly speaking, a **talus** is a slope built up by an accumulation of rock fragments at the foot of a cliff or a ridge. The rock fragments are sometimes referred to as **rock waste** or **slide rock**. In practice, however, talus is widely used as a synonym for the rock debris itself.

In the development of a talus, rock fragments are loosened from the cliff and clatter downward in a series of free falls, bounces, and slides. As time passes, the rock waste finally builds up a heap or sheet of rock rubble. An individual talus resembles a half cone with its apex resting against the cliff face in a small gulch. A series of these half cones often forms a girdle around high mountains, completely obscuring their lower portions (see Figure 9.12). Eventually, if the rock waste accumulates more rapidly than it can be destroyed or removed, even the upper cliffs become buried, and the growth of the talus stops. The slope angle of the talus varies with the size and shape of the rock fragments. Although angular material can maintain slopes up to 50°, rarely does a talus ever exceed angles of 40°.

A talus is subject to the normal process of chemical weathering, particularly in a moist climate. The rock waste is decomposed, especially toward its lower limit, or toe, which may grade imperceptibly into a soil.

9.3 SLOW MOVEMENTS

Slow mass movements of unconsolidated material are more difficult to recognize and less fully understood than rapid movements; yet they are extremely important in the sculpturing of the land surface. Because they operate over long periods of time, they are probably responsible for the transportation of more material than

FIGURE 9.13

Alternating beds of sandstone and shale have been tilted until they stand vertically. A little over 1 m below the surface they begin to bend downslope under the influence of gravity. The amount of downslope movement increases toward the surface. Haymond Formation in the Marathon region of Texas. [William C. Bradley.]

are rapid and violent movements of rock and soil.

Before the end of the nineteenth century William Morris Davis aptly described the nature of slow movements.

The movement of land waste is generally so slow that it is not noticed. But when one has learned that many land forms result from the removal of more or less rock waste, the reality and the importance of the movement are better understood. It is then possible to picture in the imagination a slow washing and creeping of the waste down the land slopes; not bodily or hastily, but grain by grain, inch by inch; yet so patiently that in the course of ages even mountains may be laid low.[1]

CREEP

In temperate and tropical climates, a slow downward movement of surface material known as **creep** operates even on gentle slopes with a protective cover of grass and trees. It is hard to realize that this movement is actually taking place. Because the observer sees no break in the vegetative mat, no large scars or hummocks, he has no reason to suspect that the soil is in motion beneath his feet.

Yet this movement can be demonstrated by exposures in soil profile (see Figure 9.13) and by the behavior of tree roots, of large blocks of resistant rock, and of artificial objects such as fences and telephone poles (see Figure 9.14). Figure 9.15 shows a section through a hillside underlain by flat-lying beds of shale, limestone, clay, sandstone, and coal. The slope is covered with rock debris and soil. But notice that the beds near the base of the soil bend downslope and thin out rapidly. These beds are being pulled downslope by gravity and are strung out in ever-thinning bands that may extend for hundreds of meters. Eventually they approach the

[1]William Morris Davis, *Physical Geography*, p. 261, Ginn and Company, Boston, 1898.

FIGURE 9.14

Telephone poles set in a talus slope south of Yellowstone National Park record movement of the slope material. Originally the poles stood vertically, but over a period of about 10 years the talus has moved and tilted the two nearest poles 5° and the most distant pole 8°. [Sheldon Judson, 1968.]

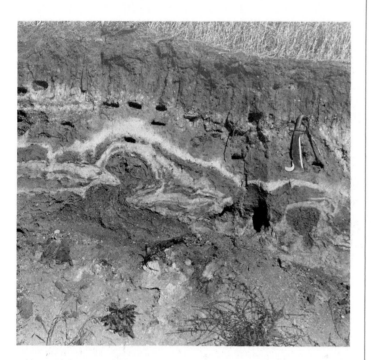

FIGURE 9.16

Beds of silt, sand, and clayey gravel have been contorted by differential freezing and thawing during the more rigorous climates of glacial times. The gravel at the base of the exposure is clayey, gravelly till deposited directly by glacier ice. Above this are the highly contorted beds of water-laid sand (dark) and silt (light), which were originally lying flat. The modern soil has developed across the contortions. The white of the silt bands is due to precipitation of calcium carbonate from soil water. The brush hook is about 0.5 m long. Exposure north of Devils Lake, east-central North Dakota. [U.S. Geological Survey, Saul Aronow.]

FIGURE 9.15

The partially weathered edges of horizontal sedimentary rocks are dragged downslope by soil creep. The tree is also moving slowly downslope, as is evidenced by the root system spread out behind the more rapidly moving trunk. [Redrawn from C. F. S. Sharpe and E. F. Dosch, "Relation of Soil-creep to Earthflow in the Appalachian Plateaus," *J. Geomorphol.*, vol. 5, p. 316, 1942. By permission of Columbia University Press.]

surface and lose their identity in the zone of active chemical weathering.

Figure 9.15 also shows other evidence that the soil is moving. Although when viewed from the surface the tree appears to be growing in a normal way, it is actually creeping slowly down the slope. Because the surface of the soil is moving more rapidly than is the soil beneath it, the roots of the tree are unable to keep up with the trunk.

We can find other evidence of the slow movement of soil in displaced fences and tilted telephone poles and gravestones. On slopes where resistant rock layers crop up through the soil, fragments are sometimes broken off and distributed down the slope by the slowly moving soil.

Many other factors cooperate with gravity to produce creep. Probably the most important is moisture in the soil, which works to weaken the soil's resistance to movement. In fact any process that causes a dislocation in the soil brings about an adjustment of the soil downslope under the pull of gravity. Thus the burrows of animals tend to fill downslope, and the same is true of cavities left by the decay of organic material, such as the root system of a dead tree. The prying action of swaying trees and the tread of animals and even of human beings may also aid in the motion. The result of all these processes, aided by the influence of gravity, is to produce a slow and inevitable downslope creep of the surface cover of debris and soil.

SOLIFLUCTION

The term **solifluction** (from the Latin *solum*, "soil," and *fluere*, "to flow") refers to the downslope movement of debris under saturated conditions. Solifluction is most pronounced in high latitudes, where the soil is strongly affected by alternate freezing and thawing and where the ground freezes to great depths; but even moderately deep seasonal freezing promotes solifluction.

Solifluction takes place during periods of thaw. Because the ground thaws from the surface downward, the water that is released cannot percolate into the subsoil and adjacent bedrock, which are still frozen and therefore impermeable to water. As a result, the surface soil becomes sodden and tends to flow down even the gentlest slopes. Solifluction is an important process in the reduction of land masses in Arctic climates, where it transports great sheets of debris from higher to lower elevations.

During the glacier advances of the Pleistocene, a zone of intense frost action and solifluction bordered the southward-moving ice. In some places we can still find the evidence of these more rigorous climates pre-

served in distorted layers of earth material just below the modern soil (see Figure 9.16).

Frost action plays queer tricks in the soils of the higher elevations and latitudes. Strange polygonal patterns made up of rings of boulders surrounding finer material, stripes of stones strewn down the face of hillsides, great tabular masses of ice within the soil, and deep ice wedges that taper downward from the surface—all are found in areas where the ground is deeply frozen (see Figure 9.17). The behavior of frozen ground is one of the greatest barriers to the settlement of Arctic regions. The importance of these regions has increased in recent years, and studies begun by Scandinavian and Soviet investigators are now being intensively pursued by United States and Canadian scientists. The present-day distribution of permanently frozen ground is shown in Figure 9.18.

FIGURE 9.17

Aerial photograph of polygonal patterns developed by ice wedges in the coastal plain of northern Alaska. Area shown is about 650 m wide. [U.S. Coast and Geodetic Survey.]

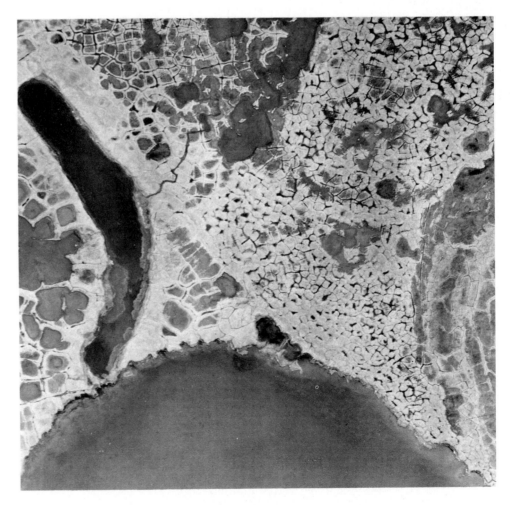

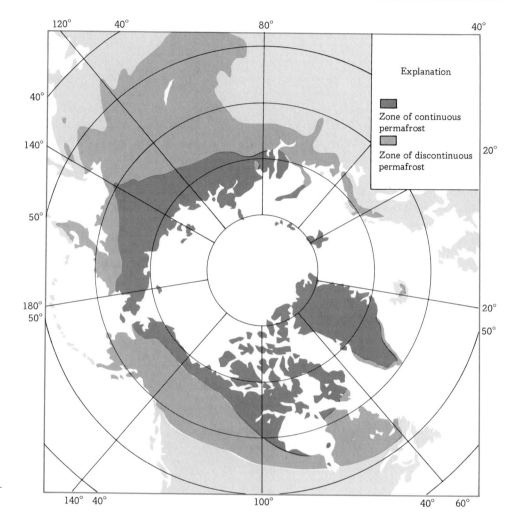

FIGURE 9.18

Distribution of permafrost in the northern hemisphere. [Troy L. Pewe (ed.), *The Periglacial Environment: Past and Present*, Arctic Institute of North America, McGill–Queen's University Press, Montreal, 1969.]

ROCK GLACIERS

Rock glaciers are tongues of rock waste that form in the valleys of certain mountainous regions. Although they consist almost entirely of rock, they bear a striking resemblance to ice glaciers. A typical rock glacier is marked by a series of rounded ridges, suggesting that the material has behaved as a viscous mass.

Observations on active rock glaciers in Alaska indicate that movement takes place through flow of interstitial ice within the mass. Favorable conditions for the development of rock glaciers include a climate cold enough to keep the ground continuously frozen, steep cliffs to supply debris, and coarse blocks that allow for large interstitial spaces.

9.4 DIRECT ACTION OF WATER ON HILLSLOPES

We have referred to the role of water in promoting mass movement. It also operates directly on slopes to move material downward to the channels of conventional streamways.

IMPACT OF RAINDROPS

W. D. Ellison has made a study of the effectiveness of individual **raindrops in erosion**. A drop striking a wet,

FIGURE 9.19

A raindrop splashes onto a water-soaked surface, scattering fine particles of earth. [Wide World.]

muddied field will splash water and suspended solid particles into the air. If there is any slope at all to the land, the material splashed uphill will not travel as far as will that splashed downhill. There will, then, be a net downslope motion of material by rain splash (see Figure 9.19).

RAIN WASH

Rain falling on a slope may soak into the ground or run off the slope. Vegetation will slow the runoff, promote infiltration, and generally reduce the erosive effectiveness of water flowing on the slopes; but if vegetation is lacking or sparse, rainwater will rapidly flow down the slope. This flow, called **rain wash,** can become an effective agent of erosion.

A thin sheet of water may accumulate on a slope and become thicker as it proceeds downslope. This is sometimes referred to as **sheet wash,** or **sheet flow.** Theoretically, a film of water would increase at a constant rate as it moved down a uniform slope, gathering more and more water. This increase occurs but not in a uniform sheet, for the simple reason that no slope is completely uniform. Initial irregularities, differing resistance to erosion, and varying lengths of flow may all contribute to the concentration of sheet wash into broad shallow troughs, or channels, which determine the location of miniature stream channels called **rills.** The rills are eroded along the axes of broad troughs; for here the water has its greatest velocity and hence its greatest erosive power. Most of us have seen the effect of such erosion on artificially created exposures along a roadway or on a slope newly prepared for planting.

SOLUTION

It is clear that, as water on the surface comes into contact with earth material, it will begin to take mineral matter into solution. This will be true whether the water flows off the surface or soaks into the ground. The effect is to move material from the slopes, and we therefore must add solution as a third way in which hillslopes are reduced by the direct action of water.

OUTLINE

Erosion of hillslopes occurs as a direct result of gravity and by the action of water.

Mass movement of bedrock and unconsolidated material is in response to the pull of gravity. The material may behave as an elastic solid, as a plastic substance, or as a liquid. Movements are classified as either rapid or slow and may be influenced by saturation of material by water, by steepening of slopes by erosion, by earthquakes, and by the activity of animals, including human beings.

Rapid movements include landslides, mudflows, earthflows, and talus formation.

Slow movements include creep, solifluction, and rock glaciers.

Water moves material on slopes by the direct impact of raindrops, by rain wash, and by solution.

SUPPLEMENTARY READINGS

Carson, M. A., and M. J. Kirby *Hillslope Form and Process,* Cambridge University Press, New York, 1972. *An advanced text on the subject.*

Embleton, Clifford, and Cuchlaine A. M. King *Periglacial Geomorphology,* 2nd ed., John Wiley & Sons, Inc., New York, 1975. *A short but authoritative treatment of surface processes found in Arctic areas.*

Washburn, A. Lincoln *Periglacial Processes and Environments,* Edward Arnold (Publishers) Ltd., London, 1973. *A handsomely illustrated treatment of periglacial activities.*

Zaruba, Quido, and Vojtech Mencl *Landslides and their control,* Elsevier Publishing Company, Amsterdam, 1969. *This volume gives many examples of mass movement, most of them drawn from central Europe.*

10

RUNNING WATER

Through millions of years of earth history, agents of erosion have been working constantly to reduce the landmasses to the level of the seas. Of these agents **running water** is the most important. Year after year the streams of the earth (Figure 10.1) move staggering amounts of debris and dissolved material through their valleys to the great settling basins, the oceans.

10.1 WORLD DISTRIBUTION OF WATER

The distribution of the earth's estimated water supply is given in Appendix C. The great bulk of this water, over 97 percent, is in the oceans, and less than 3 percent is on the land. Atmospheric moisture is surprisingly low, about 0.001 percent of the total. Even less is in the stream channels of the world at any one moment. We see, then, that the earth's water is in varying amounts at different places; yet it moves from one place to another. As is written in Ecclesiastes 1:7, "All the rivers run into the sea; yet the sea is not full; unto the place from whence the rivers come, thither they return again." This is part of the circulation of water from land to ocean and back again to land that we call the **hydrologic cycle.**

10.2 HYDROLOGIC CYCLE

The streams of the world flow to the sea at a rate that would fill the ocean basins in 40,000 years. These streams, flowing as they do, must have a continuing supply of water. Furthermore, it is clear that the oceans must themselves somehow lose water in order to make room for the continuous supply that comes to them.

The identification of the source of stream water is historically recent. Well into the eighteenth century, it

FIGURE 10.1

Stream valleys are the single most characteristic feature of the world's landscape. Even in the desert, as here on the slopes leading down into Death Valley, streams have fashioned an intricate system of valleys. [U.S. Army Air Corps.]

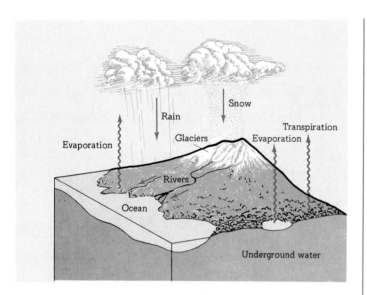

FIGURE 10.2

In the hydrologic cycle water evaporated into the atmosphere reaches the land as rain or snow. Here it may be temporarily stored in glaciers, lakes, or the underground before returning by the rivers to the sea. Or some may be transpired or evaporated directly back into the atmosphere before reaching the sea.

was the general belief that streams were replenished by springs that drew their water by some complex system through the underground from the oceans. This belief was nourished by the assumption that rainfall was inadequate to account for the flow observed in rivers; for rivers ran continuously even though rain was intermittent. Furthermore, it was generally held that rainwater could not soak into the ground and replenish springs, a conclusion that seems to owe its origin to Seneca (3 B.C.–A.D. 65), Roman statesman and philosopher, who based his conclusions on inconclusive observations while he tended his vineyards.

In the seventeenth century three different types of observation laid the base for the hydrologic cycle as we accept it, which is diagrammatically represented in Figure 10.2. In 1674 Pierre Perrault (1611–1680), a French lawyer and sometime hydrologist, presented the results of his measurements in the upper portion of the drainage basin of the Seine River. Over a 3-year span he collected data on the amount of precipitation in this portion of the basin. At the same time he kept track of the amount of water discharged by the river below the portion of the river basin where he had data on precipitation. The results showed that the flow of the river was surprisingly low when compared with the total amount of water available from precipitation. In fact the annual precipitation was six times the total volume of river flow. There was enough rainfall, then, to account for the flow of the Seine in this part of its basin.

At about the same time, the French physicist Edmé Mariotte (1620–1684) made more exact studies of discharge in the Seine basin. His publications, appearing posthumously in 1684, verified Perrault's conclusions. Mariotte further demonstrated, by experimentation at the Paris Observatory, that seepage through the earth cover was less, but not much less, than the amount of rainfall. He also demonstrated the increase in the flow of springs during rainy weather and the decrease during time of drought. It was then evident that the earth permits penetration of moisture.

In 1693 Edmund Halley (1656–1742), of comet fame, provided data on evaporation in relation to rainfall. He roughly calculated the amount of water discharged annually by rivers into the Mediterranean Sea and added this to the amount that fell directly on the sea's surface. He was then able to compute the approximate amount of water that was being evaporated back into the air each year from the surface of the Mediterranean. He found that there was more than enough water being pumped into the atmosphere to feed all the rivers coming into the sea.

10.3 PRECIPITATION AND STREAM FLOW

After water has fallen on the land as precipitation, it follows one of the many paths that make up the hydrologic cycle. By far the greatest part is evaporated directly into the air or is taken up by plants and transpired ("breathed" back) to the atmosphere. A smaller amount follows the path of **runoff,** the water that flows off the land, and the smallest amount of precipitation soaks into the ground through **infiltration.**

Figure 10.3 shows how infiltration, runoff, and **evaporation-transpiration** vary in six widely separated localities in the United States. In the examples given, between 54 and 97 percent of the total precipitation travels back to the atmosphere through transpiration and evaporation. About 2 to 27 percent drains into streams and oceans as runoff, and between 1 and 20 percent penetrates the ground through infiltration.

In Chapters 11 and 12 we shall consider water at various stages of the hydrologic cycle. In Chapter 11 we look at the water that infiltrates into the ground; in Chapter 12 we study water that has been impounded on the land as glacier ice, a condition that represents a temporary halting of the water's progress through the cycle. In this chapter we shall concentrate on the na-

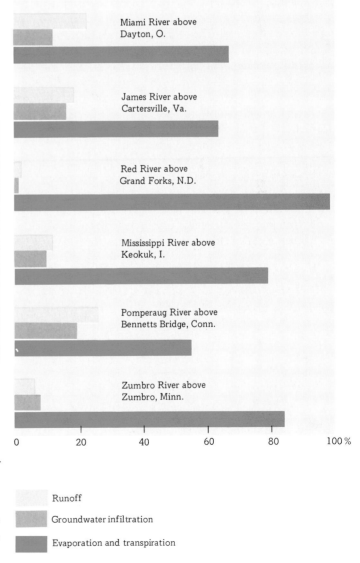

FIGURE 10.3

Distribution of precipitation in selected drainage basins. Notice that in all cases 50 percent or more of all moisture that falls is returned to the atmosphere by evaporation and transpiration. Runoff from the surface is comparatively small, and infiltration of water into the underground is still less. [Data from W. G. Hoyt and others, "Studies of Relation of Rainfall and Run-off in the United States, *U.S. Geol. Surv. Water Supply Paper* 772, 1936.]

ture and effects of runoff. Bearing in mind the ways in which water that falls as precipitation proceeds through the hydrologic cycle, we can express the amount of runoff by the following generalized formula:

Runoff = precipitation − (infiltration + evaporation and transpiration)

L. B. Leopold and W. B. Langbein of the United States Geological Survey have reported some figures on the water budget of the hydrologic cycle in the United States. There is an average of around 75 cm of precipitation per unit of area over the country each year. Of this total, about 22.5 cm are carried as runoff by the streams to the ocean. Another 52.5 cm go back into the atmosphere by evaporation and transpiration. Of course, some water filters into the ground, but all but a very small amount—about 0.25 cm—is returned to the surface, where it disappears either as runoff or as evapotranspiration. These figures indicate that across the United States there is an apparent loss of 22.5 cm annually from the atmosphere. This is the runoff that goes to the ocean, but the debit is made up by the evaporation of moisture from the ocean and its transfer by the atmosphere to the continents.

Figures on worldwide water budget are in Figure 10.4: Average precipitation for the lands of the world is variously estimated and approximates about 80 cm/ year. Estimates of runoff from the continents to the oceans vary. One estimate is given in Appendix C. This does not include water that may get to the oceans in

the form of ice. The figures indicate that 32.4×10^{15} l (liters) of water annually flow to the sea. This is a runoff of 24.9 cm/unit area. The difference between runoff and precipitation is the evapotranspiration, in this instance about 55 cm. It follows that the oceans contribute about 25 cm to the yearly continental precipitation budget.

LAMINAR AND TURBULENT FLOW

Consider now the manner in which water moves. When it moves slowly along a smooth channel or through a tube with smooth walls, it follows straight-line paths that are parallel to the channel or walls. This type of movement is called **laminar flow.**

If the rate of flow increases, however, or if the confining channel becomes rough and irregular, this smooth, streamlined movement is disrupted. The water in contact with the channel is slowed down by friction, whereas the rest of the water tends to move along as before. As a result (see Figures 10.5 and 10.6), the water is deflected from its straight paths into a series of eddies and swirls. This type of movement is known as **turbulent flow.** Water in streams usually flows along in this way, its turbulent flow being highly effective both in eroding a stream's channel and in transporting materials.

When a stream reaches an exceptionally high velocity along a sharply inclined stretch or over a waterfall, the water moves in plunging, jetlike surges. This type of flow, closely related to turbulent flow, is called **jet,** or **shooting, flow.**

VELOCITY, GRADIENT, AND DISCHARGE

The **velocity** of a stream is measured in terms of the distance its water travels in a unit of time. A velocity of 15 cm/s is relatively low, and a velocity of about 625 to 750 cm/s is relatively high.

A stream's velocity is determined by many factors, including the amount of water passing a given point, the nature of the stream banks, and the **gradient,** or slope, of the stream bed. In general a stream's gradient decreases from its headwaters toward its mouth; as a result, a stream's longitudinal profile is more or less concave toward the sky (Figure 10.7). We usually express the gradient of a stream as the vertical distance a stream descends during a fixed distance of horizontal flow. The Mississippi River from Cairo, Illinois, to the mouth of the Red River in Arkansas has a low gradient;

FIGURE 10.4

Water budget for the world. [Data from U.S. Geological Survey.]

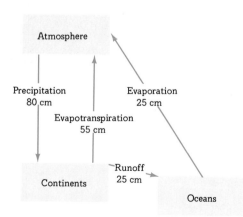

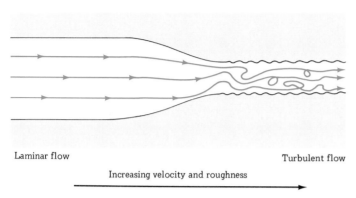

Laminar flow

Turbulent flow

Increasing velocity and roughness

FIGURE 10.5

Diagram showing laminar and turbulent flow of water through a section of pipe. Individual water particles follow paths depicted by the colored lines. In laminar flow the particles follow paths parallel to the containing walls. With increasing velocity or increasing roughness of the confining walls laminar flow gives way to turbulent flow. The water particles no longer follow straight lines but are deflected into eddies and swirls. Most water flow in streams is turbulent.

FIGURE 10.6

Most stream flow being turbulent, the Lewis River in Wyoming tumbles over falls and swirls and boils through a rough, irregular channel. [Sheldon Judson.]

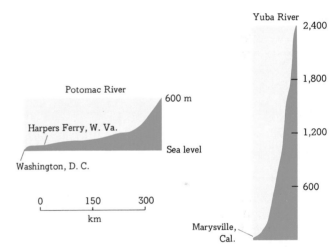

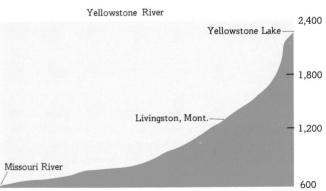

FIGURE 10.7

In longitudinal profile (from mouth to headwaters), a stream valley is concave to the sky. Irregularities along the profile indicate variations in rates of erosion. [Redrawn from Henry Gannett, "Profiles of Rivers in the United States," *U.S. Geol. Surv. Water Supply Paper 44*, 1901.]

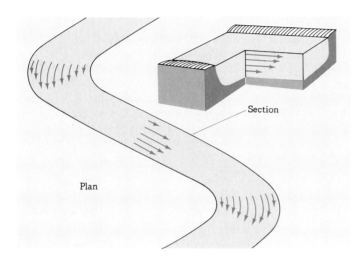

Plan

Section

FIGURE 10.8

Velocity variations in a stream. Both in plan view and in cross section the velocity is slowest along the stream channel, where the water is slowed by friction. On the surface it is most rapid at the center in straight stretches and toward the outside of a bed where the river curves. Velocity increases upward from the river bottom.

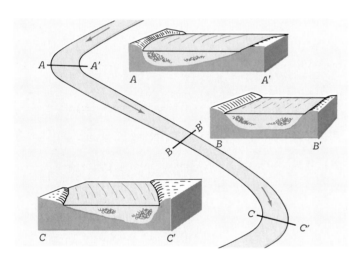

A   A'

B   B'

C   C'

A   A'

B   B'

C   C'

FIGURE 10.9

Zones of maximum turbulence in a stream are shown by the curly symbols in the sections through a river bed. They occur where the change between the two opposing forces—the forward flow and the friction of the stream channel—is most marked. Note that the maximum turbulence along straight stretches of the river is located where the stream banks join the stream floor. On bends the two zones have unequal intensity; the greater turbulence is located on the outside of a curve.

for along this stretch the drop varies between 2 and 10 cm/km. On the other hand the Arkansas River, in its upper reaches through the Rocky Mountains in central Colorado, has a high gradient; for there the drop averages 7.5 m/km. The gradients of other rivers are even higher. The upper 20 km of the Yuba River in California, for example, have an average gradient of 42 m/km; and in the upper 6.5 km of the Uncompahgre River in Colorado the gradient averages 66 m/km.

The velocity of a stream is checked by the turbulence of its flow, friction along the banks and bed of its channel, and to a much smaller extent by friction with the air above. Therefore, if we were to study a cross section of a stream, we should find that the velocity would vary from point to point. Along a straight stretch of a channel the greatest velocity is achieved toward the center of the stream at, or just below, the surface, as shown in Figure 10.8.

We then have two opposing forces: the **forward flow** of the water under the influence of gravity and the **friction** developed along the walls and bed of the stream. These are the two forces that create different velocities. Zones of maximum turbulence occur where the different velocities come into closest contact. These zones are very thin because the velocity of the water increases very rapidly as we move into the stream away from its walls and bed; but within these thin zones of great turbulence a stream shows its highest potential for erosive action (see Figure 10.9).

There is one more term that will be helpful in our discussion of running water: **discharge,** the quantity of water that passes a given point in a unit of time. (In this book we shall measure it in cubic meters per second.) Discharge varies not only from one stream to another but also within a single stream from time to time and from place to place along its course. Discharge usually increases downstream as more and more tributaries add their water to the main channel. Spring floods may so greatly increase a stream's discharge that its normally peaceful course becomes a raging torrent.

10.4 ECONOMY OF A STREAM

Elsewhere we have seen that earth processes tend to seek a balance, to establish an equilibrium, and that there is, in the words of James Hutton, "a system of beautiful economy in the works of nature." We found, for example, that weathering is a response of earth materials to the new and changing conditions they meet as they are exposed at or near the earth's surface. On a larger scale we found that the major rock groups—igneous, sedimentary, and metamorphic—reflect certain environments and that, as these environments change, members of one group may be transformed into members of another group. These changes were traced in what we called the rock cycle (Figure 1.8). Water running off the land in streams and rivers is no exception to this universal tendency of nature to seek equilibrium.

ADJUSTMENTS OF DISCHARGE, VELOCITY, AND CHANNEL

Just a casual glance tells us that the behavior of a river during its flood stage is very different from its behavior during the low-water stage. For one thing a river carries more water and moves more swiftly in flood time. Furthermore, the river is generally wider during flood; its level is higher; and we should guess, even without measuring, that it is also deeper. We can relate the discharge of a river to its width, depth, and velocity:

Discharge (m³/s) = channel width (m) × channel depth (m) × water velocity (m/s)

In other words, if the discharge at a given point along a river increases, then the width, depth, velocity, or some combination of these factors must also increase. We now know that variations in width, depth, and velocity are neither random nor unpredictable. In most streams, if the discharge increases, then the width, depth, and velocity each increase at a definite rate. The stream maintains a balance between the amount of water it carries, on the one hand, and its depth, width, and velocity, on the other. Moreover, it does so in an orderly fashion, as shown in Figure 10.10.

Let us turn now from the behavior of a stream at a single locality to the changes that take place along its entire length. From our own observation we know that the discharge of a stream increases downstream as more and more tributaries contribute water to its main channel. We also know that the width and depth increase as we travel downstream. But if we go beyond casual observation and gather accurate data on the width, depth, velocity, and discharge of a stream from its headwaters to its mouth for a particular stage of flow—for example, flood or low water—we should find again that the changes follow a definite pattern and that depth and width increase downstream as the discharge increases (Figure 10.11). And, surprisingly enough, we should also find that the stream's *velocity* increases toward its mouth. This is contrary to our ex-

FIGURE 10.10

As the discharge of a stream increases at a given gauging station, so do its velocity, width, and depth. They increase in an orderly fashion, as shown by these graphs based on data from a gauging station in the Cheyenne River near Eagle Butte, South Dakota. [Redrawn from Luna B. Leopold and Thomas Maddock, "The Hydraulic Geometry of Stream Channels and Some Physiographic Implications," *U.S. Geol. Surv. Prof. Paper* 252, p. 5, 1953.]

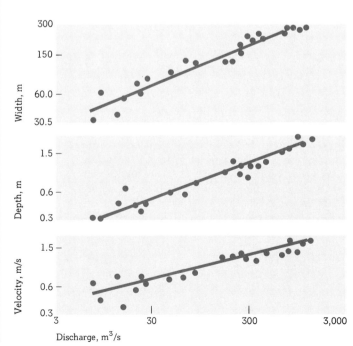

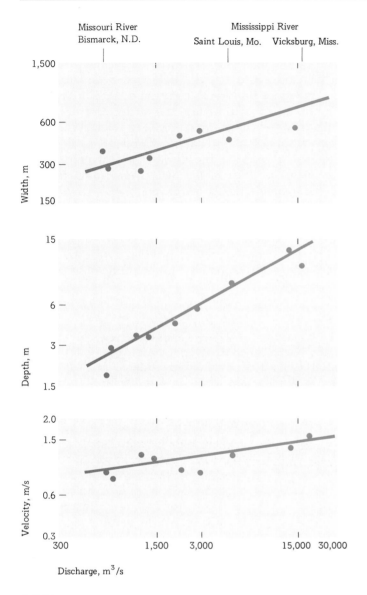

FIGURE 10.11

Stream velocity and depth and width of a channel increase as the discharge of a stream increases downstream. Measurements in this example were made at mean annual discharge along a section of the Mississippi-Missouri river system. [Luna B. Leopold and Thomas Maddock, "The Hydraulic Geometry of Stream Channels and Some Physiographic Implications," *U.S. Geol. Surv. Prof. Paper* 252, p. 13, 1953.]

pectations; for we know that the gradients are higher upstream, which suggests that the velocities in the steeper headwater areas would also be higher. But the explanation for this seeming anomaly is simple: To handle the greater discharge downstream, a stream not only must deepen and widen its channel but also must increase its velocity.

Floods   When the stream discharge exceeds the channel capacity, water must rise over the channel banks and flood the adjacent low-lying land. Floods can be catastrophic, but disastrous as they may be, we must consider them as the expression of natural stream behavior. A study of long-term stream records shows that there is some method to the apparent madness of a stream. For instance, a stream flow that just fills its channel, that is, the **bank-full stage,** occurs every 1 to 2 years. Looking at greater volumes of flow, we find that the frequency of floods, which exceed the bank-full stage, can be statistically anticipated. The procedure is simple and involves a stream-discharge record of at least 10 years: The maximum discharge for each year is listed; then these flows are ranked according to magnitude, that is, from largest to smallest; and then the flows can be plotted, as shown in Figure 10.12.

The example in Figure 10.12 is based on floods that have been recorded at Nashville, Tennessee, over a 93-year period. The largest flow, 5,750 m^3/s, is entered at 93 years because floods that large occur apparently about once in 93 years. The size of the second largest flood, 5,570 m^3/s, is plotted at $\frac{93}{2} = 46.5$ years because two floods of this size or larger have occurred in 93 years. The third largest flood is shown at $\frac{93}{3} = 31$ years, and so on through the entire record. Because the graph shows points on a roughly straight line, we can extend the line to the upper right to estimate the size of floods to be expected once in 100 years or once in 200 years. On this basis we speak of 10-year floods, 50-year floods, and even 1,000-year floods. But remember that this is a statistical prediction. We don't know precisely when a flood of a given size will occur. The historical record merely tells us to expect a flood of a given size some time within a given span.

That floods recur and that the recurrence interval of a flood of a given size is predictable should assure us that floods are a natural, expectable phenomenon. We should expect them every so often. And indeed we should look at the river's flood plain as a part of the river's domain, useful to the river in flood times and certain to be claimed by the river from time to time. Therefore, if we use the flood plain, we must remember

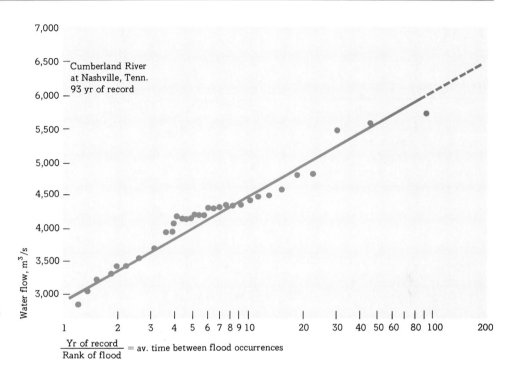

FIGURE 10.12

Average interval between floods of various sizes on the Cumberland River at Nashville, Tennessee, plotted from data in *U.S. Geol. Surv. Water Supply Paper* 771. The vertical scale is an ordinary linear scale, and the horizontal scale is known as a logarithmic scale.

that it is only on loan to us. Whatever use we put the flood plain to ought to be as compatible as possible with the use to which the river is sure to put it. Experience should teach us not only that streams will top their banks from time to time but that even the most energetic flood-control plans (including the use of dams, levees, dredging, and channel modification) sometimes fail. The problem is partly an economic one: Against what magnitude flood shall we try to protect the flood plain? The 10-year flood? The 100-year flood? The 500-year flood? Obviously the larger the flood we attempt to defend against, the greater the expense. At what point does the expense of protection outweigh the gain achieved? This is obviously a matter of public policy. An alternative to protection by engineering works is to put the flood plain to a use compatible with flooding, such as agriculture or a park system. This practice may in many instances be more appropriate than tampering with the river system.

BASE LEVEL OF A STREAM

Base level is a key concept in the study of stream activity. The **base level** is defined as the lowest point to which a stream can erode its channel. Anything that prohibits the stream from lowering its channel serves to create a base level. For example, the velocity of a stream is checked when it enters the standing, quiet waters of a lake. Consequently the stream loses its ability to erode, and it cannot cut below the level of the lake. The lake's control over the stream is actually effective along the entire course upstream; for no part of the stream can erode beneath the level of the lake—at least before the lake has been destroyed. But in a geological sense every lake is temporary, and therefore, after the lake has been destroyed (perhaps by the downcutting of its outlet), it will no longer control the stream's base level, and the stream will be free to continue its downward erosion. Because of its impermanence, the base level formed by a lake is referred to as **temporary**. But even after a stream has been freed from one temporary base level, it will be controlled by others farther downstream; and its erosive power is always influenced by the ocean, which is the **ultimate** base level. Yet, as we shall see in Chapter 13, the ocean itself is subject to changes in level; so even the ultimate base level is not fixed.

The base level of a stream may be controlled not only by lakes but also by layers of resistant rock and the level of the main stream into which a tributary drains (see Figure 10.13).

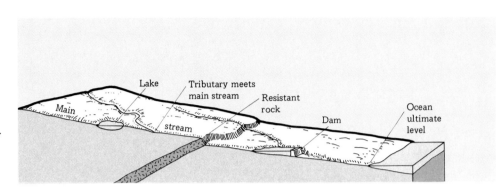

FIGURE 10.13

Base level for a stream may be determined by natural and artificial lakes, by a resistant rock stratum, by the point at which a tributary stream enters a main stream, and by the ocean. Of these the ocean is considered ultimate base level; others are temporary base levels.

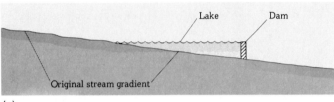

(a)

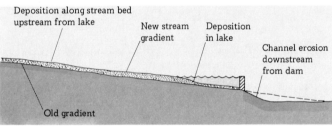

(b)

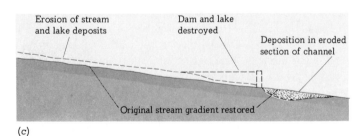

(c)

FIGURE 10.14

A stream adjusts its channel to changing base level. Construction of a dam across a stream raises its base level, imposes a lower velocity on the stream above the dam, and thus causes deposition in this section of the channel. Failure of the dam lowers base level, increases the velocity, and causes erosion of the previously deposited sediments.

Adjustment to changing base level We have defined base level as the lowest level to which a stream can erode its channel. If for some reason the base level is either raised or lowered, the stream will adjust the level of its channel to adapt to the new situation.

Let us see what happens when we *raise* the base level of a stream by building a dam and creating a lake across its course. The level of the lake serves as a new base level, and the gradient of the stream above the

dam is now less steep than it was originally. As a result, the stream's velocity is reduced. Because the stream can no longer carry all the material supplied to it, it begins to deposit sediments at the point where it enters the lake. As time passes, a new river channel is formed with approximately the same slope as the original channel but at a higher level (Figure 10.14).

What happens when we *lower* the base level by removing the dam and hence the lake? The river will now cut down through the sediments it deposited when the lake still existed. In a short time the profile of the channel will be essentially the same as it was before we began to tamper with the stream.

In general, then, we may say that a stream adjusts itself to a rise in base level by building up its channel through sedimentation and that it adjusts to a fall in base level by eroding its channel downward.

10.5 WORK OF RUNNING WATER

The water that flows through river channels does several jobs: (1) It **transports** debris, (2) it **erodes** the river channel deeper into the land, and (3) it **deposits** sediments at various points along the valley or delivers them to lakes or oceans. Running water may help to create a chasm like that of the Grand Canyon of the Colorado, or in flood time it may spread mud and sand across vast expanses of valley flats, or it may build deltas, as at the mouths of the Nile and Mississippi.

The nature and extent of these activities depend on the kinetic energy of the stream, and this in turn depends on the amount of water in the stream and the gradient of the stream channel. A stream expends its energy in several ways. By far the greatest part is used up in the friction of the water with the stream channel and in the friction of water with water in the turbulent eddies we discussed above. Relatively little of the stream's energy remains to erode and transport material. Deposition takes place when energy decreases and the stream can no longer move the material it has been carrying.

TRANSPORTATION

The material that a stream picks up directly from its own channel—or that is supplied to it by slope wash, tributaries, or mass movement—is moved downstream toward its eventual goal, the ocean. The amount of material that a stream carries at any one time, which is called its **load,** is usually less than its **capacity**—that is, the total amount it is capable of carrying under any given set of conditions (see Figure 10.15). The maximum size of particle that a stream can move measures the **competence** of a stream.

There are three ways in which a stream can transport material: (1) solution, (2) suspension, and (3) bed load.

Solution In nature no water is completely pure. We have already seen that, when water falls and filters down into the ground, it dissolves some of the soil's

compounds. Then the water may seep down through openings, pores, and crevices in the bedrock and dissolve additional matter as it moves along. Much of this water eventually finds its way to streams at lower levels. The amount of dissolved matter contained in water

FIGURE 10.15

These converging rivers in British Columbia illustrate the different loads carried by two streams. The Frazer River enters from the upper right, milky with suspended sediment derived largely from the melting of mountain glaciers. Its load is high but probably somewhat less than capacity. The Thompson River, entering from the lower right, is relatively clear and carries a very small load, much less than its capacity. [Elliott A. Riggs.]

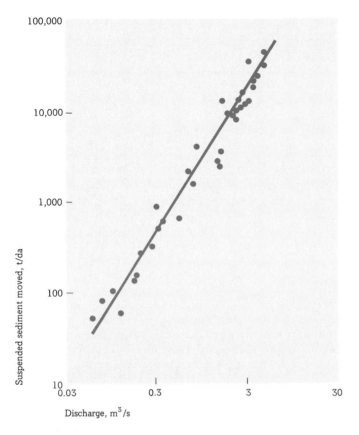

FIGURE 10.16

The suspended load of a stream increases very rapidly during floods, as illustrated by this graph based on measurements in the Rio Puerco near Cabezon, New Mexico. [Redrawn from Luna B. Leopold and John P. Miller, "Ephemeral Streams—Hydraulic Factors and Their Relation to the Drainage Net," *U.S. Geol. Surv. Prof. Paper* 282-A, p. 11, 1956.]

varies with climate, season, and geologic setting and is measured in terms of parts of dissolved matter per million parts of water. Sometimes the amount of dissolved material exceeds 1,000 ppm, but usually it is much less. By far the most common compounds found in solution in running water, particularly in arid regions, are calcium and magnesium carbonates. In addition streams carry small amounts of chlorides, nitrates, sulfates, and silica, with perhaps a trace of potassium. It has been estimated that the total load of dissolved material delivered to the seas every year by the streams of the United States is nearly 300 million t. The rivers of the world average an estimated 115 to 120 ppm

of dissolved matter, which means that annually they carry to the sea about 3.9 billion t.

Suspension Particles of solid matter that are swept along in the turbulent current of a stream are said to be in **suspension**. This process of transportation is controlled by two factors: the turbulence of the water and a characteristic known as **terminal velocity** of each individual grain. Terminal velocity is the constant rate of fall that a grain eventually attains when the acceleration caused by gravity is balanced by the resistance of the fluid through which the grain is falling. In this case the fluid is water. If we drop a grain of sand into a quiet pond, it will settle toward the bottom at an ever-increasing rate until the friction of the water on the grain just balances this rate of increase. Thereafter it will settle at a constant rate, its terminal velocity. If we can set up a force that will equal or exceed the terminal velocity of the grain, we can succeed in keeping it in suspension. Turbulent water supplies such a force. The eddies of turbulent water move in a series of orbits, and grains caught in these eddies will be buoyed up, or held in suspension, as long as the velocity of the turbulent water is equal to, or greater than, the terminal velocity of the grains.

Terminal velocity increases with the size of the particle if its general shape and density remain the same. The bigger a particle, the more turbulent the flow needed to keep it in suspension. And because turbulence increases when the velocity of stream flow increases, it follows that the greatest amount of material is moved during flood time, when velocities and turbulence are highest. The graph in Figure 10.16 shows how the suspended load of a stream increases as the discharge increases. In just a few hours or a few days during flood time a stream transports more material than it does during the much longer periods of low or normal flow. Observations of the area drained by Coon Creek, at Coon Valley, Wisconsin, over a period of 450 days showed that 90 percent of the stream's total suspended load was carried during an interval of 10 days, slightly over 2 percent of total time.

Silt and clay-sized particles are distributed fairly evenly through the depth of a stream, but coarser particles in the sand-size range are carried in greater amounts lower down in the current, in the zone of greatest turbulence.

Bed load Materials in movement along a stream bottom constitute the stream's *bed load*, in contrast to its suspended load and solution load. Because it is difficult to observe and measure the movement of the bed load, we have little data on the subject. Measurements on the Niobrara River near Cody, Nebraska, however, have shown that, at discharges between about 6 and 30 m³/s, the bed load averaged about 50 percent of the total load. Particles in the bed load move along in three ways: saltation, rolling, or sliding.

The term **saltation** has nothing to do with salt. It is derived from the Latin *saltare*, "to jump." A particle moving by saltation jumps from one point on the stream bed to another. First it is picked up by a current or turbulent water and flung upward; then, if it is too heavy to remain in suspension, it drops to the stream floor again at some spot downstream.

Some particles are too large and too heavy to be picked up, even momentarily, by water currents, but they may be pushed along the stream bed, and depending on their shape, they move forward by either **rolling** or **sliding.**

EROSION

People have not always been convinced that most of the world's stream valleys are fashioned by the streams that flow in them, and by the processes that streams have encouraged. Today, however, we know enough about running water to confirm this. In the next few paragraphs we look at the various ways a stream may remove material from its channel and its banks. But you may wish, as you continue this chapter, to keep in mind the larger question and consider what added evidence we might cite to demonstrate that a stream is responsible for its own valley.

Direct lifting In turbulent flow, as we have seen, water travels along paths that are not parallel to the bed. The water eddies and whirls, and if an eddy is powerful enough, it dislodges particles from the stream channel and lifts them into the stream. Whether or not this will happen in a given situation depends on a number of variables that are difficult to measure, but if we assume that the bed of a stream is composed of particles of uniform size, then the graph in Figure 10.17 gives us the approximate stream velocities that are needed to erode particles of various sizes, such as clay, silt, sand, granules, and pebbles. A stream bed composed of fine-sized sand grains, for example, can be eroded by a stream with a velocity of between 18 and 50 cm/s, depending on how firmly the sand grains are packed. As the fragments become larger and larger, ranging from coarse sand to granules to pebbles, increasingly higher velocities are required to move them, as we should expect.

FIGURE 10.17

The central band in this diagram shows the velocity range at which turbulent water will lift particles of differing size off the stream bed. The width is affected by the shape, density, and consolidation of the particles. [From Ake Sundborg, "The River Klarälven, a Study in Fluvial Processes," *Geografis. Ann.*, vol. 38, p. 197, 1956.]

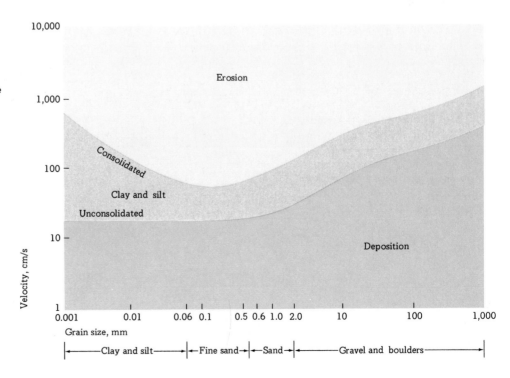

But what we might *not* expect is that, as the particles become smaller than about 0.06 mm in diameter, they do not become more easily picked up by the stream. In fact if the clay and silt are firmly consolidated, then increased velocities will be needed to erode the particles. The reason for this is generally that, the smaller the particle, the more firmly packed the deposit is and thus the more resistant to erosion. Moreover, the individual particles may be so small that they do not project sufficiently high into the stream to be swept up by the turbulent water.

Abrasion, impact, and solution   The solid particles carried by a stream may themselves act as erosive agents; for they are capable of abrading (wearing down) the bedrock itself or larger fragments in the bed of the stream. When the bedrock is worn by abrasion, it usually develops a series of smooth, curving surfaces, either convex or concave. Individual cobbles or pebbles on a stream bottom are sometimes moved and rolled about by the force of the current, and as they rub together, they become both rounder and smoother.

The impact of large particles against the bedrock or against other particles knocks off fragments, which are added to the load of the stream.

Some erosion also results from the solution of channel debris and bedrock in the water of the stream. Most of the dissolved matter carried by a stream, however, is probably contributed by the underground water that drains into it.

DEPOSITION

As soon as the velocity of a stream falls below the point necessary to hold material in suspension, the stream begins to deposit its suspended load. Deposition is a selective process. First, the coarsest material is dropped; then, as the velocity (and hence the energy) continues to slacken, finer and finer material settles out. We shall reconsider stream deposits later.

10.6 FEATURES OF VALLEYS

CROSS-VALLEY PROFILES

Earlier in this chapter we mentioned the longitudinal profile of a stream (Figure 10.7). Now let us turn to a discussion of the cross-valley profile, that is, the profile of a cross section at right angles to the trend of the stream's valley. In Figure 10.18 notice that the channel of the river runs across a broad, relatively flat **flood plain**. During flood time, when the channel can no longer accommodate the increased discharge, the

FIGURE 10.18

Cross-sectional sketches of typical stream valleys. The major features of valleys in cross section include divides, valley walls, river channel, and in some instances a flood plain. Divides may be flat topped, broadly rounded, or narrow.

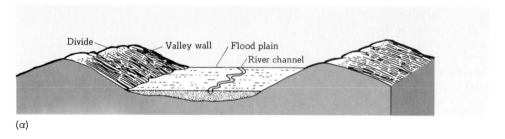

(a)

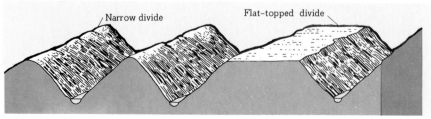

(b)

stream overflows its banks and floods this area. On either side of the flood plain **valley walls** rise to crests called **divides**, separations between the central valley and the other valleys on either side. In Figure 10.18*b* no flood plain is present; for the valley walls descend directly to the banks of the river. This diagram also illustrates two different shapes of divide. One is broad and flat; the other is narrow, almost knife edged. Both are in contrast to the broadly convex divides shown in Figure 10.18*a*.

DRAINAGE BASINS, NETWORKS, AND PATTERNS

A **drainage basin** is the entire area from which a stream and its tributaries receive their water. The Mississippi River and its tributaries drain a tremendous section of the central United States, reaching from the Rocky to the Appalachian Mountains, and each tributary of the Mississippi has its own drainage area, which forms a part of the larger basin. Every stream, even the smallest brook, has its own drainage basin, shaped differently from stream to stream but characteristically pear shaped, with the main stream emerging from the narrow end (see Figure 10.19).

Individual streams and their valleys are joined together into networks. In any single network the streams prove to have a definite geometric relationship in a way first detailed in 1945 by the United States hydraulic engineer Robert Horton. We can devise a demonstration of this relationship by first ranking the streams in a hierarchy. This ranking, shown in Figure 10.20, lists small headwater streams without tributaries as belonging to the first order in the hierarchy. Two or more first-order streams join to form a second-order stream; two or more second-order streams join to form a third-order stream; and so on. In other words a stream segment of any given order is formed by the junction of at least two stream segments of the next lower order. The main stream segment of the system always has the highest order number in the network, and the number of this stream is assigned to describe the stream basin. Thus the main stream in the basin shown in Figure 10.20 is a fourth-order stream, and the basin is a fourth-order basin.

It is apparent from Figure 10.20 that the number of stream segments of different order decreases with increasing order. When we plot the number of streams of a given order against order, the points define a straight

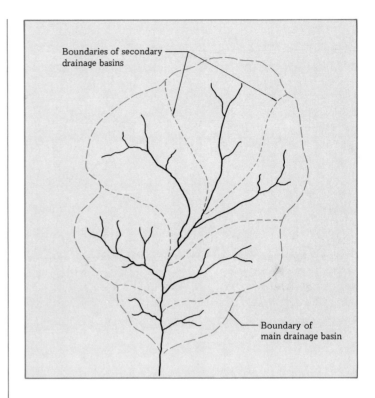

FIGURE 10.19

Each stream, no matter how small, has its own drainage basin, the area from which the stream and its tributaries receive water. This basin displays a pattern reminiscent of a tree leaf and its veins.

line on semilogarithmic paper (see Figure 10.21*a*). Likewise, if we plot order against the average length of streams of a given order or plot order against areas of drainage basins of a particular order, we find a well-defined relationship (see Figure 10.21*b* and *c*). Using these relationships, we can describe the river channels and basins of the United States as in Table 10.1.

We have seen that, when streams are arranged in a hierarchical fashion, the arrangement bears a definite relation to the number and lengths of the stream segments in each order and to the size of the basins of the stream segments of different orders. Beyond this there are overall patterns developed by stream systems that depend in part on the nature of the underlying rocks, including their arrangement, distribution, differential resistance to erosion, and even history.

The overall pattern developed by a system of streams and tributaries depends partly on the nature of

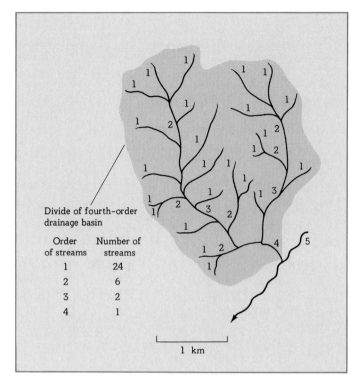

Divide of fourth-order
drainage basin

Order of streams	Number of streams
1	24
2	6
3	2
4	1

1 km

FIGURE 10.20

Method of designating stream orders. [After A. N. Strahler, "Quantitative Geomorphology in Erosional Landscapes," *Proc. Intern. Geol. Congr.*, sec. 13, pt. 3, p. 344, 1954.]

FIGURE 10.21

Relation between stream order and the number of streams, the mean length of streams, and the mean drainage area of streams of a location in central Pennsylvania. The system used to designate stream orders is that proposed by Robert Horton in 1945 and differs slightly from that illustrated in Figure 10.20. [After Lucien Brush, Jr., *U.S. Geol. Surv. Prof. Paper 282-F*, 1961.]

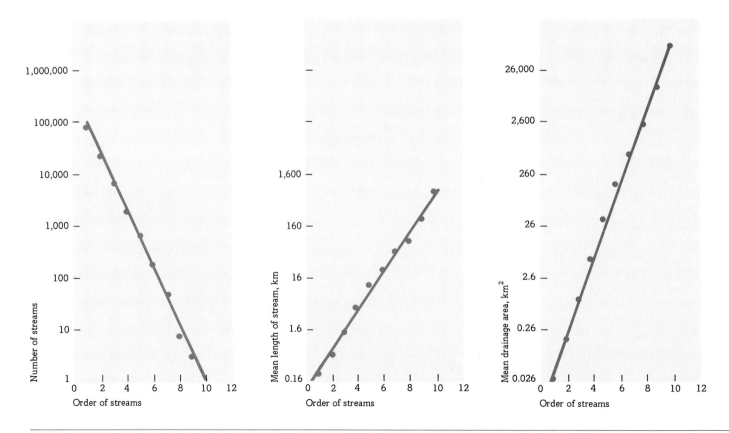

TABLE 10.1

Number and length of river channels of various sizes in the United States, excluding tributaries of smaller order[a]

Order	Number	Av. length, km	Total length, km	Mean drainage area, incl. tributaries, km²	River representative of size
1[b]	1,570,000	1.6	2,512,000	2.6	
2	350,000	3.7	1,295,000	12	
3	80,000	8.5	680,000	59	
4	18,000	19	342,000	283	
5	4,200	45	189,000	1,348	Charles
6	950	102	97,000	6,400	Raritan
7	200	219	44,000	30,300	Allegheny
8	41	541	22,000	144,000	Gila
9	8	1,243	10,000	683,000	Columbia
10	1	2,880	2,880	3,238,000	Mississippi

[a]From Luna B. Leopold, "Rivers," *Am. Sci.*, vol. 50, p. 512, 1962.
[b]The size of the order 1 channel depends on the scale of the maps used; these order numbers are based on the determination of the smallest order using maps of scale 1:62,500.

the underlying rocks and partly on the history of the streams. Almost all streams follow a branching pattern in the sense that they receive tributaries; the tributaries, in turn, are joined by still smaller tributaries, but the manner of branching varies widely (see Figure 10.22).

A stream that resembles the branching habit of a maple, oak, or similar deciduous tree is called **dendritic**, "treelike." A dendritic pattern develops when the underlying bedrock is uniform in its resistance to erosion and exercises no control over the direction of valley growth. This situation occurs when the bedrock is composed either of flat-lying sedimentary rocks or massive igneous or metamorphic rocks. The streams can cut as easily in one place as another; thus the dendritic pattern is, in a sense, the result of the random orientation of the streams.

Another type of stream pattern is **radial**: Streams radiate outward in all directions from a high central zone. Such a pattern is likely to develop on the flanks of a newly formed volcano, where the streams and their valleys radiate outward and downward from various points around the cone.

A **rectangular** pattern occurs when the underlying bedrock is crisscrossed by fractures that form zones of

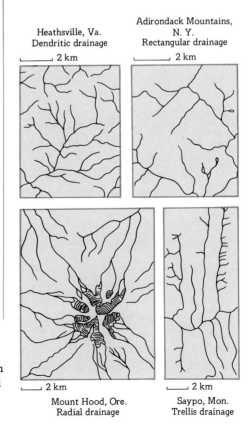

Heathsville, Va.
Dendritic drainage
2 km

Adirondack Mountains, N. Y.
Rectangular drainage
2 km

Mount Hood, Ore.
Radial drainage
2 km

Saypo, Mon.
Trellis drainage
2 km

FIGURE 10.22

The overall pattern developed by a stream system depends in part on the nature of the bedrock and in part on the history of the area. See the text for discussion.

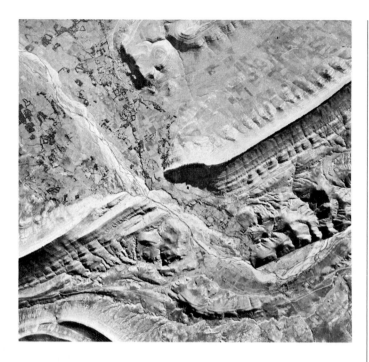

FIGURE 10.23

An aerial photograph of a water gap in North Africa. The master stream cuts through the ridge of resistant rock. [U.S. Army Air Corps.]

weakness particularly vulnerable to erosion. The master stream and its tributaries then follow courses marked by nearly right-angle bends.

Some streams, particularly in a belt of the Appalachian Mountains running from New York to Alabama, follow what is known as a **trellis** pattern. This pattern, like the rectangular one, is caused by zones in the bedrock that differ in their resistance to erosion. The trellis pattern usually, though not always, indicates that the region is underlain by alternate bands of resistant and nonresistant rock.

Particularly intriguing features of some valleys are short, narrow segments walled by steep, rocky slopes or cliffs. These are called **water gaps**. The river, in effect, flows through a narrow notch in a ridge or mountain that lies across its course, as shown in Figure 10.23. Here is one way water gaps can form: Picture an area of hills and valleys carved by differential erosion from rocks of varying resistance. Then imagine sediments covering this landscape so that the valleys and hills are buried beneath the debris. Any streams that flow over the surface of this cover will establish their own courses across the buried hills and valleys. When these streams erode downward, some of them may encounter an old ridge crest. If such a stream has sufficient erosional energy, it may cut down through the resistant rock of the hill. The course of the stream is thus superimposed across the old hill, and we call the stream a **superimposed stream**. Differential erosion may excavate the sedimentary fill from the old valleys, but the main, superimposed stream flows through the hill in a new, narrow gorge, or water gap. An example of such a gap is that followed by the Big Horn River through the northern end of the Big Horn Mountains in Montana.

ENLARGEMENT OF VALLEYS

We cannot say with assurance how running water first fashioned the great valleys and drainage basins of the continents; for the record has been lost in time. But we do know that certain processes are now at work in widening and deepening valleys, and it seems safe to assume that they also operated in the past.

If a stream were left to itself in its attempt to reach base level, it would erode its bed straight downward, forming a vertically walled chasm in the process. But because the stream is not the only agent at work in valley formation, the walls of most valleys slope upward and outward from the valley floor. In time the cliffs of even the steepest gorge will be angled away from the axis of its valley.

As a stream cuts downward and lowers its channel into the land surface, weathering, slope wash, and mass movement come into play, constantly wearing away the valley walls and pushing them farther back (see Figure 10.24). Under the influence of gravity, material is carried down from the valley walls and dumped into the stream, to be moved toward the seas. The result is a valley whose walls flare outward and upward from the stream in a typical **cross-valley profile** (see Figure 10.25).

The rate at which valley walls are reduced and the angles that they assume depend on several factors. If the walls are made up of unconsolidated material that is vulnerable to erosion and mass movement, the rate will be rapid; but if the walls are composed of resistant rock, the rate of erosion will be very slow, and the walls may rise almost vertically from the valley floor (see Figure 10.26).

In addition to cutting downward into its channel a stream also cuts from side to side, or laterally, into its

FIGURE 10.24

This gully in Wisconsin has been formed by the downward cutting of the stream combined with the slope wash and mass movement on the gully walls. [U.S. Forest Service.]

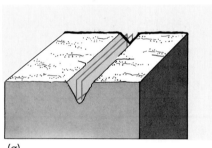

(a)

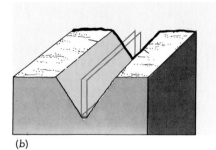

(b)

FIGURE 10.25

If a stream of water were the only agent in valley formation, we might expect a vertically walled valley no wider than the stream channel, as suggested by the colored lines in (a) and (b). Mass movement and slope wash, however, are constantly wearing away the valley walls, carving slopes that flare upward and away from the stream channel, as shown in the diagrams.

FIGURE 10.26

Erosion by a small stream has fashioned vertically walled Labyrinth Canyon, Utah, in beds of massive sandstone. The stream is the major agent in formation of the valley. Mass movement and other agents of slope modification have yet to lower the angle of valley walls. Compare Figure 10.25. [William C. Bradley.]

banks. In the early stages of valley enlargement, when the stream is still far above its base level, downward erosion is dominant. Later, as the stream cuts its channel closer and closer to base level, downward erosion becomes progressively less important. Now a larger percentage of the stream's energy is directed toward eroding its banks. As the stream swings back and forth, it forms an ever-widening flood plain on the valley floor, and the valley itself broadens.

How fast do these processes of valley formation proceed? In some instances we can measure the ages of gully formation in years and even months, but for large valleys we must be satisfied with rough approximations—and usually with plain guesses. For instance, one place where we can make an approximation of the rate of valley formation is the Grand Canyon of the Colorado in Arizona. Several lines of evidence suggest that it has taken something on the order of 10 million years for the Colorado to cut its channel downward about 1,800 m and expand its valley walls 6 to 20 km.

FEATURES OF NARROW VALLEYS

Waterfalls are among the most fascinating spectacles of the landscape. Thunderous and powerful as they are, however, they are actually short-lived features in the history of a stream. They owe their existence to some sudden drop in the river's longitudinal profile—a drop that may be eliminated with the passing of time.

Waterfalls are caused by many different conditions. Niagara Falls, for instance, is held up by a relatively resistant bed of dolomite underlain by beds of nonresistant shale (see Figure 10.27). This shale is easily undermined by the swirling waters of the Niagara River as they plunge over the lip of the falls. When the undermining has progressed far enough, the dolomite collapses and tumbles to the base of the falls. The same process is repeated over and over again as time passes, and the falls slowly retreat upstream. Historical records suggest that the Horseshoe, or Canadian, Falls (by far the larger of the two falls at Niagara) have been retreating at a rate of 1.2 to 1.5 m/year, whereas the smaller American Falls have been eroded away 5 to 6 cm/year. The 11 km of gorge between the foot of the falls and Lake Ontario are evidence of the headward retreat of the falls through time.

Yosemite Falls in Yosemite National Park, California, plunge 770 m over the Upper Falls, down an intermediate zone of cascades, and then over the Lower Falls. The falls leap from the mouth of a small valley high above the main valley of the Yosemite. The Upper Falls alone measure 430 m, nine times the height of Niagara. During the Ice Age, glaciers scoured the main valley much deeper than they did the unglaciated side valley. Then, when the glacier ice melted, the main valley was left far below its tributary, which now joins it after a drop of nearly a kilometer.

Rapids, like waterfalls, occur at a sudden drop in the stream channel. Although rapids do not plunge straight down as waterfalls do, the underlying cause of formation is often the same. In fact, many rapids have developed directly from preexisting waterfalls (Figure 10.28).

FEATURES OF BROAD VALLEYS

If conditions permit, the various agents working toward valley enlargement ultimately produce a broad valley with a wide level floor. During periods of normal or low water the river running through the valley is confined to its channel; but during high water it overflows its banks and spreads out over the flood plain.

A flood plain that has been created by the lateral erosion and the gradual retreat of the valley walls is called an **erosional flood plain** and is characterized by a thin cover of gravel, sand, and silt a few meters or a few tens of meters in thickness. On the other hand, the floors of many broad valleys are underlain by deposits of gravel, sand, and silt scores of meters thick. These thick deposits are laid down as changing conditions force the river to drop its load across the valley floor. Such a flood plain, formed by the building up of the valley floor, or aggradation, is called a **flood plain of aggradation.** Flood plains of aggradation are much more common than erosional flood plains and are found in the lower reaches of the Mississippi, Nile, Rhône, and Yellow Rivers, to name but a few.

Both erosional flood plains and flood plains of aggradation exhibit the following characteristics.

Meanders The channel of the Menderes River in Asia Minor curves back on itself in a series of broad hairpin bends, and today all such bends are called *meanders* (Figure 10.29).

Both erosion and deposition are involved in the formation of a meander. First, some obstruction swings the current of a stream against one of the banks, and then the current is deflected over to the opposite bank. Erosion takes place on the outside of each bend, where

FIGURE 10.27

Niagara Falls tumbles over a bed of dolomite, underlain chiefly by shale. As the less resistant bed is eroded, the undermined ledge of dolomite breaks off, and the lip of the falls retreats. [Redrawn from G. K. Gilbert, *Niagara Falls and Their History*, p. 213, American Book Company, New York, 1896.]

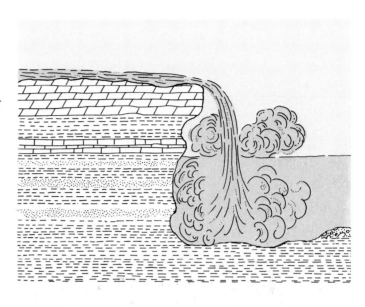

(a)

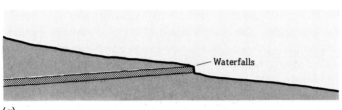

(b)

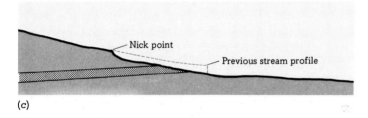

(c)

FIGURE 10.28

Rapids may represent a stage in the destruction of waterfalls, as suggested in this diagram.

FIGURE 10.29

Aerial photograph and a drawing of meanders in an Alaskan stream. Note the oxbow lakes. [U.S. Army Air Corps.]

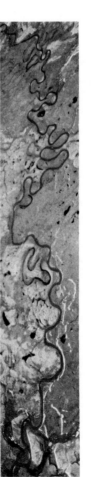

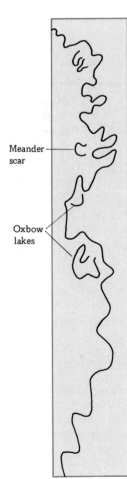

FIGURE 10.30

This oxbow lake near Weslaco, Texas, was once a part of the meandering Rio Grande, which lies in the distance. An old bend in the river was cut through at the neck, became isolated from the river, and filled with water. [Exxon Corp.]

the turbulence is greatest. The material detached from the banks is moved downstream, there to be deposited in zones of decreased turbulence—either along the center of the channel or on the inside of the next bend. As the river swings from side to side, the meander continues to grow by erosion on the outside of the bends and by deposition on the inside. Growth ceases when the meander reaches a critical size, a size that increases with an increase in the size of the stream.

Because a meander is eroded more on its downstream side than on its upstream side, the entire bend tends to move slowly down-valley. This movement is not uniform, however, and under certain conditions the downstream sweep of a series of meanders is distorted into cutoffs, meander scars, and oxbow lakes.

In its down-valley migration a meander sometimes runs into a stretch of land that is relatively more resistant to erosion. But the next meander upstream continues to move right along, and gradually the neck between them is narrowed. Finally the river cuts a new, shorter channel, called a **neck cutoff**, across the neck. The abandoned meander is called an **oxbow** because of its characteristic shape. Usually both ends of the oxbow are gradually silted in, and the old meander becomes completely isolated from the new channel. If the abandoned meander fills up with water, an **oxbow lake** results. Although a cutoff will eliminate a particular meander, the stream's tendency toward meandering still exists, and before long the entire process begins to repeat itself.

We found that a meander grows and migrates by erosion on the outside of the bend and by deposition on the inside. This deposition on the inside leaves behind a series of low ridges and troughs. Swamps often form in the troughs, and during flood time the river may develop an alternate channel through one of the troughs. Such a channel is called a **chute cutoff**, or simply a *chute* (see Figures 10.30 and 10.31).

The meandering river demonstrates a unity in ways other than the balance of erosion and deposition. The length of a meander, for example, is proportional to the width of the river, and this is true regardless of

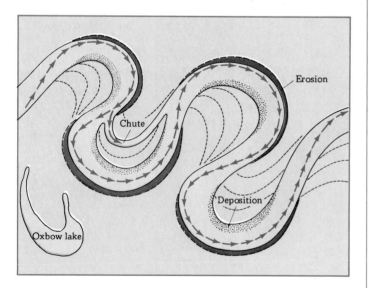

FIGURE 10.31

Erosion takes place on the outside of a meander bend, whereas deposition is most marked on the inside. If the neck of a meander is eroded through, an oxbow forms. A chute originates along the inside of a meander where irregular deposition creates ridges and troughs as the meander migrates.

FIGURE 10.32

(a) Length of the meander increases with the widening meandering stream. (b) A similar orderly relationship exists between length of the meander and the mean radius of curvature of the meander. [Redrawn from Luna B. Leopold and M. Gordon Wolman, "River Meanders," *Geol. Soc. Am. Bull.* 71, p. 773, 1960.]

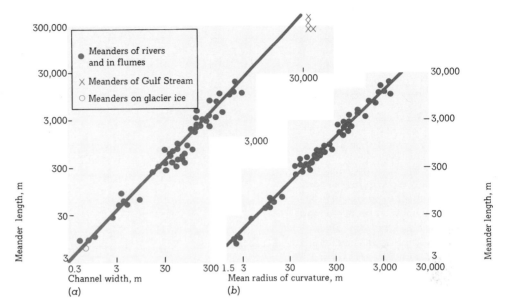

the size of the river. It holds for channels a few meters wide as well as those as large as that of the Mississippi River. As shown in Figure 10.32a, this principle is also true of the Gulf Stream even though this "river" is unconfined by solid banks. A similar relationship holds between the length of the meander and the radius of curvature of the meander (Figure 10.32b). These relationships seem to be controlled largely by the discharge and the sediments carried. Increasing discharge causes an increase in the size of characteristics such as width, length, and wavelength. For a given discharge the wavelength tends to be greater for streams carrying a high proportion of sand and gravel in their loads than for those transporting mainly fine sands.

Braided streams On some flood plains, particularly where large amounts of debris are dropped rapidly, a stream may build up a complex tangle of converging and diverging channels separated by sandbars or islands. A stream of this sort is called **braided.** The pattern generally occurs when the discharge is highly variable and the banks easily eroded to supply a heavy load. It is characteristic of alluvial fans, glacial-outwash deposits, and certain heavily laden rivers.

In general the gradient of a braided stream is higher than that for a meandering stream of the same discharge, as indicated in Figure 10.33. This is apparently indicative of a tendency of a stream to increase its efficiency in order to transport proportionally larger loads. Thus, if a stream with a single channel divides into two channels, the new channels will have a higher gradient and a greater total width than the original channel carrying the same amount of water; and turbulence increases. The two new channels can now carry a greater load, particularly bed load, than could the single-channeled stream.

Natural levees In many flood plains, the water surface of the stream is held above the level of the valley floor by banks of sand and silt known as **natural levees,** a name derived from the French *lever,* "to raise." These banks slope gently, almost imperceptibly, away from their crest along the river toward the valley wall. The low-lying flood plain adjoining a natural levee may contain marshy areas known as **back swamps.** Levees are built up during flood time, when the water spills over the river banks onto the flood plain. Because the muddy water rising over the stream bank is no longer confined by the channel, its velocity and turbulence drop immediately, and much of the suspended load is deposited close to the river; but some is carried farther along to be deposited across the flood plain. The deposit of one flood is a thin wedge tapering away from the river, but over many years the cumulative effect produces a natural levee that is considerably higher beside the river bank than away from it (see Figure 10.34). On the Mississippi delta, for instance, the levees stand 5 to 6 m above the back

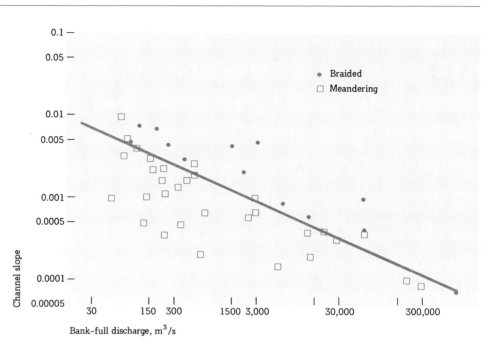

Meandering streams have lower gradients than do braided streams of the same discharge. In this diagram each symbol represents measurements at a single point on a stream. [Adapted from Luna B. Leopold, M. Gordon Wolman, and John P. Miller, *Fluvial Processes in Geomorphology*, pp. 7–39, W. H. Freeman and Company, San Francisco, copyright © 1964.]

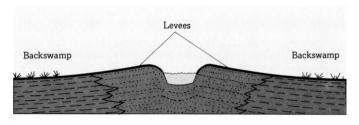

FIGURE 10.34

Natural levees characterize many aggrading streams. They build up during periods of flood as coarser material is deposited closest to the stream channel to form the levee and finer material is deposited in the back swamps. As the banks build up, the floor of the channel also rises.

swamps. Although natural levees tend to confine a stream within its channel, each time the levees are raised slightly, the bed of the river is also raised. In time the level of the bed is raised above the level of the surrounding flood plain. If the river manages to escape from its confining walls during a flood, it will assume a new channel across the lowest parts of the flood plain toward the back swamps.

A tributary stream entering a river valley with high levees may be unable to find its way directly into the main channel; so it will flow down the back-swamp zone and may run parallel to the main stream for many kilometers before finding an entrance. Because the Yazoo River typifies this situation by running 320 km parallel to the Mississippi, all rivers following similar courses are known as **yazoo-type** rivers.

Flood plains and their deposits   Virtually all streams are bordered by flood plains. These may range from very narrow (a few meters wide) to the flood plains whose widths are measured in kilometers, as is that of the lower Mississippi River. The flood plain appears very flat, particularly when viewed in relation to the slopes of the valley walls that flank it. But the plain is not without its ups and downs. A large flood plain may have differences of relief of several meters and be marked by such features as natural levees, meander scars, and oxbow lakes.

The flood plain is made up of stream-carried sediments. Two general categories of deposit are common.

One of these is made up of fine silts, clays, and sands that may be spread across the plain by a river that overflows its banks during flood. These are called **overbank deposits.** The other group of deposits is made of coarse material, gravel and sand, and is related directly to the channel of the stream. These deposits include chiefly the material that is deposited in the slack water on the inside of the bends of a winding or a meandering river. These bars of sand and gravel are called **point bars.**

We found, in discussing meanders, that the loops of a channel migrate laterally so that the river swings both across the valley and down-valley. As the river migrates laterally, it leaves behind coarse deposits in point bars and in the deeper portions of the abandoned channel. While erosion takes place on one bank of the river, deposition of sediments from upstream takes place on the point bars and, during flood, by overbank deposits—the river builds its flood plain in some places and simultaneously destroys it in others. Thus the flood plain becomes a temporary storage place for river sediments.

We can view the flood plain, then, as a depositional feature more or less in equilibrium. This equilibrium may be disturbed in several ways. The stream may lose some of its ability to erode and carry sediment. A net gain of deposits will occur, and the flood plain will build up. On the other hand, the flow of water or the gradient may increase or the supply of sediments may decrease, and there will be a net loss of sediments as erosion begins to destroy the previously developed flood plain.

Deltas and alluvial fans For centuries the Nile River has been depositing sediments as it empties into the Mediterranean Sea, forming a great triangular plain with its apex upstream. This plain came to be called a **delta** because its shape is similar to the Greek letter Δ (see Figures 10.35 and 10.36).

Whenever a stream flows into a body of standing water, such as a lake or an ocean, its velocity and transporting power are quickly stemmed. If it carries enough debris and if conditions in the body of standing water are favorable, a delta will gradually be built. An

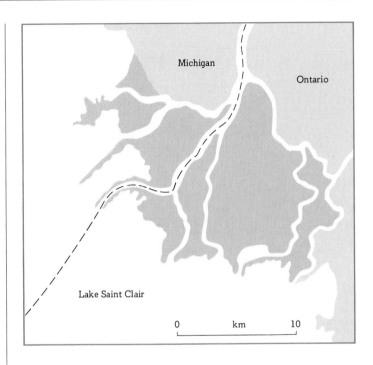

FIGURE 10.35

The delta (darker area) of the Saint Clair River in Lake Saint Clair has the classic shape as well as its distributary channels. [Redrawn from Leon J. Cole, "The Delta of the St. Clair River," *Geol. Surv. Mich.* 9, pt. 1, 1903.]

FIGURE 10.36

In this picture taken from *Gemini 4*, the delta of the Nile stands out because it is well vegetated and thus contrasts with the desert country that borders it. [NASA.]

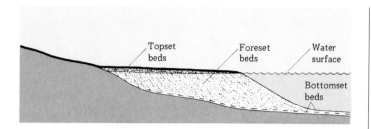

Topset beds   Foreset beds   Water surface   Bottomset beds

FIGURE 10.37

The ideal arrangement of sediments beneath a delta. Some material deposited in a lake or sea is laid on the bottom as bottomset beds. Other material is dumped in inclined foreset beds, built farther into the water and partly covering the bottomset beds. Over the foreset beds the stream lays down topset beds.

ideal delta is triangular in plan, with the apex pointed upstream and with the sediments arranged according to a definite pattern. The coarse material is dumped first, forming a series of dipping beds called **foreset beds.** But the finer material is swept farther along to settle across the sea or lake floor as **bottomset beds.** As the delta extends farther and farther out into the water body, the stream must extend its channel to the edge of the delta. As it does so, it covers the delta with **topset beds,** which lie across the top of the foreset beds (see Figure 10.37).

Very few deltas, however, show either the perfect delta shape or this regular sequence of sediments. Many factors, including lake and shore currents, varying rates of deposition, the settling of delta deposits as a result of their compaction, and the downwarping of the earth's crust—all conspire to modify the typical form and sequence.

Across the top of the delta deposits, the stream spreads seaward in a complex of channels radiating from the apex. These **distributary channels** shift their position from time to time as they seek more favorable gradients.

Deltas are characteristic of many of the larger rivers of the world, including the Nile, Mississippi, Ganges, Rhine, and Rhône. On the other hand, many rivers have no deltas either because the deposited material is swept away as soon as it is dumped or because the streams do not carry sufficient detrital material to build up a delta.

An **alluvial fan** is the land counterpart of a delta.

These fans are typical of arid and semiarid climates, but they may form in almost any climate if conditions are right. A fan marks a sudden decrease in the carrying power of a stream as it descends from a steep gradient to a flatter one—for example, when the stream flows down a steep mountain slope onto a plain. As the velocity is checked, the stream rapidly begins to dump its load. In the process it builds up its channel, often with small natural levees along its banks. Eventually, as the levees continue to grow, the stream may flow above the general level. Then during a time of flood it seeks a lower level and shifts its channel to begin deposition elsewhere. As this process of shifting continues, an alluvial fan proceeds to build up (see Figure 10.38).

Stream terraces A *stream terrace* is a relatively flat surface running along a valley, with a steep bank separating it either from the flood plain or from a lower terrace. It is a remnant of the former channel of a stream that now has succeeded in cutting its way down to a lower level.

The so-called **cut-and-fill terrace** is created when a stream first clogs a valley with sediments and then cuts its way down to a lower level (Figure 10.39). The initial aggradation may be caused by a change in climate that leads either to an increase in the stream's load or a decrease in its discharge. Or the base level of the stream may rise, reducing the gradient and causing deposition. In any event the stream chokes the valley with sediment, and the flood plain gradually rises. Now, if the equilibrium is upset and the stream begins to erode, it will cut a channel down through the deposits it has already laid down. The level of flow will be lower than the old flood plain, and at this lower level the stream will begin to carve out a new flood plain. As time passes, remnants of the old flood plain, terraces, may be left standing on either side of the new one. Terraces that face each other across the stream at the same elevation are referred to as **paired terraces.** Sometimes the downward erosion by streams creates **unpaired terraces.** If the stream swings back and forth across the valley, slowly eroding as it moves, it may encounter resistant rock beneath the unconsolidated deposits; the exposed rock will then deflect the stream and prevent further erosion, and a single terrace is left behind, with no corresponding terrace on the other side of the stream (see Figure 10.40).

Terraces, either paired or unpaired, may be cut into bedrock as well. A thin layer of sand and gravel usually rests on the beveled bedrock of these terraces.

FIGURE 10.38

An alluvial fan is the land counterpart of a delta. In this example in Death Valley, California, the streams flow only during the rare rains, carrying debris from the steep gulches along the cliff face. As the velocity of the streams is checked on the flat valley floor, material is deposited to form the alluvial fan. [Sheldon Judson.]

FIGURE 10.40

Unpaired terraces do not match across the stream that separates them. Here is one way in which they may form. The stream has cut through unconsolidated deposits within the valley. As it eroded down-ward, it also swept laterally across the valley and created a sloping surface. Lateral migration was stopped locally when the stream en-countered resistant bedrock (see ar-rows) beneath the softer valley fill. This bedrock not only deflected the stream back across the valley but also protected remnants of the valley fill from further stream erosion and allowed them to be preserved as terraces. Because they are portions of a surface sloping across the valley, however, no single remnant matches any other on the opposite side of the river.

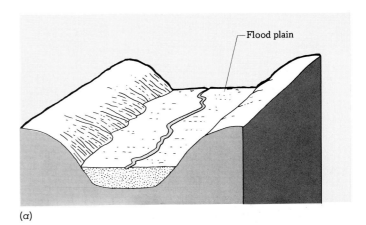

(a)

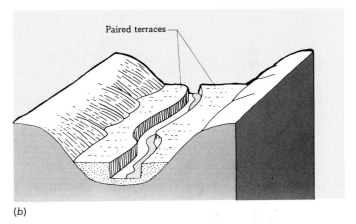

(b)

FIGURE 10.39

One example of the formation of paired terraces. (a) The stream has partially filled its valley and has created a broad flood plain. (b) Some change in conditions has caused the stream to erode into its own deposits; the remnants of the old flood plain stand above the new river level as terraces of equal height. This particular example is referred to as a cut-and-fill terrace. (See also Figure 10.40.)

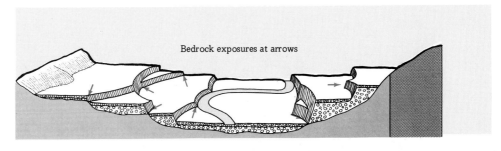

10.7 OF RIVERS AND CONTINENTS

So far we have been discussing some of the details of streams and stream action. In concluding this chapter, we shall take a more global look at some aspects of running water.

Of all the processes at work to move material from the continents to the ocean basins the most important is running water. Like great conveyor belts the rivers of the world move the weathered waste of the land to the oceans. They annually carry to the oceans an estimated 10 billion t of earth material, material delivered to them down the slopes by mass movement. This rate of erosion could in a few million years reduce the continents to base level, leaving the landmasses as flat as pancakes. But geologic studies tell us that sediments and sedimentary rocks have been forming for billions of years. We conclude, therefore, that continuous or repeated uplift of the earth's surface has for eons provided materials for erosion. The tectonics associated with plate motions have kept erosion from flattening the continents. Despite this, it *is* startling to observe on a world scale how close erosion comes to beveling the continents and reducing them to sea level: To a first approximation streams *have* leveled the continents.

A final observation about some of the major features of the world's drainage systems commands attention. If we look at a world map, we find that many of the major divides are related to plate boundaries. South America provides a spectacular example. There the Andes, formed by the collision of the Nazca and South American plates, form the continental divide. Most of the continent drains to the east into the Atlantic Ocean, and only a narrow area sheds water to the west and the Pacific Ocean. The great mountain masses of the Himalaya, the Tibetan Plateau, and the mountains of Afghanistan, Iran, the Caucasus, and on to the Alps represent a complex convergence of the great Eurasian plate with several other plates along its southern border. The continental divide of North America, from southern Mexico to Alaska, is associated with plate movements although their details are still to be worked out. And we can also note that ancient mountains, such as the Appalachians of eastern North America and the Urals of eastern Europe, mark old lines of plate convergence, and they still serve as significant drainage divides in a period that is hundreds of millions of years later.

OUTLINE

Running water has fashioned over 400 million km of stream channels, down which the bulk of material eroded from the continent moves toward the oceans.

The hydrologic cycle is the path of water circulation from oceans to atmosphere to land to oceans.

Precipitation and stream flow (runoff) are related as follows:
Runoff = precipitation − (infiltration + evaporation and transpiration)
Flow is either **turbulent** or **laminar**, but turbulent is the rule.
Water in a stream has **velocity**, flows down a **gradient**, and is measured in terms of **discharge.**

The economy of a stream is part of a dynamic equilibrium that characterizes earth processes.
It is clear that in a stream **discharge = width × depth × velocity.** The change of one characteristic will affect one or more of the others. In adjusting to an increased discharge downstream, the gradient decreases, and width, depth, and velocity increase.
Floods are a normal, but relatively rare, stage of stream flow. Flood-recurrence intervals are statistically predictable.
Base level of a stream is the point below which it cannot erode. Lowering of the base level produces erosion, and raising of the base level produces deposition.

Work of running water includes transportation, erosion, and deposition.

Transportation by a stream involves its **load** (the amount it carries at any one time), its **capacity** (the total amount it can carry under given conditions), and its **competence** (the maximum-sized particle it can move under given conditions). Material is moved in **solution**, in **suspension**, and as **bed load**.

Erosion by a stream involves **direct lifting** of material, **abrasion**, **impact**, and **solution**.

Features of valleys include those of the valley bottom, the drainage basin, and the river channel.

A **cross-valley profile** of a typical valley shows **flood plains**, **valley walls**, and **divides**.

A **drainage basin** is the area drained by a river and its tributaries. Individual streams and their valleys are joined in definite geometrical relationships. The patterns developed by stream systems include **dendritic**, **radial**, **rectangular**, and **trellis**. Some streams form **water gaps**.

Enlargement of valleys is accomplished through downward and lateral erosion by the stream and by mass movement and water erosion on the valley walls.

Features of narrow valleys include **waterfalls**, **rapids**, and **steep gradients**.

Broad valleys have **meanders**, **braided streams**, natural **levees**, **floodplains**, **deltas**, alluvial **fans**, and stream **terraces**.

Rivers are the most important agent in removing material from the continents. Movement of the earth's plates continues to supply material to be eroded. Plate margins provide some of the major stream divides of the continents.

SUPPLEMENTARY READINGS

Butzer, Karl W. *Geomorphology from the Earth*, Harper & Row, Publishers, New York, 1976. *A general text on the processes that create landforms. Chapters 6 to 9 deal with the role of running water.*

Gregory, K. J., and D. E. Walling *Drainage Basin Form and Process*, Cambridge University Press, New York, 1973. *A very good advanced text devoted to the forms of drainage basins and the processes that go on in them.*

Leopold, Luna B., M. Gordon Wolman, and John P. Miller *Fluvial Processes in Geomorphology*, W. H. Freeman and Company Publishers, San Francisco, 1964. *An advanced textbook on the subject.*

Morisawa, Marie *Streams: Their Dynamics and Morphology*, McGraw-Hill Book Company, New York, 1968. *A short book of intermediate level, which is a readable summary of the hydrology and work of streams.*

11

UNDERGROUND WATER

Tremendous quantities of water lie exposed in oceans, lakes, rivers, and—in the solid state—glaciers on the earth's surface. But beneath the surface, hidden from our sight, is another great store of water. In fact in Chapter 10 we indicated that for the earth as a whole the water beneath the surface exceeded by more than 66 times the amount of water present in streams and freshwater lakes. In the United States streams each year discharge an estimated 30,000 km³ of water into the oceans. In contrast there lies beneath the surface an estimated 7,575,000 km³ of water. It is certainly true that underground reservoirs in the United States contain far more usable water than do all our surface reservoirs and lakes combined, and the United States depends on this underground supply for about one-fifth of total water needs.

11.1 BASIC DISTRIBUTION

Underground water, subsurface water, and **subterranean water** are all general terms used to refer to water in the pore spaces, cracks, tubes, and crevices of the consolidated and unconsolidated material beneath our feet. The study of underground water is largely an investigation of these openings and of what happens to the water that finds its way into them. In Chapter 10 we found that most of the water beneath the surface comes from rain and snow that fall on the face of the earth. Part of this water soaks directly into the ground, and part of it drains away into lakes and streams and thence into the underground.

ZONES OF SATURATION AND AERATION

Some of the water that moves down from the surface is caught by rock and earth materials and is checked in its downward progress. The zone in which this water is held is known as the **zone of aeration,** and the water itself is called **suspended water.** The spaces between particles in this zone are filled partly with water and partly with air. Two forces operate to prevent suspended water from moving deeper into the earth: the molecular attraction exerted on the water by the rock and earth materials and the attraction exerted by the water particles on one another (see Figures 11.1 and 11.2).

The zone of aeration can be subdivided into three belts: **belt of soil moisture, intermediate belt,** and **capillary fringe.** Some of the water that enters the belt of soil moisture from the surface is used by plants, and some is evaporated back into the atmosphere. But some water also passes down to the intermediate belt, where it may be held by molecular attraction (as suspended water). Little movement occurs in the intermediate belt except when rain or melting snow sends a new wave of moisture down from above. In some areas the intermediate belt is missing, and the belt of soil moisture lies directly above the third belt, the capillary fringe. Water rises into the capillary fringe from below, to a height ranging from a few centimeters to 2 or 3 m.

Beneath the zone of aeration lies the **zone of saturation.** Here the openings in the rock and earth materials are completely filled with **groundwater,** and the surface between the zone of saturation and the zone of aeration is called the *groundwater table,* or simply the **water table.** The level of the water table fluctuates with variations in the supply of water coming down from the zone of aeration, with variations in the rate of discharge in the area, and with variations in the amount of groundwater that is drawn off by plants and human beings.

It is the water below the water table, within the zone of saturation, that we shall focus on in this chapter.

THE WATER TABLE

The water table is an irregular surface of contact between the zone of saturation and the zone of aeration. Below the water table lies the groundwater; above it lies the suspended water. The thickness of the zone of aeration differs from one place to another, and the level of the water table fluctuates accordingly. In general, the water table tends to follow the irregularities of the ground surface, reaching its highest elevation beneath hills and its lowest elevation beneath valleys. Although the water table reflects variations in the ground surface, the irregularities in the water table are less pronounced.

In looking at the topography of the water table, let us consider an ideal situation. Figure 11.3 shows a hill underlain by completely homogeneous material. Assume that, initially, this material contains no water at

FIGURE 11.1

A drop of water held between two fingers illustrates the molecular attraction that prevents downward movement. Water is similarly suspended within the pore spaces of the zone of aeration shown in Figure 11.2.

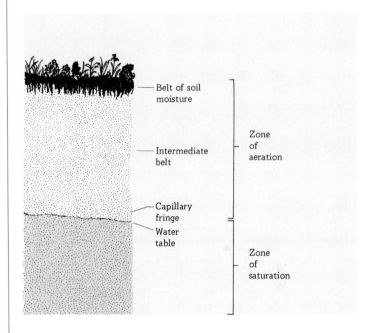

FIGURE 11.2

Underground water's two major zones: zone of aeration and zone of saturation. The water table marks the upper surface of the zone of saturation. Within the zone of aeration is a belt of soil moisture, the source of moisture for many plants. Also from here some moisture is evaporated back to the atmosphere. In many instances this belt lies above an intermediate belt where water is held by molecular attraction and little movement occurs except during periods of rain or melting snow. In the capillary fringe, just above the water table, water rises a few centimeters to a meter or more from the zone of saturation, depending on the size of the interstices.

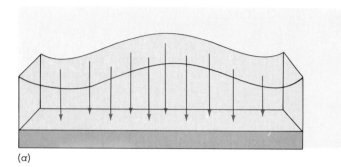

(a)

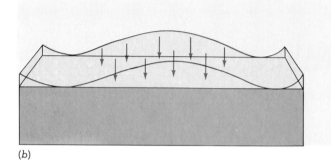

(b)

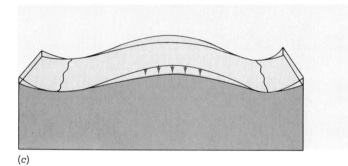

(c)

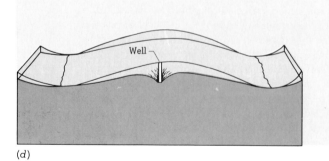

(d)

FIGURE 11.3

Ideally, the water table is a subdued reflection of the surface of the ground. In (a) and (b) the water table rises as a horizontal plane until it reaches the level of the valley bottoms on either side of the hill. Thereafter, as more water soaks into the ground, it seeks an outlet toward the valleys. Were the movement of the water not slowed down by the material making up the hill, it would remain essentially horizontal. The friction caused by the water's passing through the material (and even to some extent the internal friction of the water itself) results in a piling up of water beneath the hill; the bulge is highest beneath the crest and lowest toward the valleys (c). The shape of the water table may be altered by pumping water from a well (d). The water flows to this new outlet and forms a cone of depression.

all. Then a heavy rainfall comes along, and the water soaks slowly downward, filling the interstices at depth. In other words a zone of saturation begins to develop. As more and more water seeps down, the upper limit of this zone continues to rise. The water table remains horizontal until it just reaches the level of the two valley bottoms on either side of the hill. Then, as additional water seeps down to the water table, some of it seeks an outlet into the valleys. But this added water is "supported" by the material through which it flows, and the water table is prevented from maintaining its flat surface. The water is slowed by the friction of its movement through the interstices and even, to some degree, by its own internal friction. Consequently, more and more water is piled up beneath the hill, and the water table begins to reflect the shape of the hill. The water flows away most rapidly along the steeper slope of the water table near the valleys and most slowly on its gentler slope beneath the hill crest.

We can modify the shape of the groundwater surface by providing an artificial outlet for the water. For example, we can drill a well on the hill crest and extend it down into the saturated zone. Then, if we pumped out the groundwater that flowed into the well, we should create a dimple in the water table. The more we pumped, the more pronounced the depression—a **cone of depression**—would become.

Returning to our ideal situation, we find that, if the supply of water from the surface were to be completely stopped, the water table under the hill would slowly flatten out as water discharged into the valleys. Eventually it would almost reach the level of the water table under the valley bottoms; then the flow would stop. This condition is common in desert areas, where the rainfall is sparse.

11.2 MOVEMENT OF UNDERGROUND WATER

The preceding chapter stated that the flow of water in surface streams could be measured in terms of so many meters per second. But in dealing with the flow of underground water, we must change our scale of measurement; for here, although the water moves, it usually does so very, very slowly. Therefore we find that centimeters per day and, in some places, even centimeters or meters per year provide a better scale of measurement. The main reason for this slow rate of flow is that the water must travel through small, confined passages if it is to move at all. It will then be worthwhile for us to consider the porosity and permeability of earth materials.

POROSITY

The **porosity** of a rock is measured by the percentage of its total volume that is occupied by voids or interstices. The more porous a rock is, the greater the amount of open space it contains. Through these open spaces underground water must find its way.

Porosity differs from one material to another. What is the porosity of a rock made up of particles and grains derived from preexisting rocks? Here porosity is determined largely by the shape, size, and assortment of these rock-building units. A sand deposit composed of rounded quartz grains with fairly uniform size has a high porosity. But if mineral matter enters the deposit and cements the grains into a sandstone, the porosity is reduced by an amount equal to the volume of the cementing agent. A deposit of sand, poorly sorted with finer particles of silt and clay mixed in, has low porosity because the smaller particles fill up much of the space between the larger particles.

Even a dense massive rock, such as granite, may become porous as a result of fracturing. And a soluble massive rock, such as limestone, may have its original planes of weakness enlarged by solution.

Clearly then the range of porosity in earth materials is extremely great. Recently deposited muds (called *slurries*) may hold up to 90 percent by volume of water, whereas unweathered igneous rocks such as granite, gabbro, or obsidian may hold only a fraction of 1 percent. Unconsolidated deposits of clay, silt, sand, and gravel have porosities ranging from about 20 to as much as 50 percent. But when these deposits have been consolidated into sedimentary rocks by cementation or compaction, their porosity is sharply reduced. Average porosity values for individual rock types have little meaning because of the extreme variations within each type. In general, however, a porosity of less than 5 percent is considered low; from 5 to 15 percent represents medium porosity; and over 15 percent is considered high.

PERMEABILITY

Whether or not we find a supply of fresh groundwater in a given area depends on the ability of the earth materials to transmit water as well as on their ability to contain it. The ability to transmit underground water is termed **permeability.**

The rate at which a rock transmits water depends not only on its total porosity but also on the size of the interconnections between its openings. For example, although a clay may have a higher porosity than a sand, the particles that make up the clay are minute flakes and the interstices between them are very small. Therefore water passes more readily through the sand than through the more porous clay simply because the molecular attraction on the water is much stronger in the tiny openings of the clay. The water moves more freely through the sand because the passageways between particles are relatively large and the molecular attraction on the water is relatively low. Of course, no matter how large the interstices of a material are, there must be connections between them if water is to pass through. If they are not interconnected, the material is impermeable.

A permeable material that actually carries under-

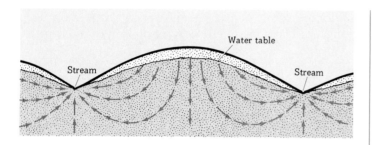

FIGURE 11.4

The flow of groundwater through uniformly permeable material is suggested here. Movement is not primarily along the groundwater table; rather, particles of water define broadly looping paths that converge toward the outlet and may approach it from below. [Redrawn from M. King Hubbert, "The Theory of Ground-water Motion," *J. Geol.,* vol. 48, p. 930, 1940.]

ground water is called an **aquifer,** from the Latin for "water" and "to bear." Perhaps the most effective aquifers are unconsolidated sand and gravel, sandstone, and some limestones. The permeability of limestone is usually due to solution that has enlarged the fractures and bedding planes into open passageways. The fractured zones of some of the denser rocks such as granite, basalt, and gabbro also act as aquifers although the permeability of such zones decreases rapidly with depth. Clay, shale, and most metamorphic and crystalline igneous rocks are generally poor aquifers.

Because the flow of underground water is usually very slow, it is largely laminar; in contrast the flow of surface water is largely turbulent. (There is one exception, however: the turbulent flow of water in large underground passageways formed in such rocks as cavernous limestone.) In laminar flow the water near the walls of interstices is presumably held motionless by the molecular attraction of the walls. Water particles farther away from the walls move more rapidly, in smooth, threadlike patterns; for the resistance to motion decreases toward the center of an opening. The most rapid flow is reached at the very center.

The energy that causes underground water to flow is derived from gravity. Gravity draws water downward to the water table; from there it flows through the ground to a point of discharge in a stream, lake, or spring. Just as surface water needs a slope to flow on, so must there be a slope for the flow of groundwater. This is the slope of the water table, the **hydraulic gradi-**

ent. It is measured by dividing the length of flow (from the point of intake to the point of discharge) into the vertical distance between these two points, a distance called **head.** Therefore, hydraulic gradient is expressed as h/l, where h is head and l is length of flow. Thus, if h is 10 m and l is 100 m, the hydraulic gradient is 0.1, or 10 percent.

An equation to express the rate of water movement through a rock was proposed by the French engineer Henri Darcy in 1856. What is now known as **Darcy's law** is essentially the same as his original equation. The law may be expressed as follows:

$$V = K\frac{h}{l}$$

where V is velocity, h head, l length of flow, and K a coefficient that depends on the permeability of the material, the acceleration of gravity, and the viscosity of water. But because h/l is simply a way of expressing the hydraulic gradient, we may say that in a rock of constant permeability the velocity of water will increase as the hydraulic gradient increases. Remembering that the hydraulic gradient and the slope of the groundwater table are the same thing, we may also say that the velocity of groundwater varies with the slope of the water table. Other things being equal, the steeper the slope of the water table, the more rapid the flow. In ordinary aquifers, the rate of water flow has been estimated as not faster than 1.5 m/day and not slower than 1.5 m/year although rates of over 120 m/day and as low as a few centimeters per year have occasionally been recorded.

The movement of underground water down the slope of the water table is only a part of the picture; for the water is also in motion at depth. Water moves downward from the water table in broad looping curves toward some effective discharge agency, such as a stream, as suggested in Figure 11.4. The water feeds into the stream from all possible directions, including straight up through the bottom of the channel. We can explain this curving path as a compromise between the force of gravity and the tendency of water to flow laterally in the direction of the slope of the water table. This tendency toward lateral flow is actually the result of the movement of water toward an area of lower pressure, the stream channel in Figure 11.4. The resulting movement is neither directly downward nor directly toward the channel but is, rather, along curving paths to the stream.

11.3 GROUNDWATER IN NATURE

So far we have assumed that groundwater is free to move on indefinitely through a uniformly permeable material of unlimited extent. Subsurface conditions actually fall far short of this ideal situation. Some layers of rock material are more permeable than others are, and the water tends to move rapidly through these beds in a direction more or less parallel to bedding planes. Even in a rock that is essentially homogeneous the groundwater tends to move in some preferred direction.

SIMPLE SPRINGS AND WELLS

Underground water moves freely downward from the surface until it reaches an impermeable layer of rock or until it arrives at the water table. Then it begins to move laterally. Sooner or later it may flow out again at the surface of the ground in an opening called a **spring.**

Springs have attracted human attention throughout history. In early days they were regarded with superstitious awe and were sometimes selected as sites for temples and oracles. To this day many persons feel that spring water possesses special medicinal and therapeutic values. Water from "mineral springs" contains salts in solution that were picked up by the water as it percolated through the ground. The same water pumped up out of a well would be regarded as merely hard and not desirable for general purposes.

Springs range from intermittent flows that disappear when the water table recedes during a dry season, through tiny trickles, to an effluence of 3.8 billion l daily, the abundant discharge of springs along a 16-km stretch of Fall River, California.

This wide variety of spring types is the result of underground conditions that vary greatly from one place to another. As a general rule, however, a spring results wherever the flow of ground water is diverted to a discharge zone at the surface (see Figure 11.5). For example, a hill made up largely of permeable rock may contain a zone of impermeable material, as shown in Figure 11.6. Some of the water percolating downward will be blocked by this impermeable rock, and a small saturated zone will be built up. Because the local water level here is actually above the main water table, it is called a **perched water table.** The water that flows laterally along this impermeable rock may emerge at the surface as a spring. Springs are not confined to

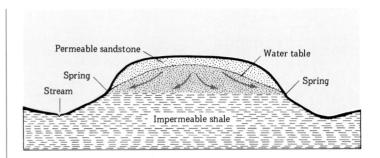

FIGURE 11.5

Nature seldom, if ever, provides uniformly permeable material. In this diagram a hill is capped by permeable sandstone and overlies impermeable shale. Water soaking into the sandstone from the surface is diverted laterally by the impermeable beds. Springs result where the water table intersects the surface at the contact of the shale and sandstone.

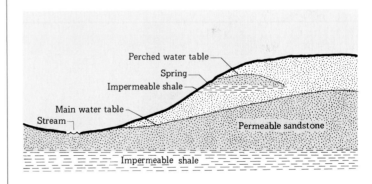

FIGURE 11.6

A perched water table results when groundwater collects over an impermeable zone and is separated from the main water table.

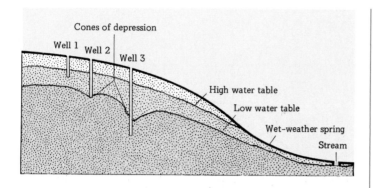

FIGURE 11.7

To provide a reliable source, a well must penetrate deep into the zone of saturation. In this diagram well 1 reaches only deep enough to tap the groundwater during periods of high water table; a seasonal drop of this surface will dry up the well. Well 2 reaches to the low water table, but continued pumping may produce a cone of depression that will reduce effective flow. Well 3 is deep enough to produce reliable amounts of water, even with continued pumping during low water-table stages.

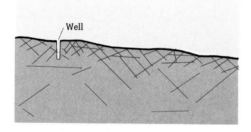

FIGURE 11.8

Wells may produce water from a fractured zone of impermeable rocks, such as granite. The supply, however, is likely to be limited not only because the size and number of fractures decrease with depth but also because the fractures do not interconnect.

points where a perched water table flows from the surface, and it is clear that, if the main water table intersects the surface at a point along a slope, a spring will form.

Even in impermeable rocks permeable zones may develop as a result of fractures or solution channels. If these openings fill with water and are intersected by the ground surface, the water will issue forth as a spring.

A spring is the result of a *natural* intersection of the ground surface and the water table, but a **well** is an *artificial* opening cut down from the surface into the zone of saturation. A well is productive only if it is drilled into permeable rock and penetrates below the water table. The greater the demands that are made on a well, the deeper it must be drilled below the water table. Continuous pumping creates the cone of depression previously described, which distorts the water table and may reduce the flow of groundwater into the well (Figure 11.7).

Wells drilled into fractured crystalline rock, such as granite, may produce a good supply of water at relatively shallow depths, but we cannot increase the yield of such wells appreciably by deepening them because the number and size of the fractures commonly decrease the farther down into the earth we go (see Figure 11.8).

Wells drilled into limestone that has been riddled by large solution passages may yield a heavy flow of water part of the time and no flow the rest of the time simply because the water runs out rapidly through the large openings. Furthermore, water soaking down from the surface flows rapidly through limestone of this sort and may make its way to a well in a very short time. Consequently water drawn from the well may be contaminated because there has been insufficient time for impurities to be filtered out as the water passes from the surface to the well.

In sandstone, on the other hand, the rate of flow is slow enough to permit the elimination of impurities even within a very short distance of underground flow. Harmful bacteria are destroyed in part by entrapment, in part by lack of food and by temperature changes, and in part by hostile substances or organisms encountered along the way, particularly in the soil.

By applying Darcy's law, we can estimate the amount of water that a well will probably yield. The quantity of water passing through a given section in a unit of time is determined by the area of the section it passes through and by the velocity of the flow. There-fore

$$V = \frac{Q}{A}$$

where V is velocity, Q the quantity of water per unit time, and A the area of the cross section through which the water flows. We then have a statement of Darcy's law that gives quantity of water in terms of permeability, hydraulic gradient, and area of cross section:

$$V = K\frac{h}{l} = \frac{Q}{A} \quad \text{or} \quad Q = AK\frac{h}{l}$$

Therefore we can calculate the rate of discharge for a given well if we know the hydraulic gradient h/l, the cross section A through which the water passes, and K, which includes the permeability of the rock material, the viscosity of the water, and the acceleration of gravity. In actual practice the most difficult problem in applying this formula is determining the hydraulic gradient.

ARTESIAN WATER

Contrary to common opinion, artesian water does not necessarily come from great depths. But other definite conditions characterize an artesian water system: (1) The water is contained in a permeable layer, the aquifer, inclined so that one end is exposed to receive water at the surface; (2) the aquifer is capped by an impermeable layer; (3) the water in the aquifer is prevented from escaping either downward or off to the sides; and (4) there is enough head to force the water above the aquifer wherever it is tapped. If the head is sufficiently great, the water will flow out to the surface either as a well or a spring (see Figure 11.9). The term **artesian** is derived from the name of a French town, Artois (originally called *Artesium* by the Romans), where this type of well was first studied.

A classic example of an artesian water system is found in western South Dakota, where the Black Hills have punched up through a series of sedimentary rocks, bending their edges up to the surface. One of these sedimentary rocks, a permeable sandstone of the Cretaceous age, carries water readily and is sandwiched between impermeable layers. Water entering the sandstone around the Black Hills moves under-

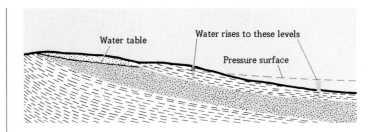

FIGURE 11.9

The wells in the diagram meet the conditions that characterize an artesian system: (1) an inclined aquifer, (2) capped by an impermeable layer, (3) with water prevented from escaping either downward or laterally, and (4) sufficient head to force the water above the aquifer wherever it is tapped. In the well at the right of the diagram the head is great enough to force water to appear at the surface.

ground eastward across the state, reaching greater and greater depths as it travels along. When we drive a well into this aquifer, we tap water that is under the pressure exerted by all the water piled up between the well and the Black Hills.

THERMAL SPRINGS

Springs that bring warm or hot water to the surface are called **thermal springs, hot springs,** or **warm springs** (Figure 11.10). A spring is usually regarded as a thermal spring if the temperature of its water is about 6.5°C higher than the mean temperature of the air. There are over 1,000 thermal springs in the western mountain regions of the United States, 46 in the Appalachian Highlands of the east, 6 in the Ouachita area in Arkansas, and 3 in the Black Hills of South Dakota.

Most of the western thermal springs derive their heat from masses of magma that have pushed their way into the crust almost to the surface and are now cooling. In the eastern group, however, the circulation of the groundwater carries it so deep that it is warmed by the normal increase in earth heat (see "Thermal gradient" in the Glossary).

The well-known spring at Warm Springs, Georgia, is heated in just this way. Long before the Civil War this spring was used as a health and bathing resort by the

FIGURE 11.10

Mammoth Hot Springs, Yellowstone National Park, are thermal springs that have built these terraces by depositing travertine. [U.S. Geological Survey.]

people of the region. Then, with the establishment of the Georgia Warm Springs Foundation for the treatment of victims of infantile paralysis, the facilities were greatly improved. Rain falling on Pine Mountain, about 3 km south of Warm Springs, enters a rock formation known as the Hollis. At the start of its journey downward the average temperature of the water is about 16.5°C. It percolates through the Hollis Formation northward under Warm Springs at a depth of around 100 m and then follows the rock as it plunges into the earth to a depth of 1,140 m, 1.6 km farther north. Normal rock temperatures in the region increase about 1.8°C/100 m of depth, and the water is warmed as it descends along the bottom of the Hollis bed. At 1,140 m the bed has been broken and shoved against an impervious layer that turns the water back. This water is now hotter than the water coming down from above, and it moves upward along the top of the Hollis Formation, cooling somewhat as it goes. Finally it emerges in a spring at a temperature of 36.5°C.

Less than 1 km away is Cold Spring, whose water comes from the same rainfall on Pine Mountain. A freak of circulation, however, causes the water at Cold Spring to emerge before it can be conducted to the depths and warmed. Its temperature is only about 16.5°C.

GEYSERS

A **geyser** is a special type of thermal spring that ejects water intermittently with considerable force. The word comes from the Icelandic name of a spring of this type, *geysir*, probably based on the verb *geysa*, "to rush forth."

Although the details of geyser action are still not understood, we know that, in general, a geyser's behavior is caused by the arrangement of its plumbing and the proximity of a good supply of heat. Here is probably what happens: Groundwater moving downward from the surface fills a natural pipe, or conduit, that opens upward to the surface. Hot igneous rocks, or the gases given off by such rocks, gradually heat the column of water in the pipe and raise its temperature toward the boiling point. Now the higher the pressure on water, the higher its boiling point; and because water toward the bottom of the pipe is under the greatest pressure, it must be heated to a higher temperature than the water above before it will come to a boil. Eventually the column of water becomes so hot that either a slight increase in temperature or a slight decrease in pressure will cause it to boil. At this critical point the water near the base of the pipe is heated to the boiling point. The water then passes to steam and,

FIGURE 11.11

Old Faithful in Yellowstone National Park is America's most widely known geyser. The periodic eruption of a geyser is due to the particular pattern of its plumbing and its proximity to a liberal source of heat and groundwater. [Barton W. Knapp.]

as it does so, expands, pushing the water above it toward the surface. But this raising of the heated column of water reduces the pressure acting on it, and it too begins to boil. The energy thus developed throws the water and steam high into the air, producing the spectacular action characteristic of many geysers. After the eruption has spent itself, the pipe is again filled with water and the whole process begins anew.

We can compare this theoretical cycle with that of Old Faithful Geyser in Yellowstone National Park (see Figure 11.11). The first indication of a coming eruption at Old Faithful is the quiet flow of water in a fountain some 1 to 2 m high. This preliminary activity lasts for a few seconds and then subsides. It represents the first upward push of the column of water described in our theoretical case. This push reduces the pressure and thereby lowers the boiling point of the water in the pipe. Consequently the water passes to steam, and less than 1 min after the preliminary fountain the first of the violent eruptions takes place. Steam and boiling water are thrown 45 to 50 m into the air. The entire display lasts about 4 min. Emptied by the eruption, the tube then gradually refills with groundwater, the water is heated, and in approximately 1 h the same cycle is repeated. The actual time between eruptions at Old Faithful averages about 65 min but may be from 30 to 90 min.

11.4 RECHARGE OF GROUNDWATER

As we have seen, the ultimate source of most underground water is precipitation that finds its way below the surface of the land either by natural or by artificial means.

Some of the water from precipitation seeps into the ground, reaches the zone of saturation, and raises the water table. Continuous measurements over long periods of time at many places throughout the United States show an intimate connection between water level and rainfall (see Figure 11.12). Because water moves relatively slowly in the zone of aeration and the zone of saturation, fluctuations in the water table usually lag a little behind fluctuations in rainfall.

Several factors control the amount of water that actually reaches the zone of saturation. For example, rain that falls during the growing season must first

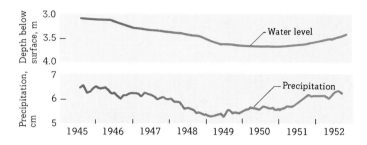

FIGURE 11.12

Relationship between the water level in an observation well near Antigo, Wisconsin, and precipitation, as shown by records from 1945 to 1952. The
water table reflects the changes in precipitation.
The graphs represent 3-year running monthly averages. For example, 5.8 cm of precipitation for
May means that precipitation averaged 5.8
cm/month from May, 1947, to May, 1950, inclusive.
[From A. H. Harder and William J. Drescher,
"Ground Water Conditions in Southwestern Lang-
lade County, Wisconsin," *U.S. Geol. Surv. Water
Supply Paper* 1294, 1954.]

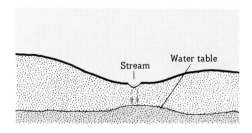

FIGURE 11.13

**The groundwater may be recharged by
water from a surface stream leaking
into the underground.**

replenish moisture used up by plants or passed off
through evaporation. If these demands are great, very
little water will find its way down to **recharge** the zone
of saturation. Then too, during a very rapid, heavy
rainfall, most of the water may run off directly into the
streams instead of soaking down into the ground. A
slow, steady rain is much more effective than a heavy,
violent rain in replenishing the supply of groundwater.
High slopes, lack of vegetation, or the presence of impermeable rock near the surface may promote runoff
and reduce the amount of water that reaches the zone
of saturation. It is true, however, that some streams are
themselves sources for the recharge of underground
water. Water from the streams leaks into the zone of
saturation, sometimes through a zone of aeration (see
Figure 11.13).

In many localities the natural recharge of the underground supplies cannot keep pace with human demands for groundwater. Consequently attempts are
sometimes made to recharge these supplies artificially.
On Long Island, New York, for example, water that has
been pumped out for air-conditioning purposes is returned to the ground through special recharging wells
or, in winter, through idle wells that are used in summer
for air conditioning. In the San Fernando Valley, California, surplus water from the Owens Valley aqueduct
is fed into the underground in an attempt to keep the
local water table at a high level.

11.5 CAVES AND RELATED FEATURES

Caves are probably the most spectacular examples of
the handiwork of underground water. In dissolving
great quantities of solid rock in its downward course,
the water fashions large rooms, galleries, and underground stream systems as the years pass. In many
caves the water deposits calcium carbonate as it drips
off the ceilings and walls, building up fantastic shapes

of *dripstone* (Figures 11.14 and 11.15).

Caves of all sizes tend to develop in highly soluble
rocks such as limestone, $CaCO_3$, and small ones occur
in the sedimentary rock dolomite, $CaMg(CO_3)_2$. Rock
salt, NaCl, gypsum, $CaSO_4 \cdot 2H_2O$, and similar rock
types are the victims of such rapid solution that underground caverns usually collapse under the weight of

FIGURE 11.14

Stalactites grow downward, some to meet stalagmites growing upward from the cave floor in Carlsbad Caverns, New Mexico. [National Park Service.]

overlying rocks before surface erosion exposes them.

Calcite, the main component of limestone, is very insoluble in pure water. But when the mineral is attacked by water containing small amounts of carbonic acid, it undergoes rapid chemical weathering; most natural water contains carbonic acid, H_2CO_3, the combination of water, H_2O, with carbon dioxide, CO_2. The carbonic acid reacts with the calcite to form calcium bicarbonate, $Ca(HCO_3)_2$, a soluble substance that is then removed in solution. If not redeposited, it eventually reaches the ocean.

Let us look more closely at underground water as it brings about the decay of calcite. Calcite contains the complex carbonate ion, $(CO_3)^{2-}$ (built by packing three oxygen atoms around a carbon atom), and the calcium ion, Ca^{2+}. These two ions are combined in much the same way as sodium and chlorine in forming salt, or halite. The weathering or solution of calcite takes place when a hydrogen ion, H^+, approaches $(CO_3)^{2-}$. Because the attraction of hydrogen for oxygen is stronger than the attraction of carbon for oxygen, the hydrogen ion pulls away one of the oxygen atoms of the carbonate ion and, with another hydrogen ion, forms water. The two other oxygen atoms remain with the carbon atom as carbon dioxide gas, CO_2. The calcium ion, Ca^{2+}, now joins with two negative bicarbonate ions, $(HCO_3)^-$, to form $Ca(HCO_3)_2$ in solution. We can express these activities as follows:

Two parts water	plus	two parts carbon dioxide	yield	two parts carbonic acid
$2H_2O$	$+$	$2CO_2$	$\rightleftharpoons$	$2H_2CO_3$

Two parts carbonic acid	yield	two hydrogen ions	plus	two bicarbonate ions
$2H_2CO_3$	$\rightleftharpoons$	$2H^+$	$+$	$2(HCO_3)^-$

Two hydrogen ions	plus	two bicarbonate ions	plus	calcite	yield
$2H^+$	$+$	$2(HCO_3)^-$	$+$	$CaCO_3$	$\rightleftharpoons$

		water	plus	carbon dioxide	plus	calcium bicarbonate in solution
		H_2O	$+$	CO_2	$+$	$Ca^{2+} + 2(HCO_3)^-$

FIGURE 11.15

Delicate corallike forms have been deposited on a massive stalagmite in the Queen's Chamber, Carlsbad Caverns, New Mexico. [National Park Service.]

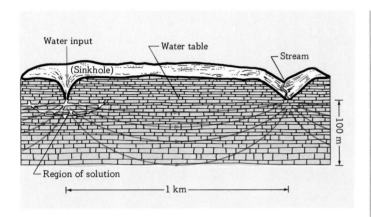

FIGURE 11.16

Diagram through a small limestone aquifer, showing approximate flow line and the location of region of solution. See the text for discussion. Vertical exaggeration, 4×. [After John Thrailkill, "Chemical and Hydrologic Factors in the Excavation of Limestone Caves," *Bull. Geol. Soc. Am.*, vol. 79, p. 39, 1968.]

The principal chemical reaction in the formation of caves is given above. Even though the reaction can be demonstrated in the laboratory and occurs in nature, its extension to the formation of caves is not so direct as it might first appear.

There is a strong difference of opinion as to whether caves form above, at (or just below), or deep beneath the water table. Those who believe that caves form below the water table call upon some mechanism that would lower the water table and drain the cave, thus filling it with air. Drilling records, in fact, show that there are large openings in limestone that lie flooded with water beneath the water table. Presumably, if land keeps rising and erosion proceeds until valleys are deep enough to drain the caves in the surrounding countryside, they will become filled with air and ready for discovery and exploration.

We have seen that water and carbon dioxide will combine to provide a solution that will corrode limestone. Such a solution, seeping into the ground, will begin to dissolve limestone as soon as it meets it just below the soil. Here the waters are rapidly neutralized and, seeping deeper, can no longer dissolve the rock. In this regard we can note that the waters seeping down from the surface into caves today are depositing rather than dissolving calcite. Yet it is also clear that

the chambers we now call caves were formed at an appreciable depth below the surface. How, then, could this happen? The details of the answer are yet to be worked out, but studies by John Thrailkill of the University of Tennessee point a probable way to the answer. He suggests mechanisms by which the flow of groundwater through limestone would allow concentrations of large amounts of fresh water undersaturated with respect to calcite. In such areas, probably close to the water table, solution of limestone would be possible.

One such mechanism is shown in Figure 11.16. In this diagram the water table is shown as being almost horizontal, a condition often encountered in limestone rock and in contrast to the sloping water table shown as an idealized case in Figure 11.4. The flow lines lead to the stream, and the water table is one of these flow lines. Water reaches the underground chiefly via an enlarged opening at the surface. This opening could be formed by differential solution from the surface downward along particularly well-developed or favorably placed fractures in the limestone rock. Its presence allows the rapid injection of large amounts of surface water into the underground. Because it is injected rapidly and in large amounts, it may pass through the zone above the water table without coming into effective contact with limestone and remain undersaturated with respect to calcite as it joins the water. If this happens, there is a zone beneath the inlet and below the water table where groundwater may still carry enough carbonic acid to effect solution in the underground.

Regardless of where and how caves are originally formed, we know that the weird stone formations characteristic of most of them must have developed above the water table when the caves were filled with air. These bizarre shapes are composed of calcite deposited by underground water that has seeped down through the zone of aeration. They develop either as **stalactites,** looking like stony icicles hanging from the cave roof (Greek *stalaktos,* "oozing out in drops"), or as **stalagmites,** heavy posts growing up from the floor (Greek *stalagmos,* "dripping"). When a stalactite and stalagmite meet, a **column** is formed. A stalactite forms as water charged with calcium bicarbonate in solution seeps through the cave roof. Drop after drop forms on the ceiling and then falls to the floor. But during the few moments that each drop clings to the ceiling a small amount of evaporation takes place; some carbon dioxide is lost, and a small amount of calcium carbonate is deposited. Over the centuries a large stalactite may gradually develop. Part of the water that falls to the

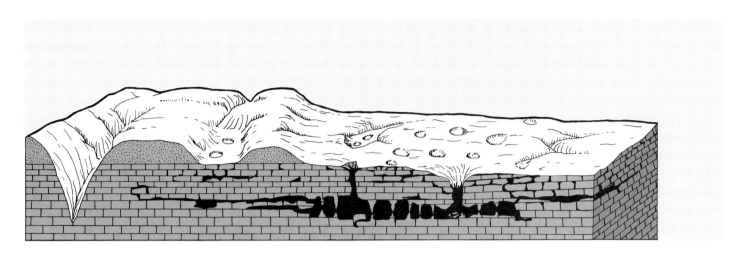

FIGURE 11.17

Underground solution of limestone not only produces caves with stalactites and stalagmites, as suggested here, but in addition may develop karst topography at the
surface with sinkholes as well as valleys that drain into the subsurface.

cave floor runs off, and part is evaporated. This evaporation again causes the deposition of calcite, and a
stalagmite begins to grow upward to meet the stalactite.

On the ground surface above soluble rock material, depressions sometimes develop that reflect areas
where the underlying rock has been carried away in
solution. These depressions, called **sinkholes** or merely
sinks, may form in one of two different ways. In one case
the limestone immediately below the soil may be dissolved by the seepage of waters downward. The process of solution may be focused by local factors, such

as a more abundant water supply or greater solubility
of the limestone. Eventually a depression—a sink—
evolves. Probably more commonly, sinks form when the
surface collapses into a large cavity below. In either
event surface water may drain through the sinkholes to
the underground; or if the sinks' subterranean outlets
become clogged, they may fill with water to form lakes.
An area with numerous sinkholes is said to display
karst topography, from the name of a plateau in Yugoslavia and northeastern Italy where this type of landscape is well developed (see Figures 11.17 and 11.18).

FIGURE 11.18

Saint Joseph's Well, a sinkhole filled
with water, is in Clark County, Kansas. [U.S. Geological Survey.]

OUTLINE

Underground water is over 60 times as abundant as the water in streams and lakes.

Distribution is in the **zone of aeration** and an underlying **zone of saturation.** The irregular surface that separates them is called the **groundwater table.**

Movement of underground water is usually by laminar flow.

Porosity is the total percentage of void space to a given volume of earth material.

Permeability is the ability of earth material to transmit water. Flow is driven by gravity according to **Darcy's law:** $V = K (h/l)$.

Groundwater in nature moves in aquifers. It comes to the surface as springs and wells.

A **spring** occurs when the ground surface intersects the water table. **Simple wells** draw water from the zone of saturation.

Artesian wells are those in which gravity drives the water level above the top of the aquifer.

Thermal springs derive their heat from the cooling of igneous rocks or by a normal increase of the earth's heat with depth.

Geysers are thermal springs marked by periodic, violent eruptions and controlled by a particular arrangement of underground passages.

Recharge of groundwater is either natural or artificial.

Caves are usually created by the solution of limestone. The chemical reaction involves carbonic acid (formed by the combination of water and carbon dioxide) with limestone.

Formation of caves apparently takes place in two stages: (1) the creation of chambers, galleries, and tunnels and (2) the decoration of these large openings by forms growing from the ceilings (**stalactites**) and forms growing from the floor (**stalagmites**). Solution of limestone develops an irregular "pitted" topography of depressions with no exterior surface drainage. This is called **karst topography** and is marked by numerous **sinkholes.**

SUPPLEMENTARY
READINGS

Davis, S. N., and R. J. M. De Wiest *Hydrogeology,* John Wiley & Sons, Inc., New York, 1966. *A somewhat advanced text on underground water.*

Sweeting, Marjorie M. *Karst Landforms,* Columbia University Press, New York, 1973. *An excellent volume on caves, karst, and associated landscape.*

Todd, D. K. *Ground Water Hydrology,* John Wiley & Sons, Inc., New York, 1959. *An old, but still good basic text on the subject.*

12

GLACIATION

The seas and rivers of moving ice known as glaciers have attracted the attention of inquisitive men deep in the Arctic, Antarctic, and mountainous regions of the world. There they have discovered that glaciers are active agents of erosion, transportation, and deposition and that these impressive masses of ice were far more widespread in the past than they are now. Geologists have learned too that the ice of the last great glacial period has modified and molded great stretches of landscape in what are now the temperate zones.

12.1 FORMATION OF GLACIER ICE

A **glacier** is a mass of ice that has been formed by the recrystallization of snow and that flows forward—or has flowed at some time in the past—under the influence of gravity. This definition eliminates the pack ice formed from seawater in polar latitudes and, by convention, icebergs even though they are large fragments broken from the seaward end of glaciers.

Like surface streams and underground reservoirs, glaciers depend on the oceans for their nourishment. Some of the water drawn up from the oceans by evaporation falls on the land in the form of snow. If the climate is right, part of the snow may last through the summer without melting. Gradually, as the years pass, the accumulation may grow deeper and deeper until at last a glacier is born. In areas where the winter snowfall exceeds the amount of snow that melts away during the summer, stretches of perennial snow known as **snowfields** cover the landscape. At the lower limit of a snowfield lies the **snow line.** Above the snow line glacier ice may collect in the more sheltered areas of the snowfields. The exact position of the snow line varies from one climatic region to another. In polar regions, for example, it reaches down to sea level, but near the equator it recedes to the mountain tops. In the high mountains of East Africa it ranges from elevations of 4,500 to 5,400 m. The highest snow lines in the world are in the dry regions known as the "horse latitudes," between 20° and 30° north and south of the equator. Here the snow line reaches higher than 6,000 m.

Fresh snow falls as a feathery aggregate of com-

FIGURE 12.1

Snowflakes exhibit a wide variety of patterns, all hexagonal and all reflecting the internal arrangement of hydrogen and oxygen. It is from snowflakes that glacier ice eventually forms.

plex and beautiful crystals with a great variety of patterns. All the crystals are basically hexagonal, however, and all reflect their internal arrangement of hydrogen and oxygen atoms (see Figure 12.1). Snow is not frozen rain; rather, it forms from the condensation of water vapor that is at temperatures below the freezing point.

After snow has been lying on the ground for some time, it changes from a light, fluffy mass to a heavier, granular material called **firn,** or *névé. Firn* derives from a German adjective meaning "of last year," and *névé* is a French word from the Latin for "snow." Solid remnants of large snowbanks, those tiresome vestiges of winter, are largely firn. Several processes are at work in the transformation of snow into firn. The first is **sublimation,** a general term for the process by which a solid material passes into the gaseous state without first becoming liquid. In sublimation molecules of water vapor escape from the snow, particularly from the edges of the flakes. Some of the molecules attach themselves to

the center of the flakes, where they adapt themselves to the structure of the snow crystals. Then, as time passes, one snowfall follows another, and the granules that have already begun to grow as a result of sublimation are packed tighter and tighter together under the pressure of the overlying snow.

Water has the rare property of increasing in volume when it freezes; conversely, it decreases in volume as the ice melts. But the cause and effect may be interchanged: If added pressure on the ice squeezes the molecules closer together and reduces its volume, the ice may melt. In fact, if the individual granules are in contact, they begin to melt with only a slight increase in pressure. The resulting meltwater trickles down and refreezes on still lower granules at points where they are not yet in contact. And all through this process the basic hexagonal structure of the original snow crystals is maintained.

Gradually, then, a layer of firn granules, ranging from a fraction of a millimeter to approximately 3 or 4 mm in diameter, is built up. The thickness of this layer varies, but the average seems to be 30 m on many mountain glaciers.

The firn itself undergoes further change as continued pressure forces out most of the air between the granules, reduces the space between them, and finally transforms the firn into **glacier ice,** a true solid composed of interlocking crystals. In large blocks it is usually opaque and takes on a blue-gray color from the air and the fine dirt it contains.

The ice crystals that make up glacier ice are minerals; the mass of glacier ice, made up of many interlocking crystals, is a metamorphic rock; for it has been transformed from snow into firn and eventually into glacier ice. Later we shall see that glacier ice itself undergoes further metamorphism.

CLASSIFICATION OF GLACIERS

The glaciers of the world fall into three principal classifications: valley glaciers, piedmont glaciers, and ice sheets.

Valley glaciers are streams of ice that flow down the valleys of mountainous areas (see Figure 12.2). Like streams of running water, they vary in width, depth, and length. A branch of the Hubbard glacier in Alaska is 120 km long, whereas some of the valley glaciers that dot the higher reaches of our western mountains are only a few hundred meters in length. Valley glaciers that are nourished on the flanks of high mountains and that flow down the mountain sides are sometimes called **mountain glaciers,** or **Alpine glaciers.** Very small mountain glaciers are referred to as **cliff glaciers, hanging glaciers,** or **glacierets.** A particular type of valley glacier sometimes grows up in areas where large masses of ice are dammed by a mountain barrier along the coast. Some of the ice escapes through valleys in the mountain barrier to form an **outlet glacier,** as it has done along the coasts of Greenland and Antarctica.

Piedmont glaciers form when two or more glaciers emerge from their valleys and coalesce to form an apron of moving ice on the plains below.

Ice sheets are broad, moundlike masses of glacier ice that tend to spread radially under their own weight. The Vatna glacier of Iceland is a small ice sheet measuring about 120 by 160 km and 225 m in thickness. A localized sheet of this sort is sometimes called an **ice cap** (see Figure 12.3). The term **continental glacier** is usually reserved for great ice sheets that obscure the mountains and plains of large sections of a continent, such as those of Greenland and Antarctica. On Greenland ice exceeds 3,000 m in thickness near the center of the ice caps. Ice in Antarctica averages about 2,300 m in thickness, and in some places this ice is over 4,000 m thick.

DISTRIBUTION OF MODERN GLACIERS

Modern glaciers cover approximately 10 percent of the land area of the world. They are found in widely scattered locations in North and South America, Europe, Asia, Africa, Antarctica, and Greenland and on many of the north polar islands and the Pacific islands of New Guinea and New Zealand. A few valley glaciers are located almost on the equator. Mount Kenya in East Africa, for instance, only 0.5° from the equator, rises over 5,100 m into the tropical skies and supports at least 10 valley glaciers.

The total land area covered by existing glaciers is estimated as 17.9 million km², of which the Greenland and Antarctic ice sheets account for about 96 percent. The Antarctic ice sheet covers approximately 15.3 million km², and the Greenland sheet covers about 1.8 million km². Small ice caps and numerous mountain glaciers scattered around the world account for the remaining 4 percent.

FIGURE 12.2

A valley glacier in the Fairweather Range of British Columbia. [Austin Post.]

FIGURE 12.3

A small ice cap on Axel Heiberg Island, Northwest Territories, Canada. [University of Washington, Austin Post.]

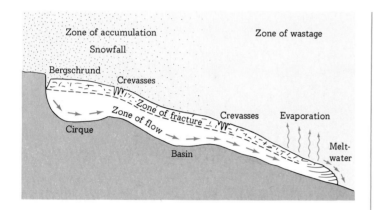

FIGURE 12.4

A glacier is marked by a zone of accumulation and a zone of wastage. Within a glacier, ice may lie either in the zone of fracture or deeper in the zone of flow. A valley glacier originates in a basin, the *cirque*, and it is separated from the headwall of the cirque by a large crevasse, known as the *bergschrund*.

NOURISHMENT AND WASTAGE OF GLACIERS

When the weight of a mass of snow, firn, and ice above the snow line becomes great enough, movement begins and a glacier is created. The moving stream flows downward across the snow line until it reaches an area where the loss through evaporation and melting is so large that the forward edge of the glacier can push no farther. A glacier, then, can be divided into two zones: a **zone of accumulation** and a **zone of wastage**. Both are illustrated in Figure 12.4.

The position of the front of a glacier depends on the relationship between the glacier's rate of nourishment and its rate of wastage. When nourishment just balances wastage, the front becomes stationary, and the glacier is said to be in equilibrium. This balance seldom lasts for long, however; for a slight change in either nourishment or wastage will cause the front to advance or recede.

Today most of the glaciers of the world are receding. With only a few exceptions this process has been going on since the latter part of the nineteenth century, although at varying rates. A striking feature of modern glaciers is that they follow the same general pattern of growth and wastage the world over and serve as indicators of widespread climatic changes.

Valley glaciers are nourished not only in the zone of accumulation but also by great masses of snow that avalanche down the steep slopes along their course. In fact, according to one interpretation, avalanches caused by earthquakes have enabled certain glaciers to advance in a single month as far as they would have if fed by the normal snowfall of several years.

Below the snow line wastage takes place through a double process of evaporation and melting known as **ablation**. If a glacier terminates in a body of water, great blocks of ice break off and float away in a process called **calving**. This is the action that produces the **icebergs** of the polar seas.

GLACIER MOVEMENT

Except in rare cases glaciers move only a few centimeters or at most a few meters per day. That they actually move, however, can be demonstrated in several ways. The most conclusive test is to measure the movement directly, by emplacing a row of stakes across a valley glacier. As time passes, the stakes move down-valley with the advancing ice, the center stakes more rapidly than those near the valley walls. A second source of evidence is provided by the distribution of rock material on the surface of a glacier. When we examine the boulders and cobbles lying along a valley glacier, we find that many of them could not have come from the walls immediately above and that the only possible source lies up-valley. We can infer, then, that the boulders must have been carried to their present position on the back of the glacier. Another indication of glacier movement is that, when a glacier melts, it often exposes a rock floor that has been polished, scratched, and grooved. It is simplest to explain this surface by assuming that the glacier actually moved across the floor, using embedded debris to polish, scratch, and groove it.

Clearly, then, a glacier moves (see Figure 12.5). In fact different parts of it move at different rates. But although we know a good bit about how a glacier flows forward, certain phases are not yet clearly understood. In any event we can distinguish two zones of movement: (1) an upper zone, between 30 and 60 m thick, which reacts like a brittle substance—that is, it breaks sharply rather than undergoing gradual, permanent distortion; and (2) a lower zone, which, because of the pressure of the overlying ice, behaves like a plastic substance. The first is the **zone of fracture**; the second is the **zone of flow**.

As plastic deformation takes place in the zone of flow, the brittle ice above is carried along. But the zone of flow moves forward at different rates—faster in some parts, more slowly in others—and the rigid ice in the zone of fracture is unable to adjust itself to this irregular advance. Consequently the upper part of the glacier cracks and shatters, giving rise to a series of deep, treacherous **crevasses** (see Figures 12.4 and 12.6).

A glacier attains its greatest velocity somewhere above the valley floor, in midstream; for the sides and bottom are retarded by friction against the valley walls and beds. In this respect the movement of an ice stream resembles that of a stream of water. The mechanics of ice flow, however, are still a matter of study—a study made difficult by the fact that we cannot actually observe the zone of flow because it lies concealed within the glacier. Yet the ice from the zone of flow eventually emerges at the snout of the glacier, and there it can be studied. We find that, by the time it has emerged, it is brittle; but it retains the imprint of movement by flow. The individual ice crystals are now several centimeters in size; in contrast crystals in ice newly formed from firn measure but a fraction of a centimeter. We can conclude that the ice crystals have grown by recrystallization as they passed through the zone of flow. The ice at the snout is also marked by bands that represent shearing and differential movement within the glacier. Recrystallization has taken place along many of the old shear planes; and along others the debris carried forward by the ice has been concentrated. These observations suggest that some movement in the zone of flow has taken place as a result of shearing.

Measurements of the direction of flow of ice in modern glaciers show that in the zone of wastage the flow of ice is upward toward the ice surface at a low angle. In the zone of accumulation the direction of movement is downward from the ice surface at a low angle (see Figure 12.4).

However glacier ice may move in the interior of a glacier, there is also a movement involving the entire ice tongue or sheet. The glacier literally slips along its base, moving across the underlying ground surface. This **basal slip** is added to the internal flow of the glacier to give the total movement of the ice body (see Figure 12.7).

Although most glacier motion is slow, some glaciers at times display a very rapid motion called a **surge.** Surges, or catastrophic advances, are now known from certain mountain and valley glaciers around the world (Figure 12.8). The largest surge so far reported seems to

FIGURE 12.5

Long sinuous bands of debris outline the flow pattern of the Lowell glacier, British Columbia. [Austin Post.]

FIGURE 12.6

Crevassed surface of the Columbia glacier, Chugach Mountains, Alaska. [University of Washington, Austin Post.]

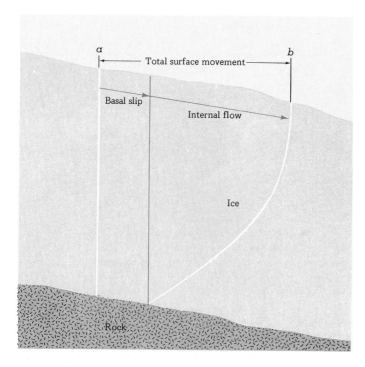

FIGURE 12.7

Forward motion of a glacier consists in part of internal flow of the ice and in part of the slippage of the entire glacier over the base. [After Robert P. Sharp, *Glaciers*, Oregon System of Higher Education, Eugene, 1960.]

be that of the Bråsellbreen, a portion of a roughly circular ice cap on Spitzbergen. Here the ice advanced up to 20 km along a 21-km front some time between 1935 and 1938, as revealed by aerial photographs taken in those two separate years. The most rapid advance authenticated by actual observation is 110 m/day on the Kutiah glacier, which is located in the mountains of northern India.

Glaciers surge because of an instability that allows gravity to tear them loose from the ground surface on which they lie. The cause of this instability is not yet completely understood, but it involves the accumulation of an excess of water at the glacier's base. One way this water may accumulate is through a thickening of the ice and thus an increase of pressure in the glacier. At the base this increase may be sufficient to allow some melting there.

Surges are as yet unrecorded from the Antarctic ice cap. Nevertheless it has been estimated that, should a surge occur there, it might well involve the movement into the Antarctic Ocean of enough ice to raise the world's sea level between approximately 10 and 30 m during a period of a few years or a few tens of years. The results would make some science fiction tame reading by comparison.

TEMPERATURES OF GLACIERS

All glaciers are cold, but some are colder than others. Therefore students of glaciers speak of "warm" and "cold" glaciers. A **cold glacier** is one in which no surface melting occurs during the summer months; its temperature is always below freezing. A **warm glacier** is

FIGURE 12.8

Rapid surges along the Yanert glacier in the Alaska Range have displaced morainal loops down-valley. The photograph was taken in 1964, and the latest surge at that time had been in 1942. [University of Washington, Austin Post.]

one that reaches the melting temperature throughout its thickness during the summer season.

Obviously no glacier can exist above the melting point of ice; yet there are glaciers that hover at this temperature. All glaciers must form at a subfreezing temperature, and therefore we must ask how a glacier warms from its formation temperature to the melting point. The sun's heat cannot penetrate more than a few meters into the poorly conducting ice. Some other mechanism must operate to warm a glacier. One in-volves the downward movement of water from the sur-face of the glacier. There the heat of the sun melts surface and near-surface ice, and the meltwater perco-lates downward into the glacier. When eventually it freezes, it gives up heat at the rate of 80 cal/g (gram) of water. (Because each gram of ice requires 0.5 cal to rise 1°C, these 80 cal increase the temperature of 160 g by 1°C.) In addition there is the normal heat flow from the earth to the surface and the heat generated by friction of the moving ice.

12.2 RESULTS OF GLACIATION

MOVEMENT OF MATERIAL

Glaciers have special ways of eroding, transporting, and depositing earth materials. A valley glacier, for example, acquires debris by means of frost action, landsliding, and avalanching. Fragments pried loose by frost action clatter down from neighboring peaks and come to rest on the back of the glacier. And great snowbanks, unable to maintain themselves on the steep slopes of the mountainsides, avalanche downward to the glacier, carrying along quantities of rock debris and rubble. This material is either buried beneath fresh snow or avalanches or tumbles into gaping crevasses in the zone of fracture and is carried along by the glacier.

When a glacier flows across a fractured or jointed stretch of bedrock, it may lift up large blocks of stone and move them off. This process is known as **plucking,** or **quarrying.** The force of the ice flow itself may suffice to pick up the blocks, and the action may be helped along by the great pressures that operate at the bottom of a glacier. Suppose the moving ice encounters a projection or rock jutting up from the valley floor. As the glacier ice forces itself over and around the projection, the pressure on the ice is increased, and some of the ice around the rock may melt. This meltwater trickles to-ward a place of lower pressure, perhaps into a crack in the rock itself. There it refreezes, forming a strong bond between the glacier and the rock. Continued movement by the glacier may then tear the block out of the valley floor.

At the heads of valley glaciers plucking and frost action sometimes work together to pry rock material loose. Along the back walls of the collection basins of mountain glaciers great hollows called **cirques,** or am-phitheaters, develop in the mountainside (see Fig-ure 12.9). As the glacier begins its movement down-slope, it pulls slightly away from the back wall, forming a crevasse known as a **bergschrund.** One wall of the bergschrund is formed by the glacier ice; the other is formed by the nearly vertical cliff or bedrock. During the day meltwater pours into the bergschrund and fills the openings in the rock. At night the water freezes, producing pressures great enough to loosen blocks of rock from the cliff. Eventually these blocks are incorpo-rated into the glacier and are moved away from the headwall of the cirque.

The streams that drain from the front of a melting glacier are charged with **rock flour,** very fine particles of pulverized rock. So great is the volume of this mate-rial that it gives the water a characteristically grayish-blue color similar to that of skim milk. Here then is further evidence of the grinding power of the glacier mill.

Glaciers also pick up rock material by means of abrasion. As the moving ice drags rocks, boulders, pebbles, sand, and silt across the glacier floor, the bedrock is cut away as though by a great rasp or file. And the cutting tools themselves are abraded. It is this mutual abrasion that produces rock flour and gives a high **polish** to many of the rock surfaces across which a glacier has ridden. But abrasion sometimes produces scratches, or **striations,** on both the bedrock floor and on the grinding tools carried by the ice (see Fig-ure 12.10). More extensive abrasion creates deep gouges, or **grooves,** in the bedrock. The striations and grooves along a bedrock surface show the direction of the glacier's movement. At Kelleys Island, in Lake Erie north of Sandusky, Ohio, the bedrock is marked by grooves 0.3 to 0.6 m deep and 0.6 to 1 m wide. In the

FIGURE 12.9

Cirques along the crest of the Wind River Mountains, Wyoming. [University of Washington, Austin Post.]

FIGURE 12.10

Striations on bedrock about 1.5 km west of Fair Haven, Vermont. The pen points about N, 10°W and marks the general direction of the younger of two sets of striations. The older set is oriented about N, 10°E. [Paul MacClintock.]

Mackenzie Valley, west of Great Bear Lake in Canada, grooves as wide as 45 m have been described, with an average depth of 15 m and lengths ranging from several hundred meters to over 1 km.

EROSIONAL EFFECTS

The erosional effects of glaciers are not limited to the fine polish and striations mentioned above, however; for glaciers also operate on a much grander scale, producing spectacularly sculptured peaks and valleys in the mountainous areas of the world.

Cirques As we have seen, a cirque is the basin from which a mountain glacier flows, the focal point for the glacier's nourishment. After a glacier has disappeared and all its ice has melted away, the cirque is revealed as a great amphitheater, or bowl, with one side partially cut away. The back wall rises a few scores of meters to over 900 m above the floor, often as an almost vertical cliff. The floor of a cirque lies below the level of the low ridge separating it from the valley of the glacier's descent. The lake that forms in the bedrock basin of the cirque floor is called a **tarn.**

A cirque begins with an irregularity in the mountainside formed either by preglacial erosion or by a process called **nivation**, a term that refers to erosion beneath and around the edges of a snowbank. Nivation works in the following way: When seasonal thaws melt some of the snow, the meltwater seeps down to the bedrock and trickles along the margin of the snowbank. Some of the water works its way into cracks in the bedrock, where it freezes again, producing pressures that loosen and pry out fragments of the rock. These fragments are moved off by solifluction, by rill wash, and perhaps by mass wasting, forming a shallow basin. As this basin gradually grows deeper, a cirque eventually develops. Continued accumulation of snow leads to the formation of firn; if the basin becomes deep enough, the firn is transformed into ice. Finally the ice begins to flow out of the cirque into the valley below, and a small glacier is born.

The actual mechanism by which a cirque is enlarged is still a matter of dispute. Some observers claim that frost action and plucking on the cirque wall within the bergschrund are sufficient to produce precipitous walls hundreds of meters in height. Others, however, point out that the bergschrund, like all glacier crevasses, remains open only in the zone of fracture, 60 m at most. Below that depth pressures cause the ice to deform plastically, closing the bergschrund.

This debate has led to the development of the so-called **meltwater hypothesis** to explain erosion along the headwalls below the base of the bergschrund. The proponents of this theory explain that meltwater periodically descends the headwalls of cirques, melts its way down behind the ice and into crevices in the rock, and there freezes at night and during cold spells. The material thus broken loose is then removed by the glacier, and cirque erosion proceeds mainly by this form of headwall recession.

Horns, arêtes, and cols A **horn** is a spire of rock formed by the headward erosion of a ring of cirques around a single high mountain. When the glaciers originating in these cirques finally disappear, they leave a steep, pyramidal mountain outlined by the headwalls of the cirques. The classic example of a horn is the famous Matterhorn of Switzerland (Figure 12.11).

An **arête** (French for "fishbone," "ridge," or "sharp edge") is formed when a number of cirques gnaw into a ridge from opposite sides. The ridge becomes knife-edged, jagged, and serrated (see Figure 12.11).

A **col** (from the Latin *collum*, "neck"), or pass, is fashioned when two cirques erode headward into a ridge from opposite sides. When their headwalls meet, they cut a sharp-edged gap in the ridge (Figure 12.11).

Glaciated valleys Instead of fashioning their own valleys, glaciers probably follow the course of preexisting valleys, modifying them in a variety of ways; usually the resulting valleys have a **broad, U-shaped cross profile,** whereas mountain valleys created exclusively by streams have **narrow, V-shaped cross profiles.** Be-

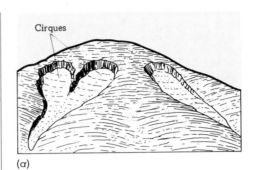

(a)

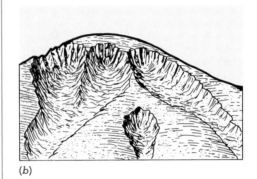

(b)

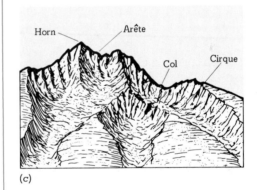

(c)

FIGURE 12.11

The progressive development of cirques, horns, arêtes, and cols. (a) Valley glaciers have produced cirques; but since erosion has been moderate, much of the original mountain surface has been unaffected by the ice. The result of more extensive glacial erosion is shown in (b). In (c) glacial erosion has affected the entire mass and has produced not only cirques but also a horn, jagged, knife-edged arêtes, and cols. [Redrawn from William Morris Davis, "The Colorado Front Range," *Ann. Assoc. Am. Geo.,* vol. 1, p. 57, 1911.]

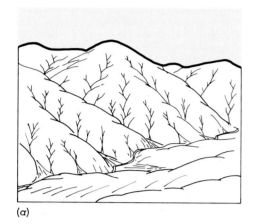

(a)

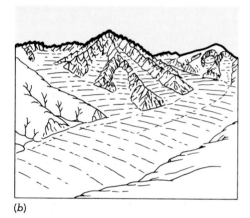

(b)

(c)

FIGURE 12.12

A mountainous area before, during, and after glaciation. [Redrawn from William Morris Davis, "The Sculpture of Mountains by Glaciers," *Scot. Geog. Mag.*, vol. 22, pp. 80, 81, 83, 1906.]

cause the tongue of an advancing glacier is relatively broad, it tends to broaden and deepen the V-shaped stream valleys, transforming them into broad, U-shaped troughs. And because the moving body of ice has difficulty manipulating the curves of a stream valley, it tends to straighten and simplify its course. In this process of straightening, the ice snubs off any spurs of land that extend into it from either side. The cliffs thus formed are shaped like large triangles, or flatirons, with their apex upward and are called **truncated spurs** (see Figures 12.12 and 12.13).

A glacier also gives a mountain valley a characteristic longitudinal profile from the cirque downward. The course of a glaciated valley is marked by a series of **rock basins**, probably formed by plucking in areas where the bedrock was shattered or closely jointed. Between the basins are relatively flat stretches of rock that was more resistant to plucking. As time passes, the rock basins may fill up with water, producing a string of lakes that are sometimes referred to as *paternoster* lakes because they resemble a string of beads.

Hanging valleys are another characteristic of mountainous areas that have undergone glaciation. The mouth of a hanging valley is left stranded high above the main valley through which a glacier has passed. As a result, streams from hanging valleys plummet into the main valley in a series of falls and plunges. Hanging valleys may be formed by processes other than glaciation, but they are almost always present in mountainous areas that formerly supported glaciers and are thus very characteristic of past valley glaciation. What has happened to leave these valleys stranded high above the main valley floor? During the time when glaciers still moved down the mountains, the greatest accumulation of ice would tend to travel along the central valley. Consequently the erosive action there would be greater than in the tributary valleys, with their relatively small glaciers, and the main valley floor would be cut correspondingly deeper. This action would be even more pronounced where the main valley was underlain by rock that was more susceptible to erosion than was the rock under the tributary valleys. Finally, some hanging valleys were probably created by the straightening and widening action of a glacier on the main valley. In any event the difference in level between the tributary valleys and the main valley does not become apparent until the glacier has melted away.

Cutting deep into the coasts of Alaska, Norway,

FIGURE 12.13

Little Cottonwood Canyon, in the Wasatch Mountains of Utah, shows the typical U-shaped profile of a glaciated valley. [William C. Bradley.]

Greenland, Labrador, and New Zealand are deep, narrow arms of the sea—**fiords** (Figure 12.14). Actually these inlets are stream valleys that were modified by glacier erosion and then partially filled by the sea. The deepest known fiord, Vanderford in Vincennes Bay, Antarctica, has a maximum depth of 2,287 m.

Some valleys have been modified by continental glaciers rather than by the valley glaciers that we have been discussing so far. The valleys occupied by the Finger Lakes of central New York State are good examples. These long, narrow lakes lie in basins that were carved out by the ice of a continental glacier. As the great sheet of ice moved down from the north, its progress seems to have been checked by the northern scarp of the Appalachian Plateau. But some of the ice moved up on the valleys that had previously drained the plateau. The energy concentrated in the valleys was so great that the ice was able to scoop out the basins that are now filled by the Finger Lakes.

Asymmetric rock knobs and hills In many places glacier erosion of bedrock produces small, rounded, asymmetric hills with gentle, striated, and polished slopes on one side and steeper slopes lacking polish and striations on the opposite side. The now-gentle slope faced the advancing glacier and was eroded by abrasion. The opposite slope has been steepened by the plucking action of the ice as it rode over the knob (Figure 12.15).

Large individual hills have the same asymmetric profiles as the smaller hills have. Here too the gentle slope faced the moving ice.

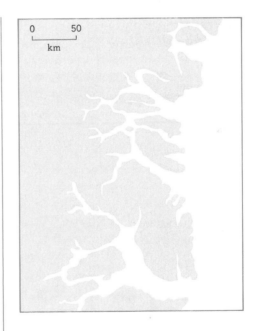

FIGURE 12.14

Glaciated valleys have been flooded by the sea to produce these fiords along the coast of Greenland. [Redrawn from Louise A. Boyd and others, "The Fiord Region of East Greenland," *Am. Geog. Soc. Spec. Publ.* 18, p. xii, 1935.]

FIGURE 12.15

The smooth, gentle slopes of these rock knobs have been produced by glacial abrasion, and the steep slopes have been produced by plucking. Ice movement was from right to left. [Geological Survey of Canada.]

FIGURE 12.16

The range in size of the particles composing till is very large. In this photograph of an exposure of till near Guilford, Connecticut, note boulders and cobbles mixed with smaller particles ranging all the way down to colloid size. Spade gives scale. [Sheldon Judson.]

FIGURE 12.17

This moraine composed of till lies along the East Rosebud River and records the last major advance of glaciers from the Beartooth Mountains, Montana. The largest boulders measure 3 m in maximum dimension. [Sheldon Judson.]

FIGURE 12.18

A sequence of diagrams to suggest the growth of a terminal moraine at the edge of a stable ice front. The progressive movement of a single particle is shown. In (a) it is moved by the ice from the bedrock floor. Forward motion of ice along a shear plane carries it ever closer to the stabilized ice margin, where it is finally deposited as a part of the moraine in (d). Diagram (e) represents the relation of the terminal moraine, ground moraine, and outwash after the final melting of the glacier.

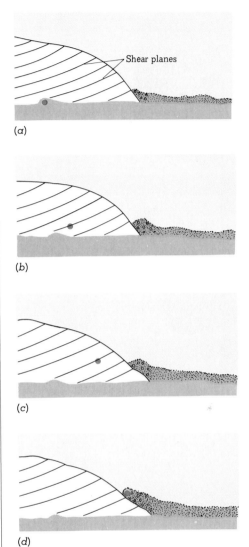

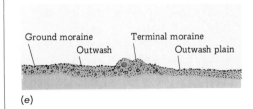

TYPES OF GLACIAL DEPOSIT

The debris carried along by a glacier is eventually deposited either because the ice that holds it melts or, less commonly, because the ice smears the debris across the land surface.

The general term **drift** is applied to all deposits that are laid down directly by glaciers or that, as a result of glacial activity, are laid down in lakes, oceans, or streams. The term dates from the days when geologists thought that the unconsolidated cover of sand and gravel blanketing much of Europe and America had been made to drift into its present position either by the sea or by icebergs. Drift can be divided into two general categories: **unstratified** and **stratified**.

Deposits of unstratified drift   Unstratified drift laid down directly by glacier ice is called **till**. It is composed of rock fragments of all sizes mixed together in random fashion, ranging from boulders weighing several tonnes to tiny clay and colloid particles (see Figures 12.16 and 12.17). Many of the large pieces are striated, polished, and faceted as a result of the wear they underwent while transported by the glaciers (see Figures 12.18 and 12.19). Some of the material picked up along the way was smeared across the landscape during the glacier's progress, but most of it was dumped when the rate of wastage began to exceed the rate of nourishment and the glacier gradually melted away.

The type of till varies from one glacier to another. Some tills, for instance, are known as **clay tills** because clay-sized particles predominate, with only a scattering of larger units. Many of the most recent tills in northeastern and eastern Wisconsin are of this type, but in many parts of New England the tills are composed for the most part of large rock fragments and boulders. Deposits of this sort are known as **boulder**, or **stony, tills**.

FIGURE 12.19

Stones from till are marked by facets and striations. The specimens are from northern Illinois. The smallest is about 5 cm across. [Willard Starks.]

Some till deposits seem to have been worked over by meltwater. The materials have begun to be sorted out according to size, and some of the finer particles may even have been washed away. This is the sort of winnowing action we should expect to find near the nose of a melting glacier, where floods of meltwater wash down through the deposits.

Till is deposited by glaciers in a great variety of topographic forms, including moraines, drumlins, erratics, and boulder trains.

Moraines **Moraine** is a general term used to describe many of the landforms that are composed largely of till.

A **terminal moraine,** or **end moraine,** is a ridge of till that marks the utmost limit of a glacier's advance. These ridges vary in size from ramparts scores of meters high to very low, interrupted walls of debris. A terminal moraine forms when a glacier reaches the critical point of equilibrium—the point at which it wastes away at exactly the same rate as it is nourished. Although the front of the glacier is now stable, ice continues to push forward; and so it will deliver a continuous supply of rock debris. As the ice melts in the zone of wastage, the debris is dumped, and the terminal moraine grows. At the same time water from the melting ice pours down over the till and sweeps part of it out in a broad flat fan that butts against the forward edge of the moraine like a giant ramp (Figure 12.18).

The terminal moraine of a mountain glacier is crescent-shaped, with the convex side extending down-valley. The terminal moraine of a continental ice sheet is a broad loop or series of loops traceable for many miles across the countryside.

Behind the terminal moraine and at varying distances from it a series of smaller ridges known as **recessional moraines** may build up. These ridges mark the position where the glacier front was stabilized temporarily during the retreat of the glacier.

Not all the rock debris carried by a glacier finds its way to the terminal and recessional moraines, however. A great deal of till is laid down as the main body of the glacier melts to form gently rolling plains across the valley floor. Till in this form, called a **ground moraine,** may be a thin veneer lying on the bedrock, or it may form a deposit scores of meters thick, partially or completely clogging preglacial valleys.

Finally, valley glaciers produce two special types of moraine. While a valley glacier is still active, large amounts of rubble keep tumbling down from the valley walls, collecting along the side of the glacier. When the ice melts, this debris is stranded as a ridge along each side of the valley, forming a **lateral moraine.** At its down-valley end, the lateral moraine may grade into a terminal moraine. The other special type of deposit produced by valley glaciers is a **medial moraine,** created when two valley glaciers join to form a single ice stream; material formerly carried along on the edges of the separate glaciers is combined in a single moraine near the center of the enlarged glacier. A streak of this kind builds up whenever a tributary glacier joins a larger glacier in the main valley (Figure 12.5). Although medial moraines are very characteristic of living glaciers, they are seldom preserved as topographic features after the disappearance of the ice.

Drumlins **Drumlins** are smooth, elongated hills composed largely of till. The ideal drumlin shape has an asymmetric profile, with a blunt nose pointing in the direction from which the vanished glacier advanced and with a gentler, longer slope pointing in the opposite direction. Drumlins range from about 8 to 60 m in

FIGURE 12.20

Map of part of the drumlin-field area south and
east of Charlevoix, Michigan. Ice moved toward
the south-southeast. [Redrawn from Frank Leverett
and F. B. Taylor, "The Pleistocene of Indiana and
Michigan and the History of the Great Lakes,"
U.S. Geol. Surv. Monog. 53, p. 311, 1915.]

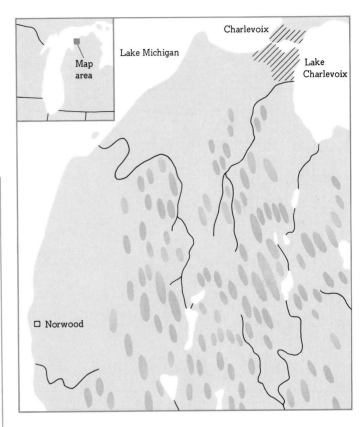

height, the average being about 30 m. Most drumlins
are between 0.5 and 1 km in length and are usually
several times longer than they are wide.

In most areas drumlins occur in clusters, or **drumlin
fields.** In the United States these are most spectacularly
developed in New England, particularly around Boston, in eastern Wisconsin, in west-central New York
State, particularly around Syracuse, in Michigan (Figure 12.20), and in certain sections of Minnesota. In
Canada extensive drumlin fields are located in western
Nova Scotia and in northern Manitoba and Saskatchewan; Figure 12.21 shows a drumlin field in British Columbia.

Just how drumlins were formed is still not clear.
Because their shape is a nearly perfect example of
streamlining, it seems probable that they were formed
deep within active glaciers in the zone of plastic flow.

Erratics and boulder trains   A stone or a boulder
that has been carried from its place of origin by a
glacier and left stranded on bedrock of different composition is called an **erratic.** The term is used whether
the stone is embedded in a till deposit or rests directly
on the bedrock. Some erratics weigh several tonnes,
and a few are even larger. Near Conway, New Hampshire, there is a granite erratic 27 m in maximum dimension, weighing close to 9,000 t. Although most erratics have traveled only a limited distance, many have
been carried along by the glacier for hundreds of kilometers. Chunks of native copper torn from the Upper
Peninsula of Michigan, for example, have been trans-

FIGURE 12.21

This aerial photograph shows the drumloidal patterns of hills and intervening grooves formed parallel to the flow of glacier ice on the Nechako Plateau, British Columbia. Weedon Lake is 9 km long.
View is northeast in the direction of ice movement.
[U.S. Army Air Corps.]

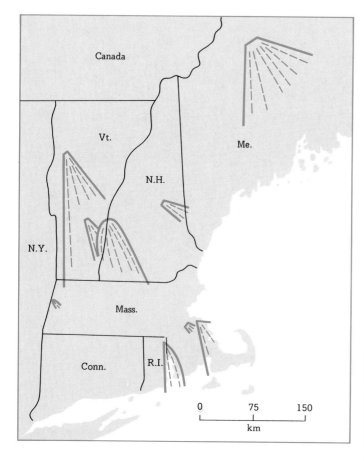

FIGURE 12.22

The boulder trains plotted on this map indicate the general direction of ice movement across New England. The apex of the fan indicates the area from which the boulders were derived; the fan itself covers the area across which they were deposited. [Redrawn from J. W. Goldthwait in R. F. Flint, "Glacial Map of North America," *Geol. Soc. Am. Spec. Paper* 60, 1945.]

ported as far as southeastern Iowa, a distance of nearly 800 km, and to southern Illinois, over 960 km.

Boulder trains consist of a series of erratics that have come from the same source, usually with some characteristic that makes it easy to recognize their common origin. The trains appear either as a line of erratics stretching down-valley from their source or in a fan-shaped pattern with the apex near the place of origin. By mapping boulder trains that have been left behind by continental ice sheets, we can obtain an excellent indication of the direction of the ice flow (see Figure 12.22).

Deposits of stratified drift Stratified drift is ice-transported sediment that has been washed and sorted by glacial meltwaters according to particle size. Because water is a much more selective sorting agent than is ice, deposits of stratified drift are laid down in recognizable layers, unlike the random arrangements of particles typical of till. Stratified drift occurs in outwash and kettle plains, eskers, kames, and varves—all discussed below.

Outwash sand and gravel The sand and gravel that are carried outward by meltwater from the front of a glacier are referred to as **outwash** (see Figure 12.23). As a glacier melts, streams of water heavily loaded with reworked till or with material *washed* directly from the ice weave a complex, braided pattern of channels across the land in front of the glacier. These streams, choked with clay, silt, sand, and gravel, rapidly lose their velocity and dump their load of debris as they flow away from the ice sheet. In time a vast apron of bedded sand and gravel is built up, which may extend for kilometers beyond the ice front. If the zone of wastage happens to be located in a valley, the outwash deposits are confined to the lower valley and compose a **valley train**. But along the front of a continental ice sheet the outwash deposits stretch out for kilometers, forming what is called an **outwash plain** (see Figure 12.24).

Kettles Sometimes a block of stagnant ice becomes isolated from the receding glacier during wastage and is partially or completely buried in till or outwash before it finally melts. When it disappears, it leaves a **kettle,** a pit, or depression, in the drift (see Figure 12.25). These depressions range from a few meters to several kilometers in diameter and from a few meters to over 30 m in depth. Many outwash plains are pockmarked with kettles and are referred to as **pitted outwash plains.** As time passes, water sometimes fills the kettles to form lakes or swamps, features found through much of Canada and the northern United States.

FIGURE 12.23 (LEFT)

Outwash forms at the snout of Brady Glacier, Fairweather Range, Alaska.
[University of Washington, Austin Post.]

FIGURE 12.24 (RIGHT)

These sand and gravel deposits represent the outwash from a now-vanished continental glacier. North of Otis Lake, Wisconsin. [U.S. Geological Survey, W. C. Alden.]

Eskers and crevasse fillings Winding, steep-walled ridges of stratified gravel and sand, sometimes branching and often discontinuous, are called **eskers** (see Figure 12.26). They usually vary in height from about 3 to 15 m although a few are over 30 m high. Eskers range from a fraction of a kilometer to over 160 km in length, but they are only a few meters wide. Most investigators believe that eskers were formed by the deposits of streams running through tunnels beneath stagnant ice. Then, when the body of the glacier finally disappeared, the old stream deposits were left standing as a ridge.

Crevasse fillings are similar to eskers in height, width, and cross profile, but unlike the sinuous and branching pattern of eskers, they run in straight ridges. As their name suggests, they were probably formed by the filling of a crevasse in stagnant ice.

Kames and kame terraces In many areas stratified drift has built up low, relatively steep-sided hills called **kames**, either as isolated mounds or in clusters. Unlike drumlins, kames are of random shape and the deposits that compose them are stratified. They were formed by

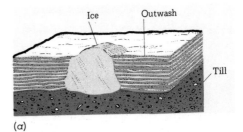

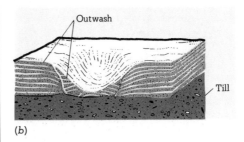

FIGURE 12.25

Kettle formation: (*a*) Stagnant ice is almost buried by outwash. (*b*) Eventual melting forms a depression.

FIGURE 12.26

An esker exposed by the retreat of the Malaspina glacier, Alaska. Note depressions in the areas of rough topography. They have formed by melting of ice blocks. [University of Washington, Austin Post.]

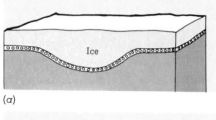

(a)

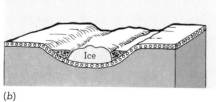

(b)

FIGURE 12.27

This isolated kame in eastern Wisconsin was formed by the partial filling of an opening in stagnant glacier ice. The melting of the ice has left this steep-sided hill of stratified material. [Raymond C. Murray.]

(c)

FIGURE 12.28

The sequence in the development of a kame terrace. (a) Ice wasting from an irregular topography lingers longest in the valleys. (b) While the ice still partially fills one of these valleys, outwash may be deposited between it and the valley walls. (c) The final disappearance of the ice leaves the outwash in the form of terraces along the sides of the valley.

the material that collected in openings in stagnant ice. In this sense they are similar to crevasse fillings but without the linear pattern (see Figure 12.27).

A **kame terrace** is a deposit of stratified sand and gravel that has been laid down between a wasting glacier and an adjacent valley wall. When the glacier disappears, the deposit stands as a terrace along the side of the valley (see Figure 12.28).

Varves A **varve** is a pair of thin sedimentary beds, one coarse and one fine (see Figure 18.5). This couplet of beds is usually interpreted as representing the deposits of a single year and is thought to form in the following way: During the period of summer thaw, waters from a melting glacier carry large amounts of clay, fine sand, and silt out into lakes along the ice margin. The coarser particles sink fairly rapidly and blanket the lake floor with a thin layer of silt and silty sand. But as long as the lake is unfrozen, the wind creates currents strong enough to keep the finer clay particles in suspension. When the lake freezes over in the winter, these wind-generated currents cease, and the fine particles sink through the quiet water to the bottom, covering the somewhat coarser summer layer. A varve is usually a few millimeters to 1 cm thick although thicknesses of 5 to 8 cm are not uncommon. In rare instances thicknesses of 30 cm or more are known.

COMPARISON OF VALLEY- AND CONTINENTAL-GLACIATION FEATURES

Some of the glacial features that we have been discussing are more common in areas that have undergone valley glaciation; others usually occur only in regions that have been overridden by ice sheets; many other features, however, are found in both types of area. Table 12.1 lists and compares the features that are characteristic of the two types.

INDIRECT EFFECTS OF GLACIATION

The glaciers that diverted rivers, carved mountains, and covered half a continent with debris also gave rise

TABLE 12.1

Valley- and continental-glaciation features compared

Features	Valley glacier	Continental ice sheet
Striations, polish, etc.	Common	Common
Cirques	Common	Absent
Horns, arêtes, cols	Common	Absent
U-shaped valley, truncated spurs, hanging valleys	Common	Rare
Fiords	Common	Absent
Till	Common	Common
Terminal moraine	Common	Common
Recessional moraine	Common	Common
Ground moraine	Common	Common
Lateral moraine	Common	Absent
Medial moraine	Common, easily destroyed	Absent
Drumlins	Rare or absent	Locally common
Erratics	Common	Common
Stratified drift	Common	Common
Kettles	Common	Common
Eskers, crevasse fillings	Rare	Common
Kames	Common	Common
Kame terraces	Common	Present in hilly country

to a variety of indirect effects that were felt far beyond the glaciers' immediate margins. Sea level fell with the accumulation of glacier ice and rose as glaciers melted. Desert and near-desert areas were better watered during maximum glaciation. In now-temperate latitudes beyond the ice margin climate also changed, wind laid down vast blankets of dust (called *loess*), and slope processes were speeded by solifluction. We defer the discussion of these subjects to Chapter 13.

12.3 DEVELOPMENT OF THE GLACIAL THEORY

Geologists have made extensive studies of the behavior of modern glaciers and have carefully interpreted the traces left by glaciers that disappeared thousands of years ago. On the basis of their studies they have developed the **glacial theory**: In the past great ice sheets covered large sections of the earth where no ice now exists, and many existing glaciers once extended far beyond their present limits.

THE BEGINNINGS

The glacial theory took many years to evolve, years of trying to explain the occurrence of erratics and the vast expanses of drift strewn across northern Europe, the British Isles, Switzerland, and adjoining areas. The exact time when inquisitive minds first began to seek an explanation of these deposits is shrouded in the past, but by the beginning of the eighteenth century explanations of what we now know to be glacial deposits and features were finding their way into print. According to the most popular early hypothesis, a great inundation had swept these deposits across the face of the land with cataclysmic suddenness or else had drifted them in by means of floating icebergs. Then, when the flood receded, the material was left where it is now.

By the turn of the nineteenth century a new theory was in the air—the theory of ice transport. We do not know who first stated the idea or when it was first proposed, but it seems quite clear that it was not hailed immediately as a great truth. As the years passed, however, more and more observers became intrigued with the idea. The greatest impetus came from Switzerland, where the activity of living glaciers could be studied on every hand.

In 1821, J. Venetz, a Swiss engineer delivering a paper before the Helvetic Society, presented the argument that Swiss glaciers had once expanded on a great scale. It has since been established that from about 1600 to the middle of the eighteenth century there was actually a time of moderate but persistent glacier expansion in many localities. Abundant evidence in the Alps, Scandinavia, and Iceland indicates that the climate was milder during the Middle Ages than it is at present, that communities existed and farming was carried on in places later invaded by advancing glaciers or devastated by glacier-fed streams. We

know, for example, that a silver mine in the valley of Chamonix was being worked during the Middle Ages and that it was subsequently buried by an advancing glacier, where it lies to this day. And the village of St. Jean de Perthuis has been buried under the Brenva glacier since about 1600.

Although Venetz's idea did not take hold immediately, by 1834 Jean de Charpentier was arguing in its support before the same Helvetic Society. Yet the theory continued to have more opponents than defenders. It was one of the skeptics, Jean Louis Rodolphe Agassiz (1807–1873), who did more than anyone else to develop the glacial theory and bring about its acceptance.

AGASSIZ

As a young zoologist, Louis Agassiz had listened to Charpentier's explanation; afterward, he undertook to demonstrate to his friend and colleague the error of his ways. During the summer of 1836 the two men made a trip together into the upper Rhône valley to the Getrotz glacier. Before the summer was over, it was Agassiz who was convinced of his error. In 1837 he spoke before the Helvetic Society championing the glacial theory and suggesting that during a "great Ice Age" not only the Alps but much of northern Europe and the British Isles were overrun by a sea of ice.

Agassiz's statement of the glacial theory was not accepted immediately, but in 1840 he visited England and won the support of leading British geologists. In 1846 he arrived in the United States, where in the following year he became professor of zoology at Harvard College and later founded the Museum of Comparative Zoology. In this country he convinced geologists of the validity of the glacial theory; by the third quarter of the nineteenth century it was firmly entrenched. The last opposition died with the turn of the twentieth century.

PROOF OF THE GLACIAL THEORY

What proof is there that the glacial theory is valid? The most important evidence is that certain features produced by glacier ice are produced by no other known process. Thus Agassiz and his colleagues found isolated stones and boulders quite alien to their present surroundings. They noticed too that boulders were actually being transported from their original location by modern ice. Some of the boulders they observed were so large that rivers could not possibly have moved them, and others were perched on high places that a river could have reached only by flowing uphill. They also noticed that, when modern ice melted, it revealed a polished and striated pavement unlike the surface fashioned by any other known process. To explain the occurrence of these features in areas where no modern glaciers exist, they postulated that the ice once extended far beyond its present limits.

Notice that the development of this theory sprang from a concept that we mentioned earlier: "The present is the key to the past." The proof of glaciation lies not in the authority of the textbook nor in the lecture. It lies in observing modern glacial activity directly and in comparing the results of this activity with features and deposits found beyond the present extent of the ice.

THEORY OF MULTIPLE GLACIATION

Even before universal acceptance of the glacial theory, which spoke of a single great Ice Age, some investigators were coming to the conclusion that the ice had advanced and retreated not just once but **several times in the recent geological past.** By the early twentieth century a broad fourfold division of the Ice Age, or Pleistocene, had been demonstrated in this country and in Europe. According to this theory, each major glacial advance was followed by a retreat and a return to climates that were sometimes even warmer than that of the present. Early in this century four stages of ice advance had been discovered in the United States and called **Nebraskan, Kansan, Illinoian,** and **Wisconsin** for the midwestern states where deposits of that particular

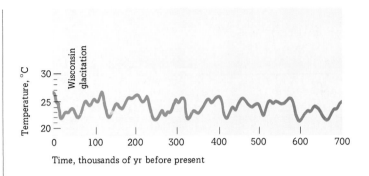

FIGURE 12.29

Analysis of deep-sea sediments shows that the average surface temperature of the ocean fluctuated many times from warm to cold in the recent geologic past. Here is the record for the last 700,000 years; the most recent cooling is labeled to represent the last major glaciation in North America. [Modified from Cesare Emeliani and Nicholas J. Shackleton, "The Brunhes Epoch: Isotopic Paleotemperatures and Geochronology," *Science,* vol. 183, p. 513, 1974.]

period were first studied or where they are well exposed. Evidence from Alaska, the Sierra Nevada, and western Europe indicates that one or two major advances preceded the Nebraskan. In central Europe successive sheets of loess suggest at least 17 periods of Pleistocene glaciation before the last major advance. Finally, analysis of sediment cores taken from the deep oceans reveals repeated cooling and warming of the oceans during the last 2 million years (Figure 12.29). About 20 of these cycles have been identified, and we believe that they represent the climatic changes accompanying repeated glaciation and deglaciation on land.

12.4 EXTENT OF PLEISTOCENE GLACIATION

During the maximum advance of the glaciers of the Wisconsin age (the last of the four great ice advances during the Pleistocene), 39 million km² of the earth's surface—about 27 percent of the present land areas—were probably buried by ice. Approximately 15 million km² of North America were covered. Greenland was also under a great mass of ice, as it is now. In Europe an ice sheet spread southward from Scandinavia across the Baltic Sea and into Germany and Poland, and the Alps and the British Isles supported their own ice caps. Eastward in Asia the northern plains of Russia were covered by glaciers, as were large sections of Siberia and of the Kamchatka Peninsula and the high plateaus of Central Asia (see Figure 12.30).

In eastern North America ice moved southward out of Canada to New Jersey, and in the Midwest it

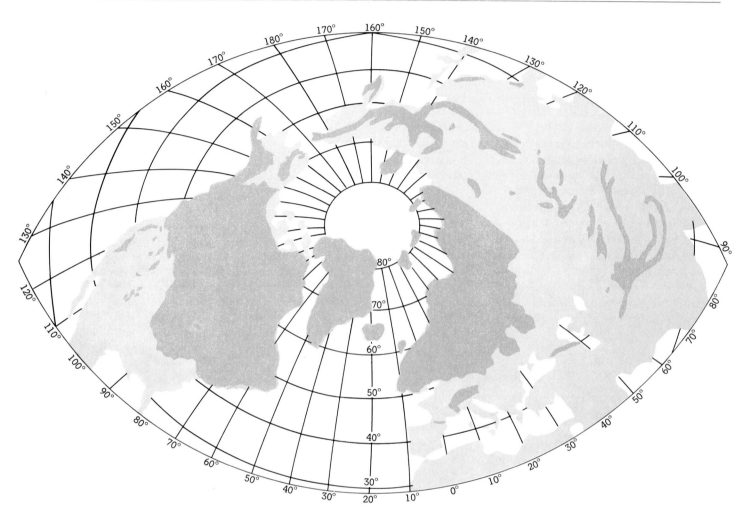

FIGURE 12.30

Extent of Pleistocene glaciation (darker areas) in the northern hemisphere. [Redrawn from Ernst Antevs, "Maps of the Pleistocene Glaciations," *Geol. Soc. Am. Bull.* 40, p. 636, 1929.]

reached as far south as Saint Louis. The western mountains were heavily glaciated by small ice caps and valley glaciers. The southernmost glaciation in the United States was in the Sierra Blanca of south-central New Mexico. The maximum extent of the Wisconsin glaciation in the United States and the maximum limit of glaciation during the Pleistocene are shown in Figure 12.31.

12.5 PRE-PLEISTOCENE GLACIATIONS

So far we have discussed only the glaciers that exist today and those which moved within the last few million years, the Pleistocene. Geologists have found evidence that glaciers appeared and disappeared in other periods as well. The record is fragmentary, as we should expect; for time tends to conceal, jumble, and

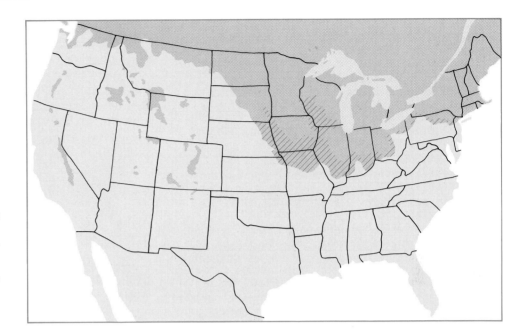

FIGURE 12.31

Extent of Pleistocene glaciation in the United States. Darker zones indicate area covered at various times during the Wisconsin glaciation. Diagonal lines represent area glaciated during pre-Wisconsin stages but not covered by the later Wisconsin ice. Tint area unglaciated. [Generalized from R. F. Flint, "Glacial Map of North America," *Geol. Soc. Am. Spec. Paper 60,* 1945.]

destroy the various manifested effects of glaciation.

Recent evidence indicates that glaciation and deglaciation, so characteristic of the Pleistocene, reaches back into the late Miocene in Alaska. Glacial deposits in Antarctica have been dated by the potassium-argon method as 10 million years in age. Farther back in the geologic time scale it seems certain that there were other extensive glaciations. Toward the end of the Paleozoic Era (230 to 290 million years ago) glaciation was widespread in what is now South Africa, South America, India, Australia, and Antarctica. There is evidence of glaciation during the Silurian and Devonian in South America, and earlier, in the Ordovician, ice spread across what is now the Sahara Desert. In the late Precambrian, a little over 600 million years ago,

glaciation affected various landmasses now in the northern hemisphere. Still earlier, perhaps 850 million years ago, large parts of the ancient continent of Gondwana (see Chapter 8) were ice covered. Some Russian geologists believe that sections of northwestern Siberia were glaciated some 1.2 billion years ago. Extensive glaciation, dating from 2.2 billion years ago, is recorded in what is now south-central Ontario, Canada. At about the same time glaciation occurred in South Africa.

Much of this glaciation took place in localities where no ice can exist today. The geologic evidence tells us that this is because the landmasses have since drifted from latitudes hospitable to glaciation to new locations (see Chapter 8).

12.6 CAUSES OF GLACIATION

As Agassiz did over 100 years ago, we can travel about the world today and observe modern glaciers at work, and we can reason convincingly that glaciers were more extensive in the past than they are at present. We can even make out a good case for the belief that glacier ice advanced and receded many times in the immediate and more remote geological past. But can we explain why glaciation takes place? The answer to

that question is a resoundingly equivocal "maybe."

The geologic record has contributed some basic data that any theory of the causes of glaciation must take into account. Among these are the following:

1 *Glaciation is related to the elevation and arrangement of continents.* During much of geologic time the continents have been lower than they are today, and shallow seas have flooded across their margins. Such conditions were unfavor-

able for widespread glaciation. But for several million years the continents have been increasing in elevation until now, in the Pleistocene, they stand on the average an estimated 450 m higher than they did in the mid-Cenozoic. We have already found that the last great Ice Age came with the Pleistocene. We believe too that the other great glaciations coincided with high continents. In addition to having adequate elevations, to allow glaciation to begin continents must be located in high latitudes, either over the pole of rotation or clustered about it.

2 *Glaciation is not due to a slow, long-term cooling off of the earth since its creation.* We have already found that extensive glaciation occurred several times during the geological past. But these glacial periods are unusual; for during most of geologic history the climate has been nonglacial.

3 *There has been a cooling of the earth's climate beginning during the Tertiary and climaxing with the glacial fluctuations of the Pleistocene.* Although the earth has been generally warm during most of its history, evidence now shows that its mean temperature had dropped an estimated 8 to 10°C from the Eocene to the end of the Pliocene and that the glacial and interglacial epochs of the Pleistocene are short-term fluctuations at the end of the long-term cooling (see Figure 12.32).

4 *The advance and retreat of glaciers have probably been broadly simultaneous throughout the world.* For instance, dating by means of radioactive carbon has demonstrated that geologically recent fluctuations of the continental glaciers in North America occurred at approximately the same time as similar fluctuations in Europe. Furthermore, observations indicate that the general retreat of mountain glaciers now recorded in North America and Europe is duplicated in South America.

One way to approach the problem is to consider

that we are dealing with two great classes of causes: One is long-term and has caused the general decline of temperatures since the Eocene; the other is shorter range and accounts for the comings and goings of the ice sheets during the Pleistocene. One operates on a scale of approximately 50 million years, and the other on the scale of perhaps 100,000 years. In considering these two different types of cause, we can envisage the long-term cause as cooling the climate to the point where one or more other causes of shorter amplitude can trigger glaciation and deglaciation.

LONG-TERM CHANGES IN CLIMATE

Geologic evidence shows that the earth's "normal" climate has been more equable than that of the present. Through most of the earth's history climate has been warmer than the present and landmasses presently in the poleward zones enjoyed average annual temperatures about 10°C higher than today. At some time during the early Tertiary, probably in the Eocene, average temperatures began to drop.

Causes of long-term climatic fluctuation What could have caused the decline of temperatures during the last 50 or 60 million years? We can make a number of suggestions.

Increasing continentality Today the average height of continents is computed to be about 875 m, as we have stated, about 450 m higher than it was in the

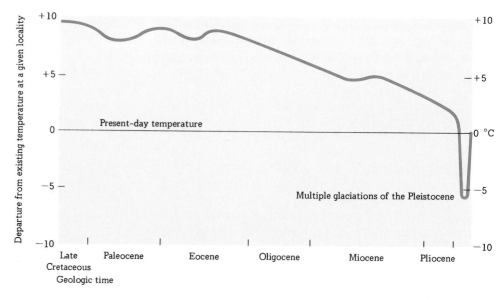

FIGURE 12.32

Temperatures trended downward from the Late Cretaceous to the Pliocene. This general decline was followed by the several glacial and interglacial episodes, the details of which cannot be shown at the scale of this diagram. The values in degrees Celsius represent approximate departures from today's temperatures of localities in the middle latitudes. [Data in part from Erling Dorf.]

beginning of the Eocene. Rising landmasses would result in a general drop in temperature. It is estimated that about one-third of the 10°C decline in average temperatures might be accounted for by the increase in the height of continents since the Eocene.

The rise of continents not only would lower global temperatures but also could interfere with the patterns of worldwide atmospheric circulation. It is conceivable that the transfer of heat from equatorial latitudes toward the poles would be hindered and that the poleward latitudes would become colder. A corollary to this would be an interruption and change in the oceanic circulation and a lessened efficiency in the transfer of heat poleward.

Continental drift It has been suggested by some geologists that continental drift has positioned the landmasses in such a manner as to favor glaciation (see Chapter 8). Thus Antarctica centers on the southern pole and is in an ideal position for the growth of ice sheets. In the northern hemisphere Europe, Asia, and North America are arranged around the Arctic Ocean, which has restricted communication with the Pacific and the Atlantic. It is argued by some that this arrangement is conducive to the glaciation of the continents. We know, however, that the present pole-continent arrangements have not changed appreciably since the early Tertiary. Therefore, if continental drift is a factor in glaciation, its effect seems to lie in bringing the continents into their present relationship to the poles and to each other, thus facilitating the slow decline in temperature that eventually ended in the glaciations of the Ice Age.

Other causes Some have suggested that an increase in the number of particles—cosmic dust—in space has decreased the amount of energy received from the sun by the earth. The suggestion is based on the fact that we know that there are great amounts of interstellar material forming vast clouds—dust clouds—in space. It is argued that perhaps the passage of the planetary system through such a cloud has caused glaciation. It is an argument that lacks proof or any forseeable way to test it.

Instead of dust in space it has been suggested that dust from volcanic eruptions has reduced the amount of solar energy that could penetrate the atmosphere and thus cause climatic cooling. The evidence indicates that there has been no appreciable variation in volcanism through a large part of geologic time for the earth as a whole. We cannot, with any confidence, assign climatic changes of large scale to volcanism.

There is, however, the possibility that the amount of heat received by the earth has varied appreciably through time. At present the amount of thermal energy received on an imaginary plane the diameter of the earth and normal to the sun's rays at the outer edge of the atmosphere is about 2 cal/cm²-min (see Section 4.1). Measurements carried on since 1918 suggest that this figure has decreased approximately 3 percent in 50 years. Assuming that the solar radiation changes, some argue that a long-range decrease might account for a cooling of the earth's climate. All we can say is that there appears to have been a measurable change in the solar radiation during historic time. Larger changes *may* have happened in the past and thus affected climate.

CLIMATIC CHANGES OF SHORTER DURATION

We have seen previously that the Ice Age was marked by the coming and going of glaciers that covered up to one-third of the earth's surface. The duration of these glacial and interglacial intervals has been much shorter than the long-term cooling of climate that we have discussed. The ratio of a glacial-interglacial cycle to the longer period of cooling that preceded the fluctuations of the Ice Age is roughly 1/500.

We are now confident that the climate had long been cooling before the first glaciation of the continents of the northern hemisphere took place. It is quite possible, however, that Antarctica supported an ice cap before this, even as it does today in a period that we can call interglacial. Whatever the cause or causes of the several glaciations of the Pleistocene, it seems reasonable that they were superimposed on the long-term cooling, which lasted some 50 million years.

Causes of short-term climatic fluctuation An impressive array of suggestions to explain the successive glaciations of the Pleistocene has emerged. These include variation in the radiant energy produced by the sun, fluctuations in the amount of volcanic ash thrown into the atmosphere, variation in the amount of cosmic dust intervening between the sun and the earth, surging of the Antarctic ice sheet, fluctuation of the carbon dioxide content of the atmosphere, variation in the patterns of oceanic circulation, and variation of the earth's position in relation to the sun, that is, variation in the geometry of the earth's orbit around the sun. Each of these theories has some support in fact, but only the last one can be extrapolated backward in time to predict the distribution of heat received at the earth's surface and thus to predict the frequency of Pleistocene glaciations. Furthermore, we can now begin to test these predicted frequencies from geologic evidence.

The changing geometry of the earth's orbit produces a variation in the geographical and seasonal distribution of solar energy on the earth. We refer to the hypothesis using this changing geometry as the **orbital hypothesis,** sometimes called the *Milankovitch hypothesis* after the Yugoslavian astronomer instrumental in its development. It holds that variations of the earth's position in relation to the sun produced climatic changes that brought on the great Ice Age. Three factors enter into the changing position of the earth in relation to the sun, each of which can be measured and its rate of change determined. One is the eccentricity of the earth's orbit around the sun. The earth is somewhat closer to the sun at some epochs than others. This motion has a period of about 93,000 years. A second factor is the obliquity of the ecliptic, which produces seasons and is determined by the angle that the earth's axis makes with the plane in which the earth circles the sun. This angle, or obliquity, changes through about 3° every 41,000 years. Finally, there is the precession of the equinoxes, which merely means that the axis of the earth wobbles because of the gravitational effect of the sun, the moon, and the planets. Like a giant top, the earth completes one wobble every 21,000 years. All these factors affect the relation of the earth to the sun—hence the amount of heat received at different places on the earth at any one time. The variation can be calculated backward in time to any desired date for any latitude. These calculations have been made, and a curve showing maxima and minima of heat received at the earth during the Pleistocene is the result. The periods of low heat receipt are said to coincide with, and be the cause of, the various ice advances.

The orbital hypothesis has been around in ever-increasing sophistication for over 100 years. In the last 25 years more and more bits of geologic history have accumulated and now suggest a tantalizing fit between the astronomical prediction and the actual earth record. Many workers have been involved, but here we cite only three: James D. Hays (Columbia University), John Imbrie (Brown University), and N. J. Shackleton (Cambridge University) have recently marshalled the climatic evidence found in deep-sea cores extending back for nearly 0.5 million years, and from these they have constructed a curve not unlike that of Figure 12.29. The correlation of this curve with the climatic fluctuations predicted from the orbital motions of the earth is striking. And their conclusion that "changes in the earth's orbital geometry are the fundamental cause of" Pleistocene ice ages is extremely persuasive. (See the Supplementary Readings at the end of this chapter.)

SUMMARY

We have touched on only a few of the theories advanced to explain glaciation, and there are many more that we have not even mentioned. To date we do not know which, if any, of these hold the right answer.

It seems reasonable, however, to expect that the cause of Pleistocene glaciation is multiple. Clearly the world climate has cooled slowly during the 50 million years or more from the early Tertiary to the Pleistocene. Whatever the cause or causes of this cooling, we can theorize that the climate reached the point at which some other cause or causes took over and produced the glaciations and deglaciations of the Pleistocene, climatic fluctuations of a much shorter wavelength.

12.7 IMPLICATIONS FOR THE FUTURE

If geology teaches us nothing else, it demonstrates that our globe is in constant change—that the face of the earth is mobile. Mountains rise only to be laid low by erosion; seas lap over the continents; entire regions progress through various stages; and glaciers come and go. We still live in the Pleistocene Ice Age, a pinpoint in time that has been preceded, and will be followed, by extensive climatic changes.

Since the height of the last glacial invasion of the northern United States some 10,000 years ago, the climate has sometimes been warmer, sometimes colder, than that of the present. Thus at one point there was a worldwide rise in temperature that produced a climate warmer than that we are used to. This interval, known as the **altithermal phase** (that is, a period of great heat), reached its height about 6,000 years ago. Plants that are now found only in more southerly latitudes began to grow in northern areas. Many small glaciers in our western mountains disappeared completely, the glaciers in the Alps retreated, and the Greenland ice cap shrank (see Figure 12.33).

The altithermal phase gave way to a period in which the climate grew cooler again. Vegetation zones were pushed southward, and extensive areas in the higher latitudes and altitudes were stripped of their forests. Today we are apparently beginning to emerge from the latter phase, which some have called the "little Ice Age," or "neoglaciation."

What will happen in the future? We can be sure that the climate will grow either hotter or colder and that glaciers will either recede or advance in the ages ahead, just as they have in the past. But we cannot tell with certainty just what the changes will be and when they will occur. If one believes the orbital-hypothesis projections, then the long-term outlook over the next several thousand years is toward widespread glaciation.

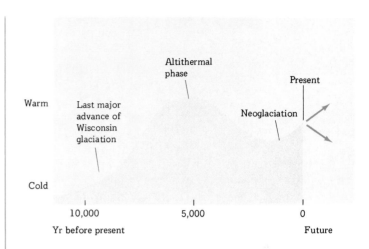

FIGURE 12.33

The climate during the last few thousand years has not been constant. Rather, it has been marked by the fluctuations generalized here. We can assume that the climate of the future will vary from that of the present.

OUTLINE

Glacier formation is a low-temperature, metamorphic process that converts snow to firn to ice.

Classification of glaciers includes **valley glaciers, piedmont glaciers,** and **ice sheets.**

Nourishment of glaciers occurs in the **zone of accumulation.** The **zone of wastage** lies below the snow line, where melting and evaporation exceed snowfall.

Glacier movement is usually a few centimeters per day, but some glaciers move rapidly in **surges.** Below a 60-m **brittle zone** glacier ice is in a zone of flow. The glacier moves by slipping along its base, by recrystallization of individual crystals, and by shearing.

Temperatures of glaciers vary. A warm glacier reaches the pressure melting point throughout the summer. In a cold glacier no melting occurs even in the summer.

Results of glaciation are both erosional and depositional.

Erosion of the ground surface takes place by **plucking** or by **abrasion.** Features formed include **cirques, horns, arêtes, cols, U-shaped valleys, hanging valleys, fiords, striations, grooves,** and **polish.**

Glacial deposits include unstratified material called **till,** which is found in **moraines** and **drumlins. Boulder trains** are special depositional features.

Stratified deposits include **outwash** of sand and gravel, which is usually found in **outwash plains, eskers, crevasse fillings, kames,** and **kame terraces. Varves** usually consist of clay and silt and form in glacial lakes.

The glacial theory was born in Switzerland and was first clearly stated by Louis Agassiz. **Proof of the glacial theory** rests on the principle of uniformitarianism.

Multiple glaciations mark the Pleistocene and four glaciations—**Nebraskan, Kansan, Illinoian,** and **Wisconsin**—are recognized in the central United States.

Pre-Pleistocene glaciations are known to have occurred several times in the Paleozoic and the Precambrian.

Causes of glaciation are not yet known, but for the Pleistocene they seem to involve a long, slow cooling of climate that began in the early Tertiary and a later set of short-term fluctuations that produced the glacial and interglacial epochs.

Implications for the future are several, and the forecast is that climatic change is certain.

SUPPLEMENTARY READINGS

Agassiz, Louis   *Studies on Glaciers*, Neuchâtel, 1840; trans. and ed. A. V. Carozzi, Harper & Row, Publishers, New York, 1967. *No better exposition of the proof of the former extent of glaciers can be found.*

Embleton, Clifford, and Cuchlaine A. M. King   *Glacial Geomorphology*, 2nd ed., John Wiley & Sons, Inc., New York, 1975. *A very good volume stressing glacier ice, erosion, and deposition.*

Flint, Richard Foster   *Glacial and Quaternary Geology*, John Wiley & Sons, Inc., New York, 1971. *A standard text in the field and a very good one.*

Hays, J. D., John Imbrie, and N. J. Shackleton   "Variations in the Earth's Orbit: Pacemaker of the Ice Ages," *Science*, vol. 194, pp. 1121–1132, 1976. *A technical report that assigns the comings and goings of Pleistocene glaciers to changes in the geometry of the earth's orbit.*

Paterson, W. S. B.   *The Physics of Glaciers*, Pergamon Press Ltd., London, 1968. *If you are interested in physics, this is the best summary available on its application to glacier ice.*

Sugden, David E., and Brian S. John   *Glaciers and Landscape*, John Wiley & Sons, Inc., New York, 1976. *Discusses (in intelligible style) glacier movement and distribution and the effects of glacial erosion and deposition. Well illustrated; good bibliography.*

13

DESERTS, WIND, AND SHORELINES

Deserts and the shores of the ocean are greatly different environments. But they have in common the important role of wind, usually a minor factor in surface processes. So we shall begin our study of the work and effects of the wind with an analysis of the desert regime.

13.1 DESERTS

Although there is no generally accepted definition of a desert, we can at least say that a desert is characterized by a lack of moisture, leading to—among other things—a restriction on the number of living things that can exist there. There may be too little initial moisture, or the moisture that occurs may be evaporated by extremely high temperatures or locked up in ice by extreme cold. Because we are not concerned here with polar deserts, we shall consider only those in the hotter climates. The distribution of middle- and low-latitude deserts is shown in Figure 13.1; they fall into two general groups.

The first are the so-called **topographic deserts**, deficient in rainfall either because they are located toward the center of continents, far from the oceans, or more commonly because they are cut off from rain-bearing winds by high mountains. Takla Makan, north of Tibet and Kashmir in extreme western China, is an example of a desert located deep inside a continental landmass. The desert climate of large sections of Nevada, Utah, Arizona, and Colorado, on the other hand, is caused by the Sierra Nevada of California, which cuts off the rain-bearing winds blowing in from the Pacific. A similar, though smaller, desert area in western Argentina has been created by the Andes.

Much more extensive than the topographic deserts are the **tropical deserts**, lying in zones that range 5° to 30° north and south of the equator. We can best understand their origin by looking at the general circulation of the earth's atmosphere.

Imagine our earth as completely covered by water. In such a situation we should find the air moved in the manner depicted in Figure 13.2. The equator, where the greatest heating by the sun takes place, would be a

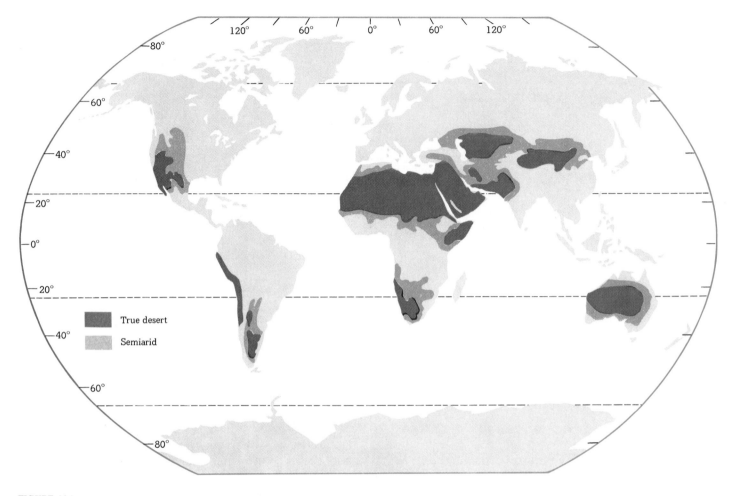

FIGURE 13.1

Deserts and near-deserts cover nearly one-third of the land surface of the earth. Middle- and low-latitude deserts, but not polar deserts, are shown here.

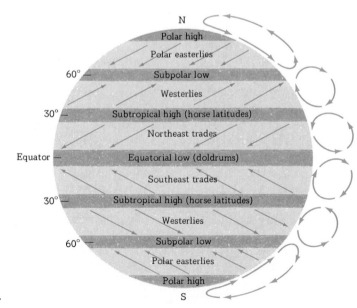

FIGURE 13.2

Idealized circulation of air on an earth presumed to have no landmasses. See the text for discussion.

belt of warm, rising air marked by low pressure. At the poles lower temperatures would cause the air to settle and create a high-pressure zone. Other high-pressure zones of descending air, the subtropical high-pressure zones (the so-called **horse latitudes**), would lie in belts at about 30° north and south of the equator. And at about 60° north and south would be two more belts of lower atmospheric pressure. At the surface of the earth air would move away from zones of high pressure toward zones of low pressure. Thus from the equatorward sides of each of the horse-latitude belts air would move toward the equator. From the poleward sides wind would blow generally toward the low-pressure belts at about 60° north and south.

If it were not for the rotation of the earth, these surface winds would blow either directly south or directly north. But the earth's rotation introduces a factor known as the **Coriolis effect,** named after G. G. Coriolis, a nineteenth-century French mathematician, who made the first extensive analysis of this phenomenon. The Coriolis effect influences everything that moves across the face of the earth: the atmosphere, ocean currents, birds in flight, aircraft, flowing streams, even an automobile speeding along a straight road. We shall not analyze the reason for the effect, but the results can be simply stated: Because of the Coriolis effect, anything moving in the *northern* hemisphere tends to veer to the *right,* and in the *southern* hemisphere any moving object tends to veer to the *left.*

Now apply this principle to the movement of air. If we stand at 30°N and face southward, in the direction toward which the air moves, the Coriolis effect will shift the air movement to our right, that is, to the west. These are the northeast **trade winds.** Standing at 30°S and facing northward, we find that the winds blowing toward the low pressure near the equator will veer to the left; they are known as the southeast trades. (Remember that, to apply the Coriolis effect, you must face in the direction *toward* which the air moves; the winds are named for the direction *from* which they move.)

Now we can return to the origin of tropical deserts. At about 30° north and south of the equator air de-scends in the subtropical high-pressure zones and spreads laterally across the surface toward both the equator and the poles. Under the influence of the Coriolis effect the air currents moving toward the poles become the **prevailing westerly winds** of the middle latitudes, and those moving toward the equator become the trade winds of the low latitudes.

Now the warmer the air becomes, the more moisture it can hold and the less likely it is to release moisture as precipitation. In the subtropical high-pressure belts the air is heated as it descends over the warm land and tends to retain the moisture it contains. Consequently the climate in these areas is relatively dry. The air that moves along in the trade winds continues to be heated as it enters warmer and warmer latitudes, and the dry belt is extended toward the equator. Finally, as the air approaches the equator itself, it becomes so heated that it rises to a higher altitude and is rapidly cooled. Then, unable to carry its great quantities of moisture, it releases the torrential rainfalls that are characteristic of the true tropics. This rising air eventually moves poleward, some of it descending in the horse latitudes to begin its path all over again.

It is this continuing circulation of the air that creates the great deserts on either side of the equator. These include the Sahara Desert of North Africa, the Arabian Desert of the Middle East, the Victoria Desert of Australia, the Kalahari Desert of Bechuanaland, in South Africa, the Sonora Desert of northwestern Mexico, southern Arizona, and California, the Atacama Desert of Peru and Chile, and the deserts of Afghanistan, Baluchistan, and northwestern India.

Some of the smaller deserts along the tropical coastlines have been created by the influence of oceanic currents bordering the continents. Winds blow in across the cool water of the ocean and suddenly strike the hot tropical landmass. There the air is heated, and its ability to retain moisture is increased. The resulting lack of precipitation gives rise to desert conditions, as along the coast of southern Peru and northern Chile, where the cool Humboldt current flows north toward the equator.

13.2 WORK OF THE WIND

In humid lands a virtually continuous vegetative cover blankets the earth's surface; but in deserts lack of moisture ensures that vegetation is absent or sparse, and the wind can then directly affect the land surface. Therefore we find good examples of erosion, transportation, and deposition by wind in deserts.

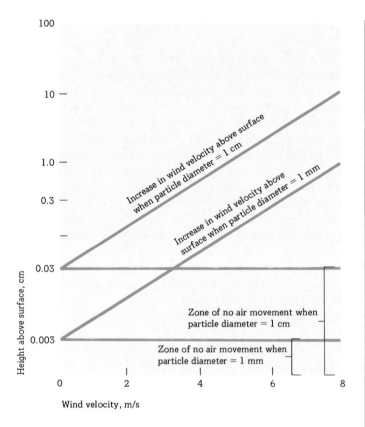

FIGURE 13.3

In a thin zone close to the ground there is little or no air movement, regardless of the wind velocity immediately above. This zone is approximately one-thirtieth the average diameter of surface particles. Two zones are shown in the graph: one for surfaces on which the particles average 1 mm in diameter and one for surfaces with 1-cm particles. Diagonal lines represent the increase in velocity of a wind of given intensity blowing over surfaces covered with particles of 1-mm and 1-cm average diameter. [Reproduced by permission from R. A. Bagnold, *The Physics of Blown Sand and Desert Dunes*, p. 54, Methuen & Co., Ltd., London, 1941.]

MOVEMENT OF MATERIAL

Wind velocities increase rapidly with height above the ground surface, just as the velocity of running water increases at levels above the channel floor. Furthermore, like running water, most air moves in turbulent flow. But wind velocities increase at a greater rate than water velocities, and the maximum velocities attained are much higher.

The general movement of wind is forward, across the surface of the land, but within this general movement the air is moving upward, downward, and from side to side. In the zone about 1 m above ground surface, the average velocity of upward motion in an air eddy is approximately one-fifth the average forward velocity of the wind. This upward movement greatly affects the wind's ability to transport small particles of earth material, as we shall see.

Right along the surface of the ground there is a thin but definite zone where the air moves very little or not at all. Field and laboratory studies have shown that the depth of this zone depends on the size of the particles that cover the surface. On the average, the depth of this "zone of no movement" is about one-thirtieth the average diameter of the surface grains (see Figure 13.3). Thus, over a surface of evenly distributed pebbles with an average diameter of 30 mm, the zone of no movement would be about 1 mm deep. This fact too has a bearing on the wind's ability to transport material.

Dust storms and sandstorms Material blown by the wind usually falls into two size groups. The diameter of wind-driven sand grains averages between 0.15 and 0.30 mm, with a few grains as fine as 0.06 mm. All particles smaller than 0.06 mm are classified as dust.

In a true **dust storm** (see Figure 13.4) the wind picks up fine particles and sweeps them upward hundreds or even thousands of meters into the air, forming a great cloud that may blot out the sun and darken the sky. In contrast a **sandstorm** is a low, moving blanket of wind-driven sand with an upper surface 1 m or less above the ground. The greatest concentration of moving sand is usually just a few centimeters above the ground surface, and individual grains seldom rise even as high as 2 m. Above the blanket of moving sand the air is quite clear, and a person on the ground appears to be partially submerged, as though standing in a shallow pond.

Often, of course, the dust and sand are mixed together in a wind-driven storm; but the wind soon

FIGURE 13.4

Silt-sized particles are swept high into the sky from the flood plain of the Knik River valley, Alaska. The Knik River is fed, in part, by active glaciers. [William C. Bradley.]

sweeps the finer particles off, and eventually the air above the blanket of moving sand becomes clear. Apparently, then, the wind handles particles of different size in different ways. A dust-sized grain is swept high into the air, and a sand-sized grain is driven along closer to the ground. The difference arises from the strength of the wind and the terminal velocity of the grain.

We defined the terminal velocity of a grain as the constant rate of fall attained by the grain when the acceleration due to gravity is balanced by the resistance of the fluid—in this case the air—through which the grain falls (see Section 10.5). Terminal velocity varies only with the size of a particle when shape and density are constant. As the particle size increases, both the pull of gravity and the air resistance increase too. But the pull of gravity increases at a faster rate than the air resistance: A particle with a diameter of 0.01 mm has a terminal velocity in air of about 0.01 m/s; a particle with a 0.2-mm diameter has a terminal velocity of about 1 m/s; and a particle with a diameter of 1 mm has a terminal velocity of about 8 m/s.

To be carried upward by an eddy of turbulent air, a particle must have a terminal velocity that is less than the upward velocity of the eddy. Close to the ground surface, where the upward currents are particularly strong, dust particles are swept up into the air and carried in suspension. Sand grains, however, have terminal velocities greater than the velocity of the up-

ward-moving air; they are lifted for a moment and then fall back to the ground. But how does a sand grain get lifted into the air at all if the eddies of turbulent air are unable to support it?

Movement of sand grains Careful observations, both in the laboratory and on open deserts, show sand grains moving forward in a series of jumps, in a process known as **saltation**. We used the same term to describe the motion of particles along a stream bed, but there is a difference: An eddy of water can actually lift individual particles into the main current, whereas wind by itself cannot pick up sand particles from the ground.

Sand particles are thrown into the air only under the impact of other particles. When the wind reaches a critical velocity, grains of sand begin to roll forward along the surface. Suddenly one rolling grain collides with another; the impact may lift either the second particle into the air or the first.

Once in the air, the sand grain is subjected to two forces. First, gravity tends to pull it down to earth again, and eventually it succeeds. But even as the grain falls, the horizontal velocity of the wind drives it forward. The resulting course of the sand grain is parabolic from the point where it was first thrown into the air to the point where it finally hits the ground. The angle of impact varies between 10° and 16° (see Figure 13.5).

When the grain strikes the surface, it may either bounce off a large particle and be driven forward once again by the wind or bury itself in the loose sand, per-

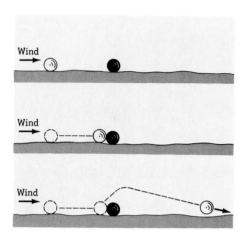

FIGURE 13.5

A sand grain is too heavy to be picked up by the wind but may be put into the air by saltation. Here a single grain is rolled forward by the wind until it bounces off a second grain. Once in the air, it is driven forward by the wind and is then pulled to the ground by gravity. It follows a parabolic path, hitting the ground at an angle between 10° and 16°.

haps throwing other grains into the air by its impact.

In any event it is through the general process of saltation that a sand cloud is kept in motion. Countless grains are thrown into the air by impact and are driven along by the wind until they fall back to the ground. Then they either bounce back into the air again or else pop other grains upward by impact. The initial energy that lifts each grain into the air comes from the impact of another grain, and the wind contributes additional energy to keep it moving. When the wind dies, all the individual particles in the sand cloud settle to earth.

Some sand grains, particularly the large ones, never rise into the air at all, even under the impact of other grains. They simply roll forward along the ground, very much like the rolling and sliding of particles along the bed of a stream of water. It has been estimated that between one-fifth and one-quarter of the material carried along in a sandstorm travels by rolling and the rest by means of saltation.

Notice that, after the wind has started the sand grains moving along the surface, initiating saltation, the wind no longer acts to keep them rolling. The cloud of saltating grains obstructs the wind and shields the ground surface from its force; thus, as soon as saltation begins, the velocity of near-surface winds drops rapidly. Saltation continues only because the impact of the grains continues. The stronger the winds blow during saltation, the heavier will be the blanket of sand and the less the possibility that surface grains will be rolled by the wind.

Movement of dust particles As we have seen, dust particles are small enough and have low enough terminal velocities to be lifted aloft by currents of turbulent air and to be carried along in suspension. But just how does the wind lift these tiny particles in the first place?

Laboratory experiments show that under ordinary conditions particles smaller than 0.03 mm in diameter cannot be swept up by the wind after they have settled to the ground. In dry country, for example, dust may lie undisturbed on the ground even though a brisk wind is blowing, but if a flock of sheep passes by and kicks loose some of the dust, a dust plume will rise into the air and move along with the wind.

The explanation for this seeming reluctance of dust particles to be disturbed lies in the nature of air movement. The small dust grains lie within the thin zone of negligible air movement at the surface. They are so small that they do not create local eddies and disturbances in the air, and the wind passes them by—or the particles may be shielded by larger particles against the action of the wind.

Some agent other than the wind must set dust particles in motion and lift them into a zone of turbulent air—perhaps the impact of larger particles or sudden downdrafts in the air movement. Irregularities in a plowed field or in a recently exposed streambed may help the wind begin its work by creating local turbulence at the surface. Also, vertical downdrafts of chilled air during a thunderstorm sometimes strike the ground with velocities of 40 to 80 km/h and churn up great swaths of dust.

EROSION

Erosion by the wind is accomplished through two processes: **abrasion** and **deflation**.

Abrasion Like the particles carried by a stream of running water, saltating grains of sand driven by the wind are highly effective abrasive agents in eroding rock surfaces. As we have seen, wind-driven sand seldom rises more than 1 m above the surface of the earth, and measurements show that most of the grains are concentrated in the 0.5 m closest to the ground. In this layer their abrasive power is greatest.

Although evidence of abrasion by sand grains is rather meager, there is enough to indicate that this erosive process takes place. For example, we sometimes find fence posts and telephone poles abraded at ground level and bedrock cliffs with a small notch along their base. In desert areas the evidence is more impressive; for here the wind-driven sand has in some places cut troughs or furrows in the softer rocks. The cross profile of one of these troughs is not unlike that of a glaciated mountain valley in miniature, the troughs ranging from a few centimeters to perhaps 8 m in depth. They run in the usual direction of the wind, and their deepening by sand abrasion has actually been observed during sandstorms.

The most common products of abrasion are certain pebbles, cobbles, and even boulders that have been eroded in a particular way. These pieces of rock are called **ventifacts,** from the Latin for "wind" and "made." They are found not only on deserts but also along modern beaches—in fact, wherever the wind blows sand grains against rock surfaces (see Figures 13.6 and 13.7). Their surface is characterized by a relatively high gloss, or sheen, and by facets, pits, gouges, and ridges.

The face of an individual ventifact may display one

FIGURE 13.6

A ventifact with three well-developed facets, on the floor of the Mojave Desert, California. The pocket knife to the right of the ventifact gives the scale. [Sheldon Judson.]

FIGURE 13.7

A facet on a ventifact is cut by the impact of grains of wind-driven sand. [Redrawn from Robert P. Sharp, "Pleistocene Ventifacts East of the Big Horn Mountains, Wyoming," *J. Geol.,* vol. 57, p. 182, 1949.]

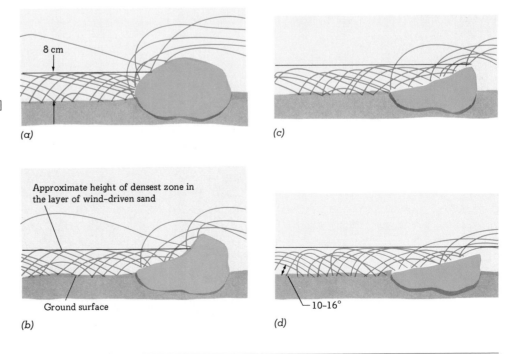

FIGURE 13.8

Wind has excavated this blowout in unconsolidated sand deposits of Terry Andrae State Park, near Lake Michigan in eastern Wisconsin. [Wisconsin Conservation Department.]

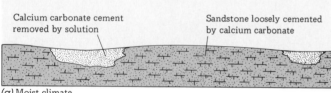

Calcium carbonate cement removed by solution

Sandstone loosely cemented by calcium carbonate

(a) Moist climate

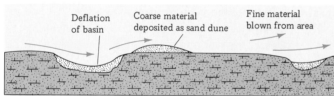

Deflation of basin

Coarse material deposited as sand dune

Fine material blown from area

(b) Dry climate

FIGURE 13.9

The High Plains of eastern New Mexico and western Texas are pockmarked with broad, shallow depressions fashioned in loosely consolidated sandstone. In this instance wind deflation has created blowouts, but only after the calcite cement of the sandstone was destroyed by downward-percolating waters. Destruction of the cement took place during moist periods in the Pleistocene, and deflation occurred in intervening dry periods. [Redrawn from Sheldon Judson, "Geology of the San Jon Site, Eastern New Mexico," *Smithsonian Misc. Coll.*, vol. 121, no. 1, p. 13, 1953.]

FIGURE 13.10

Stone-littered surfaces are common in the desert. This is a close-up view of the slopes leading down into Death Valley, California. [Sheldon Judson.]

FIGURE 13.11

The great bulk of the loess in the central United States is intimately related to the major glacier-fed valleys of the area and was probably derived from the flood plains of these valleys. In Kansas and parts of Nebraska, however, the loess is probably nonglacial in origin and has presumably been derived from local sources and the more arid regions to the west. The line of section marked *A* in Illinois refers to Figure 13.13.

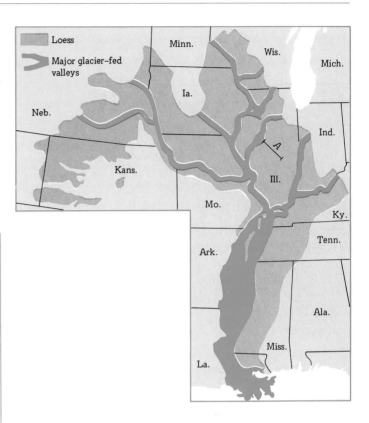

facet, or twenty facets, or more—sometimes flat but more commonly curved. Where two facets meet, they often form a well-defined ridge, and the intersection of three or more facets gives the ventifact the appearance of a small pyramid. Apparently the surface becomes pitted when it lies across the direction of wind movement at an angle of 55° or more; but it becomes grooved when it lies at angles of less than 55°.

Deflation Deflation (from the Latin for "to blow away") is the erosive process of the wind's carrying off unconsolidated material. The process creates several recognizable features in the landscape. For example, it often scoops out basins in soft, unconsolidated deposits ranging from a few meters to several kilometers in diameter. These basins are known as **blowouts,** for obvious reasons (see Figure 13.8). Even in relatively consolidated material the wind will excavate sizable basins if some other agency is at work loosening the material. We find such depressions in the almost featureless High Plains of eastern New Mexico and western Texas, where the bedrock is loosely cemented by calcium carbonate. Several times during the Pleistocene the climate in this area shifted back and forth between moist and dry. During the moist periods water dissolved some of the calcium carbonate and left the sandstone particles lying on the surface. Then during the dry periods the wind came along and removed the loosened material. Today we find the larger particles piled up in sand hills on the leeward side of the basin excavated by the wind. The smaller dust particles were carried farther along and spread in a blanket across the plains to the east (see Figure 13.9).

Deflation removes only the sand and dust particles from a deposit and leaves behind the larger particles of pebble or cobble size. As time passes, these stones form a surface cover, a **desert pavement,** that cuts off further deflation (see Figure 13.10).

DEPOSITION

Whenever the wind loses its velocity and hence its ability to transport the sand and dust particles it has picked up from the surface, it drops them back to the ground. The landscape features formed by wind-deposited materials are of various types, depending on the size of particles, the presence or absence of vegetation, the constancy of wind direction, and the amount of material available for movement by the wind. We still have a great deal to learn about this sort of deposit, but certain generalizations seem valid.

Loess Loess is a buff-colored, unstratified deposit composed of small, angular mineral fragments. Loess deposits range in thickness from a few centimeters to 10 m or more in the central United States to over 100 m in parts of China. A large part of the surface deposits across some 0.5 million km² of the Mississippi River basin is made up of loess, and this material has produced the modern fertile soils of several midwestern states, particularly Iowa, Illinois, and Missouri (see Figure 13.11).

Most geologists, though not all, believe loess to be material originally deposited by the wind. They base

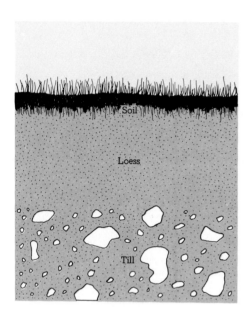

FIGURE 13.12

In many places unweathered till is overlain by loess on which a soil zone has developed. The lack of a weathering zone on the till beneath loess often indicates rapid deposition of the loess immediately after the disappearance of the glacier ice and before weathering processes could affect the till. Not until loess deposition has slowed or halted is there time available to allow weathering and organic activity capable of producing a soil.

their conclusion on several facts. The individual particles in a loess deposit are very small, strikingly like the particles of dust carried about by the wind today. Moreover, loess deposits stretch over hillslopes, valleys, and plains alike, an indication that the material has settled from the air. And the shells of air-breathing snails present in loess strongly impugn the possibility that the deposits had actually been laid down by water.

Many exposures in the north-central United States reveal that loess deposits there are intimately associated with till and outwash deposits built up during the great Ice Age. Because the loess lies directly on top of the glacial deposits in many areas, it seems likely that it was deposited by the wind during periods when glaciation was at its height rather than during interglacial

intervals. Also, because there is no visible zone of weathering on the till and outwash deposits, the loess probably was laid down on the newly formed glacial deposits before any soil could develop on them (see Figure 13.12).

Certain relationships between the loess deposits in the Midwest and the streams that drain the ancient glacial areas serve to strengthen the conclusion that there is a close connection between glaciation and the deposit of windborne materials. Figure 13.11, for example, shows that the major glacial streams cut across the loess belt. Furthermore, as seen in Figure 13.13, the thickness of the loess decreases toward the east and away from the banks of the streams. Moreover, the mean size of the particles decreases away from the glacial streams. These facts can best be explained as follows. We know that loess is not forming in this area at present; so we must look for more favorable conditions in the past. During the great Ice Age of the Pleistocene, the rivers of the Midwest carried large amounts of debris-laden meltwater from the glaciers. Consequently the flood plains of these rivers built up at a rapid rate and were broader than they are today. During periods of low water the flood plains were wide expanses of gravel, sand, silt, and clay exposed to strong westerly winds. These winds whipped the dust-sized material from the flood plains, moved it eastward, and laid down the thickest and coarsest of it closest to the rivers.

All loess, however, is not derived from glacial deposits. In one of the earliest studies of loess it was shown that the Gobi Desert has provided the source material for the vast stretches of yellow loess that blanket much of northern China and that give the characteristic color to the Yellow River and the Yellow Sea. Much of the land used for cotton growing in the eastern Sudan of Africa is thought to be made up of particles blown from the Sahara Desert to the west. We have already seen that finely divided mineral fragments are swept up in suspension during desert sandstorms and are carried by the wind far beyond the confines of the desert. Clearly, then, the large amounts of very fine material present in most deserts would make an excellent source of loess.

Sand deposits Unlike deposits of loess, which blanket whole areas, sand deposits assume certain characteristic and recognizable shapes. Wind often heaps the sand particles into mounds and ridges called **dunes,** which sometimes move slowly along in the direction of the wind. Some dunes are only a few meters in height, but others reach tremendous sizes. In south-

FIGURE 13.13

Loess related to the major glacier-fed rivers in the Midwest shows a decrease in thickness away from the rivers and a decrease in the size of individual particles. An example is shown in this diagram, based on data gathered along line *A* in Figure 13.11. [Redrawn from G. D. Smith, "Illinois Loess—Variations in Its Properties and Distributions," *Univ. Ill. Agric. Expt. Sta. Bull.* 490, 139–184, 1942.]

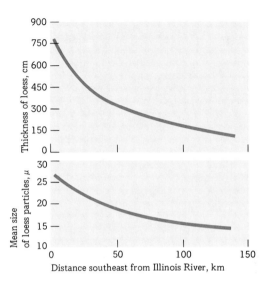

ern Iran dunes have grown to 200 m with a base 1 km wide.

In Chapter 10 we found that, as the velocity of a stream falls, so does the energy available for the transportation of material; consequently deposition of material takes place. The same relationship between decreasing energy and increasing deposition applies to the wind. But in dealing with wind-deposited sand, we need to examine the relationship more closely and to explain why sand is deposited in the form of dunes rather than as a regular, continuous blanket.

The wind shadow   Any obstacle—large or small—across the path of the wind will divert moving air and create a **wind shadow** to the leeward, as well as a smaller shadow to the windward immediately in front of the obstacle. Within each wind shadow the air moves in eddies, with an average motion less than that of the wind sweeping by outside. The boundary between the two zones of air moving at different velocities is called the **surface of discontinuity** (see Figure 13.14).

When sand particles driven by the wind strike an obstacle, they settle in the wind shadow immediately in front of it. Because the wind velocity (hence energy) is low in this wind shadow, deposition takes place, and gradually a small mound of sand builds up. Other particles move past the obstacle and cross through the surface of discontinuity into the leeward wind shadow behind the barrier. Here again the velocities are low, deposition takes place, and a mound of sand (a dune) builds up—a process aided by eddying air that tends to sweep the sand in toward the center of the wind shadow (see Figures 13.15 and 13.16). This process is familiar in dry, sandy regions, where sand piles up before and in the lee of obstacles, but it is also seen in snowy climates: The principle is illustrated by snow fences (devices that have also been used effectively against windblown sand).

Wind shadow of a dune   Actually, a sand dune itself acts as a barrier to the wind; and by disrupting

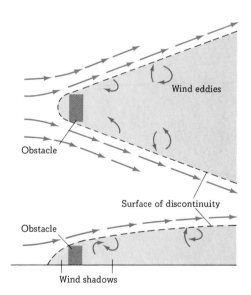

FIGURE 13.14

The shaded area indicates the wind shadow created by an obstacle. The wind is diverted over and around the obstacle. Within the wind shadow wind velocity is low and air movement is marked by eddies. A surface of discontinuity separates the air within the wind shadow from the air outside. [Reproduced by permission from R. A. Bagnold, *The Physics of Blown Sand and Desert Dunes*, p. 190, Methuen & Co., Ltd., London, 1941.]

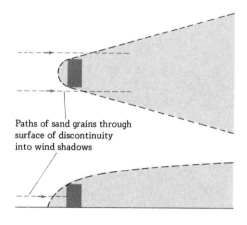

FIGURE 13.15

Because of its momentum the sand in the more
rapidly moving air outside the wind shadow either
passes through the surface of discontinuity to settle
in the wind shadow behind the obstacle or strikes
the obstacle and falls in the wind shadow in front
of the obstacle. [Reproduced by permission from
R. A. Bagnold, *The Physics of Blown Sand and
Desert Dunes*, p. 190, Methuen & Co., Ltd., London, 1941.]

FIGURE 13.16

Sand falling in the wind shadow tends to be gathered by wind eddies within the shadow to form a
shadow dune, as shown in this sequence of diagrams. [Reproduced by permission from R. A.
Bagnold, *The Physics of Blown Sand and Desert
Dunes*, p. 190, Methuen & Co., Ltd., London, 1941.]

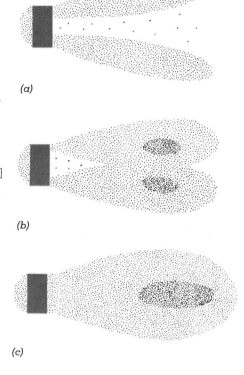

(a)

(b)

(c)

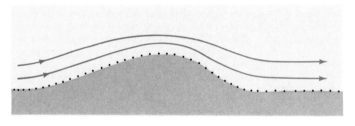

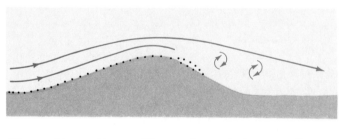

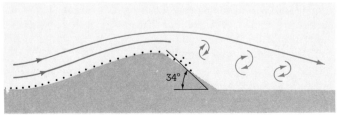

FIGURE 13.17

The development of a slip face on a dune. Wind
converges on the windward side of the dune and
over its crest and diverges to the lee of the dune.
The eventual result is the creation of a wind
shadow in the lee. In this wind shadow sand falls
until a critical angle of slope (about 34°) is
reached. Then a small landslide occurs, and the
slip face is formed. [Reproduced by permission
from R. A. Bagnold, *The Physics of Blown Sand
and Desert Dunes*, p. 202, Methuen & Co., Ltd.,
London, 1941.]

the flow of air, it may cause the continued deposition of sand. A profile through a dune in the direction toward which the wind blows shows a gentle slope facing the wind and a steep slope to the leeward. A wind shadow exists in front of the leeward slope, and it is here that deposition is active. The wind drives the sand grains up the gentle windward slope to the dune crest and then drops them into the wind shadow. The leeward slope is called the **slip face** of the dune because of the small sand slides that take place there.

The slip face is necessary for the existence of a true wind shadow. Here is how the slip face is formed. A mound of sand affects the flow of air across it, as shown in the topmost diagram of Figure 13.17. Notice that the wind flows over the mound in streamlined patterns. These lines of flow tend to converge toward the top of the mound and diverge to the leeward. In the zone of diverging air flow velocities are less than in the zone of converging flow. Consequently sand tends to be deposited on the leeward slope just over the top of the mound where the velocity begins to slacken. This slope steepens because of deposition, and eventually the sand slumps under the influence of gravity. The slump usually takes place at an angle of about 34° from the horizontal. A slip face is thus produced, steep enough to create a wind shadow in its lee. Within this shadow sand grains fall like snow through quiet air. Continued deposition and periodic slumping along the slip face account for the slow growth or movement of the dune in the direction toward which the wind blows.

Shoreline dunes Not all dunes are found in deserts. Along the shores of the ocean and of large lakes, ridges of wind-blown sand called **fore dunes** are built up even in humid climates. They are well developed along the southern and eastern shores of Lake Michigan, along the Atlantic coast from Massachusetts southward, along the southern coast of California, and at various points along the coasts of Oregon and Washington (Figure 13.18).

These fore dunes are fashioned by the influence of strong onshore winds acting on the sand particles of the beach. On most coasts the vegetation is dense enough to check the inland movement of the dunes, and they are concentrated in a narrow belt that parallels the shoreline. These dunes usually have an irregular surface, sometimes pockmarked by blowouts (see the subsection on deflation, which is on page 335).

Sometimes, however, in areas where vegetation is scanty, the sand moves inland in a series of ridges at

FIGURE 13.18

These shoreline dunes form complex patterns behind the beach at Coos Bay, Oregon. The beach serves as a source of sand, and this source is continuously renewed by longshore currents of ocean water. Onshore winds (from the left) drive the beach sand inland. The photograph shows about 450 m of shoreline.

right angles to the wind. These **transverse dunes** exhibit the gentle windward slope and the steep leeward slope characteristic of other dunes. Transverse dunes are also common in arid and semiarid country where sand is abundant and vegetation sparse.

Barchans **Barchans** are sand dunes shaped like a crescent, with their horns pointing downwind. They move slowly with the wind, the smaller ones at a rate of about 15 m/year, the larger ones about 7.5 m/year. The maximum height obtained by barchans is about 30 m, and their maximum spread from horn to horn is about 300 m (see Figure 13.19).

Just what leads to the formation of a barchan is still a matter of dispute. Certain conditions seem essential, however: a wind that blows from a fixed direction, a

FIGURE 13.19

These barchans are moving across the Pampa de Islay, Peru, in the direction in which their horns point. [Aerial Explorations, Inc.]

relatively flat surface of hard ground, a limited supply of sand, and a lack of vegetation.

Parabolic dunes Long, scoop-shaped, **parabolic dunes** look rather like barchans in reverse; that is, their horns point upwind rather than downwind. They are usually covered with sparse vegetation, which permits limited movement of the sand. Parabolic dunes are quite common in coastal areas and in various places throughout the southwestern states. Ancient parabolic dunes, no longer active, exist in the upper Mississippi valley and in central Europe.

Longitudinal dunes **Longitudinal dunes** are long ridges of sand running in the general direction of wind movement. The smaller types are about 3 m high and about 60 m long. In the Libyan and Arabian Deserts, however, they commonly reach a height of 100 m and may extend for 100 km across the country. There they are known as **seif dunes,** from the Arabic word for "sword" (see Figure 13.20).

Stratification in dunes The layers of sand within a dune are usually inclined. The layers along the slip face have an angle of about 34°, whereas the layers along the windward slope are gentler.

Because the steeper beds along the slip face are analogous to the **foreset beds** in a delta (see Figure 10.37), we can use the same term in referring to them. These beds develop if there is a continuous deposition of sand on the leeward side of the dune, as in barchans and actively moving transverse dunes.

Backset beds develop on the gentler slope to the windward. These beds constitute a large part of the total volume of a dune, especially if there is enough vegetation to trap most of the sand before it can cross over to the slip face. **Topset beds** are nearly horizontal beds laid down on top of the inclined foreset or backset beds.

FIGURE 13.20

These seif dunes are located in the Rub 'al Khali district of southern Saudi Arabia and are oriented parallel to the direction of the prevailing winds. [Arabian-American Oil Company.]

13.3 OTHER DESERT PROCESSES

WEATHERING AND SOILS

Because of the lack of moisture in the desert, the rate of weathering, both chemical and mechanical, is extremely slow. As most of the weathered material consists of unaltered rock and mineral fragments, we can conclude that mechanical weathering probably predominates.

Some mechanical weathering is simply the result of gravity, such as the shattering of rock material when it falls from a cliff. Wind-driven sand also brings about some degree of mechanical weathering. Sudden flooding of a desert by a cloudburst moves material to lower elevations, reducing the size of the rock fragments and scouring the bedrock surface in the process. In almost every desert in the world temperatures fall low enough to permit frost action. But here again the deficiency of moisture slows the process. Finally, the wide temperature variations characteristic of deserts cause rock materials to expand and contract and may produce some mechanical weathering.

This low rate of weathering is reflected in the soils of the desert. Seldom do we find extensive areas of residual soil; for the lack of protective vegetation permits the winds and occasional floods to strip away the soil-producing minerals before they can develop into true soils. Even so, soils sometimes develop in local areas, but they lack the humus of the soils in moister climates, and they contain concentrations of such soluble substances as calcite, gypsum, and even halite because there is insufficient water to carry these minerals away in solution. In the deserts of Australia rock-like concentrations of calcite, iron oxide, and even silica sometimes form a crust to be found on the surface.

One minor weathering feature of the desert is of interest because it seems to have its counterpart on the surface of Mars. A thin, shiny, reddish-brown to blackish coating called **desert varnish** occurs on some desert stones. The coating is a thin film—a fraction of a millimeter thick—made up largely of iron and manganese oxides. Interestingly enough, there has long been disagreement as to how this varnish forms on earth. Some suggest that it is the result of entirely inorganic processes. Other workers believe that some organic activity is also involved as well.

WATER

Although rainfall is extremely sparse in desert areas, there is still enough water present to act as an important agent of erosion, transportation, and deposition. In fact water is probably more effective than even the driving winds in molding desert landscape.

Very few streams flowing through desert regions ever find their way to the sea, and the few that do, such as the Colorado River in the United States and the Nile of Egypt, originate in areas from which they receive sufficient water to sustain them through their long course across the desert. Most desert stream beds, however, are dry over long periods of time and flow only when an occasional flood comes along. Even then the flow is short-lived; for the water either evaporates rapidly or vanishes into the highly permeable rubble and debris of the desert. In some places, however, such as the western United States, broad desert plains slope down from the mountain ranges toward central basins, called **playas,** where surface runoff collects from time to time. But the *playa lakes* formed in these basins usually dry up in a short time or at best exist as shallow, salty lakes, of which Great Salt Lake is the best-known example.

Desert floods are unlike floods in humid areas. The typical desert flood, like the rain that produces it, is local in extent and of short duration. In moist regions most floods arise from a general rain falling over a relatively long period of time; consequently they affect large areas. Because of the widespread vegetative cover, these floods tend to rise and fall slowly. But on the bare ground of the desert the runoff moves swiftly and floods rise and fall with great rapidity. These "flash" floods give little warning, and the experienced desert traveler has a healthy respect for them. He will never pitch camp on a dry stream floor even though the stream banks offer protection from the wind. He knows that at any moment a surging wall of debris-laden water may sweep down the stream bed, destroying everything in its path.

The amount of moisture in desert areas does change. Thus during the Ice Age, when glaciers lay across Canada and the northern United States and draped the flanks of most of the higher mountains of the West, the climate of the arid and semiarid areas of the

(a)

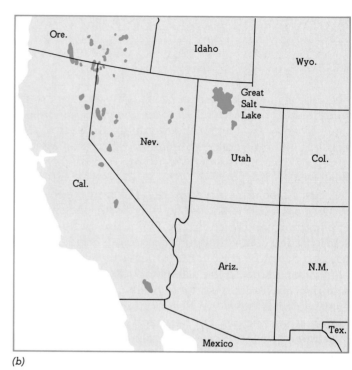

(b)

FIGURE 13.21

During glaciation many basins of the western United States held lakes where only intermittent water bodies exist today; modern saltwater lakes are remnants of basins formerly filled with lakes. Modern lakes (b) are compared with those of the glaciation of higher mountains and the western United States (a). [From O. E. Meinzer, "Map of the Pleistocene Lakes of the Basin-and-Range Province and Its Significance," *Geol. Soc. Am. Bull.* 33, pp. 543, 545, 1922.]

western United States was quite different from what it is today. Glaciations produced **pluvial periods** (from the Latin *pluvia*, "rain"), when the climate was undoubtedly not so moist as that of the eastern United States today but certainly more hospitable than we now find in the arid regions. During any single pluvial period rainfall was greater, evaporation less and vegetation more extensive. At this time large sections of the southwestern states were dotted with lakes known as *pluvial lakes* (Figure 13.21).

What was responsible for this pluvial climate? The presence of continental glaciers in the north is thought to have modified the general wind circulation of the globe. The belt of rain-bearing winds was moved to the south, and the temperatures were lowered. Consequently the rates of evaporation decreased, and at the same time the amount of precipitation increased. When the ice receded, the climate again became very much what it is today.

13.4 SHORELINES

In the desert it is the wind-driven wave of sand—the dune—that is characteristic of the dry environment. Along the coast it is the wind-driven wave of water that provides most of the energy for the modification of the shoreline.

THE PROCESSES

Wind-formed waves Most water waves are produced by the friction of air as it moves across a water surface. The harder the wind blows, the higher the water is piled up into long **wave crests** with intervening troughs, both crests and troughs at right angles to the wind. The distance between two successive wave crests is the **wave length,** and the vertical distance between the wave crest and the bottom of an adjacent trough is the **wave height** (Figure 13.22). When the wind is blowing, the waves it generates are called a **sea.** But wind-formed waves persist even after the wind that formed them dies. Such waves, or **swells,** may travel for hundreds or even thousands of kilometers from their zone of origin.

We are concerned with both the movement of the waveform and the motion of water particles in the path of the wave. Obviously the waveform itself moves forward at a measurable rate. But in deep water the water particles in the path of the wave describe a circular orbit: Any given particle moves forward on the crest of the wave, sinks as the following trough approaches, moves backward under the trough, and rises as the next crest advances. Such a motion can best be visualized by imagining a cork bobbing up and down on the water surface as successive wave crests and troughs pass by. The cork itself makes only very slight forward progress under the influence of the wind. Wave motion extends downward until at a depth equal to about one-half the wavelength it is virtually negligible. But between this level and the surface water particles move forward under the crest and backward under the trough of each wave, in orbits that decrease in diameter with depth (Figure 13.23).

As the wave approaches a shoreline and the water becomes more shallow, definite changes take place in the motion of the particles and in the form of the wave itself. When the depth of water is about half the wavelength, the bottom begins to interfere with the motion of water particles in the path of the wave, and their orbits become increasingly elliptical. As a result the length and velocity of the wave decrease, and its front becomes steeper. When the water becomes shallow enough and the front of the wave steep enough, the wave crest falls forward as a breaker, producing what we call **surf.** At this moment the water particles within the wave are thrown forward against the shoreline. The energy thus developed is then available to erode the shoreline or to set up currents along the shore that are

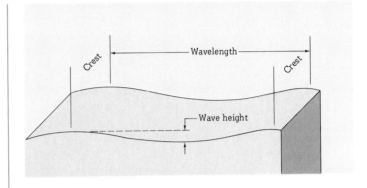

FIGURE 13.22

Diagrammatic explanation of terms used in describing water waves.

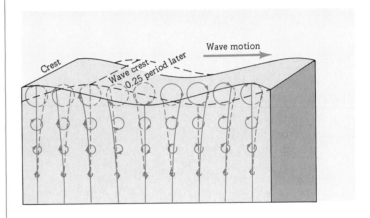

FIGURE 13.23

The motion of water particles relative to wave motion in deep water. Water particles move forward under the crest and backward under the trough, in orbits that decrease in diameter with depth. Such motion extends downward to a distance of about one-half the wavelength. [Redrawn from *U.S. Hydrog. Off. Pub.* 604, 1951.]

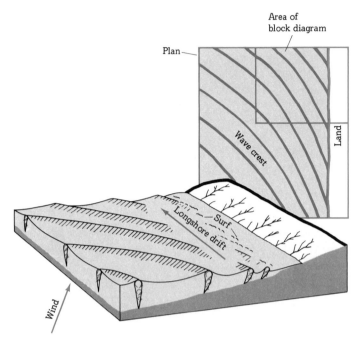

FIGURE 13.24

Wave crests that advance at an angle on a straight shoreline and across a bottom that shallows at a uniform rate are bent shoreward, as suggested in this diagram. Refraction is caused by the increasing interference of the bottom with the orbits of water-particle motion within the wave.

able to transport the sediments produced by erosion.

Wave refraction and coastal currents Although most waves advance obliquely toward a shoreline, the influence of the sea floor tends to bend or **refract** them until they approach the shore nearly head-on.

Let us assume that we have a relatively straight stretch of shoreline with waves approaching it obliquely over an even bottom that grows shallow at a constant rate. As a wave crest nears the shore, the section closest to land feels the effect of the shelving bottom first and is retarded, while the seaward part continues along at its original speed. The effect is to swing the wave around and to change the direction of its approach to the shore, as shown in Figure 13.24.

As a wave breaks, not all its energy is expended on the erosion of the shoreline. Some of the water thrown forward is deflected and moves laterally, parallel to the shore. The energy of this water movement is partly used up by friction along the bottom and partly by the transportation of material.

Refraction also helps explain why, on an irregular shoreline, the greatest energy is usually concentrated on the headland and the least along the bays. Figure 13.25 shows a bay separating two promontories and a series of wave crests sweeping in to the shore across a bottom that is shallow off the headland and deep off

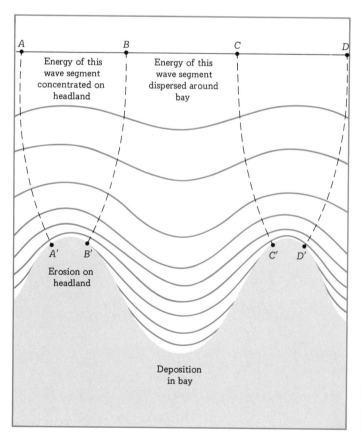

FIGURE 13.25

Refraction of waves on an irregular shoreline. It is assumed that the water is deeper off the bay than off the headlands. Consider that the original wave is divided into three equal segments, *AB*, *BC*, and *CD*. Each segment has the same potential energy. But observe that, by the time the wave reaches the shore, the energy of *AB* and *CD* has been concentrated along the short shoreline of headlands *A'B'* and *C'D'*, whereas the energy of *BC* has been dispersed over a greater front (*B'C'*) around the bay. Energy for erosion per unit of shoreline is therefore greater on the headlands than it is along the bay.

the mouth of the bay. Where the depth of the water is greater than one-half the wavelength, the crest of the advancing wave is relatively straight. Closer to shore, off the headlands, however, the depth becomes less than half the wavelength, and the wave begins to slow down. In the deeper water of the bay it continues to move rapidly shoreward until there too the water grows shallow and the wave crest slows. This differential bending of the wave tends to make it conform in a general way to the shoreline. In so doing, the wave energy is concentrated on the headland and dispersed around the bay, as suggested in Figure 13.25.

A composite profile of a shoreline from a point above high tide seaward to some point below low tide reveals features that change constantly as they are influenced by the nature of waves and currents along the shore. Not all features are present on all shorelines, but several are present in most shore profiles. The **offshore** section extends seaward from low tide. The **shore, or beach,** section reaches from low tide to the foot of the **sea cliff** and is divided into two segments: In front of the sea cliff is the **backshore,** characterized by one or more **berms,** resembling small terraces with low ridges on their seaward edges built up by storm waves; seaward from the berms to low tide is the **foreshore.** Inland from the shore lies the **coast.** Deposits of the shore may veneer a surface cut by the waves on bedrock and known as the **wave-cut terrace.** In the offshore section too there may be an accumulation of unconsolidated deposits comprising a **wave-built terrace** (Figure 13.26).

The shoreline profile is ever changing. During great storms the surf may pound in directly against the sea cliff, eroding it back and at the same time scouring down through the beach deposits to abrade the wave-cut terrace. As the storm (and hence the available energy) subsides, new beach deposits build up out in front of the sea cliff. The profile of a shoreline at any one time, then, is an expression of the available energy: It changes as the energy varies. This relation between profile and available energy is similar to the changing of a stream's gradient and channel as the discharge (and therefore the energy) of the stream varies (see Chapter 10).

SHORELINE FEATURES

Erosion and deposition work hand in hand to produce most of the features of the shoreline. An exception to this generalization is an offshore island that is merely

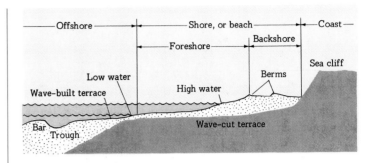

FIGURE 13.26

Some of the features along a shoreline and the nomenclature used in referring to them. [After F. P. Shepard, *Submarine Geology,* 2nd ed., p. 168, Harper & Row, Publishers, New York, 1963.]

the top of a hill or a ridge that was completely surrounded by water as the sea rose in relation to the land. But even islands formed in this way are modified by erosion and deposition.

Features caused by erosion Wave-cut cliffs are common erosional features along a shore, particularly where it slopes steeply down beneath the sea (see Figure 13.27). Here waves can break directly on the shoreline and thus can expend the greatest part of their energy in eroding the land. Wave erosion pushes the wave-cut cliff steadily back, producing a wave-cut terrace or platform at its foot. Because the surging water of the breaking waves must cross this terrace before reaching the cliff, it loses a certain amount of energy through turbulence and friction. Therefore the farther the cliff retreats and the wider the terrace becomes, the less effective are the waves in eroding the cliff. If sea level remains constant, the retreat of the cliffs becomes slower and slower.

Waves pounding against a wave-cut cliff produce various features as a result of the differential erosion of the weaker sections of the rock. Wave action may hollow out cavities, or **sea caves,** in the cliff, and if this erosion should cut through a headland, a **sea arch** is formed. The collapse of the roof of a sea arch leaves a mass of rock, a **stack,** isolated in front of the cliff (Figure 13.28).

Features caused by deposition Features of deposition along a shore are built of material eroded by the waves from the headlands and of material brought down by the rivers that carry the products of weathering and erosion from the landmasses. For example, part

of the material eroded from a headland may be drifted
by currents into the protection of a neighboring bay,
where it is deposited to form a sandy beach.

The coastline of northeastern New Jersey (Figure 13.29) illustrates some of the features caused by
deposition. Notice that the Ashbury Park–Long Branch
section of the coastline is a zone of erosion that has
been formed by the destruction of a broad headland
area. Erosion still goes on along this part of the coast,
where the soft sedimentary rocks are easily cut by the
waves of the Atlantic. The material eroded from this
section is moved both north and south along the coastline. Sand swept northward is deposited in Raritan Bay
and forms a long, sand beach projecting northward, a
spit known as Sandy Hook.

Just south of Sandy Hook the flooded valleys of the
Navesink River and of the Shrewsbury River are bays
that have been almost completely cut off from the open
ocean by sandy beaches built up across their mouths.
These beaches are called **bay barriers.**

Sand moved southward from the zone of erosion
has built up another sand spit. Behind it lies a shallow
lagoon, Barnegat Bay, that receives water from the sea
through a **tidal inlet,** Barnegat Inlet. This passage
through the spit was probably first opened by a violent
storm, presumably of hurricane force. Just inside the
inlet a delta has been formed of material deposited
partly by the original breakthrough of the bar and
partly by continued tidal currents entering the lagoon.

Long stretches of the shoreline from Long Island to
Florida and from Florida westward around the Gulf
Coast are marked by shallow, often marshy lagoons

separated from the open sea by narrow sandy beaches. Many of these beaches are similar to those that enclose Barnegat Bay, apparently elongated spits attached to broad headlands. Others, such as those that enclose Pamlico Sound at Cape Hatteras, North Carolina, have no connection with the mainland. These sandy beaches are best termed **barrier islands**.

CHANGING SEA LEVEL

Sea level is not constant. We find the evidence of marine animals in the rocks of our highest mountains, and deep drilling has recorded shallow marine sediments hundreds and even thousands of meters below the modern sea level. In Section 6.4 we found that large earthquakes can change the relative position of land and sea in local areas. In the following paragraphs we look at some geologically recent changes of sea level felt over wide areas.

Isostasy (see Section 1.3) accompanying continental glaciation and deglaciation has led to some shoreline changes. The weight of a continental glacier is so great as to cause a downward warping of the land it covers. When the glacier melts and the load of ice is removed, the land slowly recovers and achieves its original balance (see Figure 13.30).

We have good evidence of the recoil of land following glacial retreat along the shores of the Gulf of Bothnia and the Baltic Sea. Accurate measurements show that from 1800 to 1918 the land rose at rates ranging from 0.0 cm/year at the southern end to 1.1 cm/year at the northern end. Studies indicate further that the land has been rising at a comparable rate for 5,000 years. Areas that were obviously sea beaches are now elevated from a few meters to 240 m above sea level. A comparison of precise measurements along railroads and highways in Finland from 1892 to 1908 with measurements from 1935 to 1950 shows that uplift there has proceeded at the rate of from 0.3 to 0.9 m/century. The greatest change has been on the Gulf of Bothnia, at about 64°N. Most of this movement has been caused by the recoil of the land following the retreat of the Scandinavian ice cap.

Similar histories of warping and recoil have been established in other regions, including the Great Lakes area in the United States. At one stage in the final retreat of the North American ice sheet the present Lakes Superior, Michigan, and Huron had higher water levels than they do now and were joined together to

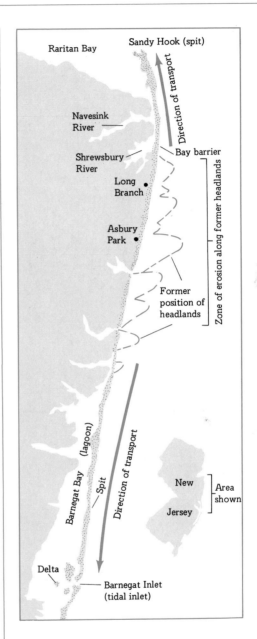

FIGURE 13.29

Erosion by the sea has pushed back the New Jersey coastline as indicated on this map. Some of the material eroded from the headlands has been moved northward along the coast to form Sandy Hook, a spit. To the south a similar but longer feature encloses Barnegat Bay, a lagoon with access to the open ocean through a tidal inlet. [After an unpublished map by Paul MacClintock.]

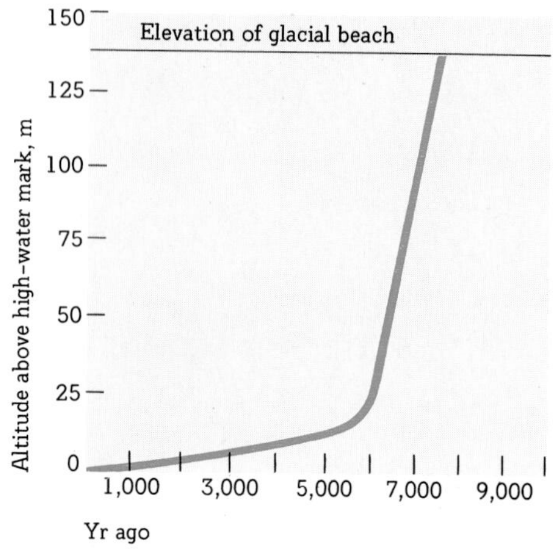

FIGURE 13.30

Change between land and sea in northern Ungava, Quebec, Canada. Land has risen over 125 m since the formation of the last glacial beach in the area more than 7,000 years ago. This change was very rapid until about 6,000 years ago, when it slowed abruptly and continued at a much lower rate to the present. The change in the relationship between land and sea has been due to the isostatic recovery of the land after the disappearance of the ice sheet that had depressed it. [After B. Matthews, "Late Quaternary Land Emergence in Northern Ungava, Quebec," *Arctic*, vol. 20, no. 3, p. 186, 1967.]

form glacial Lake Algonquin. Beaches around the borders of Lake Algonquin were horizontal at the time of their formation; today they are still horizontal south of Green Bay, Wisconsin, and Manistee, Michigan. But 290 km north, at Sault Sainte Marie, Michigan, the oldest Algonquin beach is 108 m higher than it is at Manistee.

The water that is now locked up in the ice of glaciers originally came from the oceans. It was transferred landward by evaporation and winds, precipitated as snow, and finally converted to firn and ice. If all this ice were suddenly to melt, it would find its way back to the ocean basins and would raise the sea level an estimated 60 m. A rise of this magnitude would transform the outline of the earth's landmasses and would submerge towns and cities along the coasts. For the last several thousand years melting glaciers have been raising the sea level (see Figure 13.31), and modern records of sea level indicate that the sea is still rising at about 1 mm/year.

During most of the glacial periods of the past water impounded on the land in the form of ice was more extensive than it is at present. Consequently sea level must have been lower than it is now. It has in fact been estimated that during the maximum extent of Pleistocene glaciation the sea level was from 105 to 120 m lower. Most geologists accept this estimate although some feel that it is too conservative. During the height of the Wisconsin glaciation, the last of the major ice advances, sea level is usually estimated as having been between 70 and 100 m lower than at present. On the other hand during the interglacial periods, when glaciers were somewhat less widespread than they are today, the sea level must have been higher.

FIGURE 13.31

Sea level has been rising for the last several thousand years. Here radiocarbon dates on samples of wood, shell, and peat originally deposited at or close to sea level have been combined to produce these curves of sea-level rise at various points along the eastern coast of the United States. [Redrawn from David Scholl and Minze Stuirer, "Recent Submergence of Southern Florida," *Geol. Soc. Am. Bull.*, vol. 78, p. 448, 1967.]

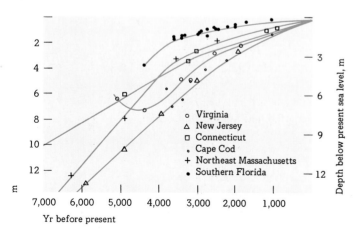

Virginia
New Jersey
Connecticut
Cape Cod
Northeast Massachusetts
Southern Florida

OUTLINE

Deserts cover one-third of the earth's surface. A **topographic** desert owes its aridity to the distance from a source of moisture or to protection from rain-bearing winds by a mountain mass. A **tropical** desert lies between 5° and 30° north or south of the equator in one of two zones of subtropical high pressure developed by the planetary atmospheric circulation.

Work of the wind is more effective in the desert than anywhere else on the lands. Even here, however, it is subordinate to the work of running water in shaping the landscape.

> **Movement of material** by wind depends in part on the size of the particle: Dust-sized particles move differently than do sand-sized particles. Dust particles are carried high into the atmosphere in **suspension**. Sand moves along in continuous contact with the ground or moves a few centimeters above the surface in **saltation**.
>
> **Erosion** by wind consists of **abrasion**, of which **ventifacts** are the most common example, and **deflation**, of which **blowouts** are examples.
>
> **Deposition** consists of dust deposits (**loess**) and sand deposits (usually **dunes**). Sand deposition in dunes begins in the **wind shadow** in the lee of an obstacle. Once established, the dune provides its own wind shadow. In the dune, material is moved up the windward slope and is deposited on the steep **slip face** in the dune's wind shadow. Dune types include **fore dunes, transverse dunes, barchans, parabolic dunes,** and **longitudinal dunes.**

Other desert processes are also important.

> **Weathering,** both mechanical and chemical, is slow because of lack of water.
>
> **Soils** are rubbly and sometimes marked by crustlike accumulations of calcite, iron oxide, and silica.
>
> **Water,** despite its scarcity, is abundant enough to create streamways in even the driest desert. Flow of streams is short-lived but often catastrophic.

Shorelines acquire characteristics generated by the energy of wind-driven waves.

> **Shoreline features** include those formed by **erosion** and those formed by **deposition.** Erosional features include cliffs, caves, arches, stacks, and tidal inlets. Depositional features include beaches, spits, bay barriers, and barrier islands.
>
> **Sea level** is not constant. Isostasy attendant upon glaciation and deglaciation changes sea level. And sea level rises and falls as glaciers melt or expand.

SUPPLEMENTARY READINGS

Bagnold, R. A. *The Physics of Blown Sand and Desert Dunes,* Methuen & Co., Ltd., London, 1941. *Although this book appeared nearly four decades ago, it is yet to be replaced by a better one on the subject. The study of the movement of sand and dust by wind and of sand dunes begins here.*

Bird, E. C. F. *Coasts,* The M.I.T. Press, Cambridge, Mass., 1968. *A good brief introductory volume on the subject.*

Butzer, Karl W. *Geomorphology from the Earth,* Harper & Row, Publishers, New York, 1976. *Chapters 11, 12, and 19 provide a treatment of coasts and deserts in somewhat greater detail than this chapter can provide.*

Cooke, Ronald V., and Andrew Warren *Geomorphology in Deserts,* B. T. Batsford, Ltd., London, 1973. *Discusses all the various physical aspects of deserts including, of course, wind action.*

King, Cuchlaine A. M. *Beaches and Coasts,* 2nd ed., St. Martin's Press, Inc., New York, 1972. *A longer and somewhat more technical treatment than the volume by Bird.*

14

ENERGY

By about 1750 the coal deposits in England began to be fully exploited, to displace water, wind, and wood, and to fuel the beginnings of the industrial revolution; for industrial society is dependent upon the availability of very large supplies of **energy.** Industrialism spread rapidly throughout the western world and then to some eastern nations and moves on to as-yet-unindustrialized countries. This change in social organizations has been driven by energy, at first in the form of coal and later as petroleum and natural gas (Figure 14.1). Now, more than three quarters of the way through the twentieth century, industrial society finds itself in a crisis, seeking ways to maintain reliable and economically feasible supplies of energy. We make no attempt here to offer solutions to this crisis, but as a basis for dealing with solutions we look at the various forms of energy that are available on earth. These include: (1) the **fossil fuels** (coal, petroleum, and natural gas), the remains of plants or animals that gathered their energy from the sun millions of years ago; (2) **nuclear fuels,** chiefly uranium at present but possibly forms of hydrogen in the future; (3) **geothermal energy** (from the earth), generated in part by the decay of radioactive elements; and (4) **solar energy,** as expressed in river-driven hydroelectric plants, or in wind, or in the direct capture of the sun's energy by artificial devices.

By far the most important sources of energy are the fossil fuels: Thus in the United States they account for about 95 percent of the energy produced. The other 5 percent is predominantly in hydroelectric and nuclear power (Figure 14.2).

14.1 NATURAL RESOURCES

Before proceeding to a consideration of energy sources, we should pause briefly to make a few general observations about natural resources. The world's natural resources are concentrated in relatively small areas and are unevenly distributed throughout the crust; deposits that have been located and mined may

FIGURE 14.1

Petroleum and natural gas are the major sources of the world's energy supply today. In recent years explorations for oil and gas have expanded from dry land to the submerged borders of the continents. Offshore reserves are increasingly important and are tapped by wells drilled from platforms set at sea, as is this one in the Arabian Gulf. [Arabian-American Oil Company.]

soon be exhausted unless there are huge reserves. They have been concentrated by the geologic processes we have been studying, such as weathering, running water, igneous activity, and metamorphism; but only a few geologic environments have favored their formation. And after they are used, they are gone forever: Unlike our practice with plant resources, we cannot grow another crop.

To put this another way, most of the natural resources that we exploit for energy and minerals are **nonrenewable**. Once they have been used up, they cannot be replenished—at least in our lifetime. Coal and oil, once burned, are vanished into water and carbon dioxide. A copper vein, once mined out, cannot be redeposited. In contrast some resources are replen-

FIGURE 14.2

Energy consumption in the United States has expanded dramatically from 1850 to 1976. During this time the major source of energy has shifted from wood to coal to oil and gas. [Data from U.S. Bureau of Mines.]

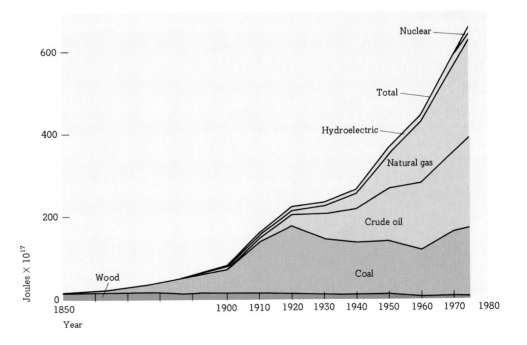

ishable. Water can be pumped from an underground supply as long as the pumpage does not exceed the infiltration of water into the underground; and we can grow a new forest to replace one that we harvest. **The sun is our ultimate source of renewable energy;** on our human time scale the atomic fires that send us light and heat across nearly 150 million km are limitless.

14.2 COAL

Coal, as we found in Section 4.8, is a sedimentary and metamorphic rock. It represents plants that grew in ancient swamps. Conditions favored preservation, which means that the remains of the plants accumulated in a nonoxidizing environment and were eventually buried by other sediments, usually sand or mud, which are now the sandstone and shale typically associated with coal beds.

The coal deposits start off as organic materials made chiefly of carbon, oxygen, and hydrogen. With rising temperature and pressure, due to the burial of the deposits, the hydrogen and oxygen are gradually lost. The higher the percentage of carbon in the remaining deposit, the higher the rank of the coal. We find, therefore, that the lowest grade of coal, **lignite,** has about 70 percent carbon; that **bituminous,** or **soft,** coal has about 80 percent carbon; and that the highest rank of coal, **anthracite,** contains from 90 to 95 percent carbon (Figure 14.3). In addition to carbon, oxygen, and hydrogen, coal contains many other elements, even though in small amounts. One, sulfur, has made some coal a dangerous pollutant of both air and water. It enters the air as sulfur dioxide gas, SO_2, which can be harmful to life if it reaches certain concentrations. The drainage from mines may be contaminated with sulfur in the form of sulfuric acid, H_2SO_4.

The geology of coal is well understood, and the search for coal is governed by the relatively simple characteristics of flat-lying to moderately folded sedimentary rocks. Coal occurs widely but unevenly around the world, with the United States being particularly well endowed (Figure 14.4). Most of the major coal-bearing basins have been identified, and we have a pretty good idea of both the known reserves as well as potential resources yet to be proved. Estimates of the total recoverable coal in the world vary, but something like 7.1×10^{12} t seems to be a reasonable figure. For the United States total recoverable coal is about 1.5×10^{12} t. These are impressive figures and represent a resource that can last for a few hundred years, a much longer time than we can project for oil and natural gas, to which we now turn.

14.3 OIL AND NATURAL GAS

Oil and **natural gas** have proved to be economical, efficient, and relatively clean fuels. As a result, by 1950 they had overtaken coal as the primary source of energy in the United States, and today some four times as much energy is produced in this country by burning oil and natural gas as is produced from coal.

ORIGIN

Almost without exception **petroleum and natural gas are associated with sedimentary rocks of marine origin.** They are mixtures of hydrocarbon compounds (composed largely of hydrogen and carbon) with minor amounts of sulfur, nitrogen, and oxygen. Virtually all geologists are now convinced that oil and natural gas were derived originally from organic remains—both plants and animals.

The hydrocarbons found in our oil and gas fields, however, differ somewhat from those we know in living things. So some changes take place between the organic remains and the end product. The first step is the accumulation of an oceanic sediment rich in remains of plant and animal life. This is not so simple as it sounds; for most sea environments provide oxygen to destroy the organic remains before they can be buried by subsequent sediments. We, however, have some examples of basins in which the circulation of water is so slow that the bottom waters become depleted in oxygen and organic-rich sediments can accumulate. Even

FIGURE 14.3

Increasing calorific value of coal with increasing rank. [From Brian J. Skinner, *Earth Resources*, p. 26, Prentice-Hall, Inc., Englewood Cliffs, N.J., 1976.]

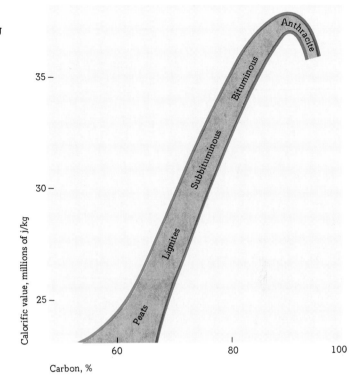

FIGURE 14.4

The coal fields of the United States are large. Extensions of deposits in the western United States are found beneath the plains of Canada, and coals typical of the eastern United States are mined in Nova Scotia. Deposits of lignitic through bituminous rank occur widely in Alaska. ["Strippable reserves of Bituminous Coal and Lignite in the United States," *U.S. Bur. Mines Info. Circ. 8531*, 1971.]

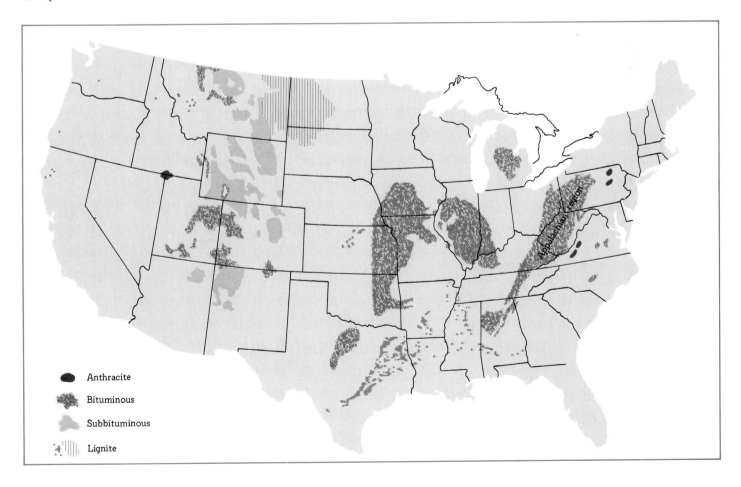

Anthracite

Bituminous

Subbituminous

Lignite

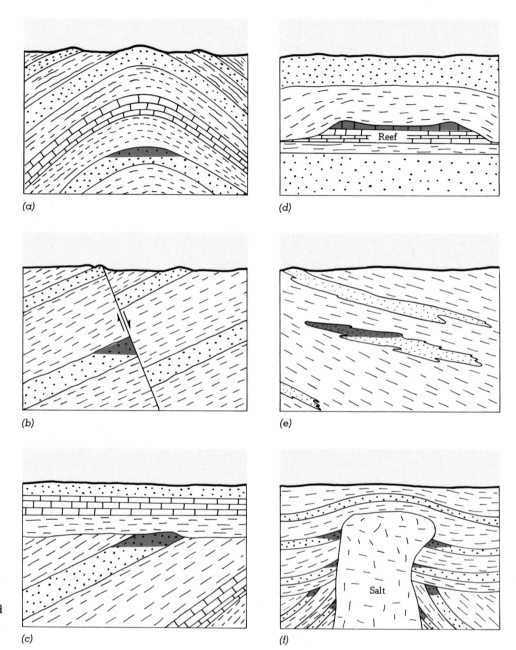

The common types of oil traps, drawn here in cross section, include (a) anticlines, (b) faults, (c) unconformities, (d) reefs, (e) sand lenses, and (f) salt domes. Oil accumulation is shown in color.

this oxygen-poor environment may not suffice to protect the organic sediments because some bacteria can utilize the sulfates of seawater to oxidize organic matter. Therefore we need to have a marine basin that accumulates organic sediments in an oxygen- *and* sulfate-deficient environment.

Once deposited, the organic-rich sediments must be buried beneath subsequent layers. With increasing burial the temperature and pressure increase; and when the depth reaches about 2.5 km, changes begin to take place in the organic portion of the sediments. The original molecules are broken down into new molecules, which begin to form hydrocarbon liquid and gas within the rock. When this happens, there exist hydrocarbons that can shift; and whether they are gathered into an oil and/or gas field depends upon a number of additional factors discussed in the following paragraphs.

ACCUMULATION

We have found that a sedimentary unit rich in organic material and buried to a depth of at least 2.5 km is needed for the creation of oil and natural gas. This sedimentary layer we call a **source bed** because it serves as a hydrocarbon source to create the oil and gas. These products may move from their original location and usually do: In the first place there is the weight of the overlying rocks, which tends to squeeze the oil and gas through pores and cracks in the rock; in the second place there is usually water in the source bed as well as in adjacent rocks, and if the pores are large enough, the oil and gas tend to move upward through the water-saturated layers. Eventually the oil and gas reach the surface and are lost there unless something intervenes to collect and hold them in the underground.

An impermeable rock unit will block or divert oil and gas in their migration. If the impermeable unit is properly positioned, it will halt the hydrocarbons on their way toward the surface and keep them underground. Associated with this impermeable **capping bed** is a permeable rock unit into which the hydrocarbons have moved. We call this the **reservoir rock;** for it contains the accumulated oil and gas. The capping bed and related reservoir rock together form a **trap** and serve as a natural tank. The traps can be of several types, as illustrated in Figure 14.5.

Once the seal provided by the capping bed is broken by the drill, the oil and natural gas move from the pores of the reservoir rock to the drill hole and then can be brought to the surface for processing and distribution. Of course the trap may be broken by natural means as well. For instance, earth movements after the formation of the trap may so disturb it that fractures develop, and the hydrocarbons will escape either to another trap or to the surface. Or, again, surface erosion may eventually reach and destroy the trap. Therefore one would expect that, the older the rock, the less chance there is that it still contains oil, and in a general way this is true. The greatest amount of production comes from rocks of Cenozoic age, followed by fields of oil and gas in Mesozoic and then Paleozoic rocks. There is essentially no production from Precambrian rocks.

THE SEARCH FOR PETROLEUM

The search for petroleum is carried on largely by indirect methods. Most of the geologic structures that contained oil and gave surficial indication of their presence have long since been drilled. So now we must resort to means other than surface observation to discover potential traps of oil and gas buried beneath kilometers of sedimentary rocks and contained in the beds submerged beneath the oceans along the continental margins.

Geologists are now using several methods in their search, including (1) core drilling, (2) seismic prospecting, and (3) gravity prospecting. These methods reveal whether or not there is a buried structure that is likely to trap oil and gas, but they do not give direct evidence of the presence of an actual reservoir.

1 In *core drilling* several closely spaced holes are drilled into the surface to reveal the structure of the underlying sedimentary beds. On the basis of the core samples the beds are matched from hole to hole, and the height of each above sea level is determined. Then each of the beds is carefully plotted, and a map of the entire structure is built up.

2 *Seismic prospecting* is based on knowledge of earthquake waves. Small dynamite blasts are set off in shallow holes about 50 m deep, and the waves generated travel into the interior and are reflected back to the surface, where they are recorded (see Figures 14.6 and 14.7). If they originate in a zone of shale, for example, and encounter a bed of sandstone, some of them bounce back to the surface, where instruments pick them up and register the time of their arrival. The depth to which they have traveled can then be computed, and through a series of such measurements an entire structure can be plotted.

FIGURE 14.6

Seismic prospecting for oil-bearing
structures. Bar-S Ranch, West Texas.
(See also Figure 14.7.) [Humble Oil
and Refining Company.]

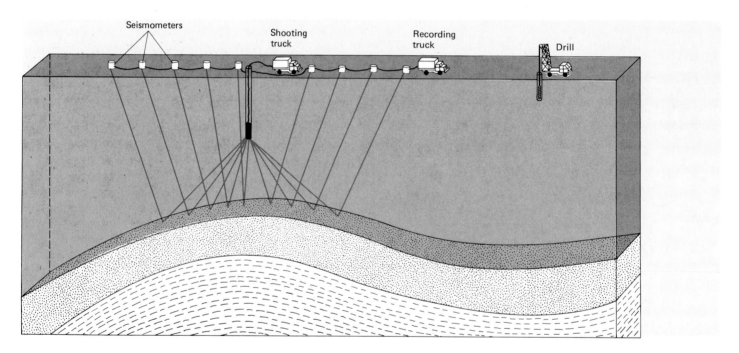

FIGURE 14.7

The operation of a seismic-exploration effort on land is shown in this diagram. First
a drill prepares a hole about 50 m deep. A shooting truck then loads the bottom of
the hole with explosives, plugs the hole, and explodes the charge electrically on a
command from the recording truck. Sound waves from the explosion follow paths
shown by the colored lines from the explosion to a reflecting layer at depth and
back up to miniature seismometers connected by cable to the recording truck. Com-
puter processing of the seismic records produces a diagram that resembles a geologic cross section. [After Sheldon Judson, Kenneth S. Deffeyes, and Robert B. Hargraves, *Physical Geology*, p. 443, Prentice-Hall, Inc., Englewood Cliffs, N.J., 1976.]

3 *Gravity prospecting* makes use of variations in the specific gravity of rocks underlying the surface. If a bed lies in a horizontal position beneath the surface, sensitive gravity meters will give a constant reading for the force of gravity all along the surface above the bed. But if the bed dips or rises, the gravity-meter readings reflect the changing structure. When the readings suggest the presence of anticlines, faults, or other structures in which oil might accumulate, test wells are drilled to determine whether reservoirs actually exist.

Special techniques have been developed for using seismic methods to locate potential traps under water-covered areas. Figure 14.8 illustrates one of these, in which energy sources are towed in water and signals are picked up on equipment towed in a streamer cable.

Figure 14.9 shows the areas in the world favorable to the exploration for oil and natural gas, both onshore and offshore. It also shows the producing areas. The coterminous United States has been very effectively explored, as have the Canadian provinces of Alberta and Saskatchewan. Probably half the oil and gas in these areas has been found and used, and all the giant fields have been discovered. The new petroleum frontiers for the North American continent lie in the arctic lands of Canada and Alaska and along the offshore margins of the continent.

How long will the supply of oil and natural gas last? That is a question we cannot answer with precision. We know that there is a finite amount of these liquid and gaseous hydrocarbons in the ground, and also know that they are not renewable resources like trees or rivers. We have several estimates of the supply duration, and each has roughly the same conclusion: namely, that early in the twenty-first century we shall run out of oil and natural gas. Two estimates for crude-oil production in the United States are given in Figure 14.10, one made in 1956 and the other in 1976. The 1956 estimate suggested that the United States would reach its peak production of crude oil between 1960 and 1970. By 1976 experience told us that the United States production of crude oil peaked in 1968. Furthermore it became apparent that by the year 2000 we shall have produced 80 percent of the crude oil available to us in the United States, exclusive of Alaska.

On a worldwide scale for both oil and gas the picture is similar to that in the United States although over the world there is a larger percentage of the original hydrocarbons yet to be found than is true for the extensively explored United States. In fact, when we consider the "big three" of our present sources of energy—coal, oil, and natural gas—we find that the world supply will be consumed over a very short time span in terms of human history (see Figure 14.11). So the fossil

FIGURE 14.8

Search for oil-bearing formations beneath the ocean floor, using marine vibrations. [Courtesy of Continental Oil Company and Westinghouse Air Brake Company, Drilling Equipment Division's *Core Driller*.]

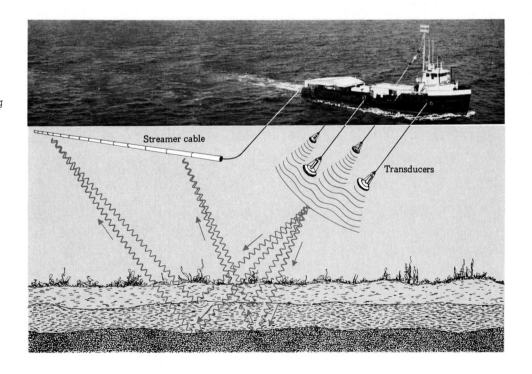

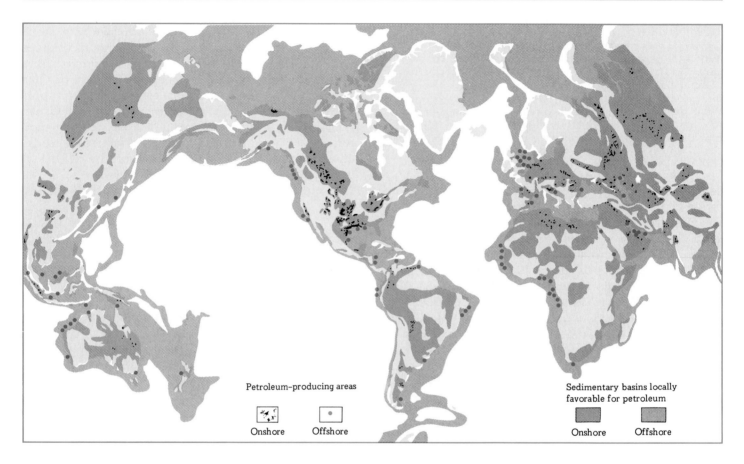

FIGURE 14.9

**Areas favorable to the accumulation of petroleum both on the continents and subsea.
Only small portions of the areas shown as favorable actually contain producible ac-
cumulations.** [After V. E. McKelvey and others, "World Subsea Mineral Resources,"
U.S. Geol. Surv. Misc. Geol. Invest. Map I-632, sheet 3, 1969.]

fuels as we currently know them will become exhausted. What is to replace them?

OIL SHALE AND TAR SANDS

Oil shale and tar sands are potential sources of additional fossil fuels. In fact the potential resources outstrip
the known and postulated supplies of natural gas.

Oil shale Most sedimentary rocks contain at least
some organic material, and one type, shale, often contains a great deal. Some shales indeed are so rich in
organic material that they give off petroleum if heated.
Thus far the production of petroleum from shale is at

best on a small scale, with some production in China,
Estonia, and other parts of the Soviet Union.

The richest known deposits of high-grade oil shale
are in the United States. During the Eocene Epoch
large freshwater lakes spread across portions of what
are now Colorado, Utah, and Wyoming (Figure 14.12).
In them accumulated great masses of organic-rich sediments today collectively known as the Green River oil
shales. Some shale is rich enough to produce nearly
240 l/t, and the potential reserves are large. Just how
large is not presently clear, but they certainly are equal
to all proved and predicted reserves of oil and natural
gas still left in the United States. For many years pilot
plants have been experimenting with the technology of

FIGURE 14.10

Crude-oil production histories and predictions for the United States and adjacent continental shelves (exclusive of Alaska) made in 1956 and 1976. Past production is shown in tint color, reserves in gray, and oil predicted to be discovered in white. [Earlier diagram from M. King Hubbert, "Nuclear Energy and the Fossil Fuels," Publication 95, p. 23, Shell Development Company, Houston, 1956; 1976 diagram updated from M. King Hubbert, "Energy Resources," in *Resources and Man*, W. H. Freeman and Company, San Francisco, 1969. Copyright National Academy of Sciences.]

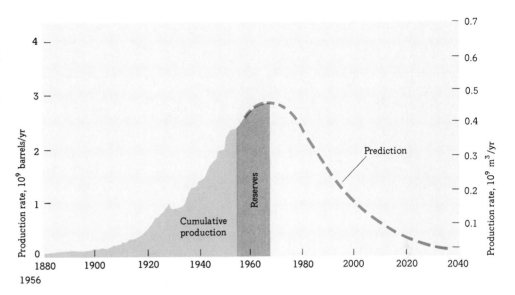

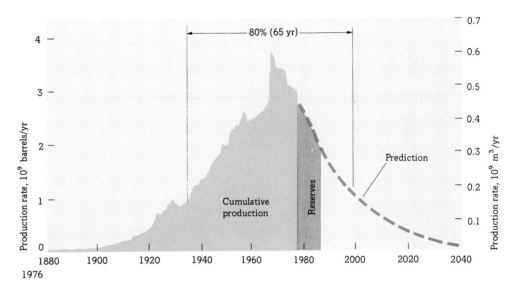

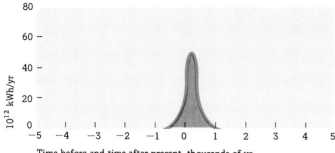

FIGURE 14.11

The period during which most of the fossil fuels will be consumed is short compared to human history and is an instant in terms of geologic time. This graph, showing the use of fossil fuels as a pimple-like bump on the scale of earth's history, has become known as "Hubbert's pimple" after M. King Hubbert, who originally published this graph in 1962. [From M. K. Hubbert, "Energy Resources: A Report to the Committee on Natural Resources," *Natl. Res. Council Pub.* 1000-D, National Academy of Sciences, Washington, 1962.]

processing the Green River shales. So far large-scale exploitation of the resource has stalled in part because the oil from processed shale is more expensive than the traditional oil-field petroleum and in part because of the environmental problems associated with extensive shale-oil operations. The problem of what to do with the shale left over after the oil extraction has yet to be satisfactorily solved. And the extraction process demands large amounts of water, already a scarce commodity in the western states.

In addition to the Green River shales the United States possesses vast amounts of additional organic-rich shales although their concentration of organic matter by no means compares with that in the Green River shales. Those shales close enough to the surface for mining are shown in Figure 14.12. Certainly, however, present technology is far from making these deposits useful to us as an energy source even though they may be eventually exploited.

Tar sands Earlier in this section we pointed out that hydrocarbons formed in the underground work their way upward and, if not contained by a trap, will

FIGURE 14.12

Organic-rich shales can be a source of oil and gas. The richest in the United States are found in the Green River Formation of Colorado, Utah, and Wyoming. Additional shales, accessible to mining and capable of yielding 40 l/t or more, are widespread in the United States. [Data from U.S. Geological Survey.]

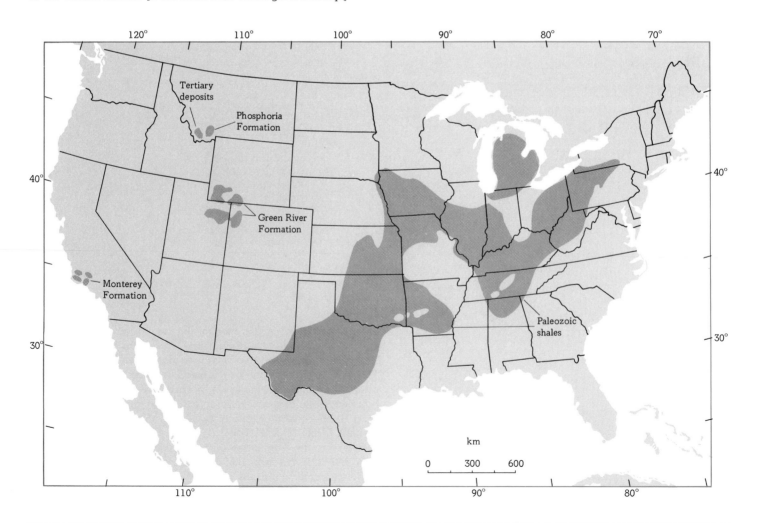

eventually reach the surface and there be lost. But in some situations certain hydrocarbons may be brought to, or close to, the surface and survive for a considerable time. These hydrocarbons are the dark, heavy, low-volatility, high-viscosity ones. We know them as **tar** or **asphalt.** And in some localities they are preserved as the dark, sticky cement of a sandstone. It is as if we were dealing with a reservoir bed from which all the flowing oil had been removed and in which only the heavy tar remained.

Three very large deposits of tar sands are known, of which the Canadian Athabascan tar sands are probably the largest (Figure 14.13). Estimates of recoverable oil from the Canadian deposits range between 250 and 500 billion barrels. Presently there is a small amount of production from the Athabascan tar sands at Fort MacMurray, where about 50,000 barrels of oil are extracted per day. Two other deposits, comparable in size to the Canadian sands, are now known. One is in Venezuela along the Orinoco River; the other, in the Soviet Union.

When these large deposits plus several small deposits are considered together, they appear to have the potential of yielding as much energy in the form of crude oil as the total energy represented by oil and natural gas originally available to us in the traditional petroleum fields of the world.

SOME CONCLUSIONS

Before leaving this discussion of oil and gas, we may find it worthwhile to draw some conclusions about the future of fossil fuels in general. As we have stated, we know that all fossil fuels are nonrenewable, and once they are used up, they cannot be replaced. We know that the traditional sources of oil and natural gas are very limited and will be essentially gone in a few decades. Coal, being more abundant, will extend the life of fossil fuels as an energy source, and so may oil shales and tar sands. But whatever we do, the prediction in Figure 14.11 will not be modified much if we continue to

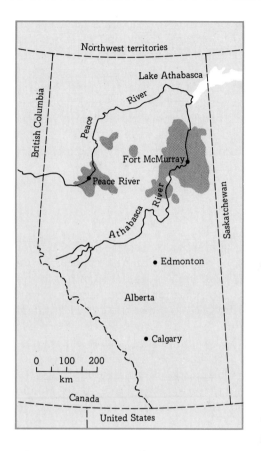

FIGURE 14.13

The areas of Alberta known to be underlain by tar sands. Some production now comes from deposits near Fort McMurray.

use fossil fuels at the ever-increasing rates that we do now. Somewhere 100 to 200 years down the line we shall have used up half the fossil fuels that were originally available to us.

What then are the alternatives to our dependence on the fossil fuels? We look at some of them in the following sections.

14.4 WATER POWER

Water power has been used as a source of energy for at least 2,000 years and perhaps longer. Water has the advantage of being a renewable resource since it is the basis of the hydrologic cycle: The sun pumps it by evaporation from the sea, and the atmosphere delivers it over the landmasses, where it falls as precipitation and flows off in rivers and streams back to the sea. Water power is clean and relatively cheap. It has the drawback that the dams generally necessary for the storage of water have limited lifetimes and usefulness because sediments silt up the lakes formed behind them. There is also the objection that damming floods out a great deal of prized scenery by drowning valleys and their associated rivers.

About 25 percent of the water-power capacity of the United States has already been developed. World-wide about 6 percent of the capacity has been developed. And it is estimated that, if the world's water power were fully developed, it would produce energy equal to about 30 percent of the total energy now consumed. Africa and South America are the continents with the greatest power potential; thus far very little, about 1 percent, has been developed there.

14.5 NUCLEAR POWER

The nucleus of an atom is, as we discussed in Chapter 2, made up of protons or protons and neutrons. It is the binding energy, holding these particles together in the nucleus, that forms the basis for nuclear power. It can be released either by breaking the nucleus of an atom into nuclei of lighter-weight elements, in a process known as **fission,** or by joining the nuclei of two elements together to form the nucleus of a heavier element, a process called **fusion.** Both of these processes have been achieved at the explosive level in the atomic bomb (fission) and the hydrogen bomb (fusion). We have been able to develop technology that controls fission so that the slow release of the energy can be converted to the generation of electric power. We cannot yet control the fusion process.

FISSION

People have been able to tap only a small fraction of the atom's potential energy. At present we are using only that available from the heavy element uranium. It is not the common isotope of **uranium,** with an atomic weight of **238,** that fissions under neutron bombardment but rather the lighter isotope, **uranium 235.** This is found in nature intermixed with ^{238}U in the proportion of about 1 part in 140. Because the two isotopes are chemically identical, the ^{235}U can be separated only by a physical process capitalizing on the difference in mass. By making uranium into a gaseous compound and then pumping it against plates pierced with billions of tiny holes less than 0.000001 cm in diameter, the process causes the lighter ^{235}U to move through the holes to concentrate a product more than 90 percent.

In nuclear fission a slow-moving neutron smashes into the nucleus of the ^{235}U atom. It splits the nucleus into two new atomic nuclei of less mass, plus several neutrons. Because the mass of the fission products is less than that of the original atom, some of the mass has been converted into energy. For nuclear fission to take place spontaneously, the piece of material must exceed a certain critical mass. That is, there must be a sufficient number of nuclei so that neutrons bombarding the material cannot fail to hit a nucleus and split it, producing more active neutrons. After the process is started, it thus proceeds by a chain reaction. This reaction, so explosive in the atomic bomb, is controllable and used in generating electrical energy. Scores of nuclear reactors are now operating throughout the world on a commercial basis.

The fuel used in the nuclear plant is the ^{235}U extracted from the more abundant ^{238}U. It is a nonrenewable resource, and so the question arises as to how much ^{238}U (and hence ^{235}U) is available and how inexpensively it can be obtained. The evidence points to a limited supply of fissionable uranium. If this is true, then we are facing a difficulty in respect to nuclear power similar to that we face with the fossil fuels.

There is, however, a particular situation that may help to solve the declining reserves of uranium. We can, in effect, replace ^{235}U with artificially produced fissionable atoms. The fairly abundant ^{238}U and ^{232}Th (thorium) not naturally fissionable, can be converted into fissionable elements. Thus ^{238}U may absorb a neutron in its nucleus and convert into ^{239}Pu (plutonium), which is fissionable and can sustain the chain reaction necessary to our purposes. In the same way neutrons will convert ^{232}Th into fissionable ^{236}U. We then need to use the neutrons of decaying ^{235}U to create radioactive plutonium or ^{236}U at a rate equal to, or faster than, the fissioning of ^{235}U. This creation of new radioactive isotopes is called **breeding,** and the reactors in which they are used are called **breeder reactors**. There are still technical problems to be solved before the process can become commercial, but solutions are hoped for.

In the United States our richest uranium deposits occur in sedimentary rocks in Utah, New Mexico, Wyoming, and Arizona. Lower-grade deposits are found with phosphate deposits, lignite beds, or shales.

In addition to the technological problems that continue to surround fission as a source of energy, the political and health-related factors concern many for this type of power generation.

FUSION

The fusion process relies upon the fusion of the nuclei of hydrogen to form helium, the same process that produces the energy in our sun and the other stars.

Shortly after the first hydrogen bomb was exploded in 1952, research began on the possibility of controlling the fusion process for peaceful uses. Most experiments now rely on **deuterium,** a heavy isotope of hydrogen (Section 2.1 and Figure 2.6), as a fuel. If controlled hydrogen fusion can be achieved (and there is little assurance that it can be), then fuel for the process presents no problem because there is an essentially limitless source of heavy hydrogen in the ocean. Even if the very large technological problems can be solved, fusion as a usable source of energy is not expected to be available before the first decade of the coming century.

14.6 GEOTHERMAL ENERGY

The volcanoes and earthquakes that plague sections of our earth are expressions of geothermal energy, as are the geysers and hot springs in such areas as Yellowstone National Park, Iceland, and New Zealand. It would be useful if this energy could be extensively harnessed. Our use of geothermal energy is already a reality on a small scale: For instance, most of the space heat for Reykjavik, the capital of Iceland, comes from hot springs on the island; the geyser basin north of San Francisco is generating electricity on a commercial scale; and the geothermal field of Larderello in central Italy has been producing power since before World War II.

The decay of radioactive elements within the earth continues to produce heat and to support a slow heat flow toward the surface (Section 3.4). In trying to harness this energy, we run into a number of problems, the first of which is that the heat flow in the earth is so diffuse that general exploitation is impractical. We look, therefore, for those areas where heat flow is unusually high. These areas are of two kinds: the plate boundaries and hot spots away from the boundaries of plates. Within such areas we can consider two possibilities in the use of geothermal power to generate electricity.

The first possibility is already being exploited where the source of energy is pools of geothermally produced hot water and steam. The energy comes from hot, igneous rocks in still active or quiescent igneous areas. The geothermal pool is not unlike an oil trap in some of its characteristics, but instead of hydrocarbons we are dealing with heated water. The permeable-rock units containing the water are sealed off, at least in part, from the surface by impermeable rocks. The water in the aquifer is heated by the igneous rocks below. Some of the heated water may leak to the surface in the form of hot springs or geysers (Figure 14.14). That which remains trapped may be tapped by drill holes and brought to the surface to produce electrical power. Commercial installations are of this type, and the bulk of them rely upon geothermal steam as the source of energy.

A second possibility can be considered but as yet is only in the experimental stage. It may be that there are hot igneous rocks at shallow depth, but the overlying rocks lack water usually because they are too impermeable. In this situation it may be possible to drill to some depth where the rocks are hot enough and then create an artificial permeability by subsurface explosions. The next step is to introduce water through the drill hole into the fractured zone where the water is to be heated. A recovery hole is drilled, and the heated water is brought to the surface to power electric generators. In such a system the water, after being used for power generation, is recycled into the ground, heated again, and once more used to produce electrical power.

The drawbacks to geothermal power include the fact that its use will be restricted to those areas where

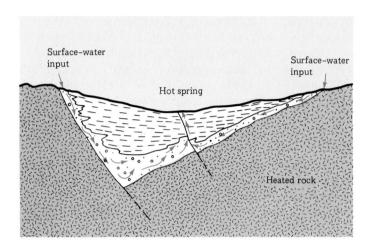

FIGURE 14.14

A geothermal system involves the heating of water by hot, igneous rocks at some depth below the surface. The system shown here involves a down-faulted basin in which sediments have accumulated. Surface water percolates downward as groundwater in the coarse sediments at the margins of the basin and is held in by finer, impermeable sediments in the center of the basin. Some heated water is shown as escaping back to the surface as a hot spring. Wells into the basin would be able to tap the heated water at depth and bring it to the surface as an energy source.

geothermal flow is particularly high. A major technological problem involves the mineral-forming elements carried in the hot water or steam. These can precipitate out in the generating installation, rapidly reduce effectiveness, and increase maintenance costs. Also, considerable amount of noise is generated in the process of using geothermal steam. And finally geothermal pools of steam are not a renewable resource, or at least we have so far been unable to develop a renewal process.

14.7 SOLAR ENERGY

We have already discussed the fossil fuels, which are in effect stored sunshine, and we have mentioned water power, which is the indirect effect of solar energy. In the United States we have seen a renewed interest in devices for directly collecting solar energy. So we can now buy solar hot-water heaters and construct houses that can be heated, at least in part, by sunlight. The big question is whether we can harness the sun's energy for large-scale power production.

Two techniques can be used for the generation of electricity. The first is the use of solar-thermal systems that employ mirrors or lenses to focus sunlight onto receivers. These in turn heat a fluid, which in its turn is used to drive a turbine and generate electricity. The second approach is the use of photosensitive material to convert some of the sunlight directly into electrical energy. The problems of both these approaches lie in the enormous area needed for the installation of the collectors and the cost not only of installation but also—perhaps more importantly—of maintenance of the collector units. Authorities are in disagreement as to whether the direct use of sunshine as an energy source will satisfy a major part of our needs in the future.

14.8 OTHER SOURCES OF ENERGY

We should not close this chapter without mentioning some other possibilities.

Wind continues to have a small place in the energy-supply picture. Windmills on land can supply small energy needs of individual homes or farms, and it is conceivable that we can occasionally return to sail for supplementary transportation over the oceans.

The **tides** of the world represent the gravitational pull of moon and sun, the moon exerting the greatest influence. Using the rise and fall of tides to generate electricity has often seemed attractive. The Rance River estuary in northern France is producing tidally generated electricity, as is a plant in the northern Soviet Union along the coast of the White Sea. The Passamaquoddy area on the Maine-Canadian border has long been cited as a possible place to install a tidal plant although thus far it has not proved economically or politically feasible.

Finally, the **thermal gradient** in the ocean has been suggested by some as a large-scale source of energy. A difference of about 20°C, for instance, is not uncommon between the cold bottom waters and the warm surface waters. Theoretically this difference in temperature should be convertible into power; but the development of this source of energy is still in the future.

OUTLINE

Fossil fuels (oil, gas, and coal) are presently the most important sources of energy in the industrialized world.

Coal is a sedimentary rock that represents ancient swamps and associated peat deposits. Coal increases in carbon content from **lignite** through **bituminous coal** to **anthracite**.

Oil and natural gas are today the primary sources of energy.
> The **origin** of oil and natural gas lies in the breakdown of organic matter. Oil and gas are associated with sedimentary rocks of marine origin and of Paleozoic age or younger.
>
> **Accumulation** of oil and natural gas in the underground demands a **source bed**, a permeable **reservoir rock**, an impermeable **capping bed**, and a **trap** to contain the hydrocarbons.
>
> The **search** for petroleum relies largely on indirect methods of geophysics and geology. In the coterminous United States we appear to have found and used over half of the hydrocarbons originally available. Worldwide the supplies of oil and natural gas will be close to exhaustion in the early part of the twenty-first century.
>
> **Oil shale** and **tar sands** are potential sources of large amounts of hydrocarbons and can extend somewhat the supply from conventional petroleum fields.

Water Power is renewable, clean, and relatively inexpensive. One drawback is that the dams needed to impound water have a limited useful life because of siltation.

Nuclear Power involves either the splitting of an atomic nucleus (**fission**) or the joining of two nuclei (**fusion**). Controlled fission is in commercial use. We do not yet know if controlled fusion is feasible. Artificial breeding of fissionable material (plutonium) could alleviate the supply problem for fission but does not answer questions of health, environment, or politics.

Geothermal energy can locally supply some energy needs by the use of steam and hot water generated at shallow depths by still-hot igneous rocks.

Solar energy: In addition to the stored solar energy in the form of fossil fuels and water power, sunlight may be used to heat water, which in turn may drive electrical generators. Another method uses photosensitive material to create electricity directly from sunlight.

Other sources of energy include the **wind** (another indirect effect of the sun), **tides**, and the **thermal gradient in the oceans**.

SUPPLEMENTARY READINGS

Landes, Kenneth K. *Petroleum Geology of the United States*, John Wiley & Sons, Inc., New York, 1970. *A state-by-state survey of the petroleum geology of the United States.*

Levorsen, A. I. *Geology of Petroleum*, 2nd ed., W. H. Freeman and Company Publishers, San Francisco, 1967. *A standard textbook on the subject.*

Skinner, Brian J.   *Earth Resources*, 2nd ed., Prentice-Hall, Inc., Englewood Cliffs, N.J., 1976. *Two of the chapters in this excellent and concise book focus on energy resources.*

Williamson, I. A.   *Coal Mining Geology*, Oxford University Press, Inc., New York, 1967. *Summarizes principles of coal accumulation and exploitation and provides a well-illustrated description of typical coal occurrences throughout the world.*

15

USEFUL MATERIALS

In the previous chapter we focused on sources of energy used in our modern technology. Here we look at the variety of minerals and rocks that we extract from the earth and with which we fabricate the vast number of things so characteristic of twentieth-century civilization. Roads and buildings, cameras and furnaces, automobiles and planes, belt buckles and pins—an almost unending list of things depends upon our ability to find and extract raw materials from the ground.

The processes by which nature has produced minerals useful to us are not unique; they are rather the same processes involved in all other geologic phenom-ena. For example, the sand shoveled from a sand deposit to mix with cement to form concrete was laid down by some agent such as a stream or the seawater sloshing on a beach. The gold of a brooch may have concentrated in veins of quartz during the final stages of igneous activity. The ore that provided the aluminum for your soft-drink can may have been mined from a deposit formed by long-continued weathering that produced a soil-related deposit high in bauxite, an oxide of aluminum. What is unique about these deposits is that we find them of value.

15.1 MINERAL DEPOSITS

Mineral deposits are generally classed as **metallic** or **nonmetallic.** Metallic deposits are mined for such elements as gold, silver, iron, copper, aluminum, lead, and zinc. Nonmetallic, or earthy, deposits include gems, salt, coal, limestone, phosphate, and even sand, gravel, and rock. The quality of mineral deposits varies from mine to mine and even within a given mine. Mineral deposits very rarely come out of the ground ready to be used; the materials must be processed after mining.

MINING

Mining involves the removal of materials from the earth. It is done in several ways, depending on the size and type of deposit. **Quarrying** is generally limited to surface removal of rock. This can be cut into dimension stone or blasted and crushed. **Placer mining** is associated with the separation of valuable metals from sand and gravel deposits (see Figure 15.1). The sluice box is

FIGURE 15.1

Panning for gold on the Anderson River, a tributary to the Fraser River in British Columbia's Sierra Cascade Mountains. The prospector partially fills the pan with water and throws in a shovelful of dirt. He picks out the pebbles and stirs the mass until clay-sized particles are dislodged and can be sloughed away in the muddied water. He partially fills the pan with water again and gives it a slightly eccentric circular motion to build up a wave that slops over the edge each time, carrying with it a little sand. He continues this process until only the specks of gold, which have greater specific gravity, remain in the pan. [Elliott A. Riggs.]

used in rich deposits. Alternatively, large quantities of sand and gravel may be worked by machines to obtain small quantities of a valuable metal. **Open-pit,** or **strip, mining** is confined to huge deposits fairly close to the surface. The operation first requires the removal of worthless overburden. Twenty-three percent of United States bituminous coal is strip mined. **Underground mining** involves thick tabular bodies of mineral deposits. Mines that were once shallow are now being worked at greater and greater depths. The deeper the mine, the more expensive the operation. However, with the development of better equipment, new techniques, better transportation, or higher prices, some abandoned mines have been reopened.

For many centuries only a few metals and precious gems were being mined. The mining depended mainly on men's muscles, and laborers toiled under extremely dangerous and harsh conditions, using only primitive tools. Despite this, mining was widespread. With a growing population and demand for a wider variety of basic minerals and metals, mining was stimulated. The greatest surge came with the advent of the industrial age. New materials were required in great quantities, and large markets were created. Miners were supplied with better tools and power to work those minerals which occur in concentrations that can be worked at a profit, that is, **ore deposits.**

Separation of valuable metal Although a metal is

more concentrated by nature in an ore deposit than in another place, it still has to be separated from the ore. This may be done in different ways.

The simplest are used in extracting metals found in an uncombined state. For example, gold-bearing quartz is pulverized, and the pulverized material is then treated with mercury, in which gold dissolves. This solution is then easily separated from the quartz because of its higher specific gravity. Heating and distilling of the mercury leaves a residue of gold. The ore after this first process may be treated again with a cyanide solution, and gold may be recovered from the solution by electrolysis or treatment with zinc.

Some metals are combined with other elements, such as sulfur, oxygen, or carbon, in the crystalline structure of a mineral. This mineral may be mixed with unwanted minerals, rock, or earthy materials called **gangue,** from which it must be separated. For example, the most common ore minerals of copper are sulfides. These are found disseminated through rocks. Concentration from this gangue begins by a process called **flotation.** The rock is finely ground and treated with a mixture of water and special oil. The water wets the silicate minerals, and the oil wets the sulfide minerals. When air is blown through the mixture, it creates a froth of the oil and sulfide minerals, which is skimmed off. The concentrated sulfide is then roasted in a furnace with air passing through. The air removes some of the sulfur as sulfur dioxide. However, some unwanted minerals such as FeO and SiO$_2$ are still left. Limestone is then mixed and heated with the roasted ore to serve as a **flux.** The iron oxide and silica combine with the limestone and form a slag, leaving the melted metal, which is drawn off.

GEOLOGIC PROCESSES INVOLVED

A convenient way of thinking about mineral deposits is to consider the **geologic processes** that have been responsible for them, as we outline below.

Igneous activity You will remember that in Chapter 3 we discussed the formation of igneous rocks from a mixture of elements in a solution called a magma. Some magmas, however, also contain elements that, because of the size of their ions, do not readily combine with the common rock-forming minerals. Sometimes these elements crystallize early in the cooling of the magma and settle out of the solution. Sometimes they form late and are trapped in the crystallized magma.

Hydrothermal activity Igneous activity often gives rise to the circulation of heated fluids that migrate beyond the immediate area of igneous activity. These solutions, some of which may carry metallic ions, may move through surrounding country rock. There, as they cool in a new chemical and physical environment, mineral deposition may result.

Weathering The homely process of weathering has been responsible for the creation of some of our most important mineral deposits. The process operates in one of two ways: It may leach out undesirable elements from the weathering earth material and leave the desirable elements concentrated in the upper soil zones; or it may move the desirable elements in low concentration near the surface to deeper zones, where they are redeposited in an enriched zone.

Placer deposits Some minerals are heavy enough and resistant enough to abrasion that they can be concentrated in selected locales in stream-bed deposits. Being heavier than the usual sand and gravel of a stream, the minerals may collect in pockets, many of which prove rich enough to mine. Such deposits are called **placers.**

Rock formation In many instances the mineral deposits comprise the whole rock and not just a small enriched fraction of it. The processes that create, for example, limestone, or marble, or basalt are creating an often-useful material, the rock itself.

15.2 METALS

When we think of mining, we think of the search for the metallic elements—from gold to lead. Here we examine some of the metallic deposits.

GOLD

Gold is a rare element used principally in coinage and jewelry. It usually occurs in the uncombined state (Figure 15.2) widely distributed in small amounts throughout other materials. It is found most commonly in hydrothermal deposits in sialic igneous rocks, particularly those that are rich in quartz (see Figure 15.3). It is also found concentrated in placer deposits. The gold, freed by weathering from the igneous rock in which it had formed, was later transported by streams to be deposited in favorable locations.

The gold discovered on the western slopes of the Sierra Nevada in California in 1848 was concentrated in placers so rich that fortunes were made simply by panning it by hand. Deposits of lower grade are uncovered by modern hydraulic giants that wash away the barren material that overlies pay dirt and sluice the gold-bearing gravels into mercury-lined boxes, where the gold is trapped. Deposits may also be worked by dredges where the gravel is below the water level of ponds. Some deposits contain only a few cents' worth in each cubic meter, but even these have been dredged at a profit.

The world's greatest gold deposits are in the Witwatersrand District of South Africa, on a plateau standing 1,800 m above sea level about 1,300 km northeast of Cape Town. (*Witwatersrand* means "white water divide," so called because of a prominent white quartzite that resists erosion and stands forth as a ridge, or divide.) The deposits in this rich area occur in conglomerates, themselves formed from ancient placer deposits, according to some geologists. Others think that permeable channels in the original rock were invaded by gold-bearing hydrothermal solutions. Approxi-

FIGURE 15.2

Native gold. A unique hornlike mass, about 8 cm long, from California. [Harvard Mineralogical Collection, Harry Groom.]

FIGURE 15.3

Working a gold-bearing quartz vein a little over 1 m wide in an underground mine at Bralorne, British Columbia.

mately 40 percent of the world's gold is produced by the Republic of South Africa.

Since 1879 the Homestake mine in the Black Hills at Lead, South Dakota, has been the leading United States gold producer. Its shafts now extend 1,600 m below the surface.

SILVER

Silver is widely distributed in small amounts in ore deposits of other metals. Native silver is principally found in the oxidized zone of ore deposits. The principal mineral containing silver is argentite, Ag_2S. Argentite is found in veins where it has been deposited by hydrothermal solutions. It may also be found in deposits concentrated through weathering processes. Significant silver production comes from ores mined primarily for lead, copper, and zinc.

Since 1885 over $1 billion of silver, lead, and zinc have been mined in the area around Coeur d'Alene, Idaho. This district is first in silver, second in lead, and fourth in zinc production in the United States. The Sunshine mine is the second largest individual producer of silver in the world; Canada as a whole is the world's largest producer, followed closely by Peru, the Soviet Union, Mexico, and of course the United States.

IRON

Iron manufacture in the United States began on the eastern seaboard. Bogs and swamps surrounding the coastal settlements of the New World contained iron that had been leached from the rocks by streams. The water carried the iron down into lowlands, where it accumulated in the marshes. Through natural chemical change this formed an iron ore of commercial value for that era. It could be extracted by baking out the moisture. Within 1 year after the founding of the colony at Jamestown, Virginia, in 1607, the settlers had sent back to England enough of this crude bog ore to produce 17 t of iron. In 1644, at Saugus, Massachusetts, the first usable iron product was manufactured. It was a 1.5-kg cast-iron cooking pot.

The iron and steel industry has grown tremendously since then. It is the basis of all our other great industries. It depends on many natural resources, of which iron is the most important. This is supplemented by other elements, which are used as alloys to harden or soften steel. Fuels such as coal and oil are required in great quantities to fire the furnaces that purify the iron and also cook the steel. Limestone is quarried as a flux for purifying iron in the refining process. When heated, it removes the impurities sulfur and phosphorus by absorbing them.

With industrial growth the bog deposits were soon exhausted, and 80 years ago the increased need of the iron and steel industry for new sources of iron ore focused interest on the Lake Superior region. Here Precambrian sedimentary rocks rich in iron-bearing minerals provided most of the ore for the major steel mills of the country through the first half of the century. The ore originally mined was very rich in iron, 50 to 60 percent. Weathering had converted the lower-grade iron formations to a residuum much richer in iron than the original iron formation. In effect the mines exploited old soil zones. With the exhaustion of the high-grade ores the Lake Superior district has been using a material called **taconite**, which contains chert, hematite, magnetite, Fe_3O_4, siderite, $FeCO_3$, and hydrous iron silicates (see Figure 15.4).

The most recently opened district in North America is on the Labrador-Quebec boundary. Here again iron formations of Precambrian age have been enriched by weathering, as in the case of the Lake Superior district. Other deposits are being worked in sedimentary formations of Silurian age, known as the Clinton beds, which crop out across Wisconsin and New York and along the southern Appalachians. These beds are being mined extensively in the Birmingham district of Alabama (Figure 15.5). The primary unleached ores from the Clinton beds are often high in $CaCO_3$ and contain 35 to 40 percent iron. But after the $CaCO_3$ has been leached out by weathering, they may contain as much as 50 percent iron.

FIGURE 15.4

Mining taconite in an open pit at Babbitt, Minnesota. The ore, containing 25 percent iron, is one of the hardest rocks in the world. The operation is like that of a rock quarry that produces crushed stone. [Hercules Powder Company, courtesy of Reserve Mining Company.]

FIGURE 15.5

More than 750 m beneath the surface of the Grace mine in eastern Pennsylvania a miner operates a load-haul-dump vehicle to transport iron ore. Predominantly magnetite, the ore is the result of contact metamorphism caused by a gabbroic intrusion into limestone. [Courtesy of Bethlehem Steel Corporation.]

FIGURE 15.6

Copper mine at Bingham Canyon, Utah. The benches are 15 to 20 m high and not less than 20 m wide. [Rotkin, P.F.I.]

COPPER

Copper occurs both uncombined and in combination with other elements in minerals such as *chalcopyrite*, $CuFeS_2$, *bornite*, Cu_5FeS_4, *chalcocite*, Cu_2S, and *enargite*, Cu_3AsS_4. Most copper deposits consist of concentrations created by hydrothermal solutions. Chalcocite deposits, however, are usually the result of secondary enrichment. In some regions igneous activity has built up deposits of copper but not in concentrations adequate for working. Groundwater has operated on some of these, dissolved the copper, and carried it down to be deposited in an enriched zone.

At Bingham Canyon, Utah (see Figure 15.6), is a spectacular open-pit mining operation that recovers at a profit an ore containing as little as 0.4 percent of copper and averaging about 1 percent. The ore is a sialic porphyry that contains finely disseminated sulfides concentrated by secondary enrichment. The benches of the mine are from 15 to 20 m high and not less than 20 m wide. The operation covers 3.5 km² and contains about 250 km of standard-gauge railroad track, most of which is moved continually to meet operating needs. In 1952 this mine was producing more than 250 million kg of copper per year, about 30 percent of all the copper mined in the United States.

Copper occurs in the pure metallic state on the Keweenaw Peninsula of northern Michigan—one of the more remarkable ore deposits in the United States. The peninsula extends 160 km into Lake Superior. It consists of beds of ancient lava flows, conglomerates, and sandstones that have been folded and glaciated. The whole series dips toward the north. The copper is found in veins intersecting the igneous rock and as a cementing material in the conglomerates. This region was the world's greatest producer of copper before Butte, Montana, was developed. At Butte the total length of deep underground workings has been reported as 1,600 km. Low-grade ore has already produced over 9 billion kg of copper, silver, lead, zinc, and gold.

At Morenci, Arizona, an impoverished zone is as deep as 65 m. Beneath this an enriched zone extends about 300 m farther down. Underlying the enriched zone is unaltered bedrock, often too low-grade to mine.

URANIUM

In Section 14.5 we discussed the use of radioactive elements in the production of fission energy. Not until it was realized that uranium is a source of energy usable for military and nonmilitary purposes did the search for uranium-bearing minerals become extensive.

The highest grade is *uraninite*, UO_2, a constituent of granitic rocks and pegmatites and also a placer mineral. The principal deposits found to date are in Zaïre, in Great Bear Lake in Canada, and in Czechoslovakia. In the United States the greatest supply of uraninite so far discovered is at the Mi Vida mine, which is located 65 km southeast of Moab, Utah. This is in the center of the Colorado Plateau. Workable deposits have also been found on the Colorado Plateau in Arizona, Colorado, New Mexico, and other sections of Utah.

Another mineral that is a source for uranium is *carnotite*, a soft, yellow weathering product. It comes principally disseminated in cross-bedded sandstones found in southwest Colorado and adjoining districts of Utah, New Mexico, and Arizona as well as in certain other sections of Arizona.

Uranium prospectors use a Geiger counter, an instrument that makes an audible click every time it is hit by a particle released by the spontaneous decay of uranium. Because this is a simple, inexpensive device, much of the prospecting for this mineral has been accomplished by laymen who possess a little spare cash and time.

ALUMINUM

Aluminum, although the third most abundant element in the earth's crust, is limited in its availability to the industrial world. Aluminum constitutes an integral part of the feldspar minerals as well as many of the iron- and magnesium-bearing aluminum silicates. But in the silicate form aluminum is economically impractical to extract.

The ores of aluminum, therefore, are now the weathering residue formed from rocks rich in aluminum silicates. The soil that is left is an aluminum oxide, **bauxite**. It forms as a product of chemical weathering in tropical climates; and indeed, although it was originally described in more temperate latitudes, the most extensive deposits now known are to be found in the tropics.

OTHER METALS

The leading producer of *lead* and *zinc* is the tristate district of Missouri, Kansas, and Oklahoma. The ore deposits are low-grade. Galena, PbS, is the ore of lead, and sphalerite, ZnS, of zinc. These are often found together, in irregular veins and pockets in limestone and chert. Lead and zinc valued at over \$1 billion have been mined from a 5,200 km² area. In southeastern Missouri galena disseminated through flat-lying dolomitic limestone beds has made the world's greatest deposits of lead.

The world's largest deposit of *molybdenum* is in the Rocky Mountains at Climax, Colorado. Owing largely to this deposit, the United States produces about 90 percent of the world's output. Molybdenum is derived from the mineral molybdenite, MoS_2, which forms as an accessory mineral in certain granites and pegmatites. At Climax it occurs in quartz veinlets in granite.

Deposits of *nickel*, *platinum*, and *chromium* occur in formations of simatic rock all around the world. At Sudbury, Ontario, for example, there are valuable deposits of nickel in rock of this sort. Apparently these minerals somehow became concentrated in the cooling magma; and since they were heavier than the rock-forming minerals, they settled out during crystallization. Although the nickel and chromium that make up the deposits have combined with other elements to form compounds, the platinum occurs in an uncombined state.

The chief source of chromium is *chromite*, $FeCr_2O_4$. Most of the chromium produced in the United States comes from Shasta County, California, but much greater quantities are imported from Oceania, Southern Rhodesia, and Turkey. Chromium is used chiefly to form an alloy with steel that has extreme hardness, toughness, and resistance to chemical attack. It is also used for plating hardware, plumbing fixtures, and automobile accessories.

Although nickel is a relatively rare element in the earth's crust, it is extremely important in modern industry. It is used in the manufacture of a strong, tough alloy known as *nickel steel* (2.5 to 3.5 percent nickel) and in the preparation of *Monel Metal* (68 percent nickel) and *Nichrome* (35 to 85 percent nickel). It is also used in various plating processes, and it forms 25 percent of the United States 5-cent coin. Finally, its low expansion makes it an ideal metal for watch springs and other delicate instruments.

An important ore of nickel is pentlandite,

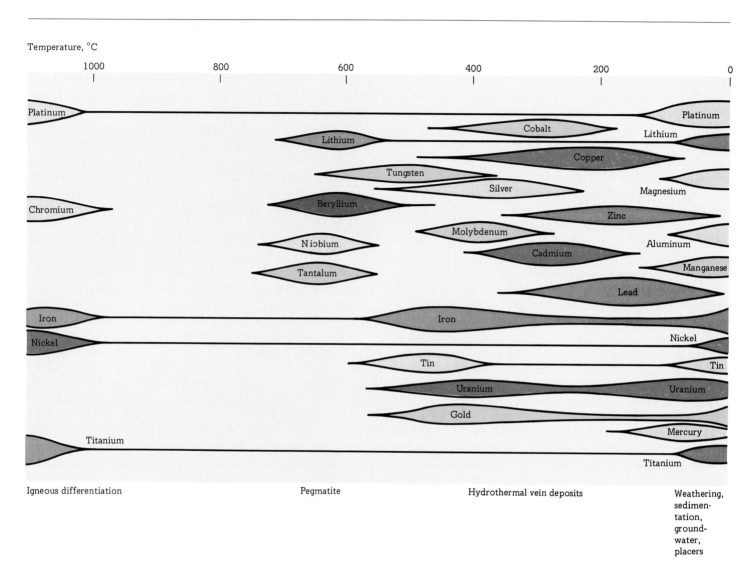

Temperature, °C

FIGURE 15.7

Most metals are associated with a particular temperature range within the overall temperature pattern of ore deposits. Igneous differentiation is the process of fractional crystallization discussed in Chapter 3. Pegmatites, also discussed in Chapter 3, are coarse-grained igneous rocks, some of which may contain unusual minerals. [Modified from Sheldon Judson, Kenneth S. Deffeyes, and Robert B. Hargraves, *Physical Geology*, p. 424, Prentice-Hall, Inc., Englewood Cliffs, N.J., 1976.]

$(Fe,Ni)_9S_8$. The Sudbury deposits of this mineral and deposits elsewhere make it the world's most valuable source of nickel.

The value of platinum in industry results from its high melting point, 1755°C, and its resistance to chemical attack. These properties make it especially useful in laboratory equipment such as crucibles, dishes, and spoons and for the contact points of bells, magnetos, and induction coils. Platinum also finds special uses in the manufacture of jewelry, in dentistry, and in photography.

PROSPECTING FOR ORE DEPOSITS

A search for ore deposits is always guided by a hypothesis of some kind. This may be based on dreams or phases of the moon. It may be an assumption that, because a certain species of spoofberry bush once grew on the surface above a copper mine, anybody who finds a spoofberry bush can expect to find a copper deposit below it. Or it may be based on the latest ideas in geophysics. But the hypothesis is there.

On the other hand, luck has often played a large role in even the most carefully planned prospecting. It did so in Nevada, where a painstaking program laid out to pan for gold was rendered ineffective by a soft, sticky, bluish mineral. When somebody recovered from the disappointment of not finding gold and had the bluish "clay" assayed, it turned out to be an important silver ore, argentite. This discovery led to the development of the fabulous Comstock lode. Many mining camps have a story of a prospector who threw a hammer at a fox and had it bounce off an outcrop of rock. When he went to retrieve the hammer, he found ore in the outcrop. Another popular account of accidental finding tells of a prospector's tracking down his stray mule and finding the mule grazing near an outcrop of ore. And rich deposits of gold and copper were discovered in rock cuts made for westward-driving railroad lines in the early days.

Known occurrences are the best guide to conditions under which to look for a particular ore. Even the most highly instrumented methods of modern geophysics or geology are best adapted to extending old ore-deposit outlines or looking for new deposits under geological conditions similar to those where old ones were formed.

Geophysical methods of prospecting for ore deposits are indirect, for the most part: They determine geological structures *favorable* for the presence of desired minerals. With the use of magnetic measurements, however, the mineral magnetite can be located *directly* because of its magnetism. But even this method is more often used indirectly, to locate minerals that commonly have magnetite associated with them; iron ores, such as hematite and limonite, though themselves nonmagnetic, have enough magnetite associated with them to permit detection and mapping. Magnetic measurements aided in finding gold-bearing beds in the Witwatersrand. The gold is scattered through steeply dipping beds of Precambrian quartz conglomerate that extend for 160 km from east to west. Parallel to these conglomerates are some shales rich in magnetite. When the Witwatersrand ran into barren ground some years ago, a geologist and geophysicist worked together to see whether the gold-bearing zone could be found again. They measured the vertical intensity of the magnetic field over extensive areas, tracing what they believed to be the magnetite-bearing shales associated with the gold-bearing conglomerates. From their measurements they were able to map the area containing the magnetite. Drilling confirmed their location of the magnetite and then of the gold-bearing formation. These were both under a cover of 600 m.

Seismic methods have been used successfully in mapping bedrock under unconsolidated deposits. This method has paid off in some areas where low places in the bedrock were sought as possible locations for placer accumulations of heavy minerals. But seismic methods have not worked well in most ore-finding problems because they cannot distinguish one mineral from another.

We have noted that there is a wide range of environments in which minerals containing useful metals are concentrated. Indeed knowing something about the expected occurrence of economic minerals serves as a help in prospecting. Figure 15.7 relates temperature and environments of concentration for several metals.

15.3 NONMETALS

DIAMONDS, ASBESTOS, SALT, AND SULFUR

Diamonds are the most valuable of the gemstones. In general the most valuable diamonds are flawless stones that are colorless or have a "blue-white" color. A faint straw-yellow color detracts much from the value. In recent years diamonds have been put to industrial uses. Fragments of diamond crystals are used to cut glass. Fine diamond powder is used in grinding and polishing diamonds and other gemstones. Wheels are impregnated with diamond powder for cutting rocks and other hard materials. Steel bits set with the cryptocrystalline variety, carbonado, make diamond drills used in geological exploratory work. The diamond is also used in wire drawing and in tools for truing grinding wheels.

Diamonds are most commonly found in placer deposits in the sands and gravels of stream beds where they have been transported after they were freed from their original environment by weathering. They have been preserved by their chemical inertness.

Ninety-five percent of the world's output of diamonds comes from the African continent. The first African diamonds were found in the gravels of the Vaal River, South Africa, in 1867. Four years later, diamonds were found in soils weathered from peridotite near the present town of Kimberley, south of the Vaal River. Most mines in the Kimberley region were originally open pits, working surface deposits of the altered peridotite in ancient volcanic necks. Later, underground methods were used in the same mine. The Kimberley mine (Figure 15.8) was developed to a depth of over 1,000 m before it was abandoned. Zaïre furnishes over 50 percent of the world's supply of diamonds from placer deposits. Diamonds also come from India and Brazil, where they are produced from gravel deposits.

Materials made from *asbestos* are extremely versatile. They can withstand fire, insulate against heat and sound, are light in weight, and can be made into pliable fabrics that resist soil, corrosion, and vermin. The most common asbestos mineral is *chrysotile*, which is an alteration product of some magnesium silicates, especially olivine, pyroxene, and amphibole. Chrysotile deposits are found in both igneous and metamorphic rocks, sometimes in such concentration as to make up the entire rock mass. Chrysotile forms soft, silky flexible fibers (see Figure 15.9).The long fibers are woven into yarn for use in brake linings and heat-resistant tapes and cloth. The United States, which is the largest user of asbestos, imports 90 percent of its needs from Quebec, Canada. In the United States chrysotile can be found in parts of Vermont, New York, New Jersey, and Arizona.

Salt, NaCl, is essential to life, and fortunately it is one of the most abundant substances in the world. It is produced commercially from brine, salt beds, and salt domes. Salt beds were formed by the natural evaporation of water from enclosed saltwater bodies. Subsequently they were covered by sediments and sometimes buried to great depths. Salt beds vary in thickness from 1 m to over 30 m and are mined at various depths.

In Louisiana and Texas salt is produced from salt domes. These are great masses of salt that appear to have pushed their way up through overlying sedimentary strata while in a plastic state. More than 100 salt domes have been located. Details of shape vary somewhat from one dome to another, but they all tend toward a cylindrical shape with a top diameter of 2 km. Some have pushed up to within a few hundred meters of the surface, but others stopped several thousand meters down. Reservoirs of oil and gas are often associated with salt domes (see Figures 14.5 and 15.10).

Salt, limestone, coal, and oil are four of the five basic raw materials of the chemical industry. The fifth is *sulfur*. The largest known reserves are in sedimentary deposits. Among the most useful and familiar compounds of sulfur are hydrogen sulfide, H_2S, sulfur dioxide, SO_2, sulfuric acid, H_2SO_4, and gypsum plaster, $CaSO_4$. About 87 percent of the sulfur produced goes into the making of sulfuric acid, which is the most used intermediate compound in chemical processing.

ROCK DEPOSITS

Many rocks are valuable in their natural condition and are usable without having to undergo changes. Stone has been used as a building material for several thousand years and is still being so used. Dimension-stone quarries cut blocks of granite, slabs of marble, and various sizes and shapes of limestone.

Rock has grown in importance during the past half century as its uses have expanded. In our modern world every mode of transportation depends in some degree on crushed rock. It provides the base for thou-

FIGURE 15.9

Chrysotile asbestos, a fibrous variety of serpentine from Thetford, Quebec. [Benjamin M. Shaub.]

FIGURE 15.8

Abandoned pit of Kimberley diamond mine in the rock of a volcanic neck near Kimberley, South Africa. Water now stands in the pit to within 200 m of ground level. Before it was abandoned, the mine was developed to a depth of 1,200 m. [Cornelius S. Hurlbut, Jr.]

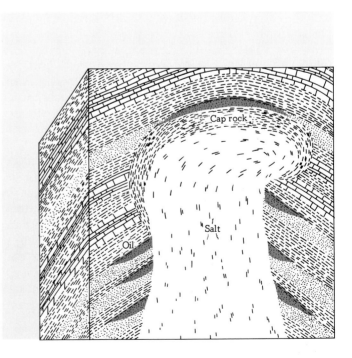

FIGURE 15.10

Schematic diagram of a salt dome.

FIGURE 15.11

Modern methods break over 200,000 t of rock from a quarry face with 25,000 kg (kilograms) of explosive. [Courtesy of New York Trap Rock Corp.]

sands of kilometers of modern highways, ballast for railways, bases for landing fields, and jetty stone for harbor facilities. With increasing demands techniques have been developed for removing it from the ground by blasting and crushing it into practical sizes (see Figure 15.11).

Some rocks have commercial value because of their chemical properties. **Limestone,** for example, is used as a flux in purifying metals. It is also used to neutralize acids in the processing of sugar, to correct the acidity of soil, and to supply calcium to plants. Limestone that contains limited amounts of impurities serves as the raw material in the manufacture of cement; the impurities give cement its characteristic hardness. The type known as Portland cement consists of 75 percent calcium carbonate (limestone), 13 percent silica, and 5 percent aluminum oxide, along with the silica and alumina that are normally present in clays or shales. Some manufacturers add the right percentages of impurities to the limestone; others use deposits called *cement rock,* in which the impurities occur naturally. **Phosphate rock** is a popular term used for sedimentary rocks that contain high percentages of phosphate, usually in the form of the mineral *apatite,* calcium fluophosphate. Phosphate rock is derived from accumulations of animal remains and chemical precipitation from seawater. It is extremely important as a source of agricultural fertilizer. The Rocky Mountain states have high-grade phosphate deposits with reserves estimated at 6 billion t, enough to last for many centuries. Reserves in Idaho run close to 5 billion t. Other deposits are being mined in Florida and Tennessee.

OUTLINE

Mineral deposits are generally classed as **metallic** or **nonmetallic**.

> **Mining** involves the removal of materials from the earth. **Ore deposits** are mineral deposits that can be worked at a profit.
>
> **Geologic processes** responsible for the formation of ore deposits include igneous activity, hydrothermal activity, weathering, sedimentary processes, and rock formation.

Metals that we mine include gold, silver, iron, copper, uranium, aluminum, lead, zinc, molybdenum, nickel, chromium, and platinum.

> **Prospecting for ore deposits** is best guided by known deposits.

Nonmetals of importance to us include diamonds, asbestos, salt, sulfur, limestone, and phosphate.

SUPPLEMENTARY
READINGS

Brobst, D. A., and W. P. Pratt (eds.)   "United States Mineral Resources," *U.S. Geol. Surv. Prof. Paper 820*, 1973. *An authoritative compendium of what we know we have and what we hope we have in the way of mineral resources.*

Mineral Resources and the Environment, Report by the Committee on Mineral Resources and the Environment, National Academy of Sciences–National Research Council, Washington, 1975. *A thoughtful, informative, and readable discussion of the subject.*

Douglas, R. J. W. (ed.)   *Geology and Economic Minerals of Canada*, Economic Geology Report 1, Geological Survey of Canada, Ottawa, 1970. *A good place to start for the Canadian picture.*

Skinner, Brian J.   "A Second Iron Age Ahead?" *Am. Sci.*, vol. 64, pp. 258–269, 1976. *A well-documented argument that the distribution of chemical elements in the crust sets natural limits to our supply of metals and that this may be more important than the limits on energy.*

Skinner, Brian J.   *Earth Resources*, 2nd ed., Prentice-Hall, Inc., Englewood Cliffs, N.J., 1976. *A very good summary of the subject particularly Chapters 1, 2, and 4 to 8.*

Minerals Yearbook Volumes I and II: Metals, Minerals, and Fuels, U.S. Bureau of Mines, Washington (issued annually). *Here are the details on the current United States situation.*

16

OUR EFFECT ON THE ENVIRONMENT

Throughout much of this book we have been concerned with the several geologic processes that have gone on through geologic time and continue into the present. In the last two chapters, however, we examined the origin and distribution of energy resources and economically useful earth materials, both of which are increasingly essential to contemporary society. In this chapter we focus more directly on the intimate relations between people and their physical environment.

In thinking about the geological processes, we are impressed with the fact that, although their results are large, their rates are generally slow. We have thus adopted the nineteenth-century view that small changes over long periods of time have produced grand effects. In thinking thus, we naturally look at ourselves as ineffectual agents in changing the earth. In one way this is correct; in another way it is far from correct.

We are disposed to consider ourselves a part of nature and to view our actions as restricted, at least in part, by the physical environment. This is obviously true, and yet, as long as we have been a numerous species on the earth, we have also influenced the other elements in our environment. For example, as we have moved from a hunting to an agricultural society, we have changed the vegetative cover of over a third of the earth's land surface and, in so doing, have modified the nature of the soil beneath. A century ago the marriage of technology with science triggered a new series of changes. These changes include, among other things, the flow and quality of both underground and surface water and the quality of the atmosphere; they also affect large portions of the oceans. Their magnitude creates new problems, forcing us to adapt to an environment that we in part created but still cannot seem to control.

What follows is a discussion of some of the things that we have been able to do to our environment. We understand the causes and some of the effects of these changes although we have yet to find effective social answers to their control. In addition we shall look at some catastrophic natural phenomena, such as earthquakes and volcanoes, that we not only cannot control but cannot yet predict.

16.1 RUNNING WATER

In Chapter 10 we considered how water is carried in streams and rivers acting to fashion the landscape. Some of the principles discussed in that chapter apply to our examination of the ways in which we can affect streams and their channels.

CHANGES IN STREAM FLOW

The hydrograph The hydrograph of a river shows the variation of stream flow with time, as indicated in Figure 16.1. In the example given we see a period of high flow, peaking in a flood, and a generally low-water stage, which represents that portion of stream flow attributable to groundwater recharge, a flow referred to as **base flow**. In addition the figure indicates a period of rainfall responsible for the high-water, or flood, stage, which follows by an interval of time called the **lag time**. The shape of the hydrograph for different streams (and even for different places on the same stream) varies with a number of natural factors, including the infiltration rate, the relief, the geology, and the vegetative cover. We are interested here in the way human activity affects this flow in ways both planned and unplanned.

Flood control Flood control depends upon modifying the hydrograph by lowering the flood peak, increasing the lag time between the precipitation and flood crest, and spreading the flood flow over a longer period of time. A flood-control dam, then, is designed to store water during periods of high runoff and to let the excess water out slowly to downstream areas. The result is to reduce the flood crest downstream and spread the discharge of flood waters over a longer time interval, as suggested in Figure 16.2.

Urbanization and suburbanization Building of cities and their suburbs affects stream flow and changes the hydrograph although in the opposite direction from that of flood-control dams. Agglomerations of buildings with their associated roads, streets, sidewalks, and paved parking areas achieve the following effects: (1) The amount of water sinking into the underground is curtailed in proportion to the amount of area sealed off by surface veneers of buildings and pavements, and even the areas of ground left open tend to be less permeable because of their compaction by extensive

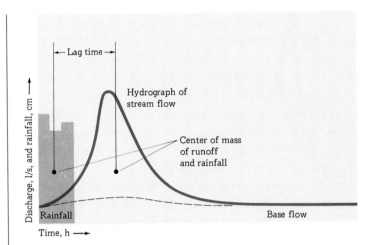

FIGURE 16.1

Hypothetical hydrograph, showing significant characteristics. [After Luna B. Leopold, "Hydrology for Urban Land Planning," *U.S. Geol. Surv. Circ.* 554, p. 3, 1968.]

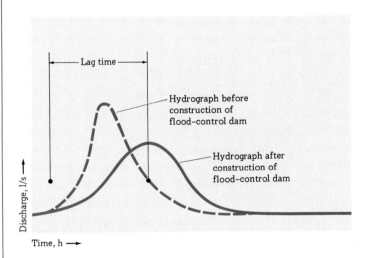

FIGURE 16.2

Hypothetical hydrograph to show the effect on stream flow of a flood-control dam upstream from the hydrograph station.

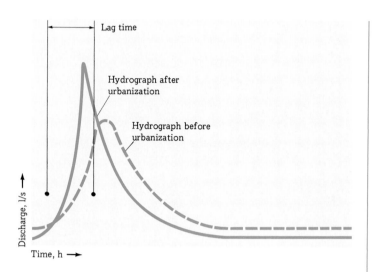

FIGURE 16.3

Hypothetical hydrograph, showing the effect of urbanization on stream flow. [After Luna B. Leopold, "Hydrology for Urban Land Planning," *U.S. Geol. Surv. Circ.* 554, p. 3, 1968.]

human activity; (2) the rate and amount of surface runoff increases; (3) the groundwater level falls because of the decreased infiltration. The effect on stream flow is shown in Figure 16.3 and can be characterized as follows: (1) The time lag between precipitation and flood peak is shortened because water is not slowed on its way to the stream channel by a vegetative cover nor does any appreciable amount sink into the ground; (2) the peak is higher because the stream must carry more water in a shorter period of time; and (3) the base flow is generally lower because of a decreased supply of groundwater on which the base flow depends. The net result of these changes is a **flashy stream,** one that has a low base flow and a high, short flood peak.

Changes in vegetative cover   Changes in vegetative cover can effect changes in the hydrograph. In a general way the decrease in the vegetative cover acts in the same direction as does an increase in urbanization.

Studies of the effect of selective forest cutting on stream flow have been carried on by the United States Forest Service in the Fernow Experimental Forest near Parsons, West Virginia. The results, not entirely unexpected, are several: The removal of the trees increases the flow of streams during the growing season chiefly

because the water ordinarily transpired into the atmosphere by trees makes its way to the streamways instead. The increase in stream flow is approximately proportional to the percentage of tree cover removed, the greatest increase—over 100 percent—occurring with complete cutting of the forest cover. Concurrently with the increase of stream flow, the flashiness of streams increases. As the tree cover replaces itself, the flow decreases and the flashiness of streams declines, and the return to stream-flow conditions similar to those before cutting is projected to be about 35 years.

CHANGES IN STREAM CHANNELS

The changes in stream channels associated with artificially induced changes in stream flow are not well documented or understood as yet, but some suggestions can be made.

In Chapter 10 we found that the stream channel is adjusted to the stream discharge, and this was expressed as discharge = width × depth × velocity. Observation shows that a stream rises to the limit of its bank on the average of once every 1.5 to 2 years. This stage of flow is called the **bank-full stage.** The stream channel is adjusted to handle the size of flow that occurs every 1.5 to 2 years. When flow exceeds this amount, flooding begins. Studies reported by Luna Leopold of the United States Geological Survey indicate that with an increase of urbanization the bank-full stage of flow increases. If an area is 50 percent urbanized, experience indicates that the bank-full stage of discharge increases by a factor of 2.7 over that of the same stream in an unurbanized condition. An example is shown in Table 16.1.

In the table the stream has increased the depth and width of its channel as the discharge for bank-full

TABLE 16.1

Stream flow and urbanization

	Before urbanization	After 50% urbanization
Discharge at bank-full stage	1.5 m³/s	4.0 m³/s
Velocity	0.75 m/s	0.75 m/s
Depth of channel	0.5 m	0.9 m
Width of channel	4.0 m	6.0 m
Area of drainage basin	2.5 km²	2.5 km²

TABLE 16.2
Pollution-caused fish kills in the United States, 1961–1968[a]

	1961	1962	1963	1964	1965	1966	1967	1968
Number of states reporting	45	37	38	40	44	46	40	42
Total estimated number of fish killed, thousands	15,910	7,118	7,860	18,387	11,784	9,115	11,591	15,236
Average size of kill, thousands	6.5	5.7	7.8	5.5	4.3	5.6	6.5	6.0
Largest kill reported, thousands	5,387	3,180	2,000	7,887	3,000	1,000	6,549	4,029

[a]Data from Federal Water Pollution Control Administration.

TABLE 16.3
Fish kill by source of pollution in the United States, 1968[a]

Source of pollution	Fish killed, thousands	Average kill	Game fish, thousands	Nongame, including commercial fish, thousands
Agriculture	422	4,240	152	270
Industry	6,398	5,675	415	5,983
Municipalities	6,952	7,585	320	6,632
Transportation	880	9,155	430	459
Other	584	1,995	19	565
Total	15,236	28,650	1,336	13,909

[a]Data from Federal Water Pollution Control Administration.

stage increased. The bank-full stage increased because of increase in runoff rates caused by urbanizing one-half of the drainage area. This erosion of the channel produces sediments that are moved downstream, where they may be deposited.

Erosion of a stream bed can be artificially induced in other ways. Thus it has been found that a stream below a dam will enlarge its channel over a considerable distance and redeposit the products of channel erosion farther downstream. This erosion is in part due to the dam's providing a settling basin in which stream deposits are stored. The stream section below the dam is deprived of its normal sedimentary load. When erosion takes place immediately below the dam, there is a lack of sediment to replace this normal loss through point-bar building or overbank deposition.

CHANGES IN WATER QUALITY

The most obvious change that we have brought to the rivers of the world is their **pollution,** a change directly caused by the planned (or unplanned) use of the rivers as a sewage system. Increasing technology, industrial-ization, and population have produced an increasing amount of refuse of all kinds. Where rivers are available, they become a handy and, at first, inexpensive way to remove our cultural refuse. This load has proved more than many can handle and still maintain their original quality. That pollution of water has its effect on the life process is shown by Tables 16.2 and 16.3.

Solid load Agriculture, more than any other human activity, increases the rate at which sediments are delivered to the streams. It is true that construction activities produce a higher rate of **sediment production,** but these are isolated events; over the long run the continuous use of land for farming is volumetrically more important in sediment production.

There is a great deal of information showing the effect of the clearing of land and of farming practices on the production of sediments. For instance, one of the effects of forest clearance in the Fernow Experimental Forest referred to earlier in this chapter was to increase the solids delivered to the stream. With complete commercial cutting of the watershed, the stream's solid load was as much as 3,700 times that of similar streams in areas in which no lumbering occurred. In Mississippi studies on sediment production from land under differ-

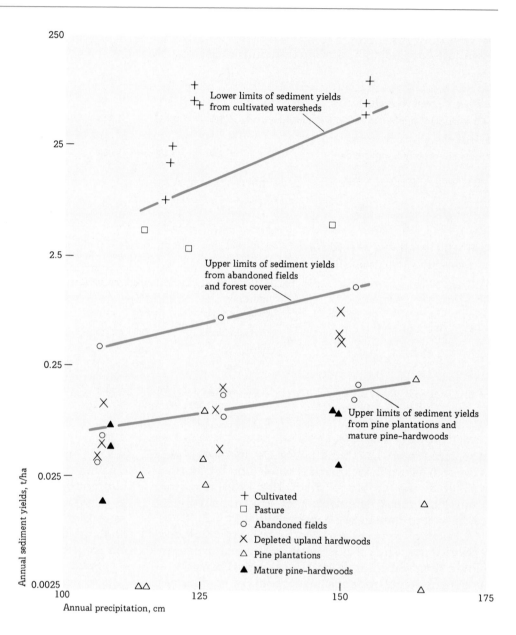

FIGURE 16.4

Variation in sediment yields from individual watersheds in northern Mississippi under different types of land use and changing amounts of precipitation. [After S. J. Ursic and F. E. Dendy, "Sediment Yields from Small Watersheds under Various Land Uses and Forest Covers," *Proc. Fed. Inter-Agency Sedimentation Conf. 1963, U.S. Dept. Agric. Misc. Pub. 970, p. 51, 1965.*]

ing uses produce the startling conclusions shown in Figure 16.4. Land under cultivation produced a thousand times more sediment than did similar land under mature stands of trees. M. Gordon Wolman has shown the effect of changing land use on the production of sediments in the Piedmont area of Maryland from about 1800 to the present (Figure 16.5). Interestingly, urbanization has reduced sediment production to the preagricultural levels.

The increased sediments move on to the streams, which then must handle them as best they can. The results vary with the stream. In general, however, increased solid loads increase the turbidity or muddiness of a stream, decrease the depth of light penetration, and can cause silting up of valley bottoms.

To agriculture, deforestation, urban construction, and grazing we must add mining and industry as producers of excess sediments. These activities may bring other undesirable features with them in addition to increased sediment. For instance, the tailings of many coal mines contain large amounts of sulfur, which can convert to undesirable compounds like sulfuric acid.

Dissolved load   We found in Chapter 10 that the average content of dissolved material carried by streams is 120 ppm on a worldwide average. Most of this comes from the chemical decomposition of earth materials during the normal processes of weathering. Some is recycled from the oceans through the atmosphere to the streams, and some is contributed by human activities.

Figure 16.6 illustrates the increase of dissolved solids in the Passaic River at Little Falls, New Jersey, from 1948 through 1963, a period of increasing industrialization and urbanization. It is not, however, just the increase of the dissolved load itself but the various factors involved in the increase that cause problems.

That part of the **dissolved load** contributed by people covers a wide range of materials. Some are toxic to river life, some serve as nutrients promoting explosive organic proliferation, and a few are merely an affront to human sensibilities. Among the most bothersome of the stream pollutants are the phosphorus and nitrogen that come from fertilizers, from sewage (both treated and raw), and, in the case of phosphorus, from detergents. These additions to streams spark a complicated set of reactions in the chemistry, biologic processes, and ecology of the stream, all usually undesirable (from the human point of view) and all leading to a stream of a different character than it was in its natural state. The list of chemical pollutants is as varied as the products of our industrial society. They are present in the streams because we continue to use the channels of water as a sewer system. Until we adopt other ways of cleaning our nests, the streams will continue to be polluted.

We shall see later that these same elements are critical in the quality of lake water as well.

Thermal pollution   The use of river water for industrial-cooling purposes, particularly by electric power plants, has led us to recognize that the discharge of waste heat into our streams can lead to unwanted effects on the aquatic life. We dignify this human activity as **thermal pollution,** a problem faced not only by rivers but by lakes, underground water, and shallow seawaters as well (see Figure 16.7).

Thus far it has been the electric-power industry that has done most to introduce us to the dangers of thermal pollution. Presently it uses about three-fourths of all waters used in the United States for industrial cooling. Up to now, however, thermal pollution has become important only in local areas. This is not to say that electric power may not pose greater problems in

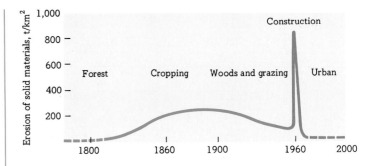

FIGURE 16.5

A sequence of land-use changes and sediment yield beginning prior to the advent of extensive farming and continuing through a period of construction and subsequent urban landscape. Based on experience in the Middle Atlantic region of the United States. [After M. G. Wolman, "A Cycle of Sedimentation and Erosion in Urban River Channels," *Geograf. Ann.*, vol. 49-A, p. 386, 1967.]

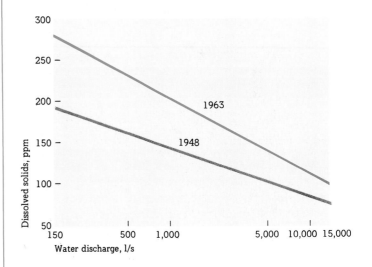

FIGURE 16.6

Increase in dissolved solids in relation to discharge in the Passaic River at Little Falls, New Jersey, from 1948 to 1963. [After Peter W. Anderson and S. D. Faust, "Quality of Water in the Passaic River at Little Falls, N.J., as Shown by Long Range Data," *U.S. Geol. Surv. Prof. Paper 525-D*, p. 215, 1965.]

FIGURE 16.7

Thermal pollution of the Connecticut River by a nuclear power plant near Haddam, Connecticut, as evidenced by infrared photography. In this color-enhanced version of a photograph made with an infrared camera, higher temperatures of the water appear as darker tones. Thus the power-plant discharge through the canal at the bottom of the picture appears dark gray: It has a temperature of about 34°C. The plume of heated water trails downstream from the mouth of the effluent canal, mixing with the river water, which has a normal temperature of 25°C and shows a lighter gray. [HRB-Singer, Inc., for U.S. Geological Survey.]

the future. One estimate suggests that by the year 2000 the industry will be producing so much waste heat that it will require, if river water is used as a coolant, approximately one-third of the country's runoff to get rid of the excess energy. This appears totally unacceptable, and some other solution must be found.

The primary effect of thermal pollution is to raise water temperatures to a point at which the native aquatic life can no longer survive. There are other effects as well. For instance, increased temperatures of river waters can produce local fog conditions as cool air sweeps across the warmed water. Such conditions could lead one to such exercises as "How many power plants will be needed on New York's East River to keep LaGuardia airfield perpetually fogbound?"

Industrial cooling is a dramatic way to change the temperature of river water. But man can bring about these changes in another way. Studies of streams in urbanized areas on Long Island have been carried out by E. J. Pluhowski, of the Johns Hopkins University. Comparing those streams in urbanized zones with a control stream still in an undeveloped area, Pluhowski found that streams most affected by human activities had summer temperatures from 5.5°C to 8°C above those of the control stream. Conversely, stretches of streams affected by human activities were found to be 2.2°C to 5.5°C colder in winter than were those unaffected.

Some of these variations are due to the nature of runoff between unurbanized and urbanized zones. During the summer months runoff from heated pavements can raise temperature of stream water appreciably, as much as 6°C as the result of a single storm, whereas an unurbanized stream shows little variation during the storm and subsequent runoff. Streams in urbanized zones are fed more by surface runoff than are streams in unurbanized zones, which depend more heavily on groundwater for their flow. In winter the surface runoff is colder than the groundwater, which tends to reflect the mean annual temperature. The stream in an urbanized area, therefore, is colder during the winter than is the stream in an unurbanized area.

16.2 LAKES

Viewed against the large span of geologic time, **lakes are only temporary features.** They are disruptions in the general drainage system of the earth's surface. Some of these disruptions may last for hundreds of thousands of years; others, for a few tens or few hundreds of years. Despite their fleeting existence in the geological record they are important to man. Here we shall consider briefly the life cycle of a lake and then look at the ways in which human beings may hasten its demise.

LIFE CYCLE OF A LAKE

The basin that holds a lake may form in one of several ways: Glaciers may scour bedrock or pile up debris across a streamway; earth movements may downdrop a portion of the earth's crust; volcanic craters and calderas may hold the waters of a lake; landslides or volcanic flows may dam valleys; the surface may subside because of solution of rock in the underground; streams may cut off meanders to form crescent-shaped oxbow lakes; lagoons may be cut off from the sea by coastal processes; wind may scour a basin later flooded by water; and human beings may build dams.

Whatever its origin, a lake begins to die as soon as it is born. Death is due primarily to two different processes, which may act concurrently: First, the outlet of a lake becomes lowered by erosion, which cuts downward and drains the lake. Second, a lake fills up with sediment. Some of the filling comes from the solid load delivered to the lake by the streams that feed it; some is due to precipitation of material from solution by either organic or inorganic processes; and some is the accumulation of the remains of plant and animal life. A third fate of a lake is to be dried up because of insufficient water, and this has been the destiny of some of the pluvial lakes of the southwestern United States, as we saw in Chapter 13.

OUR EFFECT ON LAKES

Our role in the history of lakes thus far has been one of speeding the aging process, hurrying them on to a premature end. One such example is the Caspian Sea, which we may view as a saltwater lake. The human use of water that would normally come to the Caspian is carrying that lake closer to extinction, as suggested in Figure 16.8. It is also evident that we contribute to the destruction of a lake by increasing the rates of erosion and hence the amount of sediments that streams can deliver to a lake. We have already seen that intensive land use can increase the production of sediments one to four orders of magnitude above the natural rate. This must increase the rate at which the lake basin fills.

We also contribute to lake decay by speeding up a process called **eutrophication.** *Eutrophic*, from the Greek *eutrophos*, "well-nourished," applies to lakes having abundant supplies of nutrients that support a dense growth of plant and animal life. Characteristically the decay of this organic matter depletes lake waters of oxygen, particularly during the warmer months. Lakes characteristically begin as **oligotrophic lakes** (from the Greek meaning "sparsely nourished"), which are low in accumulated nutrients and hence low in organic productivity and high in dissolved oxygen. The lake ages, and, as more nutrients can accumulate, biologic activity increases as the lake passes toward eutrophic conditions through a transitional **mesotrophic stage,** as shown in Figure 16.9. Under natural conditions the process of eutrophication proceeds at varying rates. We play a role in eutrophication by supplying large amounts of nutrient material to the lakes and bringing about an explosive increase in the biologic activity. Nitrogen and phosphorus are the most important elements contributed to the eutrophication process. Generally municipal sewage is the chief source of nutrients although rural and industrial sources are also important. Experience has shown us that in a few decades oligotrophic lakes can be converted to strongly eutrophic lakes by turning them into receptacles for municipal sewage and industrial and rural wastes.

It was long thought that the Great Lakes were so large that eutrophication would not become a major problem. This we now know was an error in judgment; for we have pushed one of them, Lake Erie, over the brink into the eutrophic stage (see Figure 16.10). This has, moreover, been an expensive misjudgment; for estimates of the cost of cleaning up Lake Erie grow ever higher. The projected price of a clean Lake Erie by the 1990's has varied from $3 to $10 billion. The other Great Lakes have yet to reach the condition of Lake Erie, but each of them shows to some extent the impact of changes induced by human activity.

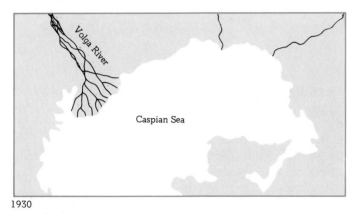

1930

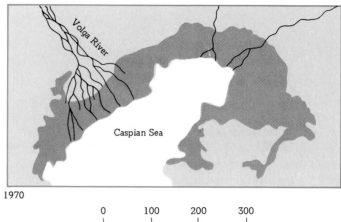

1970

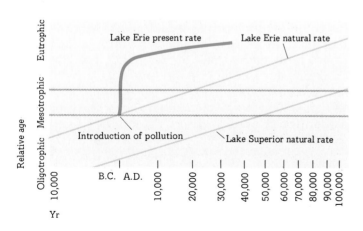

1960

FIGURE 16.8

The Caspian Sea is really a large saltwater lake. These three maps of its northern section show how it has decreased in size from 1930 to 1970. This decrease is due in part to a natural decline in rainfall and in part to human beings, who have used large amounts of water from the Volga River for irrigation, thus decreasing the river's discharge into the Caspian. [After A. E. J. Engel, "Time and the Earth," *Am. Sci.*, vol. 57, p. 499, 1969.]

FIGURE 16.9

Aging indicators as a lake progresses from oligotrophic to eutrophic. [After *Lake Erie Report*, p. 32, Federal Water Pollution Control Administration, Great Lakes Region, Washington, 1968.]

FIGURE 16.10

In two generations human-induced pollution has caused Lake Erie to advance to the eutrophic stage. [After *Lake Erie Report*, p. 32, Federal Water Pollution Control Administration, Great Lakes Region, Washington, 1968.]

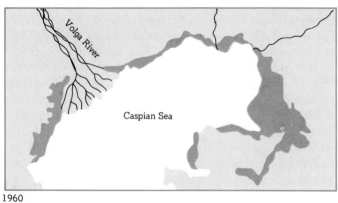

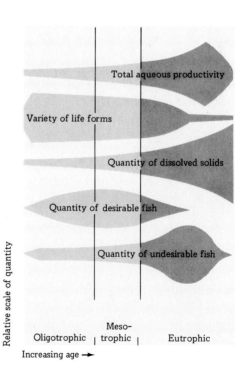

16.3 GROUNDWATER

Water underground is subject to contamination of the same type as is surface water. We have noted in Section 11.3 that water in limestone is more susceptible to pollution than is that flowing more slowly in, for example, a sandstone aquifer. Underground water may be subject to thermal pollution as well as microbial or chemical pollution: The very process of recycling water may give it the added undesirable quality of increased temperature; water used for air conditioning is withdrawn from the ground at one temperature, used in cooling, and returned to the ground at a higher temperature.

SALTWATER INVASION

A particular type of pollution to which underground freshwater supplies are subject is the **invasion of salt water.** This may be salt water of an adjacent ocean environment, or it may be salt water trapped in rocks lying beneath the freshwater aquifers.

Fresh water has a lower specific gravity than does salt water: Normal seawater has specific gravity of about 1.025, compared with 1.000 for fresh water. Therefore fresh water will float on top of salt water. If there is little or no subsurface flow, this can lead to an equilibrium in which a body of fresh water is buoyed on salt water. In a groundwater situation in which fresh water and salt water are juxtaposed we might have a situation such as that shown in Figure 16.11. Here an island (or it could as well be a peninsula) is shown underlain by a homogeneous aquifer that extends under the adjacent ocean. Fresh water falling on the land has built up a prism of fresh water in hydrostatic equilibrium with the salt water that surrounds it. The height of a column of fresh water is balanced by an equal mass of salt water. In the example given the column of fresh water (H) is equal to h_1, the height of the water table above sea level, plus h_2, the thickness of the groundwater below the sea level. A column of salt water equivalent in length to h_2 will balance a column of fresh water of length H, or $h_1 + h_2$. Working this out, we find that $h_2 = 40h_1$. This means that, if the water table is 10 m above sea level, then the freshwater-saltwater contact will be 400 m below sea level. It also means that, as we lower the water table by whatever means, we raise the elevation of the saltwater-freshwa-

ter interface at the rate of 40 m of rise for each meter of lowering of the water table. It is very possible, then, that a well that originally bottomed in fresh water could, by a lowering of the water table, turn into a saltwater well, as suggested in Figure 16.11.

LAND SUBSIDENCE

Excess pumping may deplete groundwater supplies and in certain places bring about saltwater invasion of wells. But another bothersome problem connected with heavy pumping of groundwater is **land subsidence.** It expresses itself at the surface either as broad, gentle depressions of the land or as catastrophic collapse and the formation of sinkholes.

Zones of broad subsidence have been created in California over large areas of the San Joaquin Valley and secondarily in the Santa Clara Valley south of Oakland and San Francisco (Figure 16.12). The phenomenon is related to the pumping of groundwater and to the resulting compaction of sediments. The Tulare-Wasco area of the San Joaquin Valley is a good example of what happens. In 1905 intensive pumping for

FIGURE 16.11

Cross section through an oceanic island (or a peninsula) underlain by homogeneous, permeable material. The fresh water forms a prism floating in hydrostatic balance with the neighboring salt water. Reduction of the groundwater table on the island will cause a change in the position of the saltwater-freshwater interface. See the text for discussion.

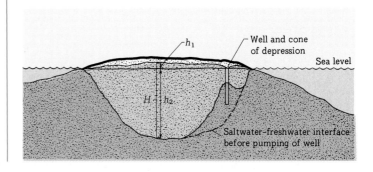

FIGURE 16.12

Principal areas in California where withdrawal of groundwater has caused subsidence of the land surface. [After B. E. Lofgren and R. L. Klausing. "Land Subsidence Due to Groundwater Withdrawal, Tulare-Wasco Area, California," *U.S. Geol. Surv. Prof. Paper 437-B,* p. 2, 1969.]

irrigation began in the area. By 1962 more than 20,000 km² of irrigable land had undergone subsidence of more than 0.3 m, and as much as 4 m of subsidence had occurred in some places. Figure 16.13 shows subsidence in the area from 1926 to 1962. Subsidence was slow through this time and hardly noticeable to residents. But the casings of many wells have been damaged, the problems of surveying and construction have been complicated, and the operation of irrigation districts endangered.

16.4 SOILS

Most arable soils are formed by weathering, which creates zonation in the surficial material of the earth. The zones (soil horizons) thus formed enlarge downward with time, and concurrently surface material is slowly removed by normal erosive processes. The introduction of farming, grazing, lumbering, and construction increases the rate of erosion. In a few years or a few decades erosion is sufficiently great to upset the balance between the downward-growing soil zones and the material removed at the surface. The rate of removal, along with human turning and disruption of the soil, may in a few years or few decades remove all or large parts of the soil profile that has taken so long to form.

The loss of the soil through humanly induced erosion can be compensated for in two ways: Either it may be artificially regenerated by the introduction of nutrient-supplying chemical fertilizers, or land may be allowed to revert to its original state and given time to recover. In the latter case the recovery time will vary with the local situation, but several generations to hundreds of years must usually be allowed.

We are conditioned to think of topsoil as a "precious" material that should not be destroyed. But there

FIGURE 16.13

Land subsidence in the Tulare-Wasco area, California, 1926 to 1962, due to withdrawal of groundwater. Lines show equal subsidence in meters and are dashed where approximate. [After B. E. Lofgren and R. L. Klausing, "Land Subsidence Due to Groundwater Withdrawal, Tulare-Wasco Area, California," *U.S. Geol. Surv. Prof. Paper* 437-B, p. 63, 1969.]

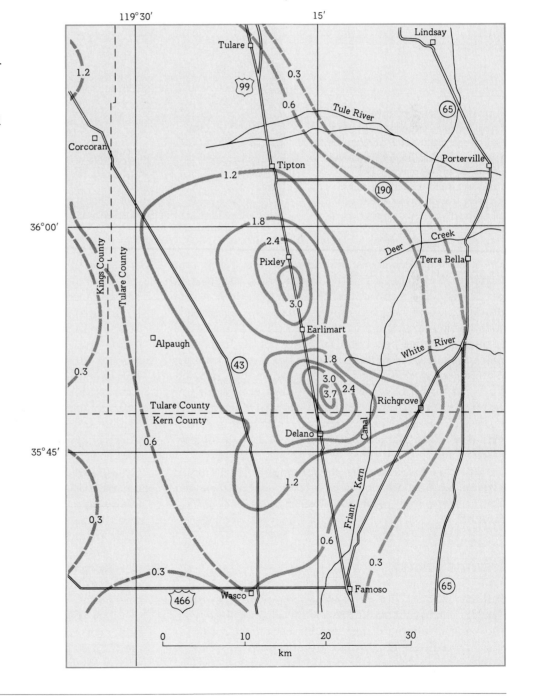

FIGURE 16.14

Strip mining of coal involves the removal of rock overlying the coal bed. Here, near Georgetown, Ohio, a giant, electric-powered shovel moves 80 m³ at a time. The bulldozer parked between the shovel's treads gives scale. [National Coal Association.]

are some situations in which one of the first things one wants to do is to get rid of the weathered material and get down to the less-weathered earth materials. A moment of thought will tell us why this is so. Some soils have been long in the process of formation. Much of the nutrient material has been leached out of them, and slow erosion at the surface has not been rapid enough to expose new unleached materials, which can provide the basis for plant growth. Under these conditions the removal of the relatively sterile upper portion and the exposure of the less-weathered material still undepleted in the elements needed for plant growth may be beneficial.

In some instances the development of a heavy *B* horizon may create not only a low-nutrient zone but also a zone that renders a soil clayey and poorly drained.

SURFACE MINING

Surface mining, the process of removing the overburden of soil and rock that covers a mineral deposit, has certain technical advantages over underground-mining methods. But it also creates problems, which include an instantaneous disruption or destruction of

the soil and a major rearrangement of the landscape.

Surface mining takes a number of forms. **Open-pit mining** is perhaps the most familiar and is represented by the gravel pit, the stone quarry, and the gigantic pits of some mines such as the vast excavations for low-grade copper in Arizona, Utah, and Nevada. **Strip mining** consists of moving to one side varying thicknesses of soil and rock until the mineral deposit is exposed for removal (see Figure 16.14). The process is most commonly used in the coal-mining industry. **Hydraulic mining** uses a strong jet of water to move deposits of sand and gravel that may contain or cover the sought-for ore. The ore is carried through concentrating equipment, where the wanted mineral is separated, usually by utilizing differences in specific gravity. **Dredging** is yet another form of surface mining. In situations where sand and gravel are sought the products of dredging are all removed for consumption. In instances where higher-priced minerals are needed, they are separated from the dredgings, and the debris that is discarded may be left as piles in the vicinity of the operation.

The primary effect of surface mining is a disturbance or, more usually, a destruction of the soil, the creation of gaping holes in the landscape, the formation of a jumbled artificial landscape unadjusted to normal surface processes, the disruption of the biota—both animal and plant—and the modification of water flow both on the surface and underground.

By 1965 a total of 12,800 km² of land had been disturbed by surface mining in the United States. This had risen by 1970 to an estimated 16,000 km². Of this total area mining for coal accounted for 41 percent; for sand and gravel, 26 percent; for stone, 8 percent; for gold, 6 percent; for phosphate, 6 percent; for iron, 5 percent; for clay, 3 percent; and for all others, 5 percent.

Since 1970 the area disturbed by surface mining has increased somewhat. But during that decade increasing attention has been paid to restoring the land disrupted by surface mining. This is generally an expensive business, but it is most economically done if it is included as an integral part of the mining project rather than as a separate operation planned and carried out after the site has been abandoned by miners.

16.5 SOLID WASTES

We have referred to some of the problems raised by chemical and organic wastes that reach our streams and lakes. We produce a very large amount of additional waste, which includes everything from paper clips to buses. Some of these wastes can be salvaged and reused, and many are. Many, however, must be disposed of, and this raises an increasingly complex problem for densely populated areas.

New York City, for instance, dumps wastes from barges onto the continental shelf beyond the New York harbor at the rate of 2 kg per person per day. As such, New York City has become the major source of sediment in the area, far outstripping the contribution of the local rivers. By the 1970s this dumping of wastes was having a deleterious effect on the marine life in the area. Human beings have, then, become a major source of sediments. We found earlier that people had also become an important factor in increasing the rates at which the processes of erosion produce sediments. The figures given in Table 16.4 suggest the magnitude of our contribution to sediments in the United States . At best they are approximations, but probably they are low. The total is staggering, particularly when we compare it with the 540 million t of material estimated to have been carried annually by the streams of the United States before we influenced the rates of erosion.

16.6 OCEANS

To anybody who has sailed the oceans or flown across them it is difficult to imagine that we could greatly affect such large bodies of water. In fact, except for local situations, the oceans have been fairly well immune to human attacks, largely because of their size. Let us, however, look at some of the local effects of human activity on the marine environment. Not surprisingly the near-shore marine environments adjacent to heavily populated areas suffer the most. We have already used as an example the dumping of New York City's solid wastes on the adjoining continental shelf.

OIL SPILLS AND BLOWOUTS

The *Torrey Canyon* disaster of March 18, 1967, awakened the world to the problem of oil pollution at sea. The *Torrey Canyon*, a 295-m tanker carrying 117,000 t of

TABLE 16.4

Major sources of unreclaimed solid wastes produced in the United States in 1970, including those estimated to result from human acceleration of the rates of erosion[a]

Source	Millions of t
Domestic, municipal, and industrial waste	325
Junked cars, trucks, and buses	10
Mining industry	1,300
Increased stream load due to human activity	560
Total	2,195

[a] Estimates compiled from various sources.

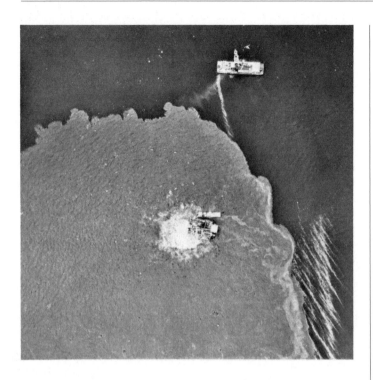

FIGURE 16.15

Oil (lighter gray) surrounds an offshore-drilling rig near Santa Barbara, California. Leakage resulted from a natural blowout of the drill hole. An estimated 750,000 l of oil escaped into the ocean before the leak could be stopped. The barge at the top in the darker, uncontaminated water is about 50 m long. [Environmental Protection Agency.]

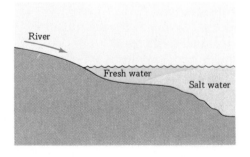

FIGURE 16.16

A saltwater wedge may develop in some estuarine situations. The position of the wedge moves as the supply of fresh water varies relative to the supply of salt water.

crude oil from Kuwait to English refineries, went aground on rocks off Land's End, at the southwestern tip of England. Over a 2-week period the stranded hulk gave up its load of petroleum, bringing a black, oily contamination to the French and English beaches along the English Channel, and either directly or indirectly caused the death of untold numbers of birds and marine organisms. The blowout in 1969 of an oil well drilled in the Santa Barbara channel off the coast of California produced similar results and emphasized to all that the process of offshore drilling can produce ecological hazards (Figure 16.15). As shipping ways become more crowded with tankers and as the pace of offshore drilling increases, we continue to witness accidents that spill oil into the oceans. Greater care will certainly reduce the number of incidents, but it seems unlikely that we can eliminate them entirely.

We have, however, learned some techniques that are useful in handling an oil spill once it occurs. Petroleum spilled at sea moves as a coherent mass 5 to 60 cm thick and parallel with the wind direction at about 3.4 percent of the wind velocity. Therefore, knowing direction and velocity of the wind, we can predict direction of movement of an oil spill and its estimated time of arrival at a selected location. Several techniques can be used to eliminate the oil before it fouls the shore. The oil can be surrounded by booms and barges and pumped from the sea's surface. In the *Torrey Canyon* disaster the French had some luck with this but were more successful in sprinkling the oil with a dust of chalk and sodium stearate. Oil adhered to the dust and then, because in combination with the dust the oil was heavier than water, it sank to the sea bottom. Over 23,000 t of oil were eliminated from the surface in this way, but their effect on the sea bottom is not known. The use of detergents to disperse the oil so that it may be more quickly oxidized or become more susceptible to degradation by bacteria is not usually successful. The British used a great volume of detergent to disperse the *Torrey Canyon* petroleum at sea and even more on the beaches. The only real result after the application of nearly 10 million l of detergent was the loss of marine life, particularly in the intertidal marine zone.

Natural processes aid in the elimination of the petroleum. Evaporation is effective, and in general about 25 percent of an oil spill will evaporate in a few days. After three months 85 percent will have volatilized. In addition photochemical reactions, degradation by bacteria, and adhesion to sedimentary particles in suspension in the water tend to eliminate the pollution.

ESTUARIES

Estuaries are long, relatively narrow incursions of the marine environment into a landmass. In reality they are valleys turned into a near-shore marine environment by invasion of the ocean waters as sea level rises in relation to the land. Estuaries are probably much more common today than they were during most of geologic time. This is because, during the Pleistocene, valley erosion of the exposed continental margins occurred during the low-water stages attending glaciations, and flooding of these valleys took place during periods of deglaciation and rising of the sea level.

The estuary has become an attractive geographic feature for us. It has provided protected anchorages and shorelines for port facilities. It permits penetration of marine shipping well inland beyond the open ocean and has provided suitable environments for various facets of the fishing industry.

The estuary is an area where fresh water from the land meets the salt water from the ocean. The ways in which these two types of water circulate and are related to each other play an important role in the use of the estuary.

As shown in Figure 16.16, the fresh water fed into the estuary by the river at its head flows above the heavier salt water, which assumes a tapering form known as the **saltwater wedge.** The position of the wedge must change with the changing supplies of salt and fresh waters. Thus during periods of drought and low river flow the saltwater wedge pushes farther and farther upstream, bringing the marine environment up the estuary with it. Conversely, in periods of heavy stream flow the saltwater wedge is pushed seaward. This bears importantly on the location of water supplies along the upper reaches of an estuary. It is possible that a freshwater intake may, in periods of low stream flow, find itself pumping brackish or salt water.

16.7 EARTHQUAKES AND VOLCANOES

In Chapter 6 we examined the general nature of earthquakes and also found that under certain circumstances we can create small earthquakes by injecting fluids into the underground and stop these tremors by halting the injection of fluids. The question for the future then is whether we can affect the frequency of small earthquakes and influence the occurrence of large quakes. It could be argued that in high-risk zones we might attempt to decrease the magnitude (and hence the damage) by creating a number of small earthquakes, thus relieving the earth stresses from time to time through artificial means. The argument would be that the potential damage from a series of small quakes would be less than that of a single large and catastrophic quake. The answer to this is that we simply don't know. And it seems probable that we will not learn the answer by actual experimentation in the field. This is simply because no one would attempt such a test while there is the good possibility that the experiment could trigger a major quake instead of stimulating a minor one.

In Chapter 3 we found that modern volcanoes have a certain distribution in the world, and we later noted that this distribution also coincides with earthquake activity. We can predict, then, that volcanoes and their activity will probably be confined to the belts where historic and present activity, both volcanic and seismic, have been concentrated. We also found in Chapter 3 that individual volcanoes have some pattern of eruptive activity. But our data on volcanic activity are not so complete as they are for seismic events, in part because the seismic events are much more numerous. Nevertheless, for some volcanoes our records are good enough and the monitoring system is sensitive enough to predict eruptions. Kilauea in Hawaii is an example. Before an eruption occurs, this volcano goes through a period of swelling related to the rise of the magma toward the surface. Measurement of this swelling by a series of delicate tiltmeters warns of impending activity.

As with earthquake prediction, the forecasting of specific volcanic activity is at best in its infancy. We saw that earthquake control is in a very early stage and may never get beyond it. When we consider control of volcanoes, we have absolutely no clues as how to proceed. It presently seems doubtful that we shall ever generate the technology, much less the social structure, to control volcanism. It is almost as doubtful that we shall ever be able to control large quakes.

OUTLINE

We affect our **physical environment** in many ways.

Streams develop a high, short-period flood peak through **urbanization** and a low, long-extended flood peak through **flood-control projects.**

Water quality is affected by solids and dissolved material and by thermal changes.

Lakes are geologically short-lived. People can hasten the death of a lake and push it into a **eutrophic** stage by addition of excess nutrients from industrial, rural, and municipal wastes.

Groundwater may be contaminated by **saltwater invasion** and by **chemical, biological, and thermal** changes. Excess pumping can lead to **collapse** and **subsidence** of the ground surface.

Soils may be destroyed by agriculture, by surface mining, and by construction activity.

Solid wastes, produced as a result of human activity, in the United States exceed the total amount of material carried annually to the oceans by the nation's rivers.

Oceans, large as they are, can be affected by human activity through, for example, **dumping of solids, oil spills,** and **blowouts.**

Estuaries are particularly easily influenced by human presence.

Earthquakes and **volcanism** can be predicted with very limited precision. We can initiate and stop small earthquakes. We cannot control large quakes or influence volcanic activity.

SUPPLEMENTARY
READINGS

Bolt, B. A., and others *Geological Hazards,* Springer-Verlag New York Inc., 1975. *The subtitle, Earthquakes—Tsunamis—Volcanoes—Avalanches—Landslides—Floods, describes the emphasis.*

Keller, Edward A. *Environmental Geology.* Charles E. Merrill Publishing Company, Columbus, Ohio, 1976. *A good introductory text on the subject.*

Man's Impact on the Global Environment, Report on the Study of Critical Environmental Problems, The M.I.T. Press, Cambridge, Mass., 1970. *An excellent overview.*

Tank, Ronald (ed.) *Environmental Geology,* Oxford University Press, Inc., New York, 1973. *A collection of case histories and readings from original sources.*

White, Gilbert (ed.) *Natural Hazards,* Oxford University Press, Inc., New York, 1974. *A number of interesting case studies of different natural hazards, with consideration of peoples' response to them.*

17

MOONS AND PLANETS

The launching of a Russian earth-orbiting satellite on October 4, 1957, began a new phase in the exploration of our planetary system and of the vast spaces beyond. In the two decades that followed, our knowledge has increased dramatically. We have mapped our moon in great detail, put astronauts on its surface, and brought them back with samples of lunar soil and rock. The Russians have landed an instrumented space vehicle on Venus, and we have received information from vehicles flying by Venus, Mars, and Jupiter. In 1976 we placed two landing vehicles on Mars and got our first close-up look at the homeland of science fiction. We turn now to some of the things we know about our planetary system.

17.1 OUR SOLAR SYSTEM

We are physically able to exist because of heat and light from our own star, the sun. As we observe it from earth, the sun rises in the east to bring daylight and sets in the west to bring night. This phenomenon anciently led to the assumption that the sun moves around a stationary earth. Likewise, the stars appear to rise in the east and set in the west. Therefore, for a long segment of human history, the earth was believed to be the center of the universe, about which everything else revolved.

This belief, supported by even noted Greek scientists, who were dominated by the teachings of Aristotle, was widely held until about 3 centuries ago. One Greek, Aristarchus, had the temerity in the third century B.C. to suggest that the planets, including the earth, circle about the sun and that the earth rotates on its axis, giving us night and day. But he failed to convince most of the people of his time, and his explanation for the movements of the planets around the sun was not to be firmly established until 1543—by Copernicus. Still another century passed, however, before there was universal acceptance of Aristarchus's idea.

We now know that the sun is the center of our physical existence. Everything within the sun's gravita-

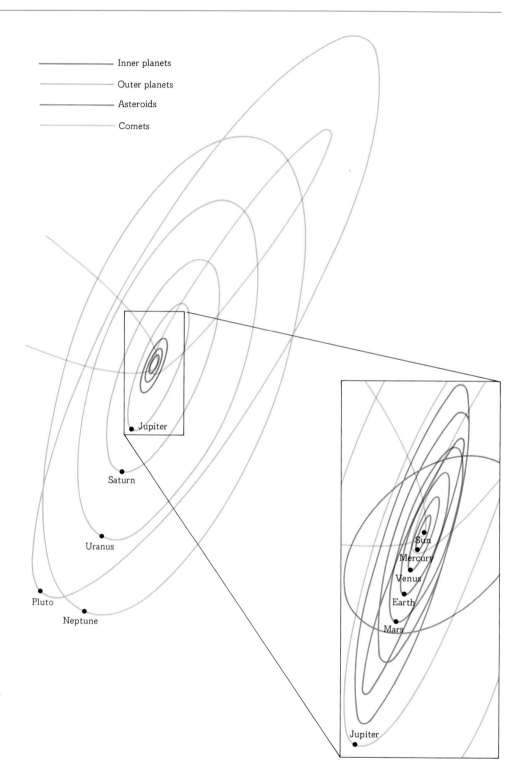

FIGURE 17.1

Orbits of the nine planets, the asteroids, and comets. Together with the 31 satellites of the planets and the uncounted millions of meteors, they make up the solar system.

tional control constitutes what we call our **solar system** (see Figure 17.1). The solar system includes nine planets, 31 satellites, or "moons," thousands of asteroids, scores of comets, and uncounted millions of meteors. All these revolve about the sun, forming a system 16 billion km in diameter.

THE SUN

The sun is a star. Like all other stars, it is a hot, self-luminous ball of gas. Located 150 million km from the earth, it has a diameter of about 1,391,000 km and a mass 332,000 times that of the earth. It is the source of the solar system's light, heat, and charged particles. Its gravitation holds the system together.

Sixty-six elements have been recognized in the spectrum of the sun, but they are in highly heated and excited atomic states and do not resemble their forms on earth. The bulk of the sun, however, is composed of two elements: Its volume is 81.76 percent hydrogen and 18.17 percent helium. All other elements total only 0.07 percent.

When the sun is at the zenith on a clear day, its light has a luminosity at the earth's surface of 10,000 candles.[1] The sun's light represents the radiant energy upon which terrestrial life depends for existence. The sun's radiant energy is believed to be generated by the conversion of matter in atomic reactions that form helium nuclei from hydrogen nuclei, in a process called the **carbon cycle** (Section 2.2). This takes place in the sun's deep interior.

When the nucleus of hydrogen $_1^1H$ collides with the nucleus of carbon $_6^{12}C$ and joins it to form the nucleus of nitrogen $_7^{13}N$, radiant energy is emitted. After five more steps the process terminates with the $_6^{12}C$ back in its original state but accompanied by a newly assembled helium nucleus that was formed along the way and by the energy converted from mass during the synthesis of helium (see Chapter 2). The carbon has acted simply as a **catalyst**; it takes part in the reaction but emerges unchanged so that it is used over and over again. The hydrogen has been converted to helium. In effect the sun is "burning" 4 million t of hydrogen per second, producing an "ash" of helium. Even at this rate the sun can keep on for 30 billion years.

These atomic reactions take place at temperatures

of several million degrees and are automatically controlled. If the temperature increases, the process operates too rapidly and expands the gas of the sun. This expansion causes cooling, which lowers the rate of energy production.

The sun's nuclear reactions create lethal gamma rays, which travel toward the sun's surface and change to X rays and ultraviolet rays on their journey as they collide with other atoms. The continuous output of solar energy has been calculated at 38×10^{16} W (watts).

Sunspots are eruptions of hot gas seen on the sun's surface, causing magnetic storms here on earth. They begin at about 20° or 30° north or south of the solar equator and move toward it. When they are within 8° of the equator, they disappear. Sunspots occur in groups, with each individual's being short-lived, rarely lasting more than 50 days. They range in size from a few hundred to over 80,000 km in diameter. Sunspots wax and wane in 11-year cycles. They seem to be a part of the larger cycle in which the sun's magnetic field reverses itself every 22 years. The last reversal occurred in 1957 and 1958, when sunspots were at their greatest intensity.

There are other eruptions from the sun's surface, which are related to magnetic discharges. The largest blasts of hot ionized gases—plasma—create the solar flares. These propel particles all the way to earth.

PLANETS AND SATELLITES

The name **planet** was given to certain celestial bodies that appear to wander about the sky, in contrast to the seemingly fixed stars. It came from the Greek *planēs*, meaning "wanderer." Our earth is a planet, one of nine that circle the sun. All travel around it from west to east in nearly the same plane, the **ecliptic**. And they rotate on their respective axes in a "forward" sense, or from west to east, with the exception of Venus and Uranus.

In order of distance from the sun the first four planets are Mercury, Venus, the earth, and Mars. These planets are about the same size and fairly dense, as though they were made of iron and stone. They are called the **terrestrial planets** because of their similarity to the earth. Next in order of distance from the sun are Jupiter, Saturn, Uranus, Neptune, and Pluto. The first four of these are of relatively large size and low density (see Figure 17.2). Little is known about Pluto, whose discovery was announced on March 12, 1930, but it is more like the terrestrial planets than the others. A uni-

---

[1] The *candle*, unit of intensity, is defined as one-sixtieth of the intensity of 1 cm² of a black-body radiator operated at the temperature of freezing platinum.

form pattern of spacing outward from the sun is broken between Mars and Jupiter. In the "gap" are the thousands of asteroids, ranging in size from 2 km to about 770 km in diameter. These asteroids are believed to be either the remains of a planet that exploded or matter that never completed the planet-forming process.

Table 17.1 lists some of the characteristics of the planets and our moon.

17.2 FEATURES OF THE MOON

The first man to see the moon with other than the naked eye was the Italian scientist Galileo Galilei (1564–1642). A Dutch lens grinder had just constructed the first telescope, and Galileo, imitating it, built his own instrument. Early in 1609 Galileo, then a mathematics professor at the University of Padua, turned his telescope on, among other celestial objects, the moon. What he saw is published in a treatise, *Sidereus nuncius (The Starry Messenger)*, one of the great landmarks of scientific discovery (Figure 17.3). For the first time man found that, as Galileo wrote, the "moon is not robed in a smooth and polished surface, but is rough and uneven" (Figure 17.4). In the last three and one-half centuries better telescopes, the camera, spacecraft and their instruments, and direct observation have given us a very good idea of what the moon is like and how it got that way (Figure 17.5).

What Galileo was able to see were dark zones, the **maria,** or "seas," and lighter-toned zones, the **highlands.** We now turn to a discussion of the nature of these zones and other features, as revealed by modern investigation methods.

MARIA

The Italian astronomer Giovanni Riccioli named most of the moon's maria back in 1651. Even as Galileo had before him, he felt that these great dark areas were actual seas of water. Today we know that they are floored with the products of volcanic activity, chiefly lava flows.

Looking at a map of the moon, one can distinguish two main types of mare. First there are the large (hundreds of kilometers across) circular maria, such as Imbrium, Serenitas, Crisium, Nectaris, Cognitum, and, partway around to the far side, Orientale. Then there is a group of irregular maria, such as Frigoris, Fecunditatis, Tranquillitatis, and the great Oceanus Procellarum.

The circular maria are outlined by discontinuous arcs of mountainous masses that stand hundreds and in many places thousands of meters above the maria floors. The Appenine Mountains on the southwestern

FIGURE 17.2

Relative sizes of the planets.

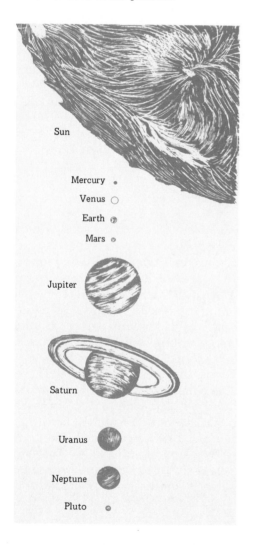

Sun

Mercury

Venus

Earth

Mars

Jupiter

Saturn

Uranus

Neptune

Pluto

TABLE 17.1
Some solar-system data

Planet	Mass, kg	Diameter, km	Av. distance from sun, thousands of km	Rotation period, earth units	Yr length, earth units	Mean density, g/cm³	Uncompressed density, g/cm³	Probable iron content, %
Mercury	3.30×10^{23}	4,864	58,000	59 da	88.0 da	5.5	5.2	57
Venus	4.87×10^{24}	12,100	108,000	243 da	224.7 da	5.2	4.0	30
Earth	5.98×10^{24}	12,756	147,598	23.9 h	365.3 da	5.52	4.0	30
[Moon]	7.35×10^{22}	3,476				3.34	3.3	10
Mars	6.44×10^{23}	6,788	229,000	24.6 h	687.0 da	3.9	3.7	26
Jupiter	1.90×10^{27}	137,400	780,000	9.8 h	11.9 yr	1.40		
Saturn	5.69×10^{26}	115,100	1,431,000	10.4 h	29.5 yr	0.71		
Uranus	8.76×10^{25}	50,000	2,880,000	10.8 h	84.0 yr	1.32		
Neptune	1.03×10^{26}	49,400	4,510,000	15.7 h	164.8 yr	1.63		
Pluto	6.6×10^{23}	5,880 ?	5,950,000	6.4 da	249.9 yr	6 ?		

FIGURE 17.3

In 1610 Galileo published *The Starry Messenger*, which contained the results of the first telescopic examination of the heavens. The title page is shown here. [Willard Starks.]

FIGURE 17.4

These reproductions from *The Starry Messenger* record part of what Galileo saw of the moon through his telescope. North is to the top, the last quarter is to the left, and the first quarter is to the right. Compare with Figure 17.5. [Willard Starks.]

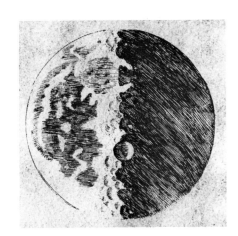
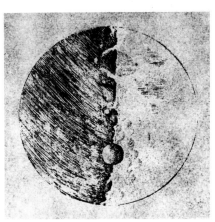

FIGURE 17.5

The moon as seen through a present-day optical telescope. North is to the top, the last quarter is to the left, and the first quarter is to the right. [Lick Observatory.]

margin of the Mare Imbrium, for instance, reach an elevation of about 4,300 m above the neighboring mare floor. The mare basins are flooded with dark, basaltic lava flows. Earthbound geologists have mapped many of these flows and see their frontal margins as 20 to 100 m in height. These mare fillings are heavily pocked with later craters. Mare Orientale looks like a great bull's-eye set in a series of concentric rings (Figure 17.6). It does not take much imagination to see the troughs and ridges that form these rings as great waves moving out from a common center. In fact this is the interpretation that has been offered by some who see these features as "frozen" waves created by the explosion that formed the main mare. Similar, though less well-displayed, shock-wave features can be seen around Mare Nectaris and most other circular maria.

We have already noted that igneous rocks appear to fill the basins of the maria and have also raised the question of what the source of the igneous melts on the moon could be. We can repeat here the observation that the lavas flooring the maria may be the result of very large impact events and not the product of heat originating in the subsurface.

The irregular maria are also lava-filled basins. In a very general way they lie outside the circular maria. For instance the continuous stretch of Mare Frigoris–Oceanus Procellarum surrounds Mare Nubium on three sides. On the west Nubium is contiguous to Serenitatis, and smaller irregular maria more or less complete the ring around these two. In many places the surfaces of the irregular maria are continuous through the broken and missing walls of the circular maria, the most conspicuous example being the connection between Nubium and Procellarum.

CRATERS

A **crater,** more than any other feature, is the trademark of the moon, and the circular maria we have just described are craters now partially filled. Here we shall look at the familiar but smaller craters. The largest of these smaller craters on the moon's near side is the crater Bailly, nearly 300 km across. Most of the craters we see on the moon, however, are a few tens of kilometers in diameter.

From analogy with topographic features on earth, we know that craterlike features can form in a number of ways. These include craters formed by impact of meteorites, volcanic craters that may be explosive or due to caldera collapse, and sinks from collapse of roof rock into voids created by the solution of sedimentary deposits such as limestone or halite in the underground. Other depressions, not necessarily with true crater forms, are the result of wind action or the melting of blocks of stagnant glacier ice. On the moon it is apparent that impact is the chief process of crater formation. Sinkholes, glacial kettle holes, and blowouts are ruled out because of the lack of atmosphere on the moon and hence the lack of moisture, sedimentary rocks, processes of solution, and the action of wind. Volcanic craters are very rare.

We now know that there is a range in the diameter of moon craters or pits from hundreds of kilometers to a few micrometers. We can refer to these craters as **primary craters,** formed by the original impact of meteorites, and as **secondary,** or **satellite, craters,** made by the falling fragments ejected from a primary crater.

Copernicus, a well-known, fairly large primary crater, shows many of the features characteristic of moon craters. It measures about 90 km in diameter and over 3 km in depth. The relatively level floor has a raised center portion, a feature that is more pronounced as a central peak in somewhat smaller craters. The walls facing into the crater are terraced or stepped and represent the slumping of large slices of the crater walls into the central depression. The rim of the crater is about 1 km high and slopes away from the crest at the crater's edge, to feather out at the edge of an irregular zone up to 150 km wide and concentric to the crater.

The material that forms the rim is zoned into three rings. The interior zone is very hummocky, with ridges concentric to the crater. Outward beyond this the material is less hummocky, and the topography is one of branching ridges more or less radial to the central crater. The outermost zone is still less hummocky, but the radial ridges of the intermediate zone still persist in subdued form. This zone is pocked by many small satellite craters thought to have been formed by fragments thrown out of the primary crater. Superimposed on the rim material and spreading far beyond it to a distance of 500 km from the crater are light-colored, discontinuous streaks, the **rays** of Copernicus (see Figure 17.7).

Copernicus is interpreted by most observers as an impact crater. The central rise on the floor of the crater is thought to result from upward adjustment following impact. The rim is thought to have been pushed up-

FIGURE 17.6

Mare Orientale is partially visible from the earth on the extreme southwestern side of the moon. Here in an *Orbiter 4* picture we see the entire mare. The outer circular scarp measures nearly 1,000 km in diameter. The concentric rings suggest waves generated by an energy source at the center of the bull's-eye. [NASA.]

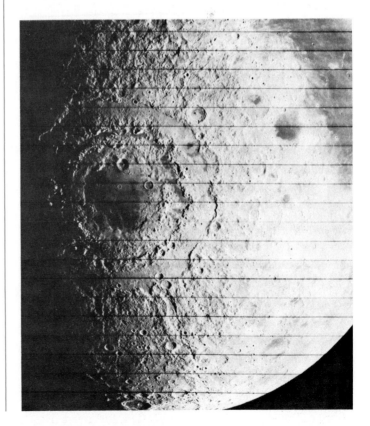

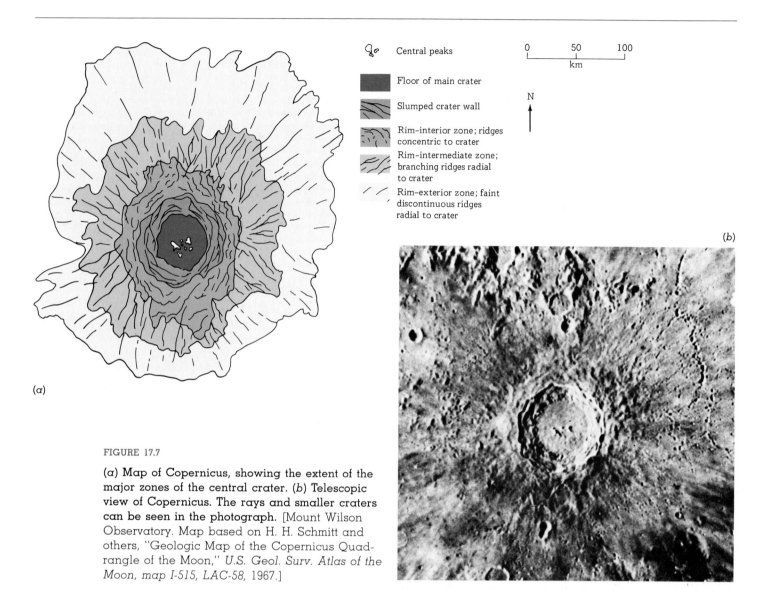

Central peaks

Floor of main crater

Slumped crater wall

Rim-interior zone; ridges concentric to crater

Rim-intermediate zone; branching ridges radial to crater

Rim-exterior zone; faint discontinuous ridges radial to crater

0   50   100
km

N

(b)

(a)

FIGURE 17.7

(*a*) **Map of Copernicus, showing the extent of the major zones of the central crater.** (*b*) **Telescopic view of Copernicus. The rays and smaller craters can be seen in the photograph.** [Mount Wilson Observatory. Map based on H. H. Schmitt and others, "Geologic Map of the Copernicus Quadrangle of the Moon," *U.S. Geol. Surv. Atlas of the Moon*, map I-515, LAC-58, 1967.]

ward and outward as a result of the impact explosion, and the hummocky ridges of the interior zone of the rim material are thought to stem from the same cause. Outward from this zone the areas of radial ridges are interpreted as material ejected from the crater and arranged in ridges subparallel to the direction of blast. The ray material is believed to be finely crushed matter thrown out from the primary crater and, to a lesser extent, similar material kicked up during formation of the satellite craters. Copernicus is one of a number of **ray craters** on the moon. Most craters, however, lack rays. Their absence is attributed to the greater age—hence longer weathering—of the nonray craters.

Some craters show only a low lip for a rim. In general these are old craters all but buried by material ejected from younger craters or by lava flows that filled the maria.

RILLES AND WRINKLE RIDGES

At various places the moon's surface is creased by long valleys called **rilles** (from the German for "furrow"). There are various forms, but they are from 10 to 150 km in length, from 0.5 to 4 km wide, and from 50 to 500 m deep, although depths of up to 1,000 m are reported.

Most rilles are related to the maria or to smaller, filled craters. There are basically two forms: One is

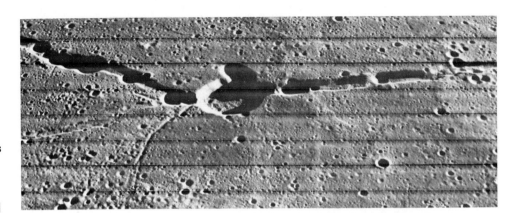

FIGURE 17.8

Rilles lead off from the crater Hyginus. The two more prominent rilles are marked by several smaller craters. Hyginus is about 10 km across, and the rille craters average about 2 km in diameter. [NASA, *Orbiter 3.*]

straight walled and, when it changes direction, does so at a well-defined angle; the second is sinuous and, in many instances, gives the impression of a meandering valley (see Figures 17.8 and 17.9).

The straight-walled types are generally thought to be downdropped strips of mare material. It is not clear whether the motion is one of collapse or whether it is caused by crustal stresses. Strings or chains of craters are located along a number of these rilles.

The sinuous rilles are particularly hard to interpret. The meandering outline suggests flowing water, a suggestion strengthened in the case of some rilles because smaller meandering channels are found in the rille floors. The similarity in form with river-formed features on earth has caused some to argue that water, in fact, caused these lunar features. The difficulty of obtaining an adequate supply of surface water has made this suggestion unacceptable to most investigators. Most students of the lunar landscape feel that the large rilles are due to the collapse of a lava tunnel, a void caused by the drainage of still-fluid lava from beneath a solidified surface zone.

Smaller rilles are found on the walls of some craters. They are different from those just described and display a border defined by levees. Very likely they represent streams of lava that flowed down into the crater, leaving levees along the cooling margins of the flows while the more molten centers drained craterward.

Across the floors of most maria and flooded craters, one sees long, low ridges, **wrinkle ridges,** comparable in scale to the large rilles. Some suggest that these are anticlinal ridges formed by laterally directed compressional stresses. Others regard them as uplifts caused by volcanic intrusions.

FIGURE 17.9

The Alpine valley creases the mountains on the northeast side of Mare Imbrium. Down its center winds a sinuous, discontinuous rille. In the upper right, toward the Mare Imbrium, are other sinuous rilles. See the text for discussion. [NASA, *Orbiter 5.*]

17.3 COMPOSITION OF THE MOON

The mean density of the moon is 3.3 g/cm³; that of the earth, 5.52. These data suggest that the mean density of the moon (and hence its composition) may be close to the composition of the earth's upper mantle. We can make another approximation, this based on the "uncompressed densities" of the moon and earth. Uncompressed density is the estimated density a body would have if all material were at the pressure of the body's surface rather than partially compressed in its interior. Such calculations give a value of 3.3 for the moon and 4.0 for the earth. These differences can be converted into estimates of the probable iron content of the two bodies. Such an exercise suggests that the moon is 10 percent iron and that the earth is 30 percent iron.

These speculations came from earth-based observations. We did not get a first-hand look at the actual material of the moon until the several *Apollo* missions returned samples from the moon's surface. What do they tell us?

MOON ROCKS

The several hundred kilograms of material returned to the earth from the moon by the *Apollo* missions have shown us that four general rock materials dominate the lunar surface.

Basaltic rocks are common and seem to form the

FIGURE 17.10 (LEFT)

A lunar gabbro as seen in a thin section magnified under a microscope. Minerals present are pyroxene (light gray), plagioclase feldspar (nearly white), and ilmenite (black). Width of section, about 1,300 μm. [NASA.]

FIGURE 17.11 (RIGHT)

A breccia from the moon as seen in a thin section magnified under the microscope. The small fragments of minerals and crystalline rock are set in a fine-grained matrix. Width of section, about 875 μm. [U.S. Geological Survey.]

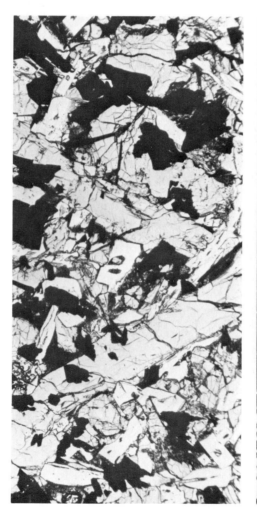

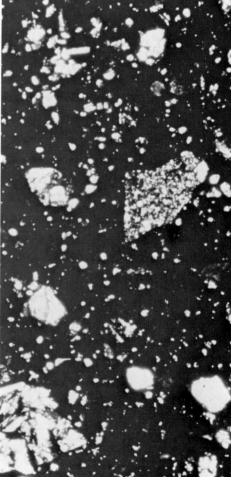

greater part of the filling of the many maria. These basalts differ slightly from those we find on earth. They contain, for instance, more titanium than our earthly basalts and are somewhat lower in silica, alumina, and sodium. Like all lunar samples they contain essentially no water.

The early *Apollo* missions brought samples of basalt and its plutonic equivalent, **gabbro** (Figure 17.10). In addition they brought us a third rock type, a **breccia** (Figure 17.11). These breccias contain fragments of basalt, gabbro, and **anorthosite.** We theorize that these rocks, composed of the fragments of other rocks, were consolidated by compaction arising from meteoritic impact.

The discovery of anorthosite fragments early in the *Apollo* program led to the speculation that there were large masses of this material on the moon. Anorthosite, a rock composed largely of the feldspar mineral anorthite, $Ca(Al_2Si_2O_8)$, usually contains a small amount of olivine and pyroxene. Anorthosite has a somewhat lower specific gravity than the basalt-gabbro group. Various lines of evidence lead us to believe that the highland areas are composed of anorthosite.

In addition to the rocks already mentioned there is a lunar "soil," which covers much of the moon's surface. This is not a soil as we think of soil on earth. It is made up of fragments of crystalline rocks, small glassy objects, and some meteoritic material. The assemblage appears to be the result of debris created by meteoritic impact.

MAGNETISM AND MASCONS

Our samples of moon rocks show a weak remanent magnetism. Furthermore an orbiting magnetometer shows that the moon's outer portion exhibits some magnetism. As we saw in Section 8.2, where we discussed the earth's magnetism, we believe that the earth's magnetism is caused by the dynamolike action of the fluid core of the earth. We infer that lunar rocks inherited their magnetism when the moon had a fluid core, a core now believed to be solidified.

One further item deserves mention. Spacecraft orbiting the moon are pulled slightly out of orbit by irregular masses embedded in various parts of the moon. These masses we call **mascons,** a word coined by shortening the descriptive words "mass concentrations." The zones of greater-than-average gravitational pull turn out to lie in the areas of the marias. We are not sure what these masses are. Were they on earth, we should expect them to sink isostatically. But this has not happened on the moon, and it suggests a crust strong enough to hold up large masses of material.

17.4 PROCESSES ON THE MOON

Much of our attention in this book has been directed to the processes that change the earth. We found that internal forces of volcanism and tectonism are countered by the surface process of water, wind, ice, and weathering, coupled with gravity. When we turn to the moon and examine the processes at work, we step into a scene at once familiar and unfamiliar when compared with the varied theater of our earthly experience.

A fundamental difference is the absence of a lunar atmosphere. There is no air, nor are there swirling clouds to distribute moisture. There is in fact no surface moisture although, as we shall see, there may be water or ice somewhere beneath the surface. Without an atmosphere there is no wind. We trace this lack of atmosphere and hydrosphere to the moon's small mass and hence its low gravitational attraction. This means that the escape velocity, the velocity that a particle must achieve to escape the lunar gravitational field, is a low 2.38 km/s. This is only a little over one-fifth of that on earth, 11.2 km/s, and lower than the escape velocity of gases and most liquids. Therefore, whatever gases and liquids may escape from the lunar interior to the surface leak off into space, as helium does from the earth. Thus, although the earth can retain nitrogen, carbon dioxide, and water at its surface and other planets retain ammonia, methane, helium, and probably even hydrogen, the moon is much too small to keep any of these elements and compounds. Without water and an atmosphere life as we know it cannot exist at all.

Another important difference between moon and earth lies in the nature of internal processes. The earth's crust is made up of a jigsaw puzzle of moving plates. The interaction of these plates produces long belts of mountains, contorted rocks, earthquakes, and

This stone at Tranquillity Base has a rounded upper surface and is set in finer material, part of which may have come from the breakdown of the stone itself. Arrows indicate location of some of the larger pits on the stone. Many of the bright spots represent reflections of pit-related glass. Reflections from the ground surface come from small fragments of glass in the granular lunar soil. Diameter of the rock is about 7 cm. [NASA, *Apollo 11.*]

volcanoes. They produce the midoceanic ridges and indeed the ocean basins themselves. The earth is, as we have seen, a dynamic system in which internal stresses are reflected in the topography. In contrast the trademarks of the lunar topography are the meteoritic craters that pock its surface. Few, if any, processes similar to the tectonic forces of earth are reflected in the lunar record.

It is not strange that the lunar surface should look different from that of the earth. And yet weathering does go on, movement does take place, and the moon's face does change although differently and at a slower pace than on the earth.

WEATHERING

In Chapter 4 we defined terrestrial weathering as a change in material at or near the earth's surface in response to atmosphere, water, and living matter. Obviously this is inapplicable to the lunar environment.

One of the first observations to be made on the moon was that many of the rocks that litter the surface are rounded on their upper, exposed portions but angular where in contact with the surface or partially buried by it. The conclusion that some sort of weathering process is affecting the exposed parts of the rocks is inescapable. Now that we have observed moon rocks at

Many of the small and intermediate-sized craters in this vertical photograph on the far side of the moon have sharp outlines. They are superposed on older craters of similar size and are identifiable by subdued outline. The longer weathering of the older craters is thought to account for the difference in crater freshness. Area shown is about 29 km across. [NASA, *Apollo 8.*]

first hand, we can see that they are marked by small pits ranging down to 20 μm or less in diameter. Those less than 4 mm in diameter are glass lined and in some instances rimmed with a glassy material. These features suggest a high-energy impact of particles with the lunar material. In those instances in which a rock fragment displays both rounded and angular sides, the pits occur more frequently on the rounded surfaces. The surface of the rocks tends to be somewhat lighter in tone than the interior, a characteristic apparently related to a thin zone of microfractures or glass attributed to impact of particles on the rock (see Figure 17.12).

Many of the younger craters of the moon are called ray craters because of the associated rays of material that spread outward as light-colored streaks created at the time of crater formation. The youth of these craters (Tycho, Kepler, and Copernicus are good examples) can be demonstrated because they and their associated features are superimposed on most other lunar features. The older craters, however, lack these rays, and the consensus is that they were originally present but have been obscured or destroyed by some type of weathering. The variation in sharpness of adjacent craters can be cited as a third example of weathering. Thus Figure 17.13 is a photograph of craters just south of the equator on the lunar far side. The small- and intermediate-sized craters show sharp, well-defined outlines in the low sun. They are superposed on a terrain pocked by craters of similar size but with much softer, more subdued outlines. Because of this superposition, we reason that the sharply outlined craters are younger than the softly outlined ones. We can then argue that the softened shapes of the older craters result from a longer period of lunar weathering than that of the younger craters. A final example is provided by the material thus far encountered on much of the moon's surface. It is a finely divided granular material, slightly cohesive but still of discrete particles (Figure 17.14). These individual particles could very

The foot of Neil Armstrong, first man on the moon, makes a clear imprint in the soil at Tranquillity Base. The soil has a granular texture, is slightly cohesive, and is made of discrete particles, many of which may represent the mechanical breakdown of larger rock masses by some processes of weathering. Depth of penetration of boot indicates that the soil here has a bearing strength between 3 to 7×10^{-4} kg/cm². [NASA, *Apollo 11.*]

well represent, at least in part, the mechanical breakdown of larger, more coherent pieces.

Now let us look at some of the ways in which weathering might take place on the moon.

Temperature changes The very great changes in temperature to which lunar materials are subjected may cause them to expand and contract so that they weaken and break. Probably the sudden changes that take place during lunar eclipses are the most effective although the slower temperature changes of the lunar day may, given time, be effective as well. Such action might either break pebble-sized fragments or spall off small bits from the surface of larger fragments.

Impact by meteorites We have examples on earth of what happens when a meteorite of large size crashes into rock. First the explosive impact creates a crater that is considerably larger than the meteorite. The rock is in part pulverized and badly shattered and broken. The rocks at the lip of the crater are bent upward and turned outward and over on themselves, and the material blasted from the crater area can be thrown for long distances. The meteorite, then, can break material into smaller and smaller pieces, truly a process of mechanical weathering. On the moon, also, we can predict that large meteorites have broken down coherent rock into smaller pieces and spread the products of impact across the lunar surface. The moon, in addition, is sub-

ject to a type of meteoritic bombardment that the earth is spared. Small meteorites do not survive the trip through the earth's atmosphere, being either melted or vaporized because of the heat generated by friction with the air. The moon, however, is not protected by an atmospheric envelope and undergoes a fairly continuous rain of small meteorites. The small pits already described on surfaces of lunar rocks are in part due to the impact of small meteorites. The process tends to break down rock material into smaller and smaller particles and to churn the uppermost portion of the rubble mantle that appears to shroud most of the moon's surface. Eugene M. Shoemaker, of the United States Geological Survey, estimates that these small meteorites—"micrometeorites," as he terms them—produce, every 100 years a complete turnover of the surface material to a depth of 1 mm.

Cosmic radiation The lack of an atmosphere exposes the lunar surface to direct attack by cosmic radiation, a process from which the earth is largely shielded. This radiation, made up chiefly of protons and secondarily of alpha particles (the nuclei of hydrogen and helium atoms, respectively), is known to cause damage to the crystal lattice of minerals and to have additional effects that, over long periods of time, could lead to the physical breakdown of minerals. Cosmic radiation may be in part responsible for the oblit-

FIGURE 17.15

This oblique photograph is of a portion of the crater Copernicus. Beyond the central peaks (over 300 m high) are the slumped blocks along the inner wall of the crater. The crater lip is about 3,000 m above the crater floor and approximately 45 km beyond the central peaks. Compare with the vertical photograph in Figure 17.16. [NASA, Orbiter 2.]

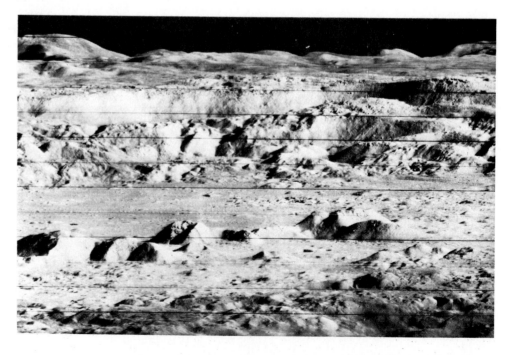

eration of the rays associated with the younger craters.

Tides We usually associate tides with our oceans. In addition, however, we know that the solid earth also reacts to the tidal pull of the moon and sun. We can measure these earth tides and find that at a maximum they amount to about 10 cm or less. The pull of the earth and sun on the moon is up to 80 times as strong as that exerted on the earth by sun and moon. The tides in lunar rock, therefore, must be correspondingly greater than those in the earth rocks. This constant kneading of the moon must contribute to the weakening and breaking down of bodies of lunar rock.

Rates of erosion Studies of materials collected on the moon have led to the conclusion that the rate of erosion of rocks on the lunar surface is about 10^{-6}mm/year, which contrasts with the average rate on earth (estimated at about 2.5×10^{-2}mm/year).

TRANSPORTATION

We must ascribe the movement of material on the moon's surface to other agents than wind and water.

Mass movement The movement of earth material under the influence of gravity was considered in Chapter 9. On the moon, as well, gravity plays a continuing role in moving material from higher to lower locations. Movements of particularly bulky masses characterize the interior margins of crater rims, especially large ones (see Figures 17.15 and 17.16). Here blocks of lunar rock material form great steplike features, each block being a slump similar to those described from the earth in Chapter 9.

Another example of gravity's action on surface material has been discovered in photography by lunar probe, particularly by the *Orbiter* missions. Boulders of varying size have left identifiable tracks hundreds of meters long as they trundled down lunar slopes.

To what extent slow, unspectacular downslope movements operate on the moon's surface is not yet known. On earth, where abundant moisture is present to facilitate movement, long-term creep of surficial material is important. Creep on lunar material remains to be demonstrated, as do the gravity movements of small individual particles although it seems reasonable to expect both.

Impact Impacting meteorites play their role in moving material across the lunar surface. Particles thrown free of the moon's surface move farther than do particles ejected with the same force on the earth's

FIGURE 17.16

A vertical view that includes a portion of the slump blocks along the inner wall of Copernicus shown in Figure 17.15. Some of the treads on the stepped slump blocks are very flat, probably the result of lakes of lava that filled the basins behind the blocks as they rotated downward and outward from the main wall. Area shown is about 53 km across. [NASA, *Orbiter 4*.]

surface. This is true for three reasons. First, there is no friction with an atmosphere to slow a particle moving across the moon's surface. Second, gravity is less than that on the earth. Third, the small radius of the moon means a greater angle of surface curvature than on the earth and thus a greater range for ejected particles. The rays of the crater Tycho, for instance, probably represent ejected material, and they extend for distances of 1,500 km from the crater.

Igneous activity Early in this chapter we noted that igneous rocks in fact exist on the moon. So we may ask not whether its rocks crystallize from a silicate melt but what the origin of the melt might be. We have two obvious choices of origin: First, the melt may be generated by internal heat of the moon in much the same way that magmas are generated on earth; second, the melt may result from the impact of meteorites on the moon. On a very small scale the glassy film associated with the pits on moon rocks reflects this impact energy. Increase the size of the impacting object to that which can create a depression the size of one of the moon's maria, and one may create a magma by impact.

The vast fillings of the maria represent lava flows. They occurred early in the history of the moon, as we shall see in Section 17.5. Whether they developed from impact or from more earthlike igneous processes is still discussed. But certainly there is little evidence that active volcanism, as known on earth, now exists.

Tectonism The absence of long, linear mountain masses on the moon similar to those of the Andean, Himalayan, and Rocky Mountain systems seems to rule out tectonic activity comparable to that on the earth. Despite this, certain features suggest movement of the moon's crust. For instance, the well-known Straight Wall in the western side of the Mare Nubium is approximately 270 m high and shows every indication of having been formed by faulting. The many rilles of the lunar surface also indicate some stress in the lunar crust and are comparable to some of the grabens and rift valleys on earth. Yet, as we have indicated, it is generally thought that these structural features are due to some cause other than tectonism: Possible sources of stress include impact, volcanic eruptions, and drainage away of molten lava beneath the surface.

Seismic instruments installed on the moon tell us that the moon experiences some small tremors. These are of two types: One is related to meteoritic impacts that generate seismic waves; the other seems to be true moonquakes. These seem to occur most commonly when the moon and earth are closest and tides are greatest. We have already noted that the tidal forces on the moon are some 80 times more powerful than those on earth. These tidally generated moonquakes occur at great depth, about 1,000 km. Some have suggested that this marks a boundary between a very rigid upper zone and a somewhat more plastic interior similar to the earth's asthenosphere, which is discussed in Section 6.12.

17.5 LUNAR HISTORY

We are not yet sure how the moon formed and came to be a satellite of the earth. Most workers currently think that the moon's origin is related in some way to the earth's origin. In this view the moon and earth formed more or less simultaneously and adjacent to each other, building up material derived from the original dust cloud from which the planetary system is thought to have evolved. Others have suggested that the earth once had rings, such as those of the planet Saturn, and that these rings eventually became the moon. Another suggestion, made originally by George Darwin, second son of Charles Darwin, is that the moon was torn from the earth at some past time. A fourth theory states that the moon once traveled in its own orbit around the sun but has since been captured by the earth and assumed its present orbit. About all we can conclude here is that knowledge about the very early history of the moon is in an unsatisfactory state.

Even though we cannot yet determine the origin and earliest history of the moon to the satisfaction of most students of that body, we have a fairly good idea of a great deal of the moon's history. Much of this story has been worked out by geologists of the United States Geological Survey working from photographs taken through earth-based telescopes and cameras carried in space probes. Rock units have been mapped on the basis of relative age, surface characteristics, and genetic types. Ages are established by cross-cutting relationships and superposition, stratigraphic principles discussed in Chapter 18 on geologic time.

Age determinations on lunar materials now give us some absolute ages that can be fitted into the relative chronology of events established by these other methods. The first samples returned to earth were of maria material, which turned out to have an age of about 3.6 billion years, at that time older than any known earth rock. Subsequent age dating has produced ages from about 3.2 billion years back to 4.6 billion years for different materials and events.

As we put together what we know about lunar history we can construct the following thumbnail biography of the moon:

The moon formed some 4.6 billion years ago, along with the earth and the other planets. Once formed, the moon heated up quickly and differentiation began. A crust of anorthosite rose to the surface, and heavier material formed a core that seems to have been molten according to the testimony provided by remanant magnetism. Massive bombardment by meteorites characterized the first 1.5 billion years. The great basins—the marias—were excavated by the impact of enormous meteorites. The undestroyed highland sections of anorthosite crust were deeply pocked by smaller meteorites. The large impact basins were eventually filled by basaltic materials, the bulk of the eruptions being between 3.2 to 3.8 billion years ago. Thereafter major volcanism ceased, and most students of the question believe that this marks the cooling of the moon and the solidification of a small liquid core. This cooling also

marks (but did not cause) a decrease in the numbers and sizes of meteorites impacting the moon. True, impressive craters have formed since then (as, for example, Copernicus and Tycho), but the period of mare-producing impacts had ceased.

Today the moon is a very quiet celestial body. There are few moonquakes (and, as we have said, these are tidally induced or represent small meteoritic impacts), little if any volcanism, no mountain building, no atmosphere, no water, no life. The planet Mars shows much greater activity, and we now turn to it.

17.6 MARS

Telescopic observation of Mars in the nineteenth century led some to believe in the presence there of long, straight "canals" that some unknown civilization had built to bring water from the poles to the arid regions of lower latitudes. Although twentieth-century spacecraft have laid this thought to rest, they have confirmed the presence of polar ice caps and of planet-obscuring dust storms, which earlier telescopic work had reported. Thus far the biological experiments conducted by martian landers have not identified past or present life.

Our first spacecraft pictures of Mars came from *Mariner 4*, which photographed the red planet in a flyby mission in 1965. By November, 1971, *Mariner 9* had

been inserted into orbit around Mars and, after the subsidence of a violent dust storm, relayed to us images that revealed a much more varied planet than we had expected. More information arrived from *Viking 1* and 2 orbiters and from their landing vehicles, which were successfully set down in July and September, 1976.

THE MARTIAN LANDSCAPE

We can divide Mars into two hemispheres: one of old, heavily cratered plains and the other of younger, lower, and lightly cratered plains (Figure 17.17). The

FIGURE 17.17

Generalized physiographic map of Mars. The clear area is lightly cratered plains. The lighter-colored area includes old, highly cratered terrain. Solid color marks young volcanoes. [Adapted from more detailed map by T. A. Mutch and R. S. Saunders.]

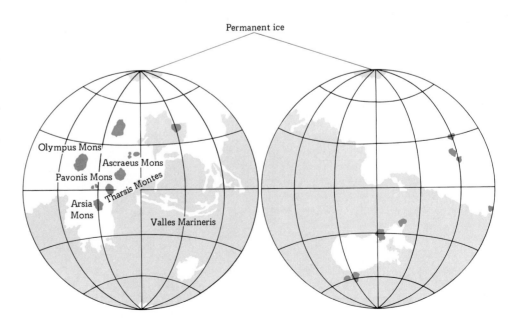

Permanent ice

Olympus Mons
Ascraeus Mons
Pavonis Mons
Tharsis Montes
Arsia Mons
Valles Marineris

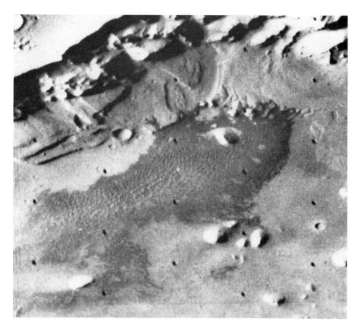

FIGURE 17.18

Particles moved by wind can totally obscure Mars, and in some places large areas of dunes occur on its surface. Here the darker, slightly mottled zone is a dune field in the bottom of the Ganges Chasm near the equator. It measures about 20 km by 50 km; sand motion seems to be from lower left to upper right. [NASA, *Viking 1.*]

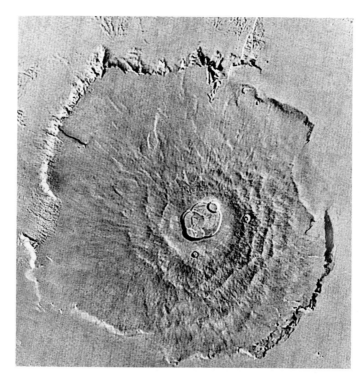

FIGURE 17.19

Olympus Mons (formerly Nix Olympica) volcano on Mars, more than 500 km in diameter and over 23 km high. This photograph is an enhanced photomosaic of five separate *Mariner 9* pictures prepared by Karl R. Blasius and James Soha of the Jet Propulsion Laboratory, Pasadena, California.

FIGURE 17.20

Part of Valles Marineris. Width of area shown is 480 km. [NASA, *Mariner 9.*]

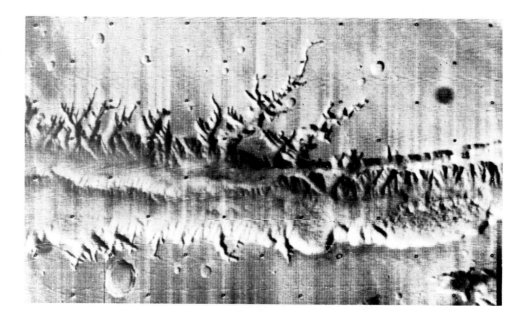

boundary between the two is inclined at about 50° to the martian equator. By extrapolating from cratered surfaces of known age on the moon, we believe that the heavily cratered surface of Mars is about 3.5 billion years old. The lightly cratered plains are estimated to be generally younger than 1.5 billion years.

Ice caps The polar ice caps, first recognized by telescope, expand and contract with the seasons. The composition of these was first thought to be all carbon dioxide (Dry Ice) but is now also known to contain water in the form of ice. Apparently the large-scale seasonal fluctuations of the caps involve the carbon dioxide component.

Wind action The atmosphere of Mars is indeed thin when compared to that of the earth. Largely carbon dioxide with some argon and nitrogen, the martian atmosphere has a pressure of less than 1 percent of the earth's. Nevertheless, dust storms occur and move material widely over the surface of the planet.

As we noted, *Mariner 9* encountered a dust storm which obscured almost all the planet and caused the spacecraft to wait for the dust to subside before imagery could be successfully carried on. We now know that windblown materials have accumulated in favorable locations on the martian surface, as shown in Figure 17.18.

Volcanism and tectonism In contrast to our moon, Mars has unquestionable and spectacular volcanic features. The largest volcano, Olympus Mons, measures more than 500 km across its base and may stand 23 to 29 km in height (Figure 17.19). As such, it is several times larger than any volcano on earth. And Olympus Mons is but one of nearly two-score volcanic centers known on Mars. Olympus Mons and several other large volcanoes are located on a broad topographic swell or dome known as Tharsis Montes, a feature that seems to indicate upwarping, perhaps similar to (but larger than) those over the hot spots known on earth. Around the margins of this uplift are fractures and faults. But we see no evidence of plate movement or of the associated mountain belts that we have on earth.

The volcanic slopes are comparatively fresh, and the sparseness of craters suggest, that the volcanism can be as young as 100 to 200 million years.

Canyons and sinuous channels Associated with the boundary between the lower, sparsely cratered plains and the higher, heavily cratered plains are great canyons and sinuous channels.

The canyons, such as Coprates Canyon and Valles Marineris, are huge. Valles Marineris, for instance, is

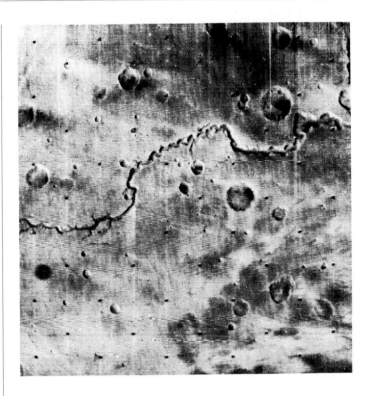

FIGURE 17.21

Sinuous valley on Mars, of which some 310 km are shown, 5 to 6 km wide. [NASA, *Mariner 9.*]

125 km wide and about 6 km deep and stretches for nearly 5,000 km. Such a feature seems to be a gigantic rift and points to vertical movements of the martian crust. Into some parts of the canyons lead intricate dendritic "drainage ways," as shown in Figure 17.20. In addition sinuous, even meandering, channels are also found (Figure 17.21), as well as wide, braided streamways (Figure 17.22). These features, taken together, suggest to many that running water has played an important role in fashioning the martian surface at some time in the planet's history.

Other features The polar caps reveal a layered pattern typical of stratified material (Figure 17.23). This has been interpreted as alternate layers of windblown material with ice.

Large areas of slump blocks similar to, but more extensive than, those found on the walls of lunar craters have been identified along the steep sides of some of the canyons (Figure 17.24). In addition, large tongues of material that resemble lobes of great mudflows are also present (Figures 17.24 and 17.25).

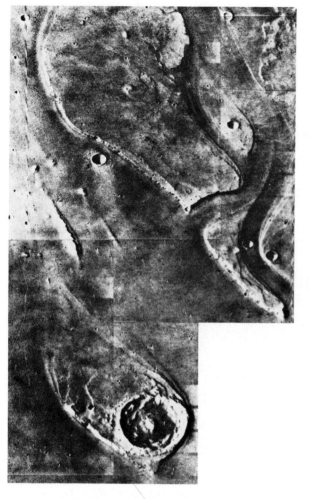

FIGURE 17.22

Braided channels of enormous size are seen in this mosaic of photographs taken over the Chryse area near the martian equator. The impact crater in the lower part of the figure measures about 15 km in diameter and has served as an obstruction around which waters have flowed to produce a streamlined pattern. More streamlined forms and interconnecting channels are seen in the upper part of the figure. [NASA, *Viking 1.*]

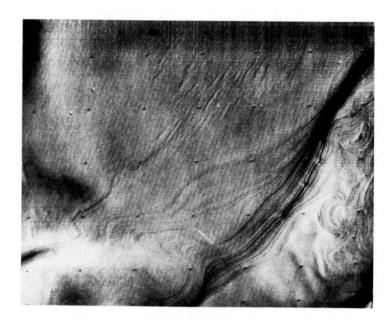

FIGURE 17.23

This layered pattern is seen in the martian north-polar region. The polar cap is made up of ice interlayered with windblown material. During the winter months the polar cap expands with the addition of extensive fields of solid carbon dioxide. The area shown is about 500 km from left to right. [NASA, *Mariner 9.*]

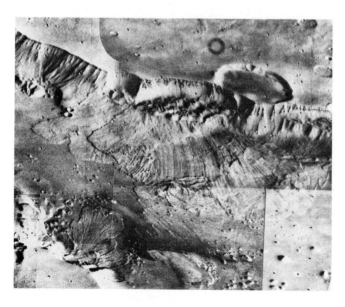

FIGURE 17.24

A view across the Valles Marineris shows a series of slump blocks along the canyon walls. Below the slumps stretch aprons of material that seem to have behaved as gigantic mudflows. At the bottom of the canyon obstacles have created wind shadows in which has accumulated wind-driven material. In the cliffs, which measure several hundreds of meters in height, differential erosion has etched out nearly horizontal layers of rock. [NASA, *Viking 1.*]

THE MARTIAN LANDERS

Viking 1 and *Viking 2* set landers down on Mars in 1976 and transmitted back the details of the planet's surface (Figure 17.26), which indeed turned out to be as red as myth, fable, and telescope would have it.

Chemical analyses of the fine material are similar for samples from both landing sites. They show an abundance of silicon and iron and significant concentrations of magnesium, aluminum, sulfur, calcium, and titanium. The material is interpreted as the weathered products of rocks of basaltic composition. Its mineral components are thought to be an iron-rich clay mineral (80 percent), magnesium oxides (10 percent), and calcite and iron oxides (5 percent each). Red oxides appear to veneer many of the larger fragments, a condition reminiscent of some of the earth's deserts. Some rocks give evidence of having been shaped in part by windblown material, and small accumulations of wind-accumulated material are visible in lander photographs.

Some have suggested that the rubbly surface, particularly at the second and more northerly landing site, may be related to glaciation dating from a time when the polar caps might have been much more extensive than they are today.

And what of life? As mentioned, the 1976 orbiters and landers have produced no positive data of present or past life. The possibility that life might have existed on Mars or even still does is small. But perhaps the evidence is to be found at an as-yet-to-be-visited site.

FIGURE 17.25

The crater Arandas, about 25 km in diameter, was formed by meteoritic impact. The ejecta from the crater seem to have moved as mudflows. It is suggested that the heat resulting from impact melted permafrost. The water thus created mixed with rock material to form the muddy ejecta. [NASA, *Viking 1.*]

FIGURE 17.26

The rubble-strewn surface at the site of Martian Lander 1. [NASA, *Viking 1.*]

FIGURE 17.27

The cratered surface of Mercury is similar to that of our moon. This detail of a portion of the planet also shows a cliff extending for more than 300 km from lower left to upper right. [NASA, *Mariner 10.*]

17.7 OTHER PLANETS

Not much is known about the geology of the other planets. But we have space probes of Venus, which have produced some tantalizing suggestions, and there are some flyby images of Mercury.

Venus is surrounded by a dense, yellowish cloud cover. Radar signals bounced off the planet's surface tell us that there is probably a cratered surface beneath the clouds and perhaps some linear mountain ranges as well. *Venera 9,* the Russian lander of 1975, sent back a panorama of a rubble-strewn surface with the suggestion of a crater rim in the distance. Radar also provides us with some evidence that the crust of Venus has rocks high in silicate content.

The temperature at the surface of Venus is very high. A Russian probe, parachuted to the surface, reported a temperature of 475°C. And it may be much higher in places. The high temperature is probably due to the atmosphere's being over 90 percent carbon dioxide, which provides a very strong "greenhouse effect." In this process the short-wavelength energy coming from the sun penetrates the atmosphere to the planet, where it warms the surface. Energy is then reradiated back into the atmosphere as long-wave radiation, which in turn is absorbed by the carbon dioxide, thus heating the atmosphere. A similar process occurs on earth, but only 0.03 percent of the atmosphere is carbon dioxide and is thus much less effective than the atmosphere of Venus in raising atmospheric and surface temperatures.

Mercury, which is the innermost planet, has an uncompressed density of 5.2, the greatest of any of the planets. This suggests a relatively large iron core. In March, 1974, *Mariner 10* sent us images from its flyby mission (Figure 17.27), and these revealed a heavily cratered surface very much like that of our moon. Although Mercury lacks maria—at least in the images we have—it seems that the planet is very similar to our moon.

As we progress outward in the planetary system, our information on the planets beyond Mars (Jupiter, Saturn, Uranus, Neptune, and Pluto) is increasingly less detailed. Our geologic knowledge is extremely inferential, and therefore we do not deal with it here. Some of the physical characteristics of these planets are listed in Figure 17.2 and Table 17.1.

OUTLINE

Our **solar system** consists of planets, satellites, asteroids, comets, and meteors held together by the sun's gravitation.

Our **sun** is a star that is the source of the solar system's light, heat, and charged particles.

Planets and **satellites** are celestial bodies that appear to wander about the sky.

Features of our moon include maria and highlands.

Maria are filled with igneous flows of basaltic material.

Craters are formed by meteoritic impact.

Rilles and **wrinkle ridges** are other features of the moon's surface.

Composition of the moon includes basalts in the maria and anorthosite in the highlands.

Magnetism of minerals suggests a once-molten core. **Mascons** underlie maria.

Processes on the moon are similar to, and different from, those on earth.

Weathering, because of lack of atmosphere, is restricted to the effect of **gravity**, **temperature, impact by meteorites, cosmic radiation,** and **tides.**

Transportation of materials is effected by **mass movement** and **meteoritic impact.**

Lunar history began with its origin 4.6 billion years ago and involved creation of maria basins by meteor impact and later filling by basalt between 3.2 and 3.8 billion years ago. Since then the moon has been quiet except for immediate meteoritic impact.

Mars has been explored by telescope, space probes, and landers.

The martian landscape is divided into a heavily cratered hemisphere and a lower, lightly cratered hemisphere. It includes **ice caps, wind deposits, volcanoes, canyons, sinuous channels,** and **landslides.** Signs of life are yet to be found on Mars.

Of the **other planets** Venus is apparently **cratered** and may have **mountain chains** and **high-silicate rocks** and very **high surface temperatures.** Mercury looks like our moon except for marias. Of the other planets we have little geologic information.

SUPPLEMENTARY
READINGS

Hartmann, **William K.**   *Moons and Planets*, Bogden & Quigley, Inc., Publishers, Tarrytown-on-Hudson, N.Y., 1972. *A readable account of our solar system.*

Mutch, Thomas A.   *Geology of the Moon: A Stratigraphic View*, 2nd ed., Princeton University Press, Princeton, 1974. *A fine summary of the geologic history of the moon.*

Mutch, Thomas A., R. E. Arvidson, K. L. Jones, J. W. Head, 3rd, and **R. S. Saunders** *Geology of Mars*, Princeton University Press, Princeton, 1976. *A good summary plus a complete photographic survey of Mars up to the Viking mission of 1976.*

Soffen, G. A., and others   "Scientific Results of the Viking Mission," *Science,* vol. 194, pp. 1274–1353, 1976. *Twenty-one short technical papers summarize the results of the* Viking *mission to November, 1976.*

18

TIME

In Chapter 1 we stressed the vastness of **geologic time** and emphasized its importance to an understanding of the physical processes of geology. In this chapter we shall look at some of the ways of measuring geologic time and at how the geologic-time scale has been worked out.

18.1 ABSOLUTE TIME

We earthlings use two basic units of time: the day, the interval required for our globe (in the present epoch) to complete one rotation on its axis, and the year, the interval required for the earth to complete one revolution around the sun. In geology, however, the problem is to determine how many of these units of time elapsed in the dim past when nobody was around to count and record them. Our most valuable clues in solving this problem are provided by the decay rates of radioactive elements.

RADIOACTIVITY

The nuclei of certain elements spontaneously emit particles, changing in size and/or form and producing new elements. This process is known as **radioactivity.**

Shortly after the turn of the twentieth century, researchers suggested that minerals containing radioactive elements could be used to determine the age of other minerals in terms of absolute time (see Figures 18.1 and 18.2 and Chapter 1).

Let us take a single radioactive element as an example of how this method works. Regardless of what element we use, we must know what products result from its radioactive decay. Let us choose uranium 238, $^{238}_{92}U$, which is known to yield helium and lead, $^{206}_{82}Pb$, as end products. We know too the rate at which uranium 238 decays. As far as we can determine, this rate is constant and unaffected by any known chemical or physical agency. The rate at which a radioactive element decays is expressed in terms of what we call its **half-life**—the time required for half of the nuclei in a sample of that element to decay. The half-life of uranium 238 is 4.51×10^9 years, which means that, if we start with 1 g of uranium 238, there will be only 0.5 g left after 4,510 million years.

The history of 1 g of uranium 238 may be recorded as shown in Table 18.1. At any instant during this process there is a unique ratio of lead 206 to uranium 238. This ratio depends on the length of time decay has been going on. Theoretically, then, we may find the age of a uranium mineral by determining how much lead 206 is present and how much uranium 238 is present. The ratio of lead to uranium then serves as an index to the age of the mineral.

One of the assumptions we have made in applying radioactivity to age determination is that the laws governing the rate of decay have remained constant over incredibly long periods of time; we are justified in wondering whether this assumption can be valid.

Geology supplies one piece of confirming evidence in the form of **pleochroic** ("many-colored") **halos**. These are minute, concentric, spherical zones of dark or colored material no more than 7.5×10^{-3} cm in diameter—the thickness of an ordinary sheet of paper—that form around inclusions of radioactive materials in biotite and in a few other minerals. If such a sphere is sliced through the center, the resulting sections show the pleochroic halos as rings. Each ring is a region in which alpha particles, shot out of the decaying radioactive mineral, came to rest and ionized the host material. The effect resembles that of light on a photographic film.

The energy possessed by an alpha particle depends on what element released it and on the stage of decay at which the particle was released. The pleochroic halos have radii that correspond to the energy of

FIGURE 18.1

The dark areas in this specimen from Grafton, New Hampshire, are made up of the uranium-bearing mineral gummite. Compare with Figure 18.2 [Specimen in the Princeton University Museum of Natural History, Willard Starks.]

FIGURE 18.2

A photograph of the same specimen as that in Figure 18.1. In this photograph the specimen was placed on a photographic plate. As the uranium and some of its radioactive daughter products decayed, the plate was exposed to the emission of particles. The white areas mark the location of the uranium-bearing minerals. [Willard Starks.]

TABLE 18.1
History of 1 g of uranium 238[a]

Age, millions of yr	$^{206}_{82}$Pb formed, g	$^{238}_{92}$U remaining, g
100	0.013	0.985
1,000	0.116	0.825
2,000	0.219	0.747
3,000	0.306	0.646
4,500 (1 half-life)	0.433	0.500
9,000 (2 half-lives)	0.650	0.250
13,500 (3 half-lives)	0.758	0.125

[a]From various sources.

present-day alpha particles. Apparently the energy of an alpha particle today is the same as it was hundreds of millions of years ago. This fact implies that the fundamental constants of nuclear physics that govern the travel of alpha particles have not changed.

This evidence, of course, does not prove explicitly that the rate of decay has always been the same. But if the laws governing the energy of the particles have not changed, it is reasonable to assume that related laws governing the rate of decay have also continued unchanged.

Uranium 238 is not the only element found useful in the age determination of rock material; there are others. But the same basic idea prevails, whatever element we choose. If we discover a mineral that contains one or more radioactive elements, we may be able (after proper chemical analysis) to determine how many years ago the mineral was formed. If the mineral was formed at the same time as the rock in which it is enclosed, the age of the mineral will also give us the age of the rock.

A radioactive element decays in one of several ways. The nucleus may lose an alpha particle—the nucleus of the helium atom, which, as we saw in Chapter 2, consists of two protons and two neutrons. This process is called **alpha decay,** and the mass of the element decreases by four (two protons plus two neutrons), and the atomic number decreases by two (two protons). An element undergoing **beta decay** loses an electron (a beta particle) from one of the neutrons of the nucleus. The neutron thereby becomes a proton, and the atomic number is increased by one. The mass—the sum of all protons and neutrons—remains the same. In a third type of decay, **electron-capture decay,** the nucleus changes by picking up an electron from its orbital electrons. This electron is added to a proton within the nucleus, converting the proton to a neutron and decreasing the atomic number by one.

Many elements, in decaying from a radioactive to a nonradioactive state, go through a series of transfor-

mations until one or more stable end products are reached. Thus uranium 238 begins to decay by alpha emissions before lead 206 is produced.

Some elements may follow two or more different paths in their decay. Potassium 40 is an example. It may decay either by beta emission to form calcium 40 or by electron capture to form argon 40.

Methods of determining age by radioactivity have produced tens of thousands of dates for events in earth history, and new ones are constantly being reported. For instance rocks from southwestern Minnesota and Greenland have been dated as approximately 3.8 billions years old. Field relations show that other, still older rocks exist. A granite about 3.2 billion years old from Pretoria, South Africa, contains inclusions of a quartzite, positive indication that older, sedimentary rocks existed before the intrusion of the granite. The exact age of the earth itself is still undetermined, but several lines of evidence converge to suggest an age of about 4.5 billion years.

Today's discoveries, then, have fully vindicated the assumptions made over a century and a half ago that geologic time *is* vast and that within earth history there *is* abundant time for slow processes to accomplish prodigious feats.

There are scores of radioactive isotopes, but only a few are useful in geologic dating (see Table 18.2). For older rocks potassium 40, rubidium 87, uranium 235, and uranium 238 have proved most important. Potassium 40 has also proved useful in dating some of the more recent events of earth history, events younger than 2 million years. For very recent events, however, the radioactive isotope of carbon, carbon 14, $^{14}_{6}C$, has proved most versatile. It is usable only on organic material that is around 50,000 years old or less than that age.

The carbon 14 method, first developed at the University of Chicago by Willard F. Libby, works as follows. When neutrons from outer space, sometimes called **cosmic rays,** bombard nitrogen in the outer atmos-

TABLE 18.2	Parent element	Half-life, yr	Daughter element	Type of decay
Some of the more useful elements for radioactive dating	Carbon 14	5,730	Nitrogen 14	Beta
	Potassium 40	1,300 million	Argon 40	Electron capture
	Rubidium 87	47,000 million	Strontium 87	Beta
	Uranium 235	713 million	Lead 207	Seven alpha and four beta
	Uranium 238	4,510 million	Lead 206	Eight alpha and six beta

phere, they knock a proton out of the nitrogen nucleus, thereby forming **carbon 14:**

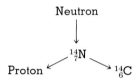

The carbon 14 combines with oxygen to form a special carbon dioxide, $^{14}CO_2$, which circulates in the atmosphere and eventually reaches the earth's surface, where it is absorbed by living matter. It has been found that the distribution of carbon 14 around the world is almost constant. Its abundance is independent of longitude, latitude, altitude, and the type of habitat of living matter.

The bulk of carbon in living material is the stable isotope, carbon 12. Nevertheless there is a certain small amount of carbon 14 in all living matter. And when the organism—whether it is a plant or an animal—dies, its supply of carbon 14 is, of course, no longer replenished by life processes (Figure 18.3). Instead, the carbon 14, with a half-life of about 5,730 years, begins spontaneously to change back to $^{14}_7N$ by beta decay. The longer the time that has elapsed since the death of the organism, the less the amount of carbon 14 that remains. So when we find carbon 14 in a buried piece of wood or in a charred bone, by comparing the amount present with the universal modern abundance, we can calculate the amount of time that has elapsed since the material ceased to take in carbon 14 dioxide, that is, since the organism died.

We have seen that there are assumptions on which we base our evaluation of the validity of radioactive dating. These assumptions are reasonable and have some support from observation. Nevertheless, the more we use radioactive methods, the more sophisticated our understanding of them—which is another way of saying that we become aware of the problems involved. Let us take carbon 14 as an example of how increased knowledge reveals increased complexity.

When Libby did his initial work on carbon 14, its half-life was determined to be 5,570 years. More refined measurements now indicate it to be about 5,730 years. By common agreement, however, 5,570 is still used as the "accepted value," and therefore all published dates should be adjusted upward 3 percent. Beyond this, Hans Suess, while working with the United States

FIGURE 18.3

This fragment of spruce log was part of a tree in a buried forest at Two Creeks, Wisconsin. Carbon 14 measurements of wood from the forest reveal that the trees died some 11,350 radiocarbon years ago, thereby establishing the date of the ice invasion. The fragment is about 20 cm long. [Willard Starks.]

Geological Survey, demonstrated that since about 1850 the amount of nonradioactive, or "dead," carbon 12 poured into the atmosphere by the combustion of coal and petroleum has diluted the amount of carbon 14 to 98 percent of its original concentration before human beings began tampering with the atmosphere. Working in the opposite direction, we have created a great amount of new carbon 14 since 1945, when we began atmospheric nuclear explosions. In fact this source of contamination has doubled the carbon 14 activity of the atmosphere.

Of more historical interest, perhaps, is the recent discovery that radiocarbon dates obtained on materials from about the time of Christ and going back for another 5,000 years, at least, are younger than we should expect when we compare them with carbon 14 dates of samples of known age dated independently of carbon 14. These samples have included historically dated material from the Mediterranean area, particularly from Egypt, and tree rings of the bristlecone pine from the western United States. This tree, which includes the world's oldest known living material, has provided us with a set of tree rings that goes back over 7,000 years and is securely tied to our modern solar calendar. When we subject tree rings older than 2,000 years to carbon 14 dating, the radiocarbon date is

FIGURE 18.4

Radiocarbon analyses of tree rings of known age show that data postulate ages that are too young for the period from the birth of Christ back at least into the sixth millennium B.C. The discrepancy increases from zero to about 750 years in 5000 B.C. [Data from E. K. Ralph and H. N. Michael, *Am. Sci.*, vol 62, pp. 553–560, 1974.]

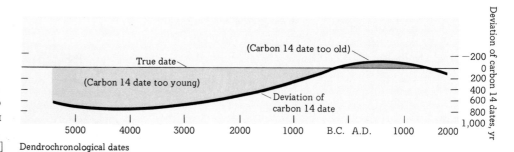

FIGURE 18.5

Varves from Puget Lowland, Washington. Strips of paper are used to record thickness and succession of the various layers. Each pair of light (summer) and dark (winter) layers is thought to represent a single year of sedimentation. (Note pocket knife holding uppermost paper.) [J. Hoover Mackin.]

younger than the actual date. The farther back we go, the greater is the divergence. For example by the time the tree ring count reaches 3200 B.C., the radiocarbon age is only 2700 B.C., (see Figure 18.4). We do not yet know the cause of this divergence or how far back into time it may continue. But it suggests that we are well-advised to use the term **radiocarbon years** to indicate that radiocarbon dates are not absolutely synonymous with dates in our solar calendar.

Despite these difficulties the nearly 20,000 radiocarbon dates now available have revolutionized many aspects of archaeology as well as the study of the last 50,000 years of earth history.

Fission-track dating One of the newest methods of absolute age dating results from the spontaneous fission of certain heavy elements, particularly the relatively abundant isotope uranium 238. As this isotope undergoes fission, the resulting parts of its nucleus leave characteristic tracks in the crystals of rocks where they occur. The number of such tracks gives us a clue to the absolute ages of those rocks, ages that can be measured from a few tens of years ago to the beginning of the solar system. This tremendous range makes the fission-track dating method extremely useful.

RADIOACTIVITY AND SEDIMENTARY ROCKS

Until recently, radioactive minerals suitable for dating geologic events were sought chiefly in the igneous rocks. These rocks were usually the uranium- and thorium-bearing minerals. Even today the great bulk of the radioactively dated rocks are igneous in origin. The development of new techniques, however, particularly the use of radioactive potassium, has extended radioactive dating to some of the sedimentary rocks.

Some sandstones and, more rarely, shales contain *glauconite*, a silicate mineral similar to biotite. Glauconite, formed in certain marine environments when the sedimentary layers are deposited, contains radioactive potassium, another geologic hourglass that reveals the age of the mineral and hence the age of the rock. The end products of the decay of radioactive potassium are argon and calcium. Age determinations are based on the ratio of potassium to argon.

Some pyroclastic rocks are composed mostly or completely of volcanic ash. Biotite in these rocks includes radioactive potassium and so offers a way of dating the biotite and sometimes the rock itself.

SEDIMENTATION AND ABSOLUTE TIME

Another way of establishing absolute dates for sedimentary strata is to determine the rate of their deposit.

Certain sedimentary rocks show a succession of thinly laminated beds. Various lines of evidence suggest that, in some instances at least, each one of these beds represents a single year of deposition. Therefore, by counting the beds, we can determine the total time it took for the rock to be deposited.

Unfortunately we have been able to link this kind of information to our modern calendar in only a very few places, such as the Scandinavian countries. Here the Swedish geologist Baron de Geer counted the annual deposits, or laminations, that formed in extinct glacial lakes. These laminations, called **varves** (see Figure 18.5 and Chapter 12), enable us to piece together some of the geologic events of the last 20,000 years or so in the countries ringing the Baltic Sea.

Much longer sequences of laminated sediments have been found in other places, but they tell us only the *total length of time* during which sedimentation took place, not *how long ago* it happened in absolute time. One excellent example of absolute-time sequence is recorded in the Green River shales of Wyoming (see Figures 18.6 and 18.7). Here each bed, interpreted as an annual deposit, is less than 0.017 cm thick, and the total thickness of the layers is about 980 m. These shales then represent approximately 6.5 million years of time.

FIGURE 18.6

The Green River Shale in Wyoming is composed of minute annual layers. Because each step in this block is 100 layers high, the entire block records 700 years of sedimentation. A portion of a fossil fish is seen on the large step on the left-hand side of the block. Counting of the layers indicates that the fish died 471 years after the formation of the lowest layer. [Specimen from Princeton University Museum of Natural History, Willard Starks.]

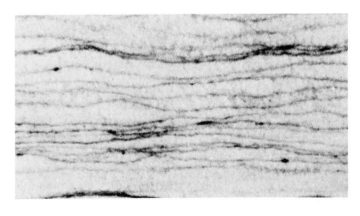

FIGURE 18.7

An enlarged section of Green River Shale, showing, greatly magnified, the layered nature of the rock. The specimen represents about 50 years of time in a piece of rock about 0.5 cm thick. At this rate of accumulation 1 m of Green River Shale represents approximately 10,000 years. [Specimen from Princeton University Museum of Natural History, Fred Anderegg.]

18.2 RELATIVE TIME

Before geologists knew how to determine absolute time, they had discovered events in earth history that convinced them of the great length of geologic time. In putting these events in chronological order, they found themselves subdividing geologic time on a relative basis and using certain labels to indicate relative time. You have probably picked up a newspaper or magazine (or indeed this textbook) and have read of the discovery of a dinosaur that lived 100 million years ago during the Cretaceous Period or of the development of a new oil field in strata formed 280 million years ago in the Pennsylvanian Period. The names of the periods are terms used by geologists to designate certain units of **relative geologic time.** In this section we shall look at how such units have been set up and how absolute dates have been suggested for them.

Relative geologic time has been determined largely by the relative positions of sedimentary rocks. Remember that a given layer of sedimentary rock represents a certain amount of time—the time it took for the original deposit to accumulate. By arranging various sedimentary rocks in their proper chronological sequence, we are in effect arranging units of time in *their* proper order. Our first task in constructing a relative-time scale, then, is to arrange the sedimentary rocks in their proper order.

THE LAW OF SUPERPOSITION

The principle used to determine whether one sedimentary rock is older than another is very simple, and it is known as the **law of superposition.** Here is an example. A deposit of mud laid down this year in, say, the Gulf of Mexico will rest on top of a layer that was deposited last year. Last year's deposit, in turn, rests on successively older deposits that extend backward into time for as long as deposition has been going on in the gulf. If we could slice through these deposits, we should expose a chronological record with the oldest deposit on the bottom and the youngest on top. This sequence would illustrate the law of superposition: **If a series of sedimentary rocks has not been overturned, the topmost layer is always the youngest and the lowermost layer is always the oldest.**

The law of superposition was first derived and used by Nicolaus Steno (1638–1687), a physician and naturalist who later turned theologian. His geologic studies were carried on in Tuscany, in northern Italy, and were published in 1669. On first glance the principle worked out by Steno is absurdly simple. For instance in a cliff of sedimentary rocks, with one layer lying on top of another, it is perfectly obvious that the oldest is on the bottom and the youngest on top. We can quickly determine the relative age of any one layer in the cliff in relation to any other layer. The difficulty, however, lies in the fact that unknown hundreds of thousands of feet of sedimentary rock have been deposited during earth time and that there is no one cliff, no one area, where *all* these rocks are exposed to view. The rocks in one place may be older or younger or of the same age as those in some other place. The task is to find out how the rocks around the world fit into some kind of relative-time scale.

CORRELATION OF SEDIMENTARY ROCKS

Because we cannot find sedimentary rocks representing all of earth time neatly arranged in one convenient area, we must piece together the rock sequence from locality to locality. This process of tying one rock sequence in one place to another in some other place is known as **correlation,** from the Latin for "together" plus "relate."

Correlation by physical features   When sedimentary rocks show rather constant and distinctive features over a wide geographic area, we can sometimes connect sequences of rock layers from different localities.

Figures 18.8 and 18.9 illustrate how this is done. Here is a series of sedimentary rocks exposed in a sea cliff. The topmost, and hence the youngest, is a sandstone. Beneath the sandstone we first find a shale, then a seam of coal, and then more shale extending down to the level of the modern beach. We can trace these layers for some distance along the cliff face, but how are they related to other rocks farther inland?

Along the rim of a canyon that lies inland from the cliff we find that limestone rocks have been exposed. Are they older or younger than the sandstone in the cliff face? Scrambling down the canyon walls, we come to a ledge of sandstone that looks very much like the sandstone in the cliff. If it *is* the same, then the limestone must be younger because it lies above it. The only

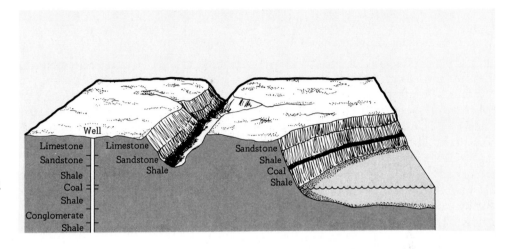

FIGURE 18.8

Diagram to illustrate the data that might be used to correlate sedimentary rocks (right) in a sea cliff with those (center) in a stream valley and with those (left) encountered in a well-drilling operation. (See also Figure 18.9.)

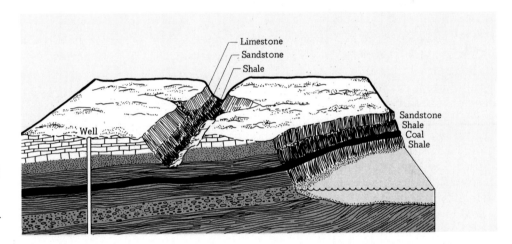

FIGURE 18.9

Similar lithologies and sequences of beds in the three different localities of Figure 18.8 suggest the correlation of rock layers shown in the diagram.

trouble is that we cannot be certain that the two sandstone beds are the same; so we continue down to the bottom of the canyon and there find some shale beds very similar to the shale beds exposed in the sea cliff beneath the sandstone. We feel fairly confident that the sandstone and the shale in the canyon are the same beds as the sandstone and upper shale in the sea cliff, but we must admit the possibility that we are dealing with different rocks.

In searching for further data, we find a well drilled still farther inland, its rig cutting through the same limestone that we saw in the canyon walls. As the bit cuts deeper and deeper, it encounters sandstone, shale, a coal seam, more shale, and then a bed of conglomerate before the drilling finally stops in another shale bed. This sequence duplicates the one we observed in the sea cliff and reveals a limestone and

an underlying conglomerate and shale that we have not seen before. We may now feel justified in correlating the sandstone in the sea cliff with that in the canyon and that in the well hole: The limestone is the youngest rock in the area; the conglomerate and lowest shale are the oldest rocks. This correlation is shown in Figure 18.9.

Many sedimentary formations are correlated in just this way, especially when physical features are our only keys to rock correlation. But as we extend the range of our correlation over a wider and wider area, physical features become more and more difficult to use. Clearly it is impossible to use physical characteristics to determine the relative age of two sequences of layered rocks in, say, England and the eastern United States. Fortunately we have another method of correlation, a method that involves the use of fossils.

Correlation by fossils Around the turn of the nine-teenth century an English surveyor and civil engineer named William Smith (1769–1839) became impressed with the relationship of rock strata to the success of various engineering projects, particularly the building of canals. As he investigated rock strata from place to place, he noticed that many of them contained fossils. Furthermore he observed that (no matter where found) some rock layers contained identical fossils, whereas the fossils in rock layers above or below were different. Eventually Smith became so skillful that, when confronted with a fossil, he could name the rock from which it had come.

At about the same time two French geologists, Georges Cuvier (1769–1832) and Alexandre Brongniart (1770–1847), were studying and mapping the fossil-bearing strata that surround Paris. They too found that certain fossils were restricted to certain rock layers, and they had also used the law of superposition to arrange the rocks of the Paris area in chronological order, just as Smith had done in England. Then Cuvier and Brongniart arranged their collection of fossils in the same order as the rocks from which the fossils had been dug. They discovered that the fossil assemblages varied in a systematic way with the chronological positions of the rocks. In comparing the fossil forms with modern forms of life, Cuvier and Brongniart discovered that the fossils from the higher rock layers bore a closer resemblance to modern forms than did the fossils from rocks lower down.

From all these observations it became evident that the **relative age** of a layer of sedimentary rock could be determined by the nature of the fossils that it contained.

This fact has been verified time and again by other workers throughout the world. It has become an axiom in geology that **fossils are a key to correlating rocks** and that **rocks containing the same fossil assemblages are similar in age.**

But when we apply this axiom to actual situations involving the use of fossils for correlation, certain complications arise. For instance it is obvious that in our modern world the distribution of living forms varies with the environment. This fact was presumably as true in the past as it is now. We found, in discussing facies of sedimentary rocks, that different sediments were laid down in different places at the same time. Plants and animals also reflect changes in environment, particularly if they happen to live on the sea bottom. Organisms living in an area where mud is being deposited are different from those living in an area where sand is being laid down. Thus we find fossils in a bed of shale (formed from mud) somewhat different from those in a bed of sandstone (formed from sand).

If both the physical features and the fossils are different, how can we correlate the rocks in two different areas? There are two possible ways: First we may actually be able to see that two different rock types, with their differing fossils, grade into each other laterally as we follow the beds along a cliff face; second we may find that a few fossils occur in both environments. In an oceanic environment, for instance, some forms float or swim over a wide geographic area that takes in more than one condition of deposition on the sea floor. The remains of these swimming or floating forms may settle to the bottom, to be incorporated in different kinds of sediment forming on the ocean floor.

18.3 THE GEOLOGIC COLUMN

Using the law of superposition and the concept that fossils are an index to time, geologists have made chronological arrangements of sedimentary rocks from all over the world. They have pictured the rocks as forming a great column, with the oldest at the bottom and the youngest at the top. The pioneer work in developing this pattern was carried out in the British Isles and in western Europe, where modern geology had its birth. There geologists recognized that the change in the fossil record between one layer and the next was not gradual but sudden. It seemed as if whole segments recording the slow change of plants and animals had

been left out of the rock sequence. These breaks, or gaps, in the fossil record served as boundaries between adjoining strata. The names assigned to the various groups of sedimentary rocks are given in the **geologic column** of Table 18.3. (See also front end paper.)

Notice that the oldest rocks are called **Precambrian,** a general term applied to all the rocks that lie beneath the Cambrian rocks. Although the Precambrian rocks represent the great bulk of geologic time, we have yet to work out satisfactory subdivisions for them because there is an almost complete absence of fossil remains, so plentiful in Cambrian and younger

TABLE 18.3

The geologic column

Era	System	Series	Aspects of the life record	
			Some major events	Dominant life form[a]
Cenozoic	Quaternary[b]	Holocene (Recent)		Man
		Pleistocene		
	Tertiary[b]	Pliocene		Mammals
		Miocene	Grasses become abundant	
		Oligocene		Flowering plants
		Eocene	Horses first appear	
		Paleocene		
Mesozoic	Cretaceous	—[c]	Extinction of dinosaurs	
	Jurassic		Birds first appear	Reptiles / Conifer and cycad plants
	Triassic		Dinosaurs first appear	
Paleozoic	Permian	—[c]		Spore-bearing land plants / Amphibia
	Pennsylvanian		Coal-forming swamps	
	Mississippian			
	Devonian			Fish
	Silurian		Vertebrates first appear (fish)	Marine invertebrates
	Ordovician			
	Cambrian		First abundant fossil record (marine invertebrates)	Marine plants
Precambrian	—[d]			Primitive marine plants and invertebrates / One-celled organisms

[a]This column does not give the complete time range of the forms listed. For example, fish are known from pre-Silurian rocks and obviously exist today; but when the Silurian and Devonian rocks were being formed, fish represented the most advanced form of animal life.
[b]Some geologists prefer to use the term Cenozoic for the Quaternary and the Tertiary.
[c]Many Mesozoic and Paleozoic series are distinguished but are not necessary here.
[d]Precambrian rocks are abundant, but worldwide subdivisions are not generally agreed upon.

rocks. Without fossils to aid them, geologists have been forced to base their correlations on the physical features of the rocks and on dates obtained from radioactive minerals. Physical features have been useful for establishing local sequences of Precambrian rocks, but these sequences cannot be extended to worldwide subdivisions. On the other hand radioactive dates are not yet numerous enough to permit subdivisions of the Precambrian rocks although such dates will become increasingly important as more are assembled.

We have said that the geologic column was originally separated into different groups of rocks on the basis of apparent gaps in the fossil record. But as geologic research progressed and as the area of investigation spread from Europe to other continents, new discoveries narrowed the gaps in the fossil record. It is now apparent that the change in fossil forms has been continuous and that the original gaps can be filled in with data from other localities. This increase in information has made it more and more difficult to draw clear boundaries between groups of rocks, and yet, despite the increasing number of "boundary" problems, the broad framework of the geologic column is still valid.

18.4 THE GEOLOGIC-TIME SCALE

The names in the geologic column refer to rock units that have been arranged in a chronological sequence from oldest to youngest. Because each of the units was formed during a definite interval of time, they provide a basis for setting up time divisions in geologic history.

In effect the terms we apply to time units are the terms that were originally used to distinguish rock units. Thus we speak either of Cambrian *time* or of Cambrian

TABLE 18.4

The geologic-time scale[a]

Era	Period	Epoch	Millions of yr	
			Duration	Before present
Cenozoic	Quaternary	Holocene	0.01	
		Pleistocene	1.5–2	
Mesozoic	Tertiary	Pliocene	5–5.5	
		Miocene	19	
Paleozoic		Oligocene	11–12	
		Eocene	15–17	
		Paleocene	11–12	
				65
	Cretaceous		71	
	Jurassic		54–59	
	Triassic		30–35	
				225
	Permian		55	
	Pennsylvanian		45	
Precambrian	Mississippian		20	
	Devonian		50	
	Silurian		35–45	
	Ordovician		60–70	
	Cambrian		70	
				570
	Precambrian time[b]			

[a]From W. B. Harland, A. Gilbert Smith, and B. Wilcock (eds.), "The Phanerozoic Time-Scale," *Quart. J. Geol. Soc. London*, vol. 120S (suppl.), 1964.

[b]The oldest rocks are about 3.8 billion years old; the earth itself is 4 to 5 billion years old.

rocks. When we speak of *time units*, we are referring to the geologic-time *scale*. When we speak of *rock units*, we are referring to the geologic *column*.

The geologic-time scale is given in Table 18.4 as well as the front end paper. Notice the terms **Era, Period,** and **Epoch** across the top of the table. These are general time terms. Thus we can speak of the Paleozoic Era, or the Permian Period, or the Pleistocene Epoch. In Table 18.3 the terms **System** and **Series,** used as general terms for rock units, correspond to the time units period and epoch, respectively. The term **Erathem** is becoming the generally accepted rock term equivalent to the era of the geologic-time scale.

ABSOLUTE DATES IN THE GEOLOGIC-TIME SCALE

We have found that the geologic-time scale is made up of units of relative time and that these units can be arranged in proper order without the use of any designations of absolute time. This relative-time scale has been constructed on the basis of sedimentary rocks. As noted earlier, geologists have only recently been able to date some sedimentary rocks by radioactive methods, and most dates have come from the igneous rocks. How can the dates obtained from igneous rocks be fitted in with the relative-time units from sedimentary rocks?

To insert the absolute age of igneous rocks into the geologic-time scale, we must know the relative-time relationships between the sedimentary and igneous rocks. The basic rule here is called the **law of crosscutting relationships,** which states that **a rock is younger than any rock that it cuts across.**

Let us consider the example given in Figure 18.10, a hypothetical section of the earth's crust with both igneous and sedimentary rocks exposed. The sedimentary rocks are arranged in three assemblages, numbered 1, 3, and 5, from oldest to youngest. The igneous rocks are 2 and 4, also from older to younger.

The sedimentary rocks labeled 1 are the oldest rocks in the diagram. First they were folded by earth forces; then a dike of igneous rock was injected into them. Because the sedimentary rocks had to be present before the dike could cut across them, they must be older than the dike. After the first igneous intrusion, erosion beveled both the sedimentary rocks and the dike, and across this surface were deposited the sedimentary rocks labeled 3. At some later time the batholith, labeled 4, cut across all the older rocks. In time this

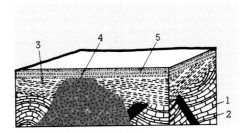

FIGURE 18.10

This diagram illustrates the law of crosscutting relationships, which states that a rock is younger than any rock that it cuts. The rock units are arranged in order of decreasing age from 1 to 5. The manner in which radioactive ages of the igneous rock (2 and 4 of the diagram) are used to give approximate ages in terms of years for the sedimentary rocks (1, 3, and 5 of the diagram) is discussed in the text and in Table 18.5.

TABLE 18.5

Relative, absolute, and approximate ages of rocks[a]

Event	Age		
	Relative	Absolute, millions of yr	Approximate, millions of yr
Sedimentary rocks	5[b]		<230
Erosion			<230
Batholith	4	230	
Sedimentary rocks	3		>230, <310
Erosion			>230, <310
Dike	2	310	
Folding			>310
Sedimentary rocks	1		>310

[a]See Figure 18.10 and the text.
[b]Youngest.

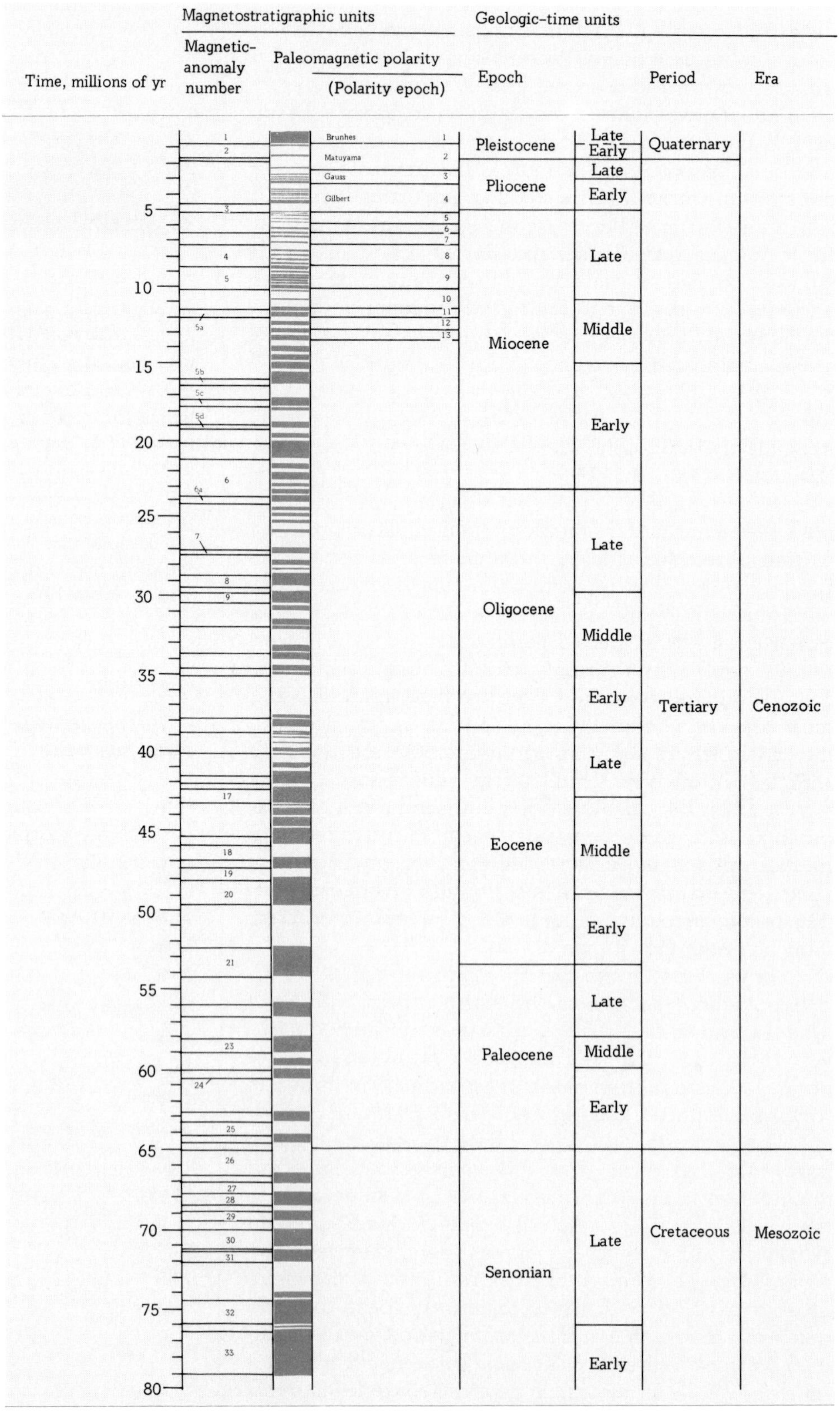

FIGURE 18.11

Chart of magnetostratigraphy, showing magnetic-anomaly numbering system (based on marine magnetic-anomaly analysis), polarity epochs (based on potassium-argon dating and paleomagnetic studies of outcropping Pliocene and Pleistocene igneous rocks), and the polarity-epoch numbering system (based on micropaleontological and paleomagnetic analysis of piston cores). See also **Figure 8.8.** [Modified from N. D. Watkins, *Geotimes*, April, 1976.]

batholith and the sedimentary rocks 3 were also beveled by erosion, and the sedimentary rocks labeled 5 were laid across this surface. We now have established the relative ages of the rocks, from oldest to youngest, as 1, 2, 3, 4, and 5.

Now, if we can date the igneous rocks by means of radioactive minerals, we can fit these dates into the relative time sequence. If we establish that the batholith is 230 million years old and that the dike is 310 million years old, the ages of the sedimentary rocks may be expressed in relation to the known dates; the final arrangement will be as shown in Table 18.5.

By this general method approximate dates have been assigned to the relative-time units of the geologic-time scale, as shown in Table 18.4. These dates may be revised and refined as new techniques for dating develop. One of the most exciting developments in this field lies in the direct application of dating methods to the radioactive minerals of sedimentary rocks, as suggested earlier.

MAGNETOSTRATIGRAPHY

The direction of the earth's magnetic field in past geologic ages can be determined by examining lava flows on land, deep-sea cores of sediments, and igneous rocks from many parts of the earth's surface (see Chapter 8). These rocks contain iron minerals (magnetite, ilmenite, hematite) that become magnetized at the time of their formation, thus creating a "magnetic fingerprint" in the rock for a certain instant of geologic time. These rock records indicate that the earth's magnetic field has changed directions many times, with the present field representing **normal** directions for north and south. During intervals of **reversed polarity** the poles apparently switch from south to north and vice versa. These changes have taken place fairly rapidly (geologically speaking) and have affected newly forming magnetic minerals. The process of reversal takes about 2,000 years according to studies conducted at the University of Pittsburgh.

Reversals have apparently occurred during at least the last billion years. Studies of magnetism in rocks from various parts of the world can be used with the geologic time scale to work out a magnetic time chart. A pattern emerges for the polarity reversals through geologic time, with **polarity epochs** lasting roughly 1 million years. Within these epochs there occur several **polarity events** in which the earth's polarity switches from one direction to another and back to the first, with durations on the order of 100,000 years for each polarity. Such a pattern has existed for the last 70 million years (see Figure 18.11). Prior to that time there appeared long intervals in which the polarity directions remained fairly constant. This latter uniform interval was preceded by a "mixed interval," when changes occurred frequently. Such polarity episodes of rapid change alternating with long intervals of little change occurred as far back as a billion years, according to present understanding. As we go farther back in geologic time, of course, the difficulties of age dating and measuring magnetic directions increase enormously. Nevertheless a distinct pattern is emerging for the use of magnetized rocks in determining a **magnetic stratigraphy (magnetostratigraphy)** for the earth.

OUTLINE

Geologic time can be expressed in either relative or absolute terms.

Absolute time is expressed in terms of years and is measured by the rate of decay of radioactive elements. Among the most useful such elements are **carbon 14, potassium 40, rubidium 87, uranium 235,** and **uranium 238.**

Relative time has been determined largely by the relative position of sedimentary rocks to each other.

The rocks are arranged in proper chronological position according to the **law of superposition.**

Correlation of sedimentary rocks from one area to another allows the extension of a chronology of relative time from region to region and continent to continent. Correlation is based on **physical features** but most importantly on **fossils.**

The geologic column is a chronological sequence of rocks, from oldest to youngest. The units are rock units.

The geologic-time scale is the chronological sequence of units of earth time represented by the rock units of the world geologic column. Absolute dates in the geologic-time scale are based on the relation of radioactively dated igneous, sedimentary, and metamorphic rocks of the geologic column. Position in the time scale of events dated from radioactive rocks is determined by the relative position of those rocks and the fossils they contain.

Magnetic reversals have occurred at intervals during the geologic past and can be used to subdivide earth history into **epochs** of **magnetostratigraphy.**

SUPPLEMENTARY READINGS

Berry, William B. N. *Growth of a Prehistoric Time Scale,* W. H. Freeman and Company Publishers, San Francisco, 1968. *Historical account of the development of the geologic time scale, with emphasis on the role of organic evolution.*

Libby, W. F. "Radiocarbon Dating," *Am. Sci.,* vol. 44, no. 1, pp. 98–112, 1956. *A summary of the method written by the man who discovered it.*

Ojakangas, R. W., and D. G. Darby *The Earth, Past and Present,* McGraw-Hill Book Company, New York, 1976. *A condensed review of the earth through time, including biologic and physical developments.*

Ralph, E. K., and H. N. Michael "Twenty-five Years of Radiocarbon Dating," *Am. Sci.,* vol. 62, no. 5, pp. 553–560, 1974. *A review of errors inherent in radiocarbon dating and the use of long-lived bristlecone pines to correct radiocarbon dates.*

Watkins, N. D. "Polarity Subcommission sets up some Guidelines," *Geotimes,* vol. 21, no. 4, pp. 18–20, 1976. *A short history of its development and presentation of data and chart for magnetostratigraphy.*

MINERALS

Many of the most common minerals may be identified in hand specimens by physical properties. Among the characteristics useful for this purpose are those discussed in Chapter 2, including crystal form, cleavage, striations, hardness, specific gravity, color, and streak. Still other properties are found to be useful in identifying certain groups of minerals. Some of these properties are summarized in the text and tables that follow.

A.1 MINERAL CHARACTERISTICS

CRYSTAL SYMMETRY AND SYSTEMS

As noted in Chapter 2, when a mineral grows without interference, it is bounded by plane surfaces symmetrically arranged, which give it a characteristic crystal form. This form is the external expression of its definite internal crystalline structure. The faces of crystals are defined by surface layers of atoms.

Every crystal consists of atoms arranged in a three-dimensional pattern that repeats itself regularly. Even in irregular mineral grains the atoms are arranged according to their typical crystalline structure.

Crystals are classified in six different systems according to the **symmetry** of their faces and the arrangement of their **axes of symmetry.** (An axis is an imaginary straight line that is drawn from the center of a face to the center of the opposite face.) The systems are described and illustrated in Table A.1.

HARDNESS

We can determine the hardness of a mineral by scratching its smooth surface with the edge of another. We must be sure that the mineral tested is actually scratched: Sometimes particles simply rub off the specimen, suggesting that it has been scratched even though it has not.

In Table A.2 ten common minerals have been arranged as examples of the degrees of the Mohs scale of relative hardness. Each of these minerals will scratch all those lower in number on the scale and will be scratched by all those higher. In other words this is a *relative scale*. In terms of absolute hardness the steps are approximately uniform up to 9; that is, number 7 is 7 times as hard as 1, and 9 is 9 times as hard as 1. But 10 is about 40 times as hard as 1. A more extensive listing is in Table A.3 (and in Table A.8).

MAGNETISM

Minerals that in their natural state are attracted to a magnet are said to be *magnetic*. Magnetite, Fe_3O_4, and pyrrhotite, $Fe_{1-x}S$, with x between 0 and 0.2, are the only common magnetic minerals although many others containing iron are drawn to a sufficiently powerful electromagnet.

TABLE A.1
Basic crystal systems[a]

Unit-cell shape[b]	Name: description	Typical forms of common-mineral crystals
	Isometric: three equal-length axes—all at right angles	Garnet Magnetite Halite, pyrite
	Tetragonal: two equal axes and a third either longer or shorter—all at right angles	Zircon
	Hexagonal: three equal axes in same plane intersecting at 60°, a fourth perpendicular to other three	Quartz Calcite, dolomite Hematite
	Orthorhombic: three unequal axes—all at right angles	Olivine Aragonite
	Monoclinic: three unequal axes—two at right angles, a third perpendicular to one but oblique to other	Pyroxene (augite) Mica, clay minerals Orthoclase Gypsum
	Triclinic: three unequal axes—all at oblique angles	Plagioclase Calcium aluminum silicate

[a]From Sheldon Judson, Kenneth S. Deffeyes, and Robert B. Hargraves, *Physical Geology*, Prentice-Hall, Inc., Englewood Cliffs, N.J., 1976.
[b]Symmetry axes in color.

TABLE A.2
Mohs scale of hardness

Scale	Mineral	Test
1	Talc	[Softest]
2	Gypsum	
2.5		Fingernail
3	Calcite	Copper coin
4	Fluorite	
5	Apatite	
5.5–6		Knife blade or plate glass
6	Orthoclase	
6.5–7		Steel file
7	Quartz	
8	Topaz	
9	Corundum	
10	Diamond	[Hardest]

PYROELECTRICITY

Pyroelectricity is the simultaneous development of positive and negative charges of electricity on different parts of the same crystal under the proper conditions of temperature change. Quartz is a good example: If it is heated to about 100° C, it will, on cooling, develop positive electric charges at three alternate prismatic edges and negative charges at the other three edges.

PIEZOELECTRICITY

Piezoelectricity is that of a charge developed in a crystallized body by pressure. Quartz is probably the most important piezoelectric mineral; for an extremely slight pressure parallel to its "electric axis" can be detected by the electric charge set up. It is used in specially oriented plates in radio equipment and in sonic sounders.

LUSTER

Luster is the way a mineral looks in reflected light. There are several kinds of luster (see page 439).

TABLE A.3
Minerals arranged according to hardness

Hardness	Mineral	Hardness	Mineral	Hardness	Mineral	Hardness	Mineral
1	Talc	3.5–4	Chalcopyrite	5–6	Tremolite	6.5–7	Jadeite
1–2	Graphite	3.5–4	Dolomite	5.5	Chromite	6.5–7	Olivine
1–3	Bauxite	3.5–4	Siderite	5.5	Enstatite	6.5–7	Spodumene
2	Gypsum	3.5–4	Sphalerite	5.5	Uraninite	6.5–7.5	Almandite
2	Stibnite	4	Azurite	5.5–6	Anthophyllite	6.5–7.5	Garnet
2–2.5	Chlorite	4	Fluorite	5.5–6.5	Hematite	7	Kyanite (across crystal)
2–2.5	Kaolinite	5	Apatite	6	Albite	7	Quartz
2–2.5	Muscovite	5	Kyanite (along crystal)	6	Anorthite	7–7.5	Staurolite
2–5	Serpentine	5–5.5	Goethite	6	Arfvedsonite	7–7.5	Tourmaline
2.5	Galena	5–5.5	Limonite	6	Magnetite	7.5	Andalusite
2.5	Halite	5–5.5	Wollastonite	6	Orthoclase	7.5	Zircon
2.5–3	Biotite	5–6	Actinolite	6–6.5	Aegirite	8	Spinel
2.5–3	Chalcocite	5–6	Augite	6–6.5	Pyrite	8	Topaz
3	Bornite	5–6	Diopside	6–7	Cassiterite	9	Corundum
3	Calcite	5–6	Hornblende	6–7	Epidote	10	Diamond
3–3.5	Anhydrite	5–6	Opal	6–7	Sillimanite		

TABLE A.4
Minerals arranged according to specific gravity

Specific gravity	Mineral	Specific gravity	Mineral	Specific gravity	Mineral	Specific gravity	Mineral
1.9–2.2	Opal	2.8–3.2	Biotite	3.3–3.5	Jadeite	4.02	Corundum
2.0–3.0	Bauxite	2.85	Dolomite	3.3–4.37	Goethite	4.1–4.3	Chalcopyrite
2.16	Halite	2.85–3.2	Anthophyllite	3.35–3.45	Epidote	4.25	Almandite
2.2–2.65	Serpentine	2.89–2.98	Anhydrite	3.4–3.55	Aegirite	4.52–4.62	Stibnite
2.3	Graphite	3.0–3.25	Tourmaline	3.4–3.6	Topaz	4.6	Chromite
2.32	Gypsum	3.0–3.3	Actinolite	3.45	Arfvedsonite	4.68	Zircon
2.57	Orthoclase	3.0–3.3	Tremolite	3.5	Diamond	5.02	Pyrite
2.6	Kaolinite	3.15–3.2	Apatite	3.5–4.1	Spinel	5.06–5.08	Bornite
2.6–2.9	Chlorite	3.15–3.2	Spodumene	3.5–4.3	Garnet	5.18	Magnetite
2.62	Albite	3.16	Andalusite	3.56–3.66	Kyanite	5.26	Hematite
2.65	Quartz	3.18	Fluorite	3.6–4.0	Limonite	5.5–5.8	Chalcocite
2.7–2.8	Talc	3.2	Hornblende	3.65–3.75	Staurolite	6.8–7.1	Cassiterite
2.72	Calcite	3.2–3.4	Augite	3.77	Azurite	7.4–7.6	Galena
2.76	Anorthite	3.2–3.5	Enstatite	3.85	Siderite	9.0–9.7	Uraninite
2.76–3.1	Muscovite	3.23	Sillimanite	3.9–4.1	Sphalerite		
2.8–2.9	Wollastonite	3.27–3.37	Olivine	4.0	Carnotite		

TABLE A.5
Scale of fusibility

Scale	Mineral	Approx. fusing point, °C	Remarks
1	Stibnite	525	Easily fusible in candle flame
2	Chalcopyrite	800	Small fragment easily fusible in Bunsen-burner flame
3	Garnet (almandite)	1050	Infusible in Bunsen flame but easily fusible in blowpipe flame
4	Actinolite	1200	Sharp-pointed splinter fuses with little difficulty in blowpipe flame
5	Orthoclase	1300	Fragment edges rounded with difficulty in blowpipe flame
6	Bronzite	1400	Only fine splinter ends rounded in blowpipe flame
7	Quartz	1470	Infusible in blowpipe flame

TABLE A.6
Plagioclase feldspars

Species	Albite, %	Anorthite, %
Albite, $Na(AlSi_3O_8)$	100–90	0– 10
Oligoclase	90–70	10– 30
Andesine	70–50	30– 50
Labradorite	50–30	50– 70
Bytownite	30–10	70– 90
Anorthite, $Ca(Al_2Si_2O_8)$	10– 0	90–100

TABLE A.7
Ions in common pyroxenes and amphiboles

Ions			
X	Y	Pyroxenes[a]	Amphiboles[b]
Mg	Mg	Enstatite	Anthophyllite
Ca	Mg	Diopside	Tremolite
Li	Al	Spodumene	
Na	Al	Jadeite	Glaucophane
Na	Fe^{3+}	Aegirite	Arfvedsonite
Ca, Na	Mg, Fe, Mn, Al, Fe^{3+}, Ti	Augite	Hornblende

[a]Basic structure: single chain, SiO_3; formula, $XY(Si_2O_6)$.
[b]Basic structure: double chains, Si_4O_{11}; formula, $X_{0-7}Y_{7-14}Z_{16}O_{44}(OH)_4$.

TABLE A.8
The common minerals[a]

Mineral	Chemical name and/or composition	Form	Cleavage and/or fracture	Hardness	Specific gravity
Actinolite (an *asbestos*; an *amphibole*)	Calcium iron silicate, $Ca_2(Mg, Fe)_5Si_8O_{22}(OH)_2$ (*tremolite* with more than 2% iron)	Slender crystals, usually fibrous	See *amphibole*	5–6	3.0–3.3
Aegirite (a *pyroxene*)	$NaFe^{3+}(Si_2O_6)$	Slender prismatic crystals	Imperfect cleavage at 87° and 93°	6–6.5	3.40–3.55

[a]From Cornelius S. Hurlbut, Jr., *Dana's Manual of Mineralogy*, 18th ed., John Wiley & Sons, Inc., New York, 1971.

Metallic Of metals
Adamantine Of diamonds
Vitreous Of a broken edge of glass
Resinous Of yellow resin
Pearly Of pearl
Silky Of silk

FLUORESCENCE AND PHOSPHORESCENCE

Minerals that become luminescent during exposure to ultraviolet light, X rays, or cathode rays are *fluorescent.* If the luminescence continues after the exciting rays are shut off, the mineral is said to be *phosphorescent.*

FUSIBILITY

Minerals can be divided into those fusible and those infusible in a blowpipe flame. Seven minerals showing different degrees of fusibility have been used as a scale to which fusible minerals can be referred. They are listed in Table A.5.

SOLUBILITY

Concentrated hydrochloric acid, HCl, diluted with three parts of water, is commonly used for the solution of minerals being tested.

Other wet reagents are for special tests to help identify minerals.

FRACTURE

Many minerals that do not exhibit cleavage (Chapter 2) break, or fracture, in a distinctive manner. Some types of fracture are:

Conchoidal Along smooth curved surfaces like the surface of a shell ("conch"); commonly observed in glass and quartz
Fibrous or splintery Along surfaces roughened by fibers or splinters
Uneven or irregular Along rough, irregular surfaces
Hackly Along a jagged, irregular surface with sharp edges

TENACITY

A mineral's cohesiveness, as shown by its resistance to breaking, crushing, bending, or tearing, is known as its *tenacity.* Various kinds of tenacity in minerals include the following:

Brittle Breaks or powders easily
Malleable Can be hammered into thin sheets
Sectile Can be cut by a knife into thin shavings
Ductile Can be drawn into wire
Flexible Bends but does not return to its original shape when pressure is removed
Elastic After being bent, will resume its original position upon release of pressure

A.2 IMPORTANT MINERALS

SILICATES

More than 90 percent of rock-forming minerals are silicates, with structures based on the $(SiO_4)^{4-}$ tetrahedron. Important classes are listed in Table 2.3 on page 29.

PLAGIOCLASE FELDSPARS

The plagioclase feldspars, also called the *soda-lime feldspars,* form a complete solid-solution series from pure albite to pure anorthite.

Calcium substitutes for sodium in all proportions, with accompanying substitution of aluminum for silicon. The series is divided into the six arbitrary species names listed in Table A.6 (also see Table A.8).

PYROXENES AND AMPHIBOLES

The pyroxene family of minerals and the amphibole family of minerals are inosilicates that parallel each other. The amphiboles contain OH^-. The pyroxenes crystallize at higher temperatures than their amphibole analogues. The two are in Table A.7 (also Table A.8)

Color	Streak	Luster	"Transparency"	Other properties
White to light green	Colorless	Vitreous	Transparent to translucent	A common ferromagnesian metamorphic mineral; a common component of green schists
Brown or green		Vitreous	Translucent	Rare rock former, chiefly in rocks rich in soda and poor in silica

Mineral	Chemical name and/or composition	Form	Cleavage and/or fracture	Hardness	Specific gravity
Albite	Sodic plagioclase *feldspar*, $Na(AlSi_3O_8)$	Tabular crystals; striations caused by twinning	Good in 2 directions at 93°34′	6	2.62
Almandite	$Fe_3Al_2(SiO_4)_3$	12- or 24-sided; massive or granular		6.5–7.5	4.25
Amphibole family	See *anthophyllite, arfvedsonite, hornblende, tremolite*		Perfect prismatic at 56° and 124°, often yielding splintery surface		
Andalusite	Aluminum silicate, Al_2SiO_5	Usually in coarse, nearly square prisms; cross section may show black cross	Not prominent	7.5	3.16
Andesine	A *plagioclase feldspar* 50–70% *albite*				
Anhydrite	Anhydrous calcium sulfate, $CaSO_4$	Commonly in massive fine aggregates not showing cleavage; crystals rate	3 directions at right angles to form rectangular blocks	3–3.5	2.89–2.98
Ankerite	Dolomite in which ferrous iron replaces more than 50% magnesium				
Anorthite	Calcic *plagioclase feldspar*, $Ca(Al_2Si_2O_8)$	Striations caused by twinning; lathlike or platy grains	Good in 2 directions at 94°12′	6	2.76
Anthophyllite (an *amphibole*)	$(Mg, Fe)_7(Si_8O_{22})(OH)_2$	Lamellar or fibrous	See *amphibole*	5.5–6	2.85–3.2
Apatite	Calcium fluophosphate, $Ca_5(F, Cl)(PO_4)_3$	Massive, granular	Poor in 1 direction; conchoidal fracture	5	3.15–3.2
Arfvedsonite (a sodium-rich amphibole)	$Na_3Mg_4Al(Si_8O_{22})(OH, F)_2$	Long prismatic crystals	See *amphibole*	6	3.45
Asbestos	See *actinolite, chrysotile, serpentine*				
Augite (a *pyroxene*)	Ferromagnesian silicate, $Ca(Mg, Fe, Al)(Al, Si_2O_6)$	Short, stubby crystals with 4- or 8-sided cross section; often in granular, crystalline masses	Perfect prismatic along 2 planes at nearly right angles, often yielding splintery surface	5–6	3.2–3.4
Azurite	Blue copper carbonate, $Cu_3(CO_3)_2(OH)_2$	Crystals complex in habit and distorted; sometimes in radiating spherical groups	Fibrous	4	3.77
Bauxite	Hydrous aluminum oxides of indefinite composition; not a mineral	In rounded grains or earthy, claylike masses	Uneven fracture	1–3	2–3
Biotite (black *mica*)	Ferromagnesian silicate, $K(Mg, Fe)_3AlSi_3O_{10}(OH)_2$	Usually in irregular foliated masses; crystals rare	Perfect in 1 direction into thin, elastic, transparent, smoky sheets	2.5–3	2.8–3.2
Bornite (peacock ore; purple copper ore)	Copper iron sulfide, Cu_5FeS_4	Usually massive; rarely in rough cubic crystals	Uneven fracture	3	5.06–5.08

Color	Streak	Luster	"Transparency"	Other properties
Colorless, white, or gray	Colorless	Vitreous to pearly	Transparent to translucent	Opalescent variety, *moonstone*
Deep red	White	Vitreous to resinous	Transparent to translucent	A garnet used to define one of the zones of middle-grade metamorphism; striking in schists
				A group of silicates with tetrahedra in double chains; *hornblende* is the most important; contrast with *pyroxene*
Flesh red, reddish brown, or olive green	Colorless	Vitreous	Transparent to translucent	Found in schists formed by middle-grade metamorphism of aluminous shales and slates; the variety chiastolite has carbonaceous inclusions in the pattern of a cross
				As grains in igneous rock; chief *feldspar* in andesite lavas of the Andes Mountains
White; may have faint gray, blue, or red tinge	Colorless	Vitreous or pearly	Transparent to translucent	Found in limestones and in beds associated with salt deposits; heavier than *calcite*, harder than *gypsum*
				Formed in low-grade regional metamorphism from conversion of *calcite* or *dolomite* or both between 80 and 120°C.
Colorless, white, gray, green, yellow, or red	Colorless	Vitreous to pearly	Transparent to translucent	Occurs in many igneous rocks
Gray to various shades of green and brown		Vitreous		From metamorphism of *olivine*
Green, brown, or red	White	Glassy	Translucent to transparent	Widely disseminated as an accessory mineral in all types of rocks; an important source of fertilizer; a transparent variety is a gem, but too soft for general use
Deep green to black		Vitreous	Translucent	Rock former in rocks poor in silica
				General term for certain fibrous minerals with similar physical characteristics though different composition; *chrysotile* is most common
Dark green to black	Greenish gray	Vitreous	Translucent only on thin edges	An important igneous rock-forming mineral found chiefly in simatic rocks
Intense azure blue	Pale blue	Vitreous to dull or earthy	Opaque	An ore of copper; a gem mineral; effervesces with HCl
Yellow, brown, gray, or white	Colorless	Dull to earthy	Opaque	An ore of aluminum; produced under subtropical to tropical climatic conditions by prolonged weathering of aluminum-bearing rocks; a component of *laterites*; clay odor when wet
Black, brown, or dark green	Colorless	Pearly or glassy	Transparent or translucent	Constructed around tetrahedral sheets; a common and important rock-forming mineral in both igneous and metamorphic rocks
Brownish bronze on fresh fracture; tarnishes to variegated purple and blue, then black	Grayish black	Metallic	Opaque	An important ore of copper

Mineral	Chemical name and/or composition	Form	Cleavage and/or fracture	Hardness	Specific gravity
Bronzite (a *pyroxene*)	*Enstatite* with 5–13% FeO				
Bytownite	A *plagioclase feldspar*, 10–30% albite				
Calcite	Calcium carbonate, $CaCO_3$	Usually in crystals or coarse to fine granular aggregates; also compact, earthy; crystals extremely varied—over 300 different forms	Perfect in 3 directions at 75° to form unique rhombohedral fragments	3	2.72
Carnotite	Potassium uranyl vanadate, $K_2(UO_2)_2(VO_4)_2$	Earthy powder	Uneven fracture	Very soft	4
Cassiterite (tinstone)	Tin oxide, SnO_2	Commonly massive granular	Conchoidal fracture	6–7	6.8–7.1
Chalcocite (copper glance)	Copper sulfide, Cu_2S	Commonly aphanitic and massive; crystals rare but small, tabular with hexagonal outline	Conchoidal fracture	2.5–3	5.5–5.8
Chalcopyrite (copper *pyrites*; yellow copper ore; fool's gold)	Copper iron sulfide, $CuFeS_2$	Usually massive	Uneven fracture	3.5–4	4.1–4.3
Chlorite	Hydrous ferromagnesian aluminum silicate, $(Mg, Fe^{2+})_5(Al, Fe^{3+})_2Si_3O_{10}(OH)_8$	Foliated massive or in aggregates of minute scales	Perfect in 1 direction, like *micas*, but into inelastic flakes	2–2.5	2.6–2.9
Chromite	Iron chromium oxide, $FeCr_2O_4$	Massive, granular to compact	Uneven fracture	5.5	4.6
Chrysotile (*serpentine asbestos*)	See *serpentine*				
Clay	See *illite*, *kaolinite*, *montmorillonite*				
Corundum (ruby, sapphire)	Aluminum oxide, Al_2O_3	Barrel-shaped crystals; sometimes deep horizontal striations; coarse or fine granular	Basal or rhombohedral parting	9	4.02
Diamond	Carbon, C	Octahedral crystals, flattened, elongated, with curved faces	Octahedral	10	3.5
Diopside (a *pyroxene*)	$CaMg(Si_2O_6)$	Prismatic crystals; also granular massive	Poor prismatic	5–6	3.2
Dolomite	Calcium magnesium carbonate, $CaMg(CO_3)_2$	Rhombohedral crystals with curved faces; granular cleavable masses or aphanitic compact	Perfect in 3 directions at 73°45′	3.5–4	2.85
Emery	Black granular *corundum* intimately mixed with *magnetite*, *hematite*, or iron *spinel*				

Color	Streak	Luster	"Transparency"	Other properties
				Primarily as grains in igneous rocks
Usually white or colorless; may be tinted gray, red, green, blue, or yellow	Colorless	Vitreous	Transparent to opaque	A very common rock mineral, occurring in masses as limestone and marble; effervesces freely in cold dilute HCl
Brilliant canary yellow		Earthy	Opaque	An ore of vanadium and uranium
Brown or black; rarely yellow or white	White to light brown	Adamantine to submetallic and dull	Translucent; rarely transparent	Principal ore of tin
Shiny lead gray; tarnishes to dull black	Grayish black	Metallic	Opaque	One of the most important ore minerals of copper; occurs principally as a result of secondary sulfide enrichment
Brass yellow; tarnishes to bronze or iridescence, but more slowly than bornite or chalcocite	Greenish black; also greenish powder in groove when scratched	Metallic	Opaque	An ore of copper; distinguished from *pyrite* by being softer than steel, distinguished from gold by being brittle; like pyrite, known as "fool's gold"
Green of various shades	Colorless	Vitreous to pearly	Transparent to translucent	A common metamorphic mineral characteristic of low-grade metamorphism
Iron black to brownish black	Dark brown	Metallic to submetallic or pitchy	Subtranslucent	The only ore of chromium; a common constituent of peridotites and *serpentines* derived from them; one of the first minerals to crystallize from cooling magma
Brown, pink, or blue; may be white, gray, green, ruby red, or sapphire blue	Colorless	Adamantine to vitreous	Transparent to translucent	Common as an accessory mineral in metamorphic rocks such as marble, mica schist, gneiss; occurs in gem form as *ruby* and *sapphire*; the abrasive *emery* is black granular corundum mixed with *magnetite*, *hematite*, or the magnesian aluminum oxide *spinel*
Colorless or pale yellow; may be red, orange, green, blue, or black	Colorless	Adamantine or greasy	Transparent	Gem and abrasive; 95% of natural diamond production is from South Africa; abrasive diamonds have been made in commercial quantities in the laboratory in the United States
White to light green		Vitreous	Transparent to translucent	Contact metamorphic mineral in crystalline limestones
Pinkish; may be white, gray, green, brown, or black	Colorless	Vitreous or pearly	Transparent to opaque	Occurs chiefly in rock masses of dolomitic limestone and marble or as the principal constituent of the rock named for it; distinguished from limestone by its less vigorous action with cold HCl (the powder dissolves with effervescence, large pieces in hot acid)

Mineral	Chemical name and/or composition	Form	Cleavage and/or fracture	Hardness	Specific gravity
Enstatite (a *pyroxene*)	Magnesium inosilicate, $Mg_2(Si_2O_6)$	Usually massive	Good at 87° and 93°	5.5	3.2–3.5
Epidote	Hydrous calcium aluminum iron silicate, $Ca_2(Al, Fe)Al_2O(SiO_4)$-$(Si_2O_7)(OH)$	Prismatic crystals striated parallel to length; usually coarse to fine granular; also fibrous	Good in 1 direction	6–7	3.35–3.45
Fayalite (an *olivine*)	$Fe_2(SiO_4)$ (see *olivine*)				
Feldspars	Aluminosilicates		Good in 2 directions at or near 90°	6	2.55–2.75
Fluorite	Calcium fluoride, CaF_2	Well-formed interlocking cubes; also massive, coarse or fine grains	Good in 4 directions parallel to faces of an octahedron	4	3.18
Forsterite (an *olivine*)	$Mg_2(SiO_4)$ (see *olivine*)				
Galena	Lead sulfide, PbS	Cube-shaped crystals; also in granular masses	Good in 3 directions parallel to faces of a cube	2.5	7.4–7.6
Garnet	$R''_3 R'''_2 (SiO_4)_3$ (R'' may be Ca, Mg, Fe, or Mn; R''' may be Al, Fe, Ti, or Cr)	Usually in 12- or 24-sided crystals; also massive granular, coarse or fine	Uneven fracture	6.5–7.5	3.5–4.3
Glaucophane (a sodium-rich *amphibole*)	Variety of *arfvedsonite*				
Goethite (bog-iron ore)	$HFeO_2$	Massive, in radiating fibrous aggregates; foliated	Perfect 010	5–5.5	3.3–4.37
Graphite (plumbago; black lead)	Carbon, C	Foliated or scaly masses common; may be radiated or granular	Good in 1 direction, folia flexible but not elastic	1–2	2.3
Gypsum	Hydrous calcium sulfate, $CaSO_4 \cdot 2H_2O$	Crystals prismatic, tabular, diamond-shaped; also in granular, fibrous, or earthy masses	Good in 1 direction, yielding flexible but inelastic flakes; fibrous fracture in another direction; conchoidal fracture in a third direction	2	2.32
Halite (rock salt; common salt)	Sodium chloride, NaCl	Cubic crystals; massive granular	Perfect cubic	2.5	2.16
Hematite	Iron oxide, Fe_2O_3	Crystals tabular; botryoidal; micaceous and foliated; massive	Uneven fracture	5.5–6.5	5.26
Hornblende (an *amphibole*)	Complex ferromagnesian silicate of Ca, Na, Mg, Ti, and Al	Long, prismatic crystals; fibrous; coarse- to fine-grained masses	Perfect prismatic at 56° and 124°	5–6	3.2
Hypersthene	*Enstatite* with more than 13% FeO				
Illite (*clay*)					

Color	Streak	Luster	"Transparency"	Other properties
Grayish, yellowish, or greenish white to olive green and brown		Vitreous	Translucent	Common in pyroxenites, peridotites, gabbros, and basalts; also in both stony and metallic meteorites
Pistachio green or yellowish to blackish green	Colorless	Vitreous	Transparent to translucent	A metamorphic mineral often associated with *chlorite;* derived from metamorphism of impure limestone; characteristic of contact metamorphic zones in limestone
				The most common igneous rock-forming group of minerals; weather to clay minerals
Variable: light green, yellow, bluish green, purple, etc.	Colorless	Vitreous	Transparent to translucent	A common, widely distributed mineral in *dolomites* and limestone; an accessory mineral in igneous rocks; used as a flux in making steel; some varieties fluoresce
Lead gray	Lead gray	Metallic	Opaque	The principal ore of lead; so commonly associated with silver that it is also an ore of silver
Red, brown, yellow, white, green, or black	Colorless	Vitreous to resinous	Transparent to translucent	Common and widely distributed, particularly in metamorphic rocks; brownish red variety, *almandite*
Yellowish brown to dark brown	Yellowish brown	Adamantine to dull	Subtranslucent	An ore of iron; one of the commonest minerals formed under oxidizing conditions as a weathering product of iron-bearing minerals
Black to steel gray	Black	Metallic or earthy	Opaque	Feels greasy; common in metamorphic rocks such as marble, schists, and gneisses
Colorless, white, or gray; with impurities, yellow, red, or brown	Colorless	Vitreous, pearly, or silky	Transparent to translucent	A common mineral widely distributed in sedimentary rocks, often as thick beds; *satin spar* is a fibrous gypsum with silky luster; *selenite* is a variety that yields broad, colorless, transparent folia; *alabaster* is a fine-grained massive variety
Colorless or white; with impurities, yellow, red, blue, or purple	Colorless	Glassy to dull	Transparent to translucent	Salty taste; permits ready passage of heat rays; a very common mineral in sedimentary rocks; interstratified in rocks of all ages to form a true rock mass
Reddish brown to black	Light to dark bloodred; blackens on heating	Metallic	Opaque	The most important ore of iron; red earthy variety known as *red ocher;* botryoidal form known as *kidney ore;* micaceous form, *specular;* widely distributed in rocks of all types and ages
Dark green to black	Colorless	Vitreous; fibrous variety often silky	Translucent on thin edges	Distinguished from *augite* by cleavage; a common rock-forming mineral that occurs in both igneous and metamorphic rocks
				A general term for clay minerals that resemble *micas;* the chief constituent in many shales

Mineral	Chemical name and/or composition	Form	Cleavage and/or fracture	Hardness	Specific gravity
Jadeite (a *pyroxene*)	$NaAl(Si_2O_6)$	Fibrous in compact massive aggregates	87° and 93°	6.5–7	3.3–3.5
Kaolinite (*clay*)	Hydrous aluminum silicate, $Al_2Si_2O_5(OH)_4$	Claylike masses	None	2–2.5	2.6
Kyanite	Aluminum silicate, Al_2SiO_5	In bladed aggregates	Good in 1 direction	5 along, 7 across, crystals	3.56–3.66
Labradorite	A *plagioclase feldspar*, 30–50% *albite*				
Limonite (brown *hematite*; bog-iron ore; rust)	Hydrous iron oxides; not a mineral	Amorphous; mammillary to stalactitic masses; concretionary, nodular, or earthy	None	5–5.5 (finely divided, apparent H as low as 1)	3.6–4
Magnetite	Iron oxide, Fe_3O_4	Usually massive granular, granular or aphanitic	Some octahedral parting	6	5.18
Mica	See *biotite, muscovite*				
Montmorillonite (*clay*)	Hydrous aluminum silicate				
Muscovite (white *mica*; potassium *mica*; common *mica*)	Nonferromagnesian silicate, $KAl_3Si_3O_{10}(OH)_2$	Mostly in thin flakes	Good in 1 direction, giving thin, very flexible, and elastic folia	2–2.5	2.76–3.1
Oligoclase	A *plagioclase feldspar*, 70–90% *albite*				
Olivine (peridot)	Ferromagnesian silicate, $(Mg, Fe)_2SiO_4$	Usually in embedded grains or granular masses	Conchoidal fracture	6.5–7	3.27–3.37
Opal	$SiO_2 \cdot nH_2O$; a mineraloid	Amorphous; massive; often botryoidal or stalactitic	Conchoidal fracture	5–6	1.9–2.2
Orthoclase	Potassium *feldspar*, $K(AlSi_3O_8)$	Prismatic crystals; most abundantly in rocks as formless grains	Good in 2 directions at or near 90°	2.57	6
Plagioclase	Soda-lime *feldspar*				
Pyrite (iron pyrites; fool's gold)	Iron sulfide, FeS_2	Cubic crystals with striated faces; also massive	Uneven fracture	6–6.5	5.02

Color	Streak	Luster	"Transparency"	Other properties
Apple green, emerald green, or white		Vitreous		Occurs in large masses in *serpentine* by metamorphism of a nepheline-*albite* rock
White	Colorless	Dull earthy	Opaque	Usually unctuous and plastic; other clay minerals similar in composition and physical properties but different in atomic structure are *illite* and *montmorillonite*; derived from weathering of *feldspars*
Blue; may be white, gray, green, or streaked	Colorless	Vitreous to pearly	Transparent to translucent	Characteristic of middle-grade metamorphism; compare with *andalusite*, which has same composition and is formed under similar conditions but has different crystal habit; contrast with *sillimanite*, which has same composition but different crystal habit and forms at highest metamorphic temperatures
				Widespread as a rock mineral; the only important constituent in large masses of rocks called *anorthosite*
Dark brown to black	Yellow brown	Vitreous	Opaque	Always of secondary origin from alteration or solution of iron minerals; mixed with fine clay, it is a pigment, *yellow ocher*
Iron black	Black	Metallic	Opaque	Strongly magnetic; may act as a natural magnet, known as *lodestone*; an important ore of iron; found in black sands on the seashore; mixed with *corundum*, it is a component of *emery*
				Unique capacity for absorbing water and expanding
Thin, colorless; thick, light yellow, brown, green, or red	Colorless	Vitreous, silky, or pearly	Thin, transparent; thick, translucent	Widespread and very common rock-forming mineral; characteristic of sialic rocks; also very common in metamorphic rocks such as gneiss and schist; the principal component of some mica schists; sometimes used for stove doors, lanterns, etc., as transparent *isinglass*; used chiefly as an insulating material
				Found in various localities in Norway with inclusions of *hematite*, which give it a golden shimmer and sparkle; this is called *aventurine* oligoclase, or *sunstone*
Olive to grayish green or brown	Pale green or white	Vitreous	Transparent to translucent	A common rock-forming mineral found primarily in simatic rocks; the principal component of peridotite; actually a series grading from *forsterite* to *fayalite*; the most common olivines are richer in magnesium than in iron; the clear-green variety *peridot* is sometimes used as a gem
Colorless; white; pale yellow, red, brown, green, gray, or blue; opalescent		Vitreous or resinous	Transparent to translucent	Many varieties; lines and fills cavities in igneous and sedimentary rocks, where it was deposited by hot waters
White, gray, or pink	White	Vitreous	Translucent to opaque	Characteristic of sialic rocks
				A continuous series varying in composition from pure *albite* to pure *anorthite*; important rock-forming minerals; characteristic of simatic rocks
Brass yellow	Greenish or brownish black	Metallic	Opaque	The most common of the sulfides; used as a source of sulfur in the manufacture of sulfuric acid; distinguished from *chalcopyrite* by its paler color and greater hardness; from gold by its brittleness and hardness

Mineral	Chemical name and/or composition	Form	Cleavage and/or fracture	Hardness	Specific gravity
Pyroxene family	Inosilicates: see *aegirite, augite, diopside, enstatite, jadeite, spodumene*				
Quartz (silica)	Silicon oxide, SiO_2, but structurally a silicate with tetrahedra sharing oxygens in 3 dimensions	Prismatic crystals with faces striated at right angles to long dimension; also massive forms of great variety	Conchoidal fracture	7	2.65
Serpentine	Hydrous magnesium silicate, $Mg_3Si_2O_5(OH)_4$	Platy or fibrous	Conchoidal fracture	2–5	2.2–2.65
Siderite (spathic iron; chalybite)	Iron carbonate, $FeCO_3$	Granular, compact, or earthy	Perfect rhombohedral	3.5–4	3.85
Sillimanite (fibrolite)	Aluminum silicate, Al_2SiO_5	Long, slender crystals without distinct terminations; often in parallel groups; frequently fibrous	Good in 1 direction	6–7	3.23
Sphalerite (zinc blende; blackjack)	Zinc sulfide, ZnS	Usually massive; crystals many-sided, distorted	Perfect in 6 directions at 120°	3.5–4	3.9–4.1
Spinel	$MgAl_2O_4$	Octahedral crystals		8	3.5–4.1

Spinel group

(XY_2O_4)			
X	Y = Al	Y = Fe	Y = Cr
Mg	Spinel, $MgAl_2O_4$	Magnesioferrite, $MgFe_2O_4$	Magnesiochromite, $MgCr_2O_4$
Fe	Hercynite, $FeAl_2O_4$	Magnetite, $FeFe_2O_4$	Chromite, $FeCr_2O_4$
Zn	Gahnite, $ZnAl_2O_4$	Franklinite, $ZnFe_2O_4$	
Mn	Galaxite, $MnAl_2O_4$	Jacobsite, $MnFe_2O_4$	

Mineral	Chemical name and/or composition	Form	Cleavage and/or fracture	Hardness	Specific gravity
Spodumene (a *pyroxene*)	Lithium aluminum inosilicate, $LiAl(Si_2O_6)$	Prismatic crystals; coarse, some large	Perfect at 87° and 93°	6.5–7	3.15–3.20
Staurolite	Iron aluminum silicate, $Fe^{+2}Al_5Si_2O_{12}(OH)$	Usually in crystals, prismatic, twinned to form a cross; rarely massive	Not prominent	7–7.5	3.65–3.75
Stibnite	Antimony trisulfide, Sb_2S_3	Slender prismatic habit; often in radiating groups	Perfect in 1 direction	2	4.52–4.62
Taconite	Not a mineral				
Talc (soapstone; steatite)	Hydrous magnesium silicate, $Mg_3Si_4O_{10}(OH)_2$	Foliated, massive	Good cleavage in 1 direction, thin, flexible, but inelastic folia	1	2.7–2.8
Topaz	Aluminum fluosilicate, $Al_2SiO_4(F, OH)_2$	Usually in prismatic crystals, often with striations in direction of greatest length	Good in 1 direction	8	3.4–3.6

Color	Streak	Luster	"Transparency"	Other properties
				A group of silicates with tetrahedra in single chains; *augite* is the most important; contrast with *amphibole*
Colorless or white; with impurities, any color	Colorless	Vitreous, greasy, or splendent	Transparent to translucent	An important constituent of sialic rocks; coarsely crystalline varieties: *rock crystal*, *amethyst* (purple), *rose quartz*, *smoky quartz*, *citrine* (yellow), *milky quartz*, *cat's eye*; cryptocrystalline varieties: *chalcedony*, *carnelian* (red chalcedony), *chrysoprase* (apple-green chalcedony), *heliotrope*, or *bloodstone* (green chalcedony with small red spots), *agate* (alternating layers of chalcedony and opal); granular varieties: *flint* (dull to dark brown), *chert* (like flint but lighter in color), *jasper* (red from hematite inclusions), *prase* (like jasper but dull green)
Variegated shades of green	Colorless	Greasy, waxy, or silky	Translucent	Platy variety, *antigorite*; fibrous variety, *chrysotile*, an asbestos; an alteration product of magnesium silicates such as *olivine*, *augite*, and *hornblende*; common and widely distributed
Light to dark brown	Colorless	Vitreous	Transparent to translucent	An ore of iron; an accessory mineral in *taconite*
Brown, pale green, or white	Colorless	Vitreous	Transparent to translucent	Relatively rare but important as a mineral characteristic of high-grade metamorphism; contrast with *andalusite* and *kyanite*, which have the same composition but form under conditions of middle-grade metamorphism
White or green; with iron, yellow to brown and black; red	White to yellow and brown	Resinous	Transparent to translucent	A common mineral; the most important ore of zinc; the red variety is called *ruby zinc*; streak lighter than corresponding mineral color
White, red, lavender, blue, green, brown, or black	White	Vitreous	Usually translucent; may be clear and transparent	A common metamorphic mineral imbedded in crystalline limestone, gneisses, and *serpentine*; when transparent and finely colored, it is a gem; the red is spinel ruby, or *balas ruby*; some are blue
White, gray, pink, yellow, or green		Vitreous	Transparent to translucent	A source of lithium, which improves lubricating properties of greases; some gem varieties
Red brown to brownish black	Colorless	Fresh, resinous or vitreous; altered, dull to earthy	Translucent	A common accessory mineral in schists and slates; characteristic of middle-grade metamorphism; associated with *garnet*, *kyanite*, *sillimanite*, *tourmaline*
Lead gray to black	Lead gray to black	Metallic	Opaque	The chief ore of antimony, which is used in various alloys
				Unleached iron formation in the Lake Superior district, consists of chert (see *quartz*) with *hematite*, *magnetite*, *siderite*, and hydrous iron silicates; an ore of iron
Gray, white, silver white, or apple green	White	Pearly to greasy	Translucent	Of secondary origin, formed by the alteration of magnesium silicates such as *olivine*, *augite*, and *hornblende*; most characteristically found in metamorphic rocks
Straw yellow, wine yellow, pink, bluish, or greenish	Colorless	Vitreous	Transparent to translucent	Represents 8 on Mohs scale of hardness; a gem stone

Mineral	Chemical name and/or composition	Form	Cleavage and/or fracture	Hardness	Specific gravity
Tourmaline	Complex boron-aluminum silicate with Na, Ca, F, Fe, Li, or Mg	Usually in crystals; common with cross section of spherical triangle	Not prominent; black variety fractures like coal	7–7.5	3–3.25
Tremolite (an *amphibole*)	$Ca_2Mg_5(Si_8O_{22})(OH)_2$	Often bladed or in radiating columnar aggregate	Good at 56°	5–6	3.0–3.3
Uraninite (pitchblende)	Complex uranium oxide with small amounts of Pb, Ra, Th, Y, N, He, and A	Usually massive and botryoidal	Not prominent	5.5	9–9.7
Wollastonite	Calcium silicate, $CaSiO_3$	Commonly massive, fibrous, or compact	Good in 2 directions at 84° and 96°	5–5.5	2.8–2.9
Zircon	Zirconium nesosilicate, $Zr(SiO_4)$	Tetragonal prism and dipyramid		7.5	4.68

Color	Streak	Luster	"Transparency"	Other properties
Varied: black or brown; red, pink, green, blue or yellow	Colorless	Vitreous to resinous	Translucent	Gem stone; an accessory mineral in pegmatites, also in metamorphic rocks such as gneisses, schists, marbles
White to light green		Vitreous	Transparent to translucent	Frequently in impure, crystalline *dolomitic* limestones where it formed on recrystallization during metamorphism; also in *talc* schists
Black	Brownish black	Submetallic or pitchy	Opaque	An ore of uranium and radium; the mineral in which helium and radium were first discovered
Colorless, white, or gray	Colorless	Vitreous or pearly on cleavage surfaces	Translucent	A common contact-metamorphic mineral in limestones
Brown; also gray, green, red, or colorless	Colorless	Adamantine	Translucent; sometimes transparent	Transparent variety is a gem stone; a source of zirconium metal, which is used in the construction of nuclear reactors

APPENDIX B | PERIODIC CHART

Legend:
- Atomic number and symbol
- Mass compared to carbon 12 (approximate in parentheses)
- Most common electric charge / Ion radius ($\times 10^{-10}$ m) with that charge
- Next most common charge / Ion radius with that charge
- Element with only radioactive species in the earth's crust
- Radioactive element not found in the crust

Example:

16 S		
32.0		
-2	1.84	
+6	0.30	
sulfur		

Z	Symbol	Mass	Charge 1	Radius 1	Charge 2	Radius 2	Name
1	H	1.01	+1	0.01			hydrogen
2	He	4.00	0	1.08			helium
3	Li	6.94	+1	0.68			lithium
4	Be	9.01	+2	0.35			beryllium
5	B	10.8	+3	0.23			boron
6	C	12.0	+4	0.16			carbon
7	N	14.0	+5	0.13	-3	1.71	nitrogen
8	O	16.0	-2	1.32			oxygen
9	F	19.0	-1	1.33			fluorine
10	Ne	20.2	0				neon
11	Na	23.0	+1	0.97			sodium
12	Mg	24.3	+2	0.66			magnesium
13	Al	27.0	+3	0.51			aluminum
14	Si	28.1	+4	0.42			silicon
15	P	31.0	+5	0.35			phosphorus
16	S	32.0	-2	1.84	+6	0.30	sulfur
17	Cl	35.5	-1	1.81			chlorine
18	Ar	40.0	0				argon
19	K	39.1	+1	1.33			potassium
20	Ca	40.1	+2	0.99			calcium
21	Sc	45.0	+3	0.73			scandium
22	Ti	47.9	+4	0.68			titanium
23	V	50.9	+5	0.59			vanadium
24	Cr	52.0	+3	0.63	+6	0.52	chromium
25	Mn	54.9	+4	0.60	+2	0.80	manganese
26	Fe	55.8	+2	0.74	+3	0.64	iron
27	Co	58.9	+2	0.72	+3	0.63	cobalt
28	Ni	58.7	+2	0.69			nickel
29	Cu	63.5	+2	0.72	+1	0.96	copper
30	Zn	65.4	+2	0.74			zinc
31	Ga	69.7	+3	0.62			gallium
32	Ge	72.6	+4	0.53			germanium
33	As	74.9	+3	0.58			arsenic
34	Se	79.0	-2	1.91			selenium
35	Br	79.9	-1	1.96			bromine
36	Kr	83.8					krypton
37	Rb	85.5	+1	1.47			rubidium
38	Sr	87.6	+2	1.12			strontium
39	Y	88.9	+3	0.89			yttrium
40	Zr	91.2	+4	0.79			zirconium
41	Nb	92.9	+5	0.69			niobium
42	Mo	95.9	+4	0.70	+6	0.62	molybdenum
43	Tc	(99)					technetium
44	Ru	101.	0	1.32			ruthenium
45	Rh	103.	0	1.34			rhodium
46	Pd	106.	0	1.37			palladium
47	Ag	108.	+1	1.26			silver
48	Cd	112.	+2	0.97			cadmium
49	In	115.	+3	0.81			indium
50	Sn	119.	+4	0.71			tin
51	Sb	122.	+3	0.76			antimony
52	Te	128.	-2	2.11			tellurium
53	I	127.	-1	2.20			iodine
54	Xe	131.					xenon
55	Cs	133.	+1	1.67			cesium
56	Ba	137.	+2	1.34			barium
57	La	139.	+3	1.02			lanthanum
72	Hf	178.	+4	0.78			hafnium
73	Ta	181.	+5	0.68			tantalum
74	W	184.	+6	0.62			tungsten
75	Re	186.	0	1.37			rhenium
76	Os	190.	0	1.33			osmium
77	Ir	192.	0	1.36			iridium
78	Pt	195.	0	1.37			platinum
79	Au	197.	0	1.44			gold
80	Hg	201.	+2	1.10			mercury
81	Tl	204.	+3	0.95			thallium
82	Pb	207.	+2	1.20			lead
83	Bi	209.	+3	0.96			bismuth
84	Po	(210)					polonium
85	At	(210)					astatine
86	Rn	(222)					radon
87	Fr	(223)					francium
88	Ra	(226)					radium
89	Ac	(227)					actinium

[a] Lanthanides, or "rare earths"

Z	Symbol	Mass	Charge 1	Radius 1	Charge 2	Radius 2	Name
58	Ce	140.	+3	1.03			cerium
59	Pr	141.	+3	1.01			praseodymium
60	Nd	144.	+3	0.99			neodymium
61	Pm	(147)					promethium
62	Sm	150.	+3	0.96			samarium
63	Eu	152.	+3	0.95	+2	1.09	europium
64	Gd	157.	+3	0.94			gadolinium
65	Tb	159.	+3	0.92			terbium
66	Dy	162.	+3	0.91			dysprosium
67	Ho	165.	+3	0.89			holmium
68	Er	167.	+3	0.88			erbium
69	Tm	169.	+3	0.87			thulium
70	Yb	173.	+3	0.86			ytterbium
71	Lu	175.	+3	0.85			lutetium

[b] Actinides

Z	Symbol	Mass	Charge 1	Radius 1	Charge 2	Radius 2	Name
90	Th	232.	+4	1.02			thorium
91	Pa	(231)					protactinium
92	U	238.	+4	0.97	+6	0.80	uranium
93	Np	(237)					neptunium
94	Pu	(242)					plutonium
95	Am	(243)					americium
96	Cm	(247)					curium
97	Bk	(247)					berkelium
98	Cf	(249)					californium
99	Es	(254)					einsteinium
100	Fm	(253)					fermium
101	Md	(256)					mendelevium
102	No	(254)					nobelium
103	Lw	(257)					lawrencium

EARTH DATA

TABLE C.1

Composition of seawater at 35 parts per thousand salinity[a]

Element	$\mu g/l$	Element	$\mu g/l$	Element	$\mu g/l$
Hydrogen	1.10×10^8	Nickel	6.6	Praesodymium	0.00064
Helium	0.0072	Copper	23	Neodymium	0.0023
Lithium	170	Zinc	11	Samarium	0.00042
Beryllium	0.0006	Gallium	0.03	Europium	0.000114
Boron	4,450	Germanium	0.06	Gadolinium	0.0006
Carbon (inorganic)	28,000	Arsenic	2.6	Terbium	0.0009
(dissolved organic)	2,000	Selenium	0.090	Dysprosium	0.00073
Nitrogen (dissolved N_2)	15,500	Bromine	6.73×10^4	Holmium	0.00022
(as NO_3^-, NO_2^-, NH_4^+)	670	Krypton	0.21	Erbium	0.00061
Oxygen (dissolved O_2)	6,000	Rubidium	120	Thulium	0.00013
(as H_2O)	8.83×10^8	Strontium	8,100	Ytterbium	0.00052
Fluorine	1,300	Yttrium	0.003	Lutetium	0.00012
Neon	0.120	Zirconium	0.026	Hafnium	<0.008
Sodium	1.08×10^7	Niobium	0.015	Tantalum	<0.0025
Magnesium	1.29×10^6	Molybdenum	10	Tungsten	<0.001
Aluminum	1	Ruthenium	—	Rhenium	—
Silicon	2,900	Rhodium	—	Osmium	—
Phosphorus	88	Palladium	—	Iridium	—
Sulfur	9.04×10^5	Silver	0.28	Platinum	—
Chlorine	1.94×10^7	Cadmium	0.11	Gold	0.011
Argon	450	Indium	—	Mercury	0.15
Potassium	3.92×10^5	Tin	0.81	Thallium	—
Calcium	4.11×10^5	Antimony	0.33	Lead	0.03
Scandium	<0.004	Tellurium	—	Bismuth	0.02
Titanium	1	Iodine	64	Radium	1×10^{-13}
Vanadium	1.9	Xenon	0.047	Thorium	0.0015
Chromium	0.2	Cesium	0.30	Protactinium	2×10^{-10}
Manganese	1.9	Barium	21	Uranium	3.3
Iron	3.4	Lanthanum	0.0029		
Cobalt	0.39	Cerium	0.0013		

[a]Adapted from Karl K. Turekian, *Oceans*, p. 92, Prentice-Hall, Inc., Englewood Cliffs, N.J., 1968.

TABLE C.2
Distribution of world's estimated supply of water[a]

	Area, thousands of		Volume, thousands of		
	km²	mi²	km³	mi³	Total volume, %
World (total area)	510,000	197,000	—	—	—
Land area	149,000	57,500	—	—	—
Water in land areas:					
Freshwater lakes	850	330	125	30	0.009
Saline lakes and inland seas	700	270	104	25	0.008
Rivers (average instantaneous volume)	—	—	1.25	0.3	0.0001
Soil moisture and vadose water	—	—	67	16	0.005
Groundwater to depth of 4,000 m	—	—	8,350	2,000	0.61
Ice caps and glaciers	19,400	7,500	29,200	7,000	2.14
Atmospheric moisture	—	—	13	3.1	0.001
World ocean	361,000	139,500	1,320,000	317,000	97.3
Total water volume (rounded)			1,360,000	326,000	100

[a]In part after R. L. Nace, *U.S. Geol. Surv. Circ.* 536, Table 1, 1967.

TABLE C.3
Runoff from the continents[a]

	Area, millions of		Annual runoff			
			Total, 10¹⁵ ×		Depth/unit area	
	km²	mi²	l	gal	cm	in
Asia	46.6	18.0	11.1	3.0	23.8	9.4
Africa	29.8	11.5	5.9	1.6	19.8	7.8
North America	21.2	8.2	4.5	1.2	21.1	8.3
South America	19.6	7.6	8.0	2.1	41.4	16.3
Europe	10.9	4.2	2.5	0.6	23.1	9.1
Australia	7.8	3.0	0.4	0.1	2.5	1.0
Total or (mean)	135.9	52.5	32.4	8.6	(24.9)	(9.8)

[a]Calculated from data from D. A. Livingstone, *U.S. Geol. Surv. Prof. Paper* 440-G, 1963.

TABLE C.4
Composition of river waters of the world[a]

Substance	Ppm
HCO_3	58.4
Ca	15
SiO_2	13.1
SO_4	11.2
Cl	7.8
Na	6.3
Mg	4.1
K	2.3
NO_3	1
Fe	0.67

[a]From Daniel A. Livingstone, "Data of Geochemistry," *U.S. Geol. Surv. Prof. Paper* 440-G, p. G-41, 1963.

TABLE C.5
Earth volume, density, and mass

	Av. thickness or radius, km	Volume, millions of km³	Mean density, g/cm³	Mass, 10²⁴ × g
Total earth	6,371	1,083,230	5.52	5,976
Oceans and seas	3.8	1,370	1.03	1.41
Glaciers	1.6	25	0.9	0.023
Continental crust	35	6,210	2.8	17.39
Oceanic crust	8·	2,660	2.9	7.71
Mantle	2,883	899,000	4.5	4,068
Core	3,471	175,500	10.71	1,881

TABLE C.8   **455**

TABLE C.6

Average composition of the crust[a]

	Av. igneous rock, %	Av. shale, %	Av. sandstone, %	Av. limestone, %	Weighted-av. crust[b], %
SiO_2	59.12	58.11	78.31	5.19	59.07
TiO_2	1.05	0.65	0.25	0.06	1.03
Al_2O_3	15.34	15.40	4.76	0.81	15.22
Fe_2O_3	3.08	4.02	1.08 ⌉		3.10 ⌉
FeO	3.80	2.45	0.30 ⌋	0.54	3.71 ⌋
MgO	3.49	2.44	1.16	7.89	3.45
CaO	5.08	3.10	5.50	42.57	5.10
Na_2O	3.84	1.30	0.45	0.05	3.71
K_2O	3.13	3.24	1.32	0.33	3.11
H_2O	1.15	4.99	1.63	0.77	1.30
CO_2	0.10	2.63	5.04	41.54	0.35
ZrO_2	0.04	—	—	—	0.04
P_2O_5	0.30	0.17	0.08	0.04	0.30
Cl	0.05	—	Tr^c	0.02	0.05
F	0.03	—	—	—	0.03
SO_3	—	0.65	0.07	0.05	—
S	0.05	—	—	0.09	0.06
$(Ce, Y)_2O_3$	0.02	—	—	—	0.02
Cr_2O_3	0.06	—	—	—	0.05
V_2O_3	0.03	—	—	—	0.03
MnO	0.12	Tr	Tr	0.05	0.11
NiO	0.03	—	—	—	0.03
BaO	0.05	0.05	0.05	0.00	0.05
SrO	0.02	0.00	0.00	0.00	0.02
Li_2O	0.01	Tr	Tr	Tr	0.01
Cu	0.01	—	—	—	0.01
C	0.00	0.80	—	—	0.04
Total	100.00	100.00	100.00	100.00	100.00

[a]After F. W. Clarke and H. S. Washington, "The Composition of the Earth's Crust," *U.S. Geol. Surv. Prof. Paper 127*, p. 32, 1924.
[b]Weighted average: igneous rock, 95%; shale, 4%; sandstone, 0.75%; limestone, 0.25%.
[c]Trace.

TABLE C.7

Earth size

	Thousands of km
Equatorial radius	6.378
Polar radius	6.357
Mean radius[a]	6.371
Polar circumference	40.009
Equatorial circumference	40.077
Ellipticity [(equatorial radius − polar radius)/ equatorial radius], 1/297	

[a]Term used by geophysicists to designate radius of a sphere of equal volume.

TABLE C.8

Earth areas

	Millions of km²
Total area	510
Land (29.22% of total)	149
Oceans and seas (70.78% of total)	361
Glacier ice	15.6
Continental shelves	28.4

TOPOGRAPHIC AND GEOLOGIC MAPS

Topography refers to the shape of the physical features of the land. A *topographic map* is a representation of the shape, size, position, and relation of the physical features of an area. In addition to mountains, hills, valleys, and rivers most topographic maps also show the culture of a region, that is, political boundaries, towns, houses, roads, and similar features.

D.1 TOPOGRAPHIC MAPS

Topographic maps are used in the laboratory for the observation and analysis of the geological processes that are constantly changing the face of the earth.

DEFINITIONS

Elevation, or altitude The vertical distance between a given point and the datum plane.

Datum plane The reference surface from which all altitudes on a map are measured. This is usually mean sea level.

Height The vertical difference in elevation between an object and its immediate surroundings.

Relief The difference in elevation of an area between tops of hills and bottoms of valleys.

Bench mark A point of known elevation and position—usually indicated on a map by the letters B.M., with the altitude given (on American maps) to the nearest foot.

Contour line A map line connecting points representing places on the earth's surface that have the same elevation. It thus locates the intersection with the earth's surface of a plane at any arbitrary elevation parallel to the datum plane. Contours represent the vertical, or third, dimension on a map, which has only two dimensions. They show the shape and size of physical features such as hills and valleys. A depression is indicated by an ordinary contour line except that hachures, or short dashes, are used on one side and point toward the center of the depression.

Contour interval The difference in elevation represented by adjacent contour lines.

Scale The ratio in a map of the distance between two points on the ground and the same two points on the map. It may be expressed in three ways:

Fractional scale If two points are 1 km apart in the field, they may be represented on the map as separated by some fraction of that distance, say, 1 cm. In this instance the scale is 1 cm to the kilometer. There are 100,000 cm in 1 km; so this scale can be expressed as the fraction, or ratio, 1:100,000. Many topographic maps of the United States Geological Survey have a scale of 1:62,500; and many recent maps have a scale of 1:31,250, and others of 1:24,000.

Graphic scale This scale is a line printed on the map and divided into units that are equivalent to some distance, such as 1 km or 1 mi.

Verbal scale This is an expression in common speech, such as "four centimeters to the kilometer," "an inch to a mile," or "two miles to the inch."

CONVENTIONAL SYMBOLS

An explanation of the symbols used on topographic maps is printed on the back of each topographic sheet, along the margin, or for newer maps on a separate legend sheet. In general, culture (artificial works) is shown in black. All water features, such as streams, swamps, and glaciers, are shown in blue. Relief is shown by contours in brown. Red may be used to indicate main highways, and green overprints may be used to designate areas of woods, orchards, vineyards, or scrub.

The United States Geological Survey distributes free of charge a single sheet, entitled "Topographic Maps," that includes an illustrated summary of topographic-map symbols. (Apply to Branch of Distribution, United States Geological Survey, 1240 South Eads Street, Arlington, Virginia 22202.)

LOCATING POINTS

Any particular point or area may be located in several ways on a topographic map. The three most commonly used are:

In relation to prominent features A point may be referred to as being so many kilometers in a given direction from a city, mountain, lake, river mouth, or other easily located feature on the map.

By latitude and longitude Topographic maps of the Geological Survey are bounded on the north and south by parallels of latitude and on the east and west by meridians of longitude. These intersecting lines form the grid into which the earth has been divided. Latitude is measured north and south from the equator, and longitude is measured east and west from the prime meridian that passes through Greenwich, England. Thus maps of the United States are within north latitude and west longitude.

By township and range Most of the area of the United States has been subdivided by a system of land survey in which a square 6 mi on a side is the basic unit, called a *township*. Not included in this system are all the states along the eastern seaboard (with the exception of Florida), West Virginia, Kentucky, Tennessee, Texas, and parts of Ohio. Townships are laid off north and south from a base line and east and west from a principal meridian. Each township is divided into 36 sections, usually 1 mi on a side. Each section may be further subdivided into half sections, quarter sections, or sixteenth sections. Thus in Figure D.1 the point *x* is located in the northeast quarter of the northwest quarter of section 3, township 9 north, range 5 west, abbreviated as NE$\frac{1}{4}$NW$\frac{1}{4}$Sec 3, T9N, R5W, or NE NW Sec 3–9N–5W.

CONTOUR SKETCHING

Many contour maps are now made from aerial photographs. Before this can be done, however, the position and location of a number of reference points, or bench marks, must be determined in the field. If the topographic map is prepared from field surveys rather than from aerial photographs, the topographer first determines the location and elevation of bench marks and a large number of other points selected for their critical position. They may be on hilltops, on the lowest point in a saddle between hills, along streams, or at places where there is a significant change in slope. On the basis of these points contours may be sketched through loci of equal elevation. The contours are preferably sketched in the field in order to include minor irregularities that are visible to the topographer.

Because contours are not ordinary lines, certain requirements must be met in drawing them to satisfy their definition. These are listed below:

1 All points on one contour line have the same elevation.

2 Contours separate all points of higher elevation than the contour from all points of lower elevation.

FIGURE D.1

Subdivision by township and range (see text for discussion).

A section

3 The elevation represented by a contour line is always a simple multiple of the contour interval. Every contour line that is a multiple of five times the contour interval is heavier than the others. (An exception is 25-unit contours, in which every multiple of four times the interval is heavier.)

4 Contours never cross or intersect one another.

5 A vertical cliff is represented by coincident contours.

6 Every contour closes on itself either within or beyond the limits of the map. In the latter case the contours will end at the edge of the map.

7 Contour lines never split.

8 Uniformly spaced contour lines represent a uniform slope.

9 Closely spaced contour lines represent a steep slope.

10 Contour lines spaced far apart represent a gentle slope.

11 A contour line that closes within the limits of the map indicates a hill.

12 A hachured contour line represents a depression. The short dashes, or hachures, point into the depression.

13 Contour lines curve up a valley but cross a stream at right angles to its course.

14 Maximum ridge and minimum valley contours always go in pairs: That is, no single lower contour can lie between two higher ones and vice versa.

TOPOGRAPHIC PROFILES

A topographic profile is a cross section of the earth's surface along a given line. The upper line of this section is irregular and shows the shape of the land along the line of profile, or section.

Profiles are most easily constructed with graph paper. A horizontal scale, usually the map scale, is chosen. Then a vertical scale sufficient to bring out the features of the surface is chosen. The vertical scale is usually several times larger than the horizontal—that is, it is exaggerated. The steps in the construction of a profile are:

1 Select a base (one of the horizontal lines on the graph paper). This may be sea level or any other convenient datum.

2 On the graph paper number each fourth or fifth line above the base, according to the vertical scale chosen.

3 Place the graph paper along the line of profile.

4 With the vertically ruled lines as guides, plot the elevation of each contour line that crosses the line of profile.

5 If great accuracy is not important, plot only every heavy contour and the tops and bottoms of hills and the bottoms of valleys.

6 Connect the points.

7 Label necessary points along the profile.

8 Give the vertical and horizontal scales.

9 State the vertical exaggeration.

10 Title the profile.

Vertical exaggeration   The profile represents both vertical and horizontal dimensions. These dimensions are not usually on the same scale because the vertical needs to be greater than the horizontal to give a clear presentation of changes in level. Thus if the vertical scale is 500 m to the centimeter and the horizontal scale is 5,000 m to the centimeter—or 1:5,000—the vertical exaggeration is 10 times, written 10 ×. This is obtained by dividing the horizontal scale by the vertical scale. Note that both horizontal and vertical scales must be expressed in the same unit (in this instance meters to the centimeter) before division.

D.2 GEOLOGIC MAPS

Geologic maps show the distribution of earth materials on the surface. In addition they indicate the relative age of these materials and suggest their arrangement beneath the surface.

DEFINITIONS

Formation   The units illustrated upon a geologic map are usually referred to as "formations." We define a formation as a rock unit with upper and lower boundaries that can be recognized easily in the field and that is large enough to be shown on the map. A formation receives a distinctive designation made up of two parts: The first part is geographic and refers to the place or general area where the formation is first described; the second refers to the nature of the rock. Thus *Trenton Limestone* is a formation dominantly composed of limestone and is named after Trenton Falls in central New York, where it was first formally described. *Wausau Granite* designates a body of granite in the Wausau, Wisconsin, area. If the lithology is so variable that no single lithologic distinction is appropriate, the word *formation* may be used. For instance, the *Raritan Formation* is named

for the area of the Raritan River and Raritan Bay in New Jersey, and its lithology includes both sand and clay.

Dip and strike   The dip and strike of a rock layer refers to its orientation in relation to a horizontal plane. In Chapter 7 we found that the dip is the acute angle that a tilted rock layer makes with an imaginary plane. We also found that the strike is the compass direction of a line formed by the intersection of the dipping surface with an imaginary horizontal plane. The direction of strike is always at right angles to the direction of dip. The dip-and-strike symbol used on a geologic map is in the form of a topheavy tee. The crossbar represents the direction of the strike of the bed. The short upright represents the direction of the dip of the bed. This sometimes, but not always, has an arrow pointing in the direction of dip. Very often the angle of dip is indicated alongside the symbol. Here is an example (the top of the page is considered to be north):

⊤ 30 = strike E–W; dip 30°S

⊀ 25 = strike N, 45°E; dip 25°SE

Contact   A contact is the plane separating two rock units. It is shown on the geologic map as a line that is the intersection of the plane between the rock units and the surface of the ground.

Outcrop An outcrop is rock material that is exposed at the surface through the cover of soil and weathered material. In areas of abundant rainfall soil and vegetation obscure the underlying rock and only a small fraction of 1 percent of the surface may be in outcrop. In dry climates, where soils are shallow or absent and the plant cover is discontinuous, bedrock usually crops out much more widely.

Legend and symbols A legend is an explanation of the various symbols used on the map. There is no universally accepted set of standard symbols, but some that are widely used are given in Figure D.2. In addition to the graphic symbols in Figure D.2 letter symbols are sometimes used to designate rock units. Such a symbol contains a letter or letters referring to the geologic column, followed by a letter or letters referring to the specific name of the rock unit. Thus in the symbol Ot the O stands for Ordovician and the t for the Trenton Limestone of central New York. The letters or abbreviations generally used for the geologic column are given in Table D.1.

Sometimes different colors are used to indicate different rock systems. There is no standardized color scheme, but many of the geologic maps of the Geological Survey use the colors given in Table D.1, combined with varying patterns, for systems of sedimentary rocks. No specific colors are designated for igneous rocks, but when colors are used, they are usually purer and more brilliant than those used for sedimentary rocks.

CONSTRUCTION OF A GEOLOGIC MAP

The basic idea of geologic mapping is simple. We are interested first in showing the distribution of the rocks at the earth's surface. Theoretically all we need to do is plot the occurrence of the different rocks on a base map, and then we have a geologic map. Unfortunately the process is not quite so simple.

In most areas the bedrock is more or less obscured in one way or another, and only a small amount of outcrop is available for observation, study, and sampling. From the few exposures available the geologist must extrapolate the general distribution of rock types. In this extrapolation his field data are obviously of prime importance. But he will also be guided by changes in soil, vegetation, and landscape as well as by patterns that can be detected on aerial photographs. Furthermore he may be aided by laboratory examination of field samples and by the records of both deep and shallow wells. The geologist may also have available to him geophysical data that help determine the nature of obscured bedrock. Eventually, when he has marshaled as many data as possible, he draws the boundaries delineating the various rock types.

In addition to the distribution of rock types the geologist is also concerned with depicting, as accurately as he can, the ages of the various rocks and their arrangement beneath the surface. These goals will also be realized in part through direct observations in the field and in part through other lines of evidence. The preparation of an accurate, meaningful, geologic map demands experience, patience, and judgment.

GEOLOGIC CROSS SECTIONS

A geologic map tells us something of how rocks are arranged in the

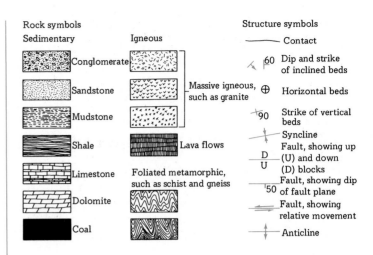

FIGURE D.2

Symbols commonly used on geologic maps.

TABLE D.1

Letter symbols and colors commonly used to designate units in the geologic column

Period	Symbol	Color
Pleistocene	Q	Yellow and gray
Pliocene	Tpl	Yellow ocher
Miocene	Tm	Yellow ocher
Oligocene	To	Yellow ocher
Eocene	Te	Yellow ocher
Paleocene	Tp	Yellow ocher
Cretaceous	K	Olive green
Jurassic	J	Blue green
Triassic	T_R	Light peacock blue or bluish gray green
Permian	P	Blue
Pennsylvanian	Cp	Blue
Mississippian	Cm	Blue
Devonian	D	Gray purple
Silurian	S	Purple
Ordovician	O	Red purple
Cambrian	$\in$	Brick red
Precambrian	P_e	Terra-cotta and gray brown

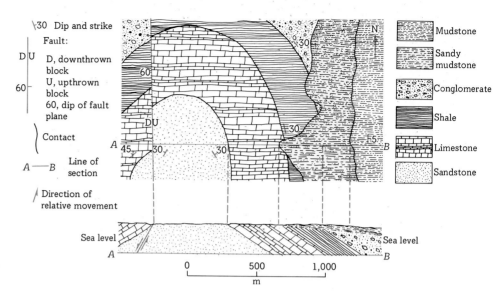

Construction of a geologic cross section from a geologic map.

underground. Often, to show these relations more clearly, we find it convenient to draw geologic cross sections. Such a section is really a diagram showing a side view of a block of the earth's crust as it would look if we could lift it up to view. We have used cross sections in many illustrations throughout this book.

A geologic cross section is drawn, insofar as possible, at right angles to the general strike of the rocks. The general manner in which a geologic cross section is projected from a geologic map is shown in Figure D.3. If the projection is made onto a topographic profile in which the vertical scale has been exaggerated, then the angle of dip of the rocks should be exaggerated accordingly.

GLOSSARY

The definitions in this glossary have been coordinated with the revised edition of *Dictionary of Geological Terms*, prepared under the direction of the American Geological Institute and published by Doubleday & Co., Inc., New York, 1976. In the definitions here words in italics are often separately defined in their alphabetic place.

Å Abbreviation for *angstrom*, a unit of length (10^{-8} cm).

aa lava *Lava* whose surface is covered with random masses of angular jagged blocks.

ablation As applied to glacier ice, process by which ice below snow line is wasted by evaporation and melting.

abrasion Erosion of rock material by friction of solid particles moved by gravity, water, ice, or wind.

absolute time Geologic time measured in years. Compare with *relative time*.

acidic lava Containing 70 percent or more of silica.

actinolite A metamorphic ferromagnesian mineral; an *asbestos*.

adiabatic rate In a body of air moving upward or downward change in temperature without gain or loss of temperature to air through which it moves.

aftershock Earthquake that follows a larger earthquake and originates at or near focus of larger earthquake. Major shallow earthquakes are generally followed by many aftershocks, which decrease in number as time goes on but may continue for days or even months.

agate Variety of chalcedony with alternating layers of chalcedony and opal.

A horizon Soil zone immediately below surface, from which soluble material and fine-grained particles have been moved downward by water seeping into soil. Varying amounts of organic matter give *A* horizon color ranging from gray to black.

Airy hypothesis Explains isostasy by assuming earth's crust has same density everywhere and differences in elevation result from differences in thickness of outer layer.

albite Feldspar in which diagnostic positive ion is Na^+; sodic feldspar, $Na(AlSi_3O_8)$. One of plagioclase feldspars.

alkali rocks Igneous rocks in which abundance of alkalies is unusually high, generally indicated by soda pyroxenes, soda amphiboles, and/or feldspathoids.

alluvial fan Land counterpart of a delta: an assemblage of sediments marking place where a stream moves from a steep gradient to a flatter gradient and suddenly loses transporting power. Typical of arid and semiarid climates but not confined to them.

almandite A deep-red garnet of iron and aluminum formed during regional metamorphism.

alpha decay Radioactive decay taking place by loss of an alpha particle from nucleus. Mass of element decreases by 4, and atomic number decreases by 2.

alpha particle A helium atom lacking electrons and therefore having a double positive charge.

Alpine, or mountain, glacier See *valley glacier*.

amorphous A state of matter in which there is no orderly arrangement of atoms.

amphibole group Ferromagnesian silicates with a double chain of silicon-oxygen tetrahedra. Common example: hornblende. Contrast with *pyroxene group*.

amphibolite A faintly foliated metamorphic rock developed during regional metamorphism of simatic rocks. Composed mainly of hornblende and plagioclase feldspars.

amphibolite facies   An assemblage of minerals formed during regional metamorphism at moderate to high pressures between 450 and 700°C.

andalusite   A silicate of aluminum built around independent tetrahedra, Al_2SiO_5. Characteristic of middle-grade metamorphism. Compare with *kyanite*, which has same composition and forms under similar conditions but which has different crystal habit. Contrast with *sillimanite*, which has same composition but different crystal habit and forms at highest metamorphic temperatures.

andesite   A fine-grained igneous rock with no quartz or orthoclase, composed of about 75 percent plagioclase feldspars, balance ferromagnesian silicates. Important as lavas; possibly derived by fractional crystallization from basaltic magma. Widely characteristic of mountain-making processes around borders of Pacific Ocean. Confined to continental sectors.

andesite line   A map line designating the petrographic boundary of Pacific Ocean. Extrusive rocks on Pacific side of line are basaltic; on other side, andesitic.

angstrom   A unit of length equal to one hundred-millionth of a centimeter (10^{-8} cm). Abbreviation, Å.

angular momentum   A vector quantity, product of mass times radius of orbit times velocity. The energy of motion of solar system.

angular unconformity   An *unconformity* in which older strata dip at different angle from that of younger strata.

anhydrite   Mineral calcium sulfate, $CaSO_4$, which is *gypsum* without water.

anion   A negatively charged atom or group of atoms produced by gain of electrons. Compare *cation*.

anorthite   Feldspar in which diagnostic positive ion is Ca^{2+}; calcic feldspar, $Ca(Al_2Si_2O_8)$. One of plagioclase feldspars.

anorthosite   A plutonic igneous rock composed of 90 percent or more of feldspar mineral anorthite. Pyroxene and some olivine usually make up balance of the rock.

Antarctic bottom water   Seawater that sinks to ocean floor off Antarctica and flows equatorward beneath North Atlantic deep water.

Antarctic intermediate water   Seawater that sinks at about 50°S and flows northward above North Atlantic deep water.

anthracite   Metamorphosed bituminous coal of about 95 to 98 percent carbon.

anticline   A configuration of folded, stratified rocks in which rocks dip in two directions away from a crest, as principal rafters of a common gable roof dip away from ridgepole. Reverse of *syncline*. The "ridgepole," or crest, is called *axis*.

aphanitic texture   Individual minerals present but in particles so small that they cannot be identified without a microscope.

aquifer   A permeable material through which groundwater moves.

arête   A narrow, saw-toothed ridge formed by cirques developing from opposite sides into a ridge.

arkose   A detrital sedimentary rock formed by cementation of individual grains of sand size and predominantly composed of quartz and feldspar. Derived from disintegration of granite.

arroyo   Flat-floored, vertically walled channel of an intermittent stream typical of semiarid climates; often applied to such features of southwestern United States. Synonymous with *wadi* and *wash*.

artesian water   Water under pressure when tapped by a well and able to rise above level at which first encountered. It may or may not flow out at ground level.

asbestos   A general term applied to certain fibrous minerals displaying similar physical characteristics although differing in composition. Some asbestos has fibers long enough to be spun into fabrics with great heat resistance, such as those for automobile brake linings. Types with shorter fibers are compressed into insulating boards, shingles, etc. Most common asbestos mineral (95 percent of United States production) is *chrysotile*, a variety of *serpentine*, a metamorphic mineral.

ash   Volcanic fragments of sharply angular glass particles, smaller than cinders.

asphalt   A brown to black, solid or semisolid bituminous substance. Occurs in nature but is also obtained as residue from refining of certain hydrocarbons ("artificial asphalt").

asteroids   Orbiting small bodies believed to be either fragments of a disintegrated planet or of matter that never completed planet-forming process.

asthenosphere   A zone within earth's *mantle* where plastic movements occur to permit *isostatic* adjustments. Begins 50 to 100 km below surface and extends perhaps to 500 km.

astrogeology   Geology of celestial bodies other than earth. Generally restricted in meaning to geology of extraterrestrial bodies of our planetary system.

asymmetric fold   A fold in which one limb dips more steeply than the other.

Atlantic Ridge   Belt of mountains under Atlantic Ocean from Iceland to Antarctica and nearly midway between continents; also called *Midatlantic Ridge*.

atoll   A ring of low coral islands arranged around a central lagoon.

atom   A building block of matter; combination of protons, neutrons, and electrons, of which 103 kinds are now known.

atomic energy   Energy associated with nucleus of an atom. It is released when nucleus is split.

atomic mass   The nucleus of an atom contains 99.95 percent of its mass. Total number of protons and neutrons in nucleus is called *mass number*.

atomic number   Number of positive charges on nucleus of an atom; number of protons in nucleus.

atomic reactor   A huge apparatus in which a radioactive core heats water under pressure and passes it to a heat exchanger.

atomic size   Radius of an atom (average distance from center to outermost electron of neutral atom). Commonly expressed in angstroms.

augite   A rock-forming, ferromagnesian silicate mineral built around single chains of silicon-oxygen tetrahedra.

aureole   A zone in which contact metamorphism has taken place.

axial plane   A plane through a rock fold that includes the axis and divides the fold as symmetrically as possible.

axis   Ridge or place of sharpest folding of an anticline or syncline.

backset beds   Inclined layers of sand developed on gentler dune slope to windward. These beds may constitute a large part of total volume of a dune, especially if there is sufficient vegetation to trap most sand before it can cross over to slip face.

bank-full stage   Stage of flow at which a stream fills its channel up to level of its bank. Recurrence interval averages 1.5 to 2 years.

barchan   A crescent-shaped dune with wings, or horns, pointing downwind. Has gentle windward slope and steep lee slope inside horns; about 30 m high and 300 m wide from horn to horn. Moves with wind about 15 m/year across flat, hard surface where limited supply of sand is available.

barrier island A low, sandy island near shore and parallel to it, on a gently sloping offshore bottom.

barrier reef A reef separated from a landmass by a lagoon of varying width and depth opening to sea through passes in reef.

basal plane Lower boundary of a zone of concentric folding.

basal slip Movement of an entire glacier over underlying ground surface.

basalt A fine-grained igneous rock dominated by dark-colored minerals, consisting of plagioclase feldspars (over 50 percent) and ferromagnesian silicates. Basalts and andesites represent about 98 percent of all extrusive rocks.

base flow That portion of stream flow attributable to groundwater flow.

base level Level below which a stream cannot erode. There may be temporary base levels along a stream's course, such as those established by lakes or resistant layers of rock. Ultimate base level is sea level.

basement complex Undifferentiated rocks underlying oldest identifiable rocks in any region. Usually sialic, crystalline, metamorphosed. Often, but not necessarily, Precambrian.

basic lava Containing less than 50 percent silica.

basin See *dome*.

batholith A *discordant pluton* that increases in size downward, has no determinable floor, and shows an area of surface exposure exceeding 100 km².

bauxite Chief ore of commercial aluminum. A *mineraloid* mixture of hydrous aluminum oxides.

bay barrier A sandy beach, built up across mouth of a bay so that it is no longer connected to main body of water.

bedding (1) A collective term used to signify existence of beds, or layers, in sedimentary rocks. (2) Sometimes synonymous with *bedding plane*.

bedding plane Surface separating layers of sedimentary rocks. Each bedding plane marks termination of one deposit and beginning of another of different character, such as surface separating a sand bed from a shale layer. Rock tends to separate, or break, readily along bedding planes.

bed load Material in movement along stream bottom or, if wind is moving agency, along surface. Contrast with material carried in *suspension* or solution.

belt of soil moisture Subdivision of *zone of aeration*. Belt from which water may be used by plants or withdrawn by soil evaporation. Some water passes down into intermediate belt, where it may be held by molecular attraction against influence of gravity.

bench mark See Appendix D.

Benioff zone A seismic zone dipping beneath a continental margin and having a deep-sea trench as surface expression.

bergschrund Gap, or crevasse, between glacier ice and headwall of a cirque.

berms In coastline terminology berms are storm-built beach features that resemble small terraces; on seaward edges are low ridges built up by storm waves.

beta decay Radioactive decay taking place by loss of a beta particle (electron) from a neutron in nucleus. Mass of element remains same, but atomic number increases by 1.

B **horizon** Soil zone of accumulation below *A* horizon. Here is deposited some material moved down from *A* horizon.

big-bang theory Theory that presently expanding universe originated as primeval cosmic fireball in very short period of time 10 to 12 billion years ago. Compare *steady-state theory*.

binding energy Amount of energy that must be supplied to break an atomic nucleus into its component fundamental particles. It is equivalent to mass that disappears when fundamental particles combine to form a nucleus.

biochemical rock A sedimentary rock made up of deposits resulting directly or indirectly from life processes of organisms.

biotite "Black mica," ranging in color from dark brown to green. Rock-forming ferromagnesian silicate mineral with tetrahedra in sheets.

bituminous coal Soft coal, containing about 80 percent carbon and 10 percent oxygen.

blowout A basin scooped out of soft, unconsolidated deposits by process of deflation. Ranges from a few meters to several kilometers in diameter.

body wave *Push-pull* or *shake* earthquake wave travelling through body of a medium; distinguished from waves travelling along free surface.

bornite A mineral, Cu_5FeS_4; an important ore of copper.

bottomset bed Layer of fine sediment deposited in a body of standing water beyond advancing edge of a growing delta, which eventually builds up on bottomset beds.

boulder size Volume greater than that of a sphere with diameter of 256 mm.

boulder train Series of glacier erratics from the same bedrock source, usually with some property that permits easy identification. Arranged across country in a fan, with apex at source and widening in direction of glacier movement.

Bowen's reaction series Series of minerals for which any early-formed phase tends to react with melt that remains to yield a new mineral further along in the series. Thus early-formed crystals of olivine react with remaining liquids to form augite crystals; these in turn may further react with liquid then remaining to form hornblende. See also *continuous reaction series* and *discontinuous reaction series*.

braided stream Complex tangle of converging and diverging stream channels separated by sand bars or islands. Characteristic of flood plains where amount of debris is large in relation to discharge.

breccia Clastic rock made up of angular fragments of such size that an appreciable percentage of rock volume consists of particles of granule size or larger.

brittle A property of material whereby strength in *tension* is greatly different from strength in *compression*.

burial metamorphism Changes resulting from pressures and temperatures in rocks buried to depths of several kilometers.

calcic feldspar Anorthite, $Ca(Al_2Si_2O_8)$.

calcite A mineral composed of calcium carbonate, $CaCO_3$.

caldera Roughly circular, steep-sided volcanic basin with diameter at least three or four times depth. Commonly at summit of a volcano. Contrast with *crater*.

caliche Whitish accumulation of calcium carbonate in soil profile.

capillary fringe Belt above *zone of saturation*, in which underground water is lifted against gravity by surface tension in passages of capillary size.

capillary size "Hairlike," or very small, such as tubes from 0.0025 to 0.25 cm in diameter.

carbohydrate Compound of carbon, hydrogen, and oxygen. Carbohydrates are chief products of life process in plants.

carbonate mineral Mineral formed by combination of complex ion $(CO_3)^{2-}$ with a positive ion. Common example: calcite, $CaCO_3$.

carbon cycle Process in sun's deep interior by which radiant energy is generated in formation of helium from hydrogen.

carbon 14 Radioactive isotope of carbon, $^{14}_6C$, which has half-life of 5,730 years. Used to date events up to about 50,000 years ago.

carbon ratio A number obtained by dividing amount of fixed carbon in coal by sum of fixed carbon and volatile matter and multiplying by 100. This is same as percentage of fixed carbon, assuming no moisture or ash.

cassiterite A mineral, tin dioxide, SnO_2. Ore of tin with specific gravity 7; nearly 75 percent of world's tin production is from placer deposits, mostly from cassiterite.

cation A positively charged atom or group of atoms produced by loss of electrons. Compare *anion*.

cementation Process by which a binding agent is precipitated in spaces between individual particles of an unconsolidated deposit. Most common cementing agents are calcite, dolomite, and quartz; others include iron oxide, opal, chalcedony, anhydrite, and pyrite.

cement rock Clayey limestone used in manufacturing hydraulic cement. Contains lime, silica, and alumina in varying proportions.

chalcedony General name applied to fibrous cryptocrystalline silica with waxy luster. Deposited from aqueous solutions and frequently found lining or filling cavities in rocks. *Agate* is a variety with alternating layers of chalcedony and opal.

chalcocite A mineral, copper sulfide, Cu_2S; sometimes called *copper glance;* one of most important ore minerals of copper.

chalcopyrite A mineral, a sulfide of copper and iron, $CuFeS_2$; sometimes called *copper pyrite* or *yellow copper ore.*

chalk Variety of limestone made up in part of biochemically derived calcite, in form of skeletons or skeletal fragments of microscopic oceanic plants and animals mixed with very fine-grained calcite deposits of biochemical or inorganic-chemical origin.

chemical rock In terminology of sedimentary rocks chemical rock is composed chiefly of material deposited by chemical precipitation, whether organic or inorganic (compare with *detrital sedimentary rock*). Chemical sedimentary rocks may have either clastic or nonclastic (usually crystalline) texture.

chemical weathering Weathering of rock material by chemical processes that transform original material into new chemical combinations. Thus chemical weathering of orthoclase produces clay, some silica, and a soluble salt of potassium.

chert Granular cryptocrystalline silica, similar to *flint* but usually light in color. Occurs as compact massive rock or as nodules.

chlorite Family of tetrahedral sheet silicates of iron, magnesium, and aluminum, characteristic of low-grade metamorphism. Green color, with cleavage like mica except that chlorite small scales are not elastic.

C horizon Soil zone that contains partially disintegrated and decomposed parent material. Lies directly under *B* horizon and grades downward into unweathered material.

chromite Mineral oxide of iron and chromium, $FeCr_2O_4$, only ore of commercial chromium. One of first minerals to crystallize from magma; is concentrated within the magma.

chrysotile Metamorphic mineral; an asbestos, fibrous variety of *serpentine*. Silicate of magnesium, with tetrahedra arranged in sheets.

chute or chute cutoff Applied to stream flow, *chute* refers to new route taken by a stream when main flow is diverted to inside of a bend, along trough between low ridges formed by deposition on inside of bend, where water velocity is reduced. Compare with *neck cutoff.*

cinder cone Structure built exclusively or predominantly of pyroclastic ejecta dominated by cinders. Parasitic to a major volcano, seldom exceeds 500 m in height. Slopes up 30° to 40°. Example: Parícutin.

cinders Volcanic fragments; small, slaglike, solidified pieces of magma 0.5 to 2.5 cm across.

cirque A steep-walled hollow in a mountainside at high elevation, formed by ice-plucking and frost action and shaped like a half bowl or half amphitheater. Serves as principal gathering ground for ice of a valley glacier.

clastic texture Texture shown by sedimentary rocks from deposits of mineral and rock fragments.

clay minerals Finely crystalline, hydrous silicates formed from weathering of such silicate minerals as feldspar, pyroxene, and amphibole. Most common clay minerals belong to *kaolinite, montmorillonite,* and *illite* groups.

clay size Volume less than that of a sphere with diameter of $\frac{1}{256}$ mm (0.004 mm).

cleavage (1) *Mineral cleavage:* Property possessed by many minerals of breaking in certain preferred directions along smooth plane surfaces. Planes of cleavage are governed by atomic pattern and represent directions in which atomic bonds are relatively weak. (2) *Rock cleavage:* Property possessed by certain rocks of breaking with relative ease along parallel planes or nearly parallel surfaces. Rock cleavage is designated as *slaty, phyllitic, schistose,* and *gneissic.*

coal Sedimentary rock composed of combustible matter derived from partial decomposition and alteration of plant cellulose and lignin.

cobble size Volume greater than that of a sphere with diameter of 64 mm and less than that of a sphere with diameter of 256 mm.

coesite High-pressure form of quartz, with density of 2.92. Associated with impact craters and cryptovolcanic structures.

col Pass through a mountain ridge. Created by enlargement of two cirques on opposite sides of ridge until headwalls meet and are broken down.

cold glacier One in which no surface melting occurs during summer months and whose temperature is always below freezing.

colloidal size Between 0.2 and 1 μm (0.0002 to 0.001 mm).

color Sensation resulting from stimulation of retina by light waves of certain lengths.

column Column, or post, of dripstone joining floor and roof of a cave; result of joining of stalactite and stalagmite.

columnar jointing Pattern of jointing that blocks out columns of rock. Characteristic of tabular basalt flows or sills.

compaction Reduction in pore space between individual grains from pressure of overlying sediments or pressures of earth movement.

composite volcanic cone Composed of interbedded lava flows and pyroclastic material and characterized by slopes of close to 30° at summit, progressively reducing to 5° near base. Example: Mayon.

compression Squeezing stress that tends to decrease volume of a material.

concentric folding Elastic bending of an originally horizontal sheet with all internal movements parallel to a *basal plane* (lower boundary of the fold).

conchoidal fracture A mineral's habit of fracturing to produce curved surfaces like interior of a shell (*conch*). Typical of glass and quartz.

concordant pluton An intrusive igneous body with contacts parallel to layering or foliation surfaces of rocks into which it has intruded.

concretion An accumulation of mineral matter formed around a center, or axis, of deposition after a sedimentary deposit has been laid down. Cementation consolidates the deposit as a whole, but the concretion is a body within host rock that represents local concentration of cementing material. Enclosing rock is less firmly cemented than the concretion. Commonly spheroidal or disk-shaped and composed of such cementing agents as calcite, dolomite, iron oxide, or silica.

cone of depression A dimple in the water table, which forms as water is pumped from a well.

cone sheet A dike, part of a concentric set dipping inward, like an inverted cone.

conglomerate Detrital sedimentary rock made up of more or less rounded fragments of such size that an appreciable percentage of volume of rock consists of particles of granule size or larger.

connate water Water trapped in a sedimentary deposit when the deposit was laid down.

contact metamorphism Metamorphism at or very near contact between magma and rock during intrusion.

continental drift Slow, lateral movement of continents; involves rigid plates that may carry both continental and oceanic areas as they move.

continental glacier An ice sheet that obscures mountains and plains of a large section of a continent. Continental glaciers exist on Greenland and Antarctica.

continental rise In some places base of continental slope is marked by somewhat gentler continental rise, which leads downward to deep ocean floor.

continental shelf Shallow, gradually sloping zone extending from sea margin to a depth at which there is marked or rather steep descent into ocean depths down continental slope. Seaward boundary of shelf averages about 130 m in depth.

continental slope Portion of ocean floor extending downward from seaward edge of continental shelves. In some places, such as south of Aleutian Islands, slopes descend directly to ocean deeps. In other places, such as off eastern North America, they grade into somewhat gentler continental rises, which in turn lead to deep ocean floors.

continuous reaction series Branch of Bowen's reaction series comprising plagioclase feldspars, in which reaction of early-formed crystals with later liquids takes place continuously—that is, without abrupt phase changes.

contour interval See Appendix D.

contour line See Appendix D.

convection Mechanism by which material moves because its density differs from that of surrounding material. Density differences are often brought about by heating.

convection cell Pair of *convection currents* adjacent to each other.

convection current Closed circulation of material sometimes developed during convection. Convection currents normally develop in pairs, each pair called *convection cell*.

convergent plate boundary Boundary between two plates moving toward each other. Compare *divergent plate boundary*.

coquina A coarse-grained, porous, friable variety of clastic limestone made up chiefly of fragments of shells.

core Innermost zone of earth, surrounded by *mantle*.

Coriolis effect Tendency of any moving body, on or starting from surface of earth, to continue in direction in which earth's rotation propels it. Direction in which the body moves because of this tendency combined with direction in which it is aimed determines ultimate course of the body relative to earth's surface. In the northern hemisphere Coriolis effect causes a moving body to veer or try to veer to right of its direction of forward motion; in the southern hemisphere, to left. Magnitude of effect is proportional to velocity of a body's motion. This effect causes cyclonic storm-wind circulation to be counterclockwise in the northern hemisphere and clockwise in the southern hemisphere and determines final course of ocean currents relative to trade winds.

correlation Process of establishing contemporaneity of rocks or events in one area with rocks or events in another area.

covalent bond Bond in which atoms combine by sharing their electrons.

crater Roughly circular, steep-sided volcanic basin with diameter less than three times depth. Commonly at summit of a volcano (contrast with *caldera*). Applied also to depressions caused by meteorites by either direct or explosive impact.

craton Relatively immobile part of the earth, generally large central portion of a continent.

creep Applied to soils and surficial material, slow downward plastic movement. As applied to elastic solids, slow permanent yielding to stresses less than yield point if applied for a short time only.

crevasse (1) Deep crevice, or fissure, in glacier ice. (2) Breach in a natural levee.

cross bedding See *inclined bedding*.

crosscutting relationships, law of A rock is younger than any rock across which it cuts.

crust Outermost shell of the earth. Continental crust averages 35 km thick, density 2.6 t/m^3; oceanic crust, about 5 km thick, density 3 t/m^3.

cryptocrystalline State of matter in which there is actually orderly arrangement of atoms characteristic of crystals but in units so small (material is so fine grained) that crystalline nature cannot be determined with an ordinary microscope.

cryptovolcanic structure Geologic structure in which rocks appear disrupted by volcanic activity but in which there are hidden no volcanic rocks.

crystal Solid with orderly atomic arrangement, which may or may not develop external faces that give it crystal form.

crystal form Geometrical form taken by a mineral, giving external expression to orderly internal atomic arrangement.

crystalline structure Orderly arrangement of atoms in a crystal. Also called *crystal structure*.

crystallization Process through which crystals separate from fluid, viscous, or dispersed state.

Curie temperature Temperature above which ordinarily magnetic material loses magnetism. On cooling below this temperature, it regains magnetism. Example: Iron loses magnetism above 760°C and regains it as it cools below this temperature. This is its Curie temperature.

current bedding See *inclined bedding*.

cutoff See *chute cutoff*, *neck cutoff*.

cyclosilicate Mineral with crystal structure containing silicon-oxygen, $(SiO_4)^{4-}$, tetrahedra arranged as rings.

datum plane See Appendix D.

debris slide Small, rapid movement of largely unconsolidated material that slides or rolls downward to produce irregular topography.

decomposition Synonymous with *chemical weathering*.

deep focus Earthquake focus deeper than 300 km. Greatest depth of focus known is 700 km.

deep-sea trenches See *island-arc deeps*.

deflation Erosive process in which wind carries off unconsolidated material.

deformation of rocks Any change in original shape or volume of rock masses; produced by mountain-building forces. *Folding, faulting,* and *plastic flow* are common modes of rock deformation.

delta Plain underlain by an assemblage of sediments accumulated where a stream flows into a body of standing water, its velocity and transporting power suddenly reduced. Originally named after Greek letter *delta* (Δ) because many are roughly triangular in plan, with apex pointing upstream.

dendritic pattern An arrangement of stream courses that, on a map or viewed from the air, resemble branching habit of certain trees, such as oaks or maples.

density Measure of concentration of matter, expressed as mass per unit volume.

density current Current due to differences in density of water from place to place caused by changes in temperature and variations in salinity or amount of material held in suspension.

depositional remanent magnetism Magnetism resulting from tendency of magnetic particles such as magnetite to orient themselves in earth's magnetic field as they are deposited. Orientation is maintained as soft sediments are lithified and thus records earth's field when particles were laid down. Abbreviation, DRM.

desert varnish Thin, shiny layer of iron and manganese oxides that coats some desert-rock surfaces.

detrital sedimentary rock Rock formed from accumulation of minerals and rocks derived from erosion of previously existing rocks or from weathered products of these rocks.

diabase Rock of basaltic composition, essentially labradorite and pyroxene, characterized by *ophitic* texture.

diamond A mineral composed of element carbon; hardest substance known. Used as a gem and in industry for cutting tools.

differential erosion Process by which different rock masses or different parts of same rock erode at different rates.

differential weathering Process by which different rock masses or different parts of same rock weather at different rates.

dike Tabular *discordant pluton*.

dike swarm Group of approximately parallel dikes.

dilatancy Tendency of rocks to expand along minute fractures immediately prior to failure; stress may be from earth movements or from controlled laboratory experiments.

diorite Coarse-grained igneous rock with composition of andesite (no quartz or orthoclase), composed of about 75 percent plagioclase feldspars and balance ferromagnesian silicates.

dip (1) Acute angle that a rock surface makes with a horizontal plane. Direction of dip is always perpendicular to *strike*. (2) See *magnetic declination*.

dipole Any object oppositely charged at two points. Most commonly refers to a molecule that has concentrations of positive or negative charge at two different points.

dipole magnetic field Portion of earth's magnetic field best described by a dipole passing through earth's center and inclined to earth's rotation axis. See also *nondipole magnetic field* and *external magnetic field*.

dip pole See *magnetic pole*.

dip-slip fault Fault in which displacement is in direction of fault's dip.

discharge With reference to stream flow, quantity of water that passes a given point in unit time. Measured in cubic meters per second or often cubic feet per second (abbreviation, cfs).

disconformity *Unconformity* in which beds on opposite sides are parallel.

discontinuity Within earth's interior, sudden or rapid changes with depth in one or more of physical properties of materials constituting the earth, as evidenced by seismic data.

discontinuous reaction series Branch of Bowen's reaction series including minerals olivine, augite, hornblende, and biotite, for which each series change represents abrupt phase change.

discordant pluton An intrusive igneous body with boundaries that cut across surfaces of layering or foliation in rocks into which it has intruded.

disintegration Synonymous with *mechanical weathering*.

divergent plate boundary Boundary between two plates moving apart. New oceanic-type lithosphere is created at the opening. Compare *convergent plate boundary*.

divide Line separating two drainage basins.

dolomite Mineral composed of carbonate of calcium and magnesium, $CaMg(CO_3)_2$. Also used as rock name for formations composed largely of mineral dolomite.

dome Anticlinal fold without clearly developed linearity of crest so that beds involved dip in all directions from a central area, like an inverted but usually distorted cup. Reverse of *basin*.

drainage basin Area from which a given stream and its tributaries receive water.

drift Any material laid down directly by ice or deposited in lakes, oceans, or streams as result of glacial activity. Unstratified glacial drift is called *till* and forms *moraines;* stratified forms *outwash plains, eskers, kames,* and *varves.*

dripstone Calcium carbonate deposited from solution by underground water entering a cave in *zone of aeration*. Sometimes called *travertine*.

DRM See *depositional remanent magnetism*.

drumlin Smooth, streamlined hill composed of *till*. Long axis oriented in direction of ice movement: Blunt nose points upstream, and gentler slope tails off downstream. In height drumlins range from 8 to 60 m, with average somewhat less than 30 m. Most drumlins are between 0.5 and 1 km in length, the length commonly several times width. Diagnostic characteristics are shape and composition of unstratified glacial drift, in contrast to *kames*, of stratified glacial drift and random shapes.

dune Mound or ridge of sand piled by wind.

dust-cloud hypotheses Hypotheses that solar system was formed from condensation of interstellar dust clouds.

dust size Volume less than that of sphere with diameter 0.06 mm; used in reference to particles carried in suspension by wind.

earthflow Combination of *slump* and *mudflow*.

earth waves Mechanism for transmitting energy from earthquake focus.

ecliptic Apparent path of the sun in the heavens; plane of planets' orbits.

eclogite facies Metamorphic rocks of *gabbroic* composition, consisting primarily of *pyroxene* and *garnet*.

elastic deformation Nonpermanent deformation, after which body returns to original shape or volume when deforming force removed.

elastic energy Energy stored within a solid during elastic deformation and released during elastic rebound.

elasticity A property of materials that defines extent to which they resist small deformations, from which they recover completely when deforming force is removed. Elasticity equals *stress* divided by *strain*.

elastic limit Maximum *stress* that produces only *elastic deformation*.

elastic rebound Recovery of elastic *strain* when material breaks or deforming force removed.

elastic solid A solid that yields to applied force by changing shape, volume, or both but returns to original condition when force is removed. Amount of yield is proportional to force.

electrical energy Energy of moving *electrons*.

electric charge Property of matter resulting from imbalance between number of *protons* and number of *electrons* in given piece of matter. The electron has negative charge: the proton, positive charge. Like charges repel each other; unlike charges attract each other.

electric current Flow of electrons.

electric turbine generator Apparatus that uses steam from heat exchanger to drive turbine and generate electricity.

electron Fundamental particle of matter, the most elementary unit of negative electrical charge. Mass, 0.00055 u (*atomic mass* unit).

electron-capture decay Radioactive decay that takes place as an orbital electron is captured by a proton in nucleus. Mass of element remains constant, but atomic number decreases by 1.

electron shell Imaginary spherical surface representing all possible paths of electrons with same average distance from *nucleus* and with approximately same energy.

element Unique combination of *protons, neutrons,* and *electrons* that cannot be broken down by ordinary chemical methods. Fundamental properties of an element are determined by number of protons, each element assigned a number corresponding to its number of protons. Combinations containing from 1 through 103 protons are now known.

end moraine Ridge or belt of *till* marking farthest advance of a glacier. Sometimes called *terminal moraine*.

energy Capacity for producing motion. Energy holds matter together and can become *mass* or be derived from *mass*. Takes such forms as kinetic, potential, heat, chemical, electrical, and atomic energy; one form of energy can be changed to another.

energy level Distance from atomic nucleus at which electrons orbit. May be thought of as shell surrounding nucleus.

epicenter Area on surface directly above *focus* of earthquake.

epidote Silicate of aluminum, calcium, and iron characteristic of low-grade metamorphism and associated with *chlorite* and *albite* in *greenschist facies*. Built around independent tetrahedra.

epidote-amphibolite facies Assemblage of minerals formed between 250 and 450°C during *regional metamorphism*.

erg Unit of energy expressing capacity for doing work. Equal to energy expended when force of 1 dyne acts through distance of 1 cm.

erosion Movement of material from one place to another on earth's surface. Agents of movement include gravity, water, ice, and wind.

erosional flood plain Flood plain created by lateral erosion and gradual retreat of valley walls.

erratic In terminology of glaciation a stone or boulder carried by ice to a place where it rests on or near bedrock of different composition.

escape velocity Minimum velocity an object must have to escape from gravitational field. For moon this is about 2.38 km/s and for earth about 11.2 km/s.

esker Winding ridge of stratified glacial *drift*, steep-sided, 3 to 15 m high, and from a fraction of a kilometer to over 160 km long.

eugeosyncline Part of *geosyncline* in which volcanism is associated with clastic sedimentation, generally located away from *craton*.

eustatic change Change in sea level produced entirely by increase or decrease in amount of water in oceans; hence of worldwide proportions.

eutrophic In lake aging pertains to an old-age lake and indicates high supply of nutrients supporting high biologic productivity.

evaporation Process by which liquid becomes vapor at temperature below boiling point.

evaporite Rock composed of minerals precipitated from solutions concentrated by evaporation of solvents. Examples: *rock salt, gypsum, anhydrite*.

exfoliation Process by which rock plates are stripped from larger rock mass by physical forces.

exfoliation dome Large, rounded domal feature produced in homogeneous coarse-grained igneous rocks and sometimes in conglomerates by process of exfoliation.

external magnetic field Component of earth's field originating from activity above earth's surface. Small when compared with dipole and nondipole components of field, which originate beneath surface.

extrusive rock Rock solidified from mass of magma poured or blown out upon earth's surface.

facies Assemblage of mineral, rock, or fossil features reflecting environment in which rock was formed. See *sedimentary facies, metamorphic facies*.

fault Surface of rock rupture along which has been differential movement.

fault-block mountain Mountain bounded by one or more faults.

feldspars Silicate minerals composed of silicon-oxygen and aluminum-oxygen tetrahedra linked together in three-dimensional networks with positive ions fitted into interstices of negatively charged framework of tetrahedra. Classed as aluminosilicates. When positive ion is K^+, mineral is *orthoclase;* when Na^+, mineral is *albite;* when Ca^{2+}, mineral is *anorthite*.

felsite General term for light-colored, fine-grained igneous rocks.

ferromagnesian silicate Silicate in which positive ions are dominated by iron, magnesium, or both.

fibrous fracture Mineral habit of breaking into splinters or fibers.

fiery cloud Avalanche of incandescent *pyroclastic debris* mixed with steam and other gases, heavier than air, and projected down a volcano's side. Also called *nuée ardente*.

fiord Glacially deepened valley now flooded by the sea to form long, narrow, steep-walled inlet.

firn Granular ice formed by recrystallization of snow. Intermediate between snow and glacier ice. Sometimes called *névé*.

fissility Property of splitting along closely spaced planes more or less parallel to bedding. Presence distinguishes shale from mudstone.

fission   Process by which atomic nucleus breaks down to form nuclei of lighter atoms.

fissure eruption   Extrusion of lava from fissure in earth's crust.

flashy stream   Stream with high flood peak of short duration, which may be caused by urbanization.

flint   Granular cryptocrystalline silica, usually dull and dark. Often occurs as lumps or nodules in calcareous rocks, such as Cretaceous chalk beds of southern England.

flood basalt   Basalt poured out from fissures in floods that tend to form great plateaus. Sometimes called *plateau basalt*.

flood frequency   Time within which a flood of a given size can be expected to occur.

flood plain   Area bordering a stream, over which water spreads in time of flood.

flood plain of aggradation   Flood plain formed by building up of valley floor by sedimentation.

flotation   Process that begins concentration of ore minerals from *gangue*.

fluid   Material that offers little or no resistance to forces tending to change its shape.

focus   Source of given set of earthquake waves.

fold   Bend, flexure, or wrinkle in rock produced when rock was in a plastic state.

fold mountains   Mountains consisting primarily of elevated, folded sedimentary rocks.

foliation   Layering in some rocks caused by parallel alignment of minerals; textural feature of some metamorphic rocks. Produces rock cleavage.

footwall   One of blocks of rock involved in fault movement. One that would be under feet of person standing in tunnel along or across fault; opposite *hanging wall*.

fore dune   Dune immediately behind shoreline of ocean or large lake.

foreset beds   Inclined layers of sediment deposited on advancing edge of growing *delta* or along lee slope of advancing sand *dune*.

foreshock   Relatively small earthquake that precedes larger earthquake by a few days or weeks and originates at or near *focus* of larger earthquake.

fossil   Evidence of past life, such as dinosaur bones, ancient clam shell, footprint of long-extinct animal, or impression of leaf in rock.

fossil fuels   Organic remains used to produce heat or power by combustion. Include *coal*, *petroleum*, and *natural gas*.

fractional distillation   Recovery—one or more at a time—of fractions of complex liquid, each of which has different density.

fractionation   Process whereby crystals that formed early from magma have time to settle appreciably before temperature drops much further. They are thus effectively removed from environment in which they formed.

fracture   As mineral characteristic, way in which mineral breaks when it does not have cleavage. May be *conchoidal* (shell-shaped), *fibrous*, *hackly*, or *uneven*.

fracture cleavage   System of joints spaced fraction of centimeter apart.

freeboard of continents   Relative elevation of continents with respect to sea level.

fringing reef   Reef attached directly to landmass.

frost action   Process of mechanical weathering caused by repeated cycles of freezing and thawing. Expansion of water during freezing cycle provides energy for process.

fumarole   Vent for volcanic steam and gases.

fundamental particles   *Protons*, *neutrons*, and *electrons*, which combine to form atoms. Each particle is defined in terms of its *mass* and its *electric charge*.

fusion   Process by which nuclei of lighter atoms join to form nuclei of heavier atoms.

gabbro   Coarse-grained igneous rock with composition of basalt.

galaxy   Family of stars grouped in space. The earth belongs to Milky Way galaxy, which contains about 100 billion stars.

galena   A mineral; lead sulfide, PbS. Principal ore of lead.

gangue   Commercially valueless material remaining after ore-mineral extraction from rock.

garnet   Family of silicates of iron, magnesium, aluminum, calcium, manganese, and chromium, which are built around independent tetrahedra and appear commonly as distinctive 12-sided, fully developed crystals. Characteristic of *metamorphic rocks*; generally cannot be distinguished from one another without chemical analysis.

gas   (1) State of matter that has neither independent shape nor volume, can be compressed readily, and tends to expand indefinitely. (2) In geology "gas" is sometimes used to refer to *natural gas*, gaseous hydrocarbons that occur in rocks, dominated by methane. Compare "oil," referring to *petroleum*.

geode   Roughly spherical, hollow or partially hollow accumulation of mineral matter from few centimeters to nearly 0.5 m in diameter. Outer layer of chalcedony lined with crystals that project toward hollow center. Crystals, often perfectly formed, usually quartz although calcite and dolomite also found and—more rarely—other minerals. Geodes most commonly found in limestone and more rarely in shale.

geographic poles   Points on earth's surface marked by ends of axis of rotation.

geologic column   Chronologic arrangement of rock units in columnar form, with oldest units at bottom and youngest at top.

geologic-time scale   Chronologic sequence of units of earth time.

geology   Organized body of knowledge about the earth, including *physical geology* and *historical geology*.

geomagnetic poles   Dipole best approximating earth's observed field is one inclined 11.5° from axis of rotation. Points at which ends of this imaginary magnetic axis intersect earth's surface are known as geomagnetic poles. They should not be confused with *magnetic*, or *dip*, poles or *virtual geomagnetic poles*.

geophysical prospecting   Mapping rock structures by methods of experimental physics. Includes measuring magnetic fields, force of gravity, electrical properties, seismic-wave paths and velocities, radioactivity, and *heat flow*.

geophysics   Physics of the earth.

geosyncline   Literally, "earth syncline." Term now refers, however, to a *basin* in which thousands of meters of sediments have accumulated, with accompanying progressive sinking of basin floor explained only in part by load of sediments. Common usage includes both accumulated sediments themselves and geometrical form of basin in which they are deposited. All folded mountain ranges were built from geosynclines, but not all geosynclines have become mountain ranges.

geothermal field   Area where wells drilled to obtain elements contained in solution in hot brines and to tap heat energy.

geyser Special type of thermal spring that intermittently ejects its water with considerable force.

glacier A mass of ice, formed by recrystallization of snow, that flows forward or has flowed at some past time under influence of gravity. By convention we exclude icebergs even though they are large fragments broken from seaward end of glaciers.

glacier ice Unique form of ice developed by compression and recrystallization of snow and consisting of interlocking crystals.

glass Form of matter that exhibits properties of a solid but has atomic arrangements, or lack of order, of a liquid.

Glossopteris flora A late-Paleozoic assemblage of fossil plants named for seed fern *Glossopteris*, a plant in the flora. Widespread in South America, South Africa, Australia, India, and Antarctica.

gneiss Metamorphic rock with *gneissic cleavage*. Commonly formed by metamorphism of granite.

gneissic cleavage Rock cleavage in which surfaces of easy breaking, if developed at all, are from a few hundredths of a millimeter to a centimeter or more apart.

goethite Hydrous iron oxide, $FeO(OH)$.

Gondwanaland Hypothetical continent thought to have broken up in Mesozoic. Resulting fragments are postulated to form present-day South America, Africa, Australia, India, and Antarctica.

graben Elongated, trenchlike, structural form bounded by parallel normal faults created when block that forms trench floor moved downward relative to blocks that form sides.

grade Term used to designate extent to which *metamorphism* has advanced. Found in such combinations as "high-grade" or "low-grade metamorphism." Compare with *rank*.

graded bedding Type of bedding shown by sedimentary deposit when particles become progressively finer from bottom to top.

gradient Slope of stream bed.

granite Coarse-grained *igneous rock* dominated by light-colored minerals, consisting of about 50 percent *orthoclase*, 25 percent *quartz*, and balance of plagioclase *feldspars* and *ferromagnesian silicates*. Granites and *granodiorites* comprise 95 percent of all intrusive rocks.

granitization Special type of *metasomatism* by which solutions of magmatic origin move through solid rocks, change ions with them, and convert them into rocks that achieve granitic character without having passed through magmatic stage.

granodiorite Coarse-grained igneous rock intermediate in composition between *granite* and *diorite*.

granular texture Composed of mineral grains large enough to be seen by unaided eye.

granulite facies *Gneissic* rocks produced by deep-seated high-grade *regional metamorphism*.

graphic structure Intimate intergrowth of *potassic feldspar* and *quartz* with long axes of quartz crystals lining up parallel to feldspar axis. Quartz part is dark and feldspar is light in color; so pattern suggests Egyptian hieroglyphs. Commonly found in *pegmatites*.

graphite "Black lead." A mineral composed entirely of carbon. Very soft because of crystalline structure; diamond, in contrast, has same composition but is hardest substance known.

gravity anomaly Difference between observed and computed values of gravity.

gravity fault Fault in which *hanging wall* appears to have moved downward relative to *footwall*. Also called *normal fault*.

gravity meter An instrument for measuring force of gravity. Also called *gravimeter*.

gravity prospecting Mapping force of gravity at different places to determine differences in *specific gravity* of rock masses and, through this, distribution of masses of different specific gravity. Done with *gravity meter* (gravimeter).

graywacke A variety of *sandstone* generally characterized by hardness, dark color, and angular grains of *quartz*, *feldspar*, and small rock fragments set in matrix of clay-sized particles. Also called *lithic sandstone*.

greenschist Schist characterized by green color. Product of *regional metamorphism* of *simatic rocks*. (Green color is imparted by mineral *chlorite*.)

greenschist facies Assemblage of minerals formed between 150 and $250°C$ during *regional metamorphism*.

groundmass Finely crystalline or glassy portion of porphyry.

ground moraine *Till* deposited from a glacier as veneer over landscape and forming gently rolling surface.

groundwater Underground water within *zone of saturation*.

groundwater table Upper surface of *zone of saturation* for underground water. An irregular surface with slope or shape determined by quantity of groundwater and permeability of earth materials. In general, highest beneath hills and lowest beneath valleys. Also referred to as *water table*.

guyot Flat-topped *seamount* rising from ocean floor like a volcano but planed off on top and covered by appreciable water depth. Synonymous with *tablemount*.

gypsum Hydrous calcium sulfate, $CaSO_4 \cdot 2H_2O$. A soft, common mineral in sedimentary rocks, where it sometimes occurs in thick beds interstratified with limestones and shales. Sometimes occurs as layer under bed of *rock salt* since it is one of first minerals to crystallize on evaporation of seawater. Alabaster is a fine-grained massive variety of gypsum.

H Symbol for mineral *hardness*.

hackly fracture Mineral habit of breaking to produce jagged, irregular surfaces with sharp edges.

half-life Time needed for one-half of nuclei in sample of radioactive element to decay.

halide Compound made from a halogen, such as chlorine, iodine, bromine, or fluorine.

halite A mineral; *rock salt*, or common salt, $NaCl$. Occurs widely disseminated or in extensive beds and irregular masses precipitated from seawater and interstratified with rocks of other types as true sedimentary rock.

hanging valley A valley that has greater elevation than the valley to which it is tributary, at point of junction. Often (but not always) created by deepening of main valley by a glacier. Hanging valley itself may or may not be glaciated.

hanging wall One of blocks involved in fault movement. One that would be hanging overhead for person standing in tunnel along or across fault; opposite *footwall*.

hardness Mineral's resistance to scratching on a smooth surface. Mohs scale of relative hardness consists of 10 minerals, each scratching all those below it in scale and being scratched by all those above it: (1) talc, (2) gypsum, (3) calcite, (4) fluorite, (5) apatite, (6) orthoclase, (7) quartz, (8) topaz, (9) corundum, (10) diamond.

head Difference in elevation between intake and discharge points for a liquid. In geology most commonly of interest in connection with movement of *underground water*.

heat energy Special manifestation of *kinetic energy* in atoms. Temperature of a substance depends on average kinetic energy of component particles. When heat is added to a substance, average kinetic energy increases.

heat exchanger Unit in atomic-power generation that uses water (heated under pressure by atomic reactor) to form steam from water in another system and to drive a turbine for electricity generation.

heat flow Product of thermal gradient and thermal conductivity of earth materials. Average over whole earth, $1.2 \pm 0.15 \ \mu cal/cm^2$-s.

height See Appendix D.

hematite Iron oxide, Fe_2O_3. Principal ore mineral for about 90 percent of commercial iron produced in United States. Characteristic red color when powdered.

hinge fault Fault in which displacement perceptibly dies out along strike and ends at definite point.

historical geology Branch of *geology* that deals with history of the earth, including record of life on earth as well as physical changes in earth itself.

horn Spire of bedrock left where *cirques* have eaten into a mountain from more than two sides around a central area. Example: Matterhorn of the Swiss Alps.

hornblende A rock-forming *ferromagnesian silicate* mineral with double chains of silicon-oxygen tetrahedra. An *amphibole*.

hornfels Dense, granular metamorphic rock. Since this term is commonly applied to metamorphic equivalent of any fine-grained rock, composition is variable.

hornfels facies Assemblage of minerals formed at temperatures greater than 700°C during *contact metamorphism*.

horst Elongated block bounded by parallel *normal faults* in such a way that it stands above blocks on both sides.

hot spot Localized melting region in mantle near base of lithosphere, a few hundred kilometers in diameter and persistent over tens of millions of years. Existence of heat is assumed from volcanic activity at surface.

hot spring *Thermal spring* that brings hot water to surface. Water temperature usually 6.5°C or more above mean air temperature.

hydration Process by which water combines chemically with other molecules.

hydraulic gradient *Head* of underground water divided by distance of travel between two points: If head 10 m for two points 100 m apart, hydraulic gradient is 0.1, or 10 percent. When head and distance of flow are same, hydraulic gradient is 100 percent.

hydraulic mining Use of strong water jet to move deposits of sand and gravel from original site to separating equipment, where sought-for mineral extracted.

hydrocarbon Compound of hydrogen and carbon that burns in air to form water and oxides of carbon. There are many hydrocarbons. The simplest, methane, is chief component of natural gas. Petroleum is a complex mixture of hydrocarbons.

hydroelectric power Conversion of energy to electricity by free fall of water. This method supplies about 4 percent of world's electrical energy.

hydrograph Graph of variation of stream flow over time.

hydrologic cycle General pattern of water movement by evaporation from sea to atmosphere, by precipitation onto land, and by return to sea under influence of gravity.

hydrothermal solution Hot, watery solution that usually emanates from *magma* in late stages of cooling. Frequently contains, and deposits in economically workable concentrations, minor elements that, because of incommensurate *ionic radii* or electronic charges,

have not been able to fit into atomic structures of common minerals of igneous rocks.

hysteresis Retardation of recovery from *elastic deformation* after *stress* is removed.

icecap Localized *ice sheet*.

ice sheet Broad, moundlike mass of glacier ice of considerable extent that has tendency to spread radially under own weight. Localized ice sheets are sometimes called *icecaps*.

igneous rock Aggregate of interlocking silicate minerals formed by cooling and solidification of magma.

illite Clay mineral family of hydrous aluminous silicates. Structure is similar to that of *montmorillonite*, but aluminum replaces 10 to 15 percent of silicon, which destroys montmorillonite's property of expanding with addition of water because weak bonds replaced by strong potassium-ion links. Structurally illite intermediate between montmorillonite and *muscovite*. Montmorillonite converts to illite in sediments; illite, to muscovite under conditions of low-grade metamorphism. Illite is commonest clay mineral in clayey rocks and recent marine sediments and is present in many soils.

ilmenite Iron titanium oxide. Accounts for much of unique abundance of titanium on moon.

inclined bedding Bedding laid down at angle to horizontal. Also referred to as *cross bedding* or *current bedding*.

index minerals *Chlorite*, low-grade metamorphism; *almandite*, middle-grade metamorphism; *sillimanite*, high-grade metamorphism.

induced magnetism In terminology of rock magnetism one of components of rock's *natural remanent magnetism*. It is parallel to earth's present field and results from it.

inertia member Central element of a seismograph, consisting of weight suspended by wire or spring so that it acts like pendulum free to move in only one plane.

infiltration Soaking into ground of water on surface.

inosilicate Mineral with crystal structure containing silicon-oxygen tetrahedra in single or double chains.

intensity Measure of effects of earthquake waves on human beings, structures, and earth's surface at particular place. Contrast with *magnitude*, which is measure of total energy released by an earthquake.

intermediate belt Subdivision of *zone of aeration*. Belt that lies between *belt of soil moisture* and *capillary fringe*.

intermediate lava Lava composed of 60 to 65 percent silica.

intrusive rock Rock solidified from mass of magma that invaded earth's crust but did not reach surface.

ion Electrically unbalanced form of an atom or group of atoms, produced by gain or loss of electrons.

ionic bond Bond in which ions are held together by electrical attraction of opposite charges.

ionic radius Average distance from center to outermost electron of an ion. Commonly expressed in *angstroms*.

island-arc deeps Arcuate trenches bordering some continents; some reach depths of 9,000 m or more below sea surface. Also called *deep-sea trenches* or *trenches*.

isoclinal folding Beds on both *limbs* nearly parallel, whether fold upright, *overturned*, or *recumbent*.

isoseismic line Line connecting all points on surface of earth where intensity of shaking from earthquake waves is same.

isostasy Ideal condition of balance that would be attained by earth materials of differing densities if gravity were the only force governing heights relative to each other.

isotope Alternative form of an element produced by variations in number of neutrons in nucleus.

jasper Granular, cryptocrystalline silica usually colored red by *hematite* inclusions.

joint Break in rock mass with no relative movement of rock on opposite sides of break.

joint system Combination of intersecting joint sets, often at approximately right angles.

juvenile water Water brought to surface or added to underground supplies from magma.

kame Steep-sided hill of stratified glacial *drift*. Distinguished from *drumlin* by lack of unique shape and by stratification.

kame terrace Stratified glacial drift deposited between wasting glacier and adjacent valley wall. When ice melts, this material stands as a terrace along valley wall.

kaolinite Clay mineral, hydrous aluminous silicate, $Al_4Si_4O_{10}(OH)_8$. Structure consists of one sheet of silicon-oxygen tetrahedra each sharing three oxygens to give ratio of $(Si_4O_{10})^{4-}$ linked with one sheet of aluminum and hydroxyl. Composition of pure kaolinite does not vary as for other clay minerals, *montmorillonite* and *illite*, in which ready addition or substitution of ions takes place.

karst topography Irregular topography characterized by *sink holes*, streamless valleys, and streams that disappear underground—all developed by action of surface and underground water in soluble rock such as limestone.

Kerguelen-Gaussberg Ridge Belt of mountains under Indian Ocean between India and Antarctica.

kettle Depression in ground surface formed by melting of a block of ice buried or partially buried by glacial *drift*, either *outwash* or *till*.

kinetic energy Energy of movement. Amount possessed by an object or particle depends on mass and speed.

kyanite A silicate mineral characteristic of temperatures of middle-grade metamorphism. Al_2SiO_5 in bladed blue crystals is softer than a knife along the crystal. Its crystalline structure is based on independent tetrahedra. Compare with *andalusite*, which has same composition and forms under similar conditions but has different crystal habit. Contrast with *sillimanite*, which has same composition but different crystal habit and forms at highest metamorphic temperatures.

L Symbol for earthquake *surface waves*.

laccolith *Concordant pluton* that has domed up strata into which it intruded.

lag time On stream *hydrograph* time interval between center of mass of precipitation and center of mass of resulting flood.

laminar flow Mechanism by which fluid (such as water) moves slowly along a smooth channel or through a tube with smooth walls with fluid particles following straight-line paths parallel to channel or walls. Contrast with *turbulent flow*.

landslide General term for relatively rapid mass movement, such as *slump*, *rock slide*, *debris slide*, *mudflow*, and *earthflow*.

lapilli *Pyroclastic debris* in pieces about walnut size.

large waves Earthquake *surface waves*.

latent heat of fusion Number of calories per unit volume that must be added to a material at melting point to complete process of melting. These calories do not raise temperature.

lateral moraine Ridge of *till* along edge of valley glacier. Composed largely of material fallen to glacier from valley walls.

laterite Tropical soil rich in hydroxides of aluminum and iron and formed under conditions of good drainage.

lava *Magma* poured out on surface of earth or rock solidified from such magma.

left-lateral fault *Strike-slip fault* where ground opposite you appears to have moved left when you face it.

levee (natural) Bank of sand and silt built by river during floods, where suspended load deposited in greatest quantity close to river. Process of developing natural levees tends to raise river banks above level of surrounding *flood plains*. Break in natural levee sometimes called *crevasse*.

lignite Low-grade coal, with about 70 percent carbon and 20 percent oxygen. Intermediate between peat and bituminous coal.

limb One of two parts of *anticline* or *syncline*, on either side of axis.

limestone Sedimentary rock composed largely of mineral *calcite*, $CaCO_3$, formed by either organic or inorganic processes. Most limestones have clastic texture, but nonclastic, particularly crystalline, textures are common. Carbonate rocks, limestone and *dolomite*, constitute estimated 12 to 22 percent of sedimentary rocks exposed above sea level.

limonite Iron oxide with no fixed composition or atomic structure; a *mineraloid*. Always of secondary origin, not a true mineral. Is encountered as ordinary rust or coloring material of yellow clays and soils.

liquefaction Process of changing soil and unconsolidated sediments into water mixture immediately following earthquake; often results in foundation failure, with sliding of ground under building structures.

liquid State of matter that flows readily so that the mass assumes form of container but retains independent volume.

lithic sandstone See *graywacke*.

lithification Process by which unconsolidated rock-forming materials are converted into consolidated or coherent state.

lithosphere Rigid outer layer of earth; includes *crust* and upper part of *mantle*. Relatively strong layer in contrast to underlying *asthenosphere*.

loess Unconsolidated, unstratified aggregation of small, angular mineral fragments, usually buff in color. Generally believed to be wind-deposited; characteristically able to stand on very steep to vertical slopes.

longitudinal dune Long ridge of sand oriented in general direction of wind movement. A small one is less than 3 m high and 60 m long. Very large ones called *seif dunes*.

longitudinal wave *Push-pull wave*.

lopolith Tabular *concordant pluton* shaped like spoon bowl, with both roof and floor sagging downward.

magma Naturally occurring silicate melt, which may contain suspended silicate crystals, dissolved gases, or both. These conditions may be met in general by a mixture containing as much as 65 percent crystals but no more than 11 percent dissolved gases.

magnetic declination Angle of divergence between geographic meridian and magnetic meridian. Measured in degrees east and west of geographic north.

magnetic inclination Angle that magnetic needle makes with surface of earth. Also called *dip of magnetic needle*.

magnetic pole North magnetic pole is point on earth's surface where north-seeking end of a magnetic needle free to move in space points directly down. At south magnetic pole the same needle points directly up. These poles are also known as *dip poles.*

magnetic reversal Shift of 180° in earth's magnetic field such that north-seeking needle of magnetic compass would point south rather than to north magnetic pole.

magnetite A mineral; iron oxide, Fe_3O_4. Black; strongly magnetic. Important ore of iron.

magnetosphere Region 1,000 to 64,000 km above earth, where magnetic field traps electrically charged particles from sun and space. First believed to consist of two bands, Van Allen belts.

magnetostratigraphy Use of magnetized rocks to determine history of events in record of changes in earth's magnetic field in past geologic ages.

magnitude Measure of total energy released by an earthquake. Contrast with *intensity,* which is measure of effects of earthquake waves at particular place.

mantle Intermediate zone of earth. Surrounded by *crust,* it rests on *core* at depth of about 2,900 km.

marble Metamorphic rock of granular texture, with no rock cleavage, and composed of *calcite, dolomite,* or both.

maria Dark-toned ''seas'' of moon. Mark moon's topographically low areas.

marl Calcareous clay or intimate mixture of clay and particles of *calcite* or *dolomite,* usually shell fragments.

marsh gas *Methane,* CH_4, simplest paraffin hydrocarbon. Predominant component of *natural gas.*

mascons Concentrations of mass located beneath surfaces of lunar *maria.*

mass A number that measures quantity of matter. It is obtained on earth's surface by dividing weight of a body by acceleration due to gravity.

mass movement Surface movement of earth materials induced by gravity.

mass number Number of protons and neutrons in atomic nucleus.

mass unit One-twelfth mass of carbon atom. Approximately mass of hydrogen atom.

matter Anything that occupies space. Usually defined by describing its states and properties: solid, liquid, or gas; possesses mass, inertia, color, density, melting point, hardness, crystal form, mechanical strength, or chemical properties. Composed of *atoms.*

meander (1) Turn or sharp bend in stream's course. (2) To turn, or bend, sharply. Applied to stream courses in geological usage.

mechanical weathering Process by which rock is broken down into smaller and smaller fragments as result of energy developed by physical forces. Also known as *disintegration.*

medial moraine Ridge of *till* formed by junction of two *lateral moraines* when two valley glaciers join to form single ice stream.

mélange Heterogeneous mixture of rock materials. Mappable body of deformed rocks that may be several kilometers in length and consists of highly sheared clayey matrix, thoroughly mixed with angular native and exotic blocks of diverse origin and geologic ages.

Mercalli intensity scale Scale to evaluate intensity of earthquake shaking on basis of effects at given place.

mesotrophic In lake aging stage between *oligotrophic* and *eutrophic.*

metal Substance fusible and opaque, good conductor of electricity, and with characteristic luster. Examples: gold, silver, aluminum. Of the elements 77 are metals.

metallic bonding Special kind of bonding in atoms of metallic elements whereby outermost electrons are not shared or exchanged but are free to move around and connect to any atoms in solid. Relative freedom of movement of electrons accounts for high level of electrical conductivity in metals.

metalloid Element of some metallic and some nonmetallic characteristics. There are nine metalloids. See also Appendix B.

metamorphic facies Assemblage of minerals that reached equilibrium during metamorphism under specific range of temperature and pressure.

metamorphic rock ''Changed-form rock.'' Any rock changed in texture or composition by heat, pressure, or chemically active fluids after original formation.

metamorphic zone Area subjected to *metamorphism* and characterized by certain metamorphic facies formed during process.

metamorphism A process whereby rocks undergo physical or chemical changes or both to achieve equilibrium with conditions other than those under which they were originally formed (weathering arbitrarily excluded from meaning). Agents of metamorphism are heat, pressure, and chemically active fluids.

metasomatism Process whereby rocks are altered when *volatiles* exchange ions with them.

meteor Transient celestial body that enters earth's atmosphere with great speed, becoming incandescent from heat generated by air resistance.

meteoric water Groundwater derived primarily from precipitation.

meteorite Stony or metallic body fallen to earth from outer space.

methane Simplest paraffin hydrocarbon, CH_4. Principal constituent of *natural gas.* Sometimes called *marsh gas.*

micas Group of silicate minerals characterized by perfect sheet or scale cleavage resulting from atomic pattern, in which silicon-oxygen tetrahedra linked in sheets. *Biotite* is ferromagnesian black mica. *Muscovite* is potassic white mica.

microseism ''Small shaking.'' Specifically limited in technical usage to earth waves generated by sources other than earthquakes and, most frequently, to waves with periods of from 1 to about 9 s, from sources associated with atmospheric storms.

Midatlantic Ridge See *Atlantic Ridge.*

midocean ridge Continuous, seismically active, median mountain range extending through North and South Atlantic, Indian, and South Pacific Oceans.

migmatite Mixed rock produced by intimate interfingering of *magma* and invaded rock.

mineral Naturally occurring solid element or compound, exclusive of biologically formed carbon components. Has definite composition or range of composition and orderly internal atomic arrangement (crystalline structure), which gives unique physical and chemical properties, including tendency to assume certain geometrical forms known as *crystals.*

mineral deposit Occurrence of one or more minerals in such concentration and form as to make possible removal and processing for use at profit.

mineraloid Substance that does not yield definite chemical formula and shows no sign of crystallinity. Examples: *bauxite, limonite,* and *opal.*

miogeosyncline That part of a geosyncline in which volcanism is absent, generally located near *craton*.

modulus of elasticity Slope of graph line relating *stress* to *strain* in *elastic deformation*.

Mohorovičić discontinuity (Moho) Base of crust marked by abrupt increases in velocities of earth waves.

molecule Smallest unit of compound that displays properties of that compound.

Monel metal Steel containing 68 percent nickel.

monocline Double flexure connecting strata at one level with same strata at another level.

montmorillonite Clay mineral family, hydrous aluminous silicate with structural sandwich of one ionic sheet of aluminum and hydroxyl between two $(Si_4O_{10})^{4-}$ sheets. Sandwiches piled on each other with water between and with nothing but weak bonds to hold them together. As result, additional water can enter lattice readily, causing mineral to swell appreciably and further weakening attraction between structural sandwiches. Consequently a lump of montmorillonite in a bucket of water slumps rapidly into a loose, incoherent mass. Compare with other clay minerals, *kaolinite* and *illite*.

moon A natural satellite.

moraine General term applied to certain landforms composed of *till*.

mountain Any part of landmass projecting conspicuously above its surroundings.

mountain chain Series or group of connected mountains having well-defined trend or direction.

mountain range Series of more or less parallel ridges, all of which formed within a single *geosyncline* or on its borders.

mountain structure Structure produced by deformation of rocks.

mudcracks Cracks caused by shrinkage of drying deposit of silt or clay under surface conditions.

mudflow Flow of well-mixed mass of rock, earth, and water that behaves like a fluid and moves down slopes with consistency similar to that of newly mixed concrete.

mudstone Fine-grained, detrital sedimentary rock made up of *silt-* and *clay-sized* particles. Distinguished from *shale* by lack of fissility.

muscovite "White mica." Nonferromagnesian rock-forming silicate mineral with tetrahedra arranged in sheets. Sometimes called *potassic mica*.

native state State in which an element occurs uncombined in nature. Usually applied to metals, as in "native copper," "native gold," etc.

natural gas Gaseous hydrocarbons that occur in rocks. Dominated by *methane*.

natural remanent magnetism Magnetism of rock. May or may not coincide with present magnetic field of earth. Abbreviation, NRM.

natural resources Energy and materials made available by geological processes.

neck cutoff Breakthrough of a river across narrow neck separating two meanders, where downstream migration of one has been slowed and next meander upstream has overtaken it. Compare with *chute cutoff*.

negative charge Condition resulting from surplus of electrons.

nesosilicate Mineral with crystal structure containing silicon-oxygen tetrahedra arranged as isolated units.

neutron *Proton* and *electron* combined and behaving like fundamental particle of matter. Electrically neutral with mass of 1.00896 u. If isolated, decays to form proton and electron.

névé Granular ice formed by recrystallization of snow. Intermediate between snow and glacier ice. Sometimes called *firn*.

Nichrome Steel alloy with 35 to 85 percent nickel.

nickel steel Steel containing 2.5 to 3.5 percent nickel.

nodule Irregular, knobby-surfaced mineral body that differs in composition from rock in which formed. Silica in form of *chert* or *flint* is major component of nodules. They are commonly found in limestone and dolomite.

nonclastic texture Applied to sedimentary rocks in which rock-forming grains are interlocked. Most sedimentary rocks with nonclastic texture are crystalline.

nonconformity *Unconformity* in which older rocks are of intrusive igneous origin.

nondipole magnetic field Portion of earth's magnetic field remaining after dipole field and external field are removed.

nonferromagnesians Silicate minerals that do not contain iron or magnesium.

nonmetal Element that does not exhibit metallic luster, conductivity, or other features of metal. Of the elements 17 are nonmetals.

normal fault Fault in which *hanging wall* appears to have moved downward relative to *footwall*; opposite of *thrust fault*. Also called *gravity fault*.

North Atlantic deep water Seawater in Arctic that sinks in North Atlantic and drifts southward as far as 60°S.

NRM See *natural remanent magnetism*.

nucleus *Protons* and *neutrons* constituting central part of an atom.

nuée ardente "Hot cloud." French term applied to highly heated mass of gas-charged lava ejected from vent or pocket at volcano summit more or less horizontally onto an outer slope, down which it moves swiftly, however slight the incline, because of its extreme mobility.

oblique slip fault Fault with components of relative displacement along both *strike* and *dip*.

obsidian Glassy equivalent of granite.

oil In geology refers to *petroleum*.

oil shale Shale containing such proportion of hydrocarbons as to be capable of yielding *petroleum* on slow distillation.

oligotrophic In lake aging pertains to a youthful lake and indicates water low in accumulated nutrients and high in dissolved oxygen.

olivine Rock-forming ferromagnesian silicate mineral that crystallizes early from magma and weathers readily at earth's surface. Crystal structure based on isolated $(SiO_4)^{4-}$ ions and positive ions of iron, magnesium, or both. General formula: $(Mg, Fe)_2SiO_4$.

oölites Spheroidal grains of sand size, usually composed of calcium carbonate, $CaCO_3$, and thought to have originated by inorganic precipitation. Some limestones largely made up of oölites.

ooze Deep-sea deposit consisting of 30 percent or more by volume of hard parts of very small, sometimes microscopic, organisms. If particular organism dominant, its name used as modifier, as in *globigerina* ooze, or *radiolarian* ooze.

opal Amorphous silica, with varying amounts of water; a mineral gel.

open-pit mining Surface mining represented by sand and gravel pits, stone quarries, and copper mines of some western states.

ophitic Rock texture in which lath-shaped plagioclase crystals are enclosed wholly or in part in later-formed augite, as commonly occurs in diabase.

order of crystallization Chronological sequence in which crystallization of various minerals of an assemblage takes place.

ore deposit Metallic minerals in concentrations that can be worked at profit.

orogeny Process by which mountain structures develop.

orthoclase Feldspar in which K$^+$ is diagnostic positive ion; K(AlSi$_3$O$_8$).

orthoquartzite Sandstone composed completely—or almost completely—of *quartz* grains. *Quartzose sandstone* is synonym.

outwash Material carried from a glacier by meltwater and laid down in stratified deposits.

outwash plain Flat or gently sloping surface underlain by outwash.

overbank deposits Sediments (usually clay, silt, and fine sand) deposited on flood plain by river overflowing banks.

overturned fold Fold with at least one *limb* rotated through more than 90°.

oxbow Abandoned *meander* caused by a *neck cutoff*.

oxbow lake Abandoned *meander* isolated from main stream channel by deposition and filled with water.

oxide mineral Mineral formed by direct union of an element with oxygen. Examples: ice, corundum, hematite, magnetite, cassiterite.

P Symbol for earthquake *primary waves.*

pahoehoe lava *Lava* whose surface is smooth and billowy, frequently molded into forms resembling huge rope coils. Characteristic of basic lavas.

paired terraces *Terraces* that face each other across stream at same elevation.

paleomagnetism Study of earth's magnetic field as has existed during geologic time.

paleosol Soil formed in past environment; often buried.

Pangaea Hypothetical continent from which all others are postulated to have originated through process of fragmentation and drifting.

parabolic dune *Dune* with long, scoop-shaped form that, when perfectly developed, exhibits parabolic shape in plan, with horns pointing upwind. Contrast *barchan*, in which horns point downwind. Characteristically covered with sparse vegetation; often found in coastal belts.

paternoster lakes Chain of lakes resembling string of beads along glaciated valley where ice plucking and gouging have scooped out series of basins.

peat Partially reduced plant or wood material, containing approximately 60 percent carbon and 30 percent oxygen. An intermediate material in process of coal formation.

pebble size Volume greater than that of a sphere with diameter of 4 mm and less than that of a sphere of 64 mm.

pedalfer Soil characterized by accumulation of iron salts or iron and aluminum salts in *B* horizon. Varieties of pedalfers include red and yellow soils of southeastern United States and *podsols* of northeastern quarter of United States.

pedocal *Soil* characterized by accumulation of calcium carbonate in its profile. Characteristic of low rainfall. Varieties include black and chestnut soils of northern Plains states and red and gray desert soils of drier western states.

pedology Science that treats of *soils*—origin, character, and utilization.

pegmatite Small *pluton* of exceptionally coarse texture, with crystals up to 12 m in length, commonly formed at margin of *batholith* and characterized by *graphic structure*. Nearly 90 percent of all pegmatites are simple pegmatites of *quartz, orthoclase*, and unimportant percentages of *micas*; others are extremely rare ferromagnesian pegmatites, and complex pegmatites. Complex pegmatites have as major components *sialic* minerals of simple pegmatites but also contain variety of rare minerals.

pelagic deposit Material formed in deep ocean and deposited there. Example: *ooze.*

pendulum Inertia member so suspended that, after displacement, restoring force will return it to starting position. If displaced and then released, oscillates, completing one to-and-fro swing in time called *period.*

perched water table Top of *zone of saturation* that bottoms on impermeable horizon above level of general *water table* in area. Is generally near surface and frequently supplies a hillside spring.

peridotite Coarse-grained igneous rock dominated by dark-colored minerals, consisting of about 75 percent *ferromagnesian silicates* and balance *plagioclase feldspars.*

period For oscillating system length of time required to complete one oscillation.

permeability For rock or earth material ability to transmit fluids. Permeability equal to velocity of flow divided by hydraulic gradient.

petroleum An oily mixture of hydrocarbons extracted from subsurface earth structures. Thought to result from physical and chemical conversion of remains of animals and plants. A fuel in natural or refined state, yielding on distillation such products as gasoline, kerosine, naphtha.

phase (1) Homogeneous, physically distinct portion of matter in physical-chemical system not homogeneous, as in three phases ice, water, and aqueous vapor. (2) Group of seismic waves of one type.

phenocryst A crystal significantly larger than crystals of surrounding minerals.

phosphate rock Sedimentary rock containing calcium phosphate.

photosynthesis Process by which carbohydrates are compounded from carbon dioxide and water in presence of sunlight and chlorophyll.

phyllite Clayey metamorphic rock with rock cleavage intermediate between *slate* and *schist*. Commonly formed by the regional metamorphism of *shale* or *tuff*. Micas characteristically impart a pronounced sheen to rock cleavage surfaces. Has phyllitic cleavage.

phyllitic cleavage Rock cleavage in which flakes are produced barely visible to unaided eye. Coarser than *slaty* and finer than *schistose cleavage.*

phyllosilicate Mineral with crystal structure containing silicon-oxygen tetrahedra arranged as sheets.

physical geology Branch of *geology* that deals with nature and properties of material composing the earth, distribution of materials throughout globe, processes by which they are formed, altered, transported, and distorted, and nature and development of landscape.

piedmont glacier Glacier formed by coalescence of *valley glaciers* and spreading over plains at foot of mountains from which valley glaciers came.

planet Heavenly body that changes position from night to night with respect to background of stars.

planetology Organized body of knowledge about planetary system.

plastic deformation Permanent change in shape or volume not involving failure by rupture and, once started, continuing without increase in deforming force.

plastic solid Solid that undergoes deformation continuously and indefinitely after stress applied to it passes a critical point.

plate Segment of earth's crust (*lithosphere*) varying in thickness from several tens to as much as 250 km and including part of upper *mantle* above *asthenosphere*.

plate tectonics Theory of worldwide dynamics involving movement and interactions of the many rigid plates of earth's *lithosphere*.

plateau basalt Basalt poured out from fissures in floods that tend to form great plateaus. Sometimes called *flood basalt*.

playa Flat-floored center of undrained desert basin.

playa lake Temporary lake formed in a *playa*.

pleochroic halo Minute, concentric-spherical zones of darkening or coloring that form around inclusions of radioactive minerals in *biotite*, *chlorite*, and a few other minerals. About 0.075 mm in diameter.

plume Pipelike convection cells thought to carry heat and *mantle* material from lower mantle up to *crust*, producing hot spots at surface.

plunge Acute angle that axis of folded rock mass makes with horizontal plane.

pluton A body of igneous rock formed beneath earth surface by consolidation from magma. Sometimes extended to include bodies formed beneath surface by metasomatic replacement of older rock.

plutonic igneous rock Rock formed by slow crystallization, which yields coarse texture. Once believed to be typical of crystallization at great depth; but not a necessary condition.

pluvial lake Lake formed during a *pluvial period*.

pluvial period Period of increased rainfall and decreased evaporation; prevailed in nonglaciated areas during time of ice advance elsewhere.

podsol Ashy-gray or gray-brown soil of *pedalfer* group. Highly bleached soil, low in iron and lime, formed under moist and cool conditions.

point bars Accumulations of sand and gravel deposited in slack waters on inside of bends of winding, or meandering, river.

Poisson's ratio Ratio of change of diameter per unit diameter to change of length per unit length in elastic stretching or compression of cylindrical specimen.

polar compound Compound, such as water, with a molecule that behaves like small bar magnet with positive charge on one end and negative charge on other.

polarity epoch Interval of time during which earth's magnetic field has been oriented dominantly in either normal or reverse direction. May be marked by shorter intervals of opposite sign, called *polarity events*.

polarity event See *polarity epoch*.

polar wandering, or migration Movement of position of magnetic pole during past time in relation to present position.

polymorphism Existence of several different morphologic kinds occurring in species or mineral.

porosity Percentage of open space or interstices in rock or other earth material. Compare with *permeability*.

porphyritic Textural term for igneous rocks in which larger crystals, called *phenocrysts*, are set in finer groundmass, which may be crystalline or glassy or both.

porphyry Igneous rock containing conspicuous *phenocrysts* in fine-grained or glassy *groundmass*.

portland cement Hydraulic cement consisting of compounds of silica, lime, and alumina.

positive charge Condition resulting from deficiency of electrons.

potassic feldspar *Orthoclase*, $K(AlSi_3O_8)$.

potential energy Stored energy waiting to be used. Energy that a piece of matter possesses because of position or because of arrangement of parts.

prairie soils Transitional soils between *pedalfers* and *pedocals*.

Pratt hypothesis Explains isostasy by assuming all portions of the crust have same total mass above certain elevation, called *level of compensation*. Higher sections would have proportionately lower density.

precipitation Discharge of water, in rain, snow, hail, sleet, fog, or dew, on land or water surface. Also, process of separating mineral constituents from solution by evaporation (*halite*, *anhydrite*) or from *magma* to form igneous rocks.

precursor Relating to earthquakes, refers to events immediately preceding actual shaking of ground. Includes changes in seismic velocities, groundwater levels, and tilt of ground surface.

pressure Force per unit area.

primary wave Earthquake body wave that travels fastest and advances by *push-pull* mechanism. Also known as *longitudinal*, compressional, or *P wave*.

proton Fundamental particle of matter with positive electrical charge of 1 unit (equal in amount, but opposite in effect, to the charge of *electron*) and mass of 1.00758 u.

proton-proton fusion Rapidly moving *protons* in hot interior of stars collide and fuse to form atoms of helium from atoms of hydrogen in continuous buildup of higher elements.

pumice Pieces of magma up to several centimeters across that have trapped bubbles of steam or other gases as they were thrown out in eruption. Sometimes they have sufficient buoyancy to float on water.

push-pull wave Wave that advances by alternate compression and rarefaction of medium, causing particles in path to move forward and backward along direction of wave's advance. In connection with earth waves, also known as *primary wave*, compressional wave, *longitudinal wave*, or *P* wave.

pyrite A sulfide mineral, iron sulfide, FeS_2.

pyroclastic debris Fragments blown out by explosive volcanic eruptions and subsequently deposited on ground. Include *ash*, *cinders*, *lapilli*, blocks, bombs, and *pumice*.

pyroxene group Ferromagnesian silicates with a single chain of silicon-oxygen tetrahedra. Common example: augite. Compare with *amphibole group* (example: hornblende), which has a double chain of tetrahedra.

pyrrhotite A mineral, iron sulfide. So commonly associated with nickel minerals that has been called "world's greatest nickel ore."

quartz A silicate mineral, SiO_2, composed exclusively of silicon-oxygen tetrahedra, with all oxygens joined in a three-dimensional network. Crystal form is six-sided prism tapering at end, with prism faces striated transversely. An important rock-forming mineral.

quartzite Metamorphic rock commonly formed by metamorphism of sandstone and composed of quartz. No rock cleavage. Breaks *through* sand grains in contrast to sandstone, which breaks *around* grains.

radar Ultrahigh-frequency electromagnetic radiation.

radial drainage Arrangement of stream courses in which streams radiate outward in all directions from central zone.

radiant energy Electromagnetic waves travelling as wave motion.

radioactivity Spontaneous breakdown of atomic nucleus, with emission of radiant energy.

rain wash Water from rain after it has fallen on ground and before concentrated in definite stream channels.

range Elongated series of mountain peaks considered to be a part of one connected unit, such as Appalachian Range or Sierra Nevada Range.

rank Term used to designate extent to which *metamorphism* has advanced. Compare with *grade*. Rank is more commonly employed in designating stage of metamorphism of *coal*.

ray craters Lunar craters marked by *rays*. Young on lunar time scale.

rays Light-toned streaks that spread outward from such lunar craters as Tycho, Kepler, and Copernicus.

reaction series See *Bowen's reaction series*.

recessional moraine Ridge or belt of *till* marking period of moraine formation, probably in period of temporary stability or slight readvance, during general wastage of a glacier and recession of its front.

recorder Part of a *seismograph* that makes record of ground motion.

rectangular pattern Arrangement of stream courses in which tributaries flow into larger streams at angles approaching 90°.

recumbent fold Fold with axial plane more or less horizontal.

reflection seismic prospecting Uses reflected waves and places seismographs at distances only a fraction of depths investigated.

refraction seismic prospecting Uses travel times of refracted waves and spreads seismographs over lines roughly four times depth being investigated.

refractory Mineral or compound that resists action of heat and chemical reagents.

regional metamorphism Metamorphism occurring over tens or scores of kilometers.

rejuvenation Change in conditions of erosion that causes a stream to begin more active erosion and a new cycle.

relative time Dating of events by place in chronologic order of occurrence rather than in years. Compare with *absolute time*.

relief See Appendix D.

reverse fault Fault in which *hanging wall* appears to have moved upward relative to *footwall*; contrast with *normal*, or *gravity*, *fault*. Also called *thrust fault*.

rhyolite Fine-grained igneous rock with composition of granite.

rift zone System of fractures in earth's crust. Often associated with lava extrusion.

right-lateral fault *Strike-slip fault* in which ground opposite you appears to have moved right when you face it.

rigidity Resistance to elastic *shear*.

rill Miniature stream channel which forms along axis of broad, shallow trough carrying *sheet wash*, or sheet flow.

rilles Trenchlike depressions on moon's surface. Some are straight walled, other sinuous.

ring dike Arcuate (rarely circular) *dike* with steep dip.

ripple marks Small waves produced in unconsolidated material by wind or water. See *ripple marks of oscillation*.

ripple marks of oscillation *Ripple marks* formed by oscillating movement of water such as may be found along sea coast outside surf zone. Symmetrical, with sharp or slightly rounded ridges separated by more gently rounded troughs.

rock Aggregate of minerals of one or more kinds in varying proportions.

rock cycle Concept of sequences through which earth materials may pass when subjected to geological processes.

rock flour Finely divided rock material pulverized by glacier and carried by streams fed by melting ice.

rock-forming silicate minerals Minerals built around framework of silicon-oxygen tetrahedra: olivine, augite, hornblende, biotite, muscovite, orthoclase, albite, anorthite, quartz.

rock glacier Tongue of rock waste found in valleys of certain mountainous regions. Characteristically lobate and marked by series of arcuate, rounded ridges that give it aspect of having flowed as viscous mass.

rock melt Liquid solution of rock-forming mineral ions.

rock salt *Halite*, or common salt, NaCl.

rock slide Sudden and rapid slide of bedrock along planes of weakness.

Rossi-Forel scale Scale for rating earthquake intensities, devised in 1878.

runoff Water that flows off land.

rupture Breaking apart or state of being broken apart.

S Symbol for *secondary wave*.

salt In geology this term usually refers to *halite*, or *rock salt*, NaCl, particularly in such combinations as salt water and *salt dome*.

saltation Mechanism by which a particle moves by jumping from one point to another.

salt dome Mass of NaCl generally of roughly cylindrical shape and with diameter of about 2 km near top. Such mass has been pushed through surrounding sediments into present position. Reservoir rocks above and alongside salt domes sometimes trap *oil* and *gas*.

saltwater wedge Body of water, found in some estuaries, which thins toward head of estuary and is overridden by fresh water from land.

sand Clastic particles of *sand size*, commonly but not always composed of mineral *quartz*.

sand size Volume greater than that of a sphere with diameter of 0.0625 mm and less than that of a sphere with diameter of 2 mm.

sandstone Detrital *sedimentary rock* formed by cementation of individual grains of sand size and commonly composed of mineral *quartz*. Sandstones constitute estimated 12 to 28 percent of sedimentary rocks.

sapropel Aquatic ooze or sludge rich in organic matter. Believed to be source material for *petroleum* and *natural gas*.

satellite crater *Crater* formed by impact of a fragment ejected during creation of a primary crater. Also called *secondary crater*.

scale See Appendix D.

schist Metamorphic rock dominated by fibrous or platy minerals. Has *schistose cleavage* and is product of *regional metamorphism*.

schistose cleavage Rock cleavage with grains and flakes clearly visible and cleavage surfaces rougher than in *slaty* or *phyllitic cleavage*.

sea-floor spreading Process by which ocean floors spread laterally from crests of main ocean ridges. As material moves laterally from ridge, new material is thought to replace it along ridge crest by welling upward from mantle.

seamount Isolated, steep-sloped peak rising from deep ocean floor but submerged beneath surface. Most have sharp peaks, but some have flat tops and are called *guyots*, or *tablemounts*. Seamounts are probably volcanic in origin.

secondary crater   See *satellite crater*.

secondary wave   Earthquake *body wave* slower than *primary wave*. *Shear*, *shake*, or *S wave*.

secular variation of magnetic field   Change in inclination, declination, or intensity of earth's magnetic field. Detectable only from long historical records.

sedimentary facies   Accumulation of deposits that exhibits specific characteristics and grades laterally into other sedimentary accumulations that were formed at same time but exhibit different characteristics.

sedimentary rock   Rock formed from accumulations of sediment, which may consist of rock fragments of various sizes, remains or products of animals or plants, products of chemical action or of evaporation, or mixtures of these. *Stratification* is single most characteristic feature of sedimentary rocks, which cover about 75 percent of land area.

sedimentation   Process by which mineral and organic matters are laid down.

seif dune   Very large *longitudinal dune*. As high as 100 m and as long as 100 km.

seismic prospecting   Method of determining nature and structure of buried rock formations by generating waves in ground (commonly by small explosive charges) and measuring length of time these waves require to travel different paths.

seismic sea wave   Large wave in ocean generated at time of earthquake. Popularly but incorrectly known as *tidal wave*. Sometimes called *tsunami*.

seismogram   Record obtained on a *seismograph*.

seismograph   Instrument for recording vibrations, most commonly employed for recording earth vibrations during earthquakes.

seismology   Scientific study of earthquakes and other earth vibrations.

serpentine   Silicate of magnesium common among metamorphic minerals. Occurs in two crystal habits: platy, known as antigorite; fibrous, known as *chrysotile*, an *asbestos*. "Serpentine" comes from mottled shades of green on massive varieties, suggestive of snake markings.

S.G.   Symbol for *specific gravity*.

shake wave   Wave that advances by causing particles in path to move from side to side or up and down at right angles to direction of wave's advance, a shake motion. Also called *shear wave*, or *secondary wave*.

shale   Fine-grained, detrital sedimentary rock made up of *silt*- and *clay-sized* particles. Contains clay minerals as well as particles of quartz, feldspar, calcite, dolomite, and other minerals. Distinguished from *mudstone* by presence of fissility.

shear   Change of shape without change of volume.

shear wave   Wave that advances by shearing displacements (which change shape without changing volume) of medium. This causes particles in path to move from side to side or up and down at right angles to direction of wave's advance. Also called *shake wave*, or *secondary wave*.

sheet flow   See *sheet wash*.

sheeting   Joints essentially parallel to ground surface. More closely spaced near surface and become progressively farther apart with depth. Particularly well developed in granitic rocks, but sometimes in other massive rocks as well.

sheet wash   Water accumulating on a slope in thin sheet of water. May begin to concentrate in *rills*. Also called *sheet flow*.

shield   Nucleus of Precambrian rocks around which a continent has grown.

shield volcano   Volcano built up almost entirely of lava, with slopes seldom as great as 10° at summit and 2° at base. Examples: five volcanoes on island of Hawaii.

sial   A term coined from chemical symbols for silicon and aluminum. Designates composite of rocks dominated by granites, granodiorites, and their allies and derivatives, which underlie continental areas of globe. Specific gravity considered to be about 2.7.

sialic rock   Igneous rock composed predominantly of silicon and aluminum, from whose chemical symbols term is constructed. Average specific gravity, about 2.7.

siderite   A mineral; iron carbonate, $FeCO_3$. An ore of iron.

silicate minerals   Minerals with crystal structure containing *silicon-oxygen tetrahedra* arranged as isolated units (nesosilicates), single or double chains (inosilicates), sheets (phyllosilicates), or three-dimensional frameworks (tectosilicates).

silicon-oxygen tetrahedron   Complex ion composed of silicon ion surrounded by four oxygen ions. Negative charge of 4 units, and represented by symbol $(SiO_4)^{4-}$. Diagnostic unit of silicate minerals, and makes up central building unit of nearly 90 percent of materials of earth's crust.

sillimanite   A silicate mineral, Al_2SiO_5, characteristic of highest metamorphic temperatures and pressures. Occurs in long slender crystals, brown, green, white. Crystalline structure based on independent tetrahedra. Contrast with *kyanite* and *andalusite*, which have same composition but different crystal habits and form at lower temperatures.

silt size   Volume greater than that of a sphere with diameter of 0.0039 mm and less than that of a sphere with diameter of 0.0625 mm.

sima   Term coined from *silicon* and *magnesium*. Designates worldwide shell of dark, heavy rocks. Sima believed to be outermost rock layer under deep, permanent ocean basins, such as Midpacific. Originally sima considered basaltic in composition, with specific gravity of about 3.0. It has been suggested also, however, that it may be *peridotitic* in composition, with specific gravity of about 3.3.

simatic rock   Igneous rock composed predominantly of ferromagnesian minerals. Average specific gravity, 3.0 to 3.3.

sink   See *sinkhole*.

sinkhole   Depression in surface of ground caused by collapse of roof over solution cavern.

slate   Fine-grained metamorphic rock with well-developed *slaty cleavage*. Formed by low-grade *regional metamorphism* of *shale*.

slaty cleavage   Rock cleavage in which ease of breaking occurs along planes separated by microscopic distances.

slip face   Steep face on lee side of a dune.

slope failure   See *slump*.

slump   Downward and outward movement of rock or unconsolidated material as unit or as series of units. Also called *slope failure*.

snowfield   Stretch of perennial snow existing in area where winter snowfall exceeds amount of snow that melts away during summer.

snow line   Lower limit of perennial snow.

soapstone   See *talc*.

sodic feldspar   Albite, $Na(AlSi_3O_8)$.

soil   Superficial material that forms at earth's surface as result of organic and inorganic processes. Soil varies with climate, plant and animal life, time, slope of land, and parent material.

soil horizon   Layer of soil approximately parallel to land surface with observable characteristics produced through operation of soil-building processes.

solar constant Average rate at which radiant energy received by earth from sun. Equal to little less than 2 cal/cm² on plane perpendicular to sun's rays at outer edge of atmosphere, when earth is at mean distance from sun.

solar system Sun with group of celestial bodies held by its gravitational attraction and revolving around it.

sole mark Cast of sedimentary structures such as cracks, tracks, or grooves formed on lower surface or underside of sandstone bed, commonly revealed after original underlying sedimentary layer has weathered away.

solid Matter with definite shape and volume and some fundamental strength. May be crystalline, glassy, or amorphous.

solid solution Single crystalline phase that may vary in composition within specific limits.

solifluction Mass movement of soil affected by alternate freezing and thawing. Characteristic of saturated soils in high latitudes.

sorosilicates Mineral with crystal structure containing *silicon-oxygen tetrahedra* arranged as double units.

space lattice In crystalline structure of mineral three-dimensional array of points representing pattern of locations of identical atoms or groups of atoms constituting a mineral's *unit cell*. There are 230 pattern types.

specific gravity Ratio between weight of given volume of material and weight of equal volume of water at 4°C.

specific heat Amount of heat necessary to raise temperature of 1 g of any material 1°C.

sphalerite A mineral; zinc sulfide, ZnS. Nearly always contains iron, (Zn, Fe)S. Principal ore of zinc. (Also known as zinc blende or blackjack.)

spheroidal weathering Spalling off of concentric shells from rock masses of various sizes as result of pressures built up during chemical weathering.

spit Sandy bar built by currents into a bay from a promontory.

spring Place where *water table* crops out at ground surface and water flows out more or less continuously.

stack Small island that stands as isolated, steep-sided rock mass just off end of promontory. Has been isolated from land by erosion and weathering concentrated behind end of a headland.

stalactite Icicle-shaped accumulation of *dripstone* hanging from cave roof.

stalagmite Post of *dripstone* growing upward from cave floor.

star A heavenly body that seems to stay in same position relative to other heavenly bodies.

staurolite Silicate mineral characteristic of middle-grade metamorphism. Crystalline structure based on independent tetrahedra with iron and aluminum. Has unique crystal habit that makes it striking and easy to recognize: six-sided prisms intersecting at 90° to form cross or at 60° to form ex.

steady-state theory Theory that universe is developing by continuous creation of matter as newly formed galaxies replace those expanding out of sight, thus keeping mass density of universe constant. Compare *big-bang theory*.

stock Discordant *pluton* that increases in size downward, has no determinable floor, and shows area of surface exposure less than 100 km². Compare with *batholith*.

stoping Mechanism by which *batholiths* have moved into crust by breaking off and foundering of blocks of rock surrounding magma chamber.

strain Change of dimensions of matter in response to *stress:* Commonly, unit strain, such as change in length per unit length (total lengthening divided by original length), change in width per unit width, change in volume per unit volume.

stratification Structure produced by deposition of sediments in layers or beds.

stratigraphic trap Structure that traps *petroleum* or *natural gas* because of variation in permeability of reservoir rock or termination of inclined reservoir formation on up-dip side.

streak Color of fine powder of mineral; may be different from color of hand specimen. Usually determined by rubbing mineral on piece of unglazed porcelain (*hardness* about 7) known as a "streak plate," which is, of course, useless for minerals of greater hardness.

stream order Hierarchy in which segments of a stream system are arranged.

stream terrace Surface representing remnants of stream's channel or flood plain when stream was flowing at higher level. Subsequent downward cutting by stream leaves remnants of old channel or *flood plain* standing as *terrace* above present stream level.

strength *Stress* at which rupture occurs or plastic deformation begins.

stress Force applied to material that tends to change dimensions: Commonly, unit stress, or total force divided by the area over which applied. Contrast with *strain*.

striations (1) Scratches, or small channels, gouged by glacial action. Bedrock, pebbles, and boulders may show striations produced when rocks trapped by ice were ground against bedrock or other rocks. Striations along bedrock surface are oriented in direction of ice flow across that surface. (2) In minerals parallel, threadlike lines, or narrow bands, on face of mineral. Reflect internal atomic arrangement.

strike Direction of line formed by intersection of a rock surface with a horizontal plane. Strike is always perpendicular to direction of *dip*.

strike-slip fault *Fault* in which movement is almost in direction of fault's *strike*.

strip mining Surface mining in which soil and rock covering sought-for commodity are moved to one side. Some coal mining is pursued in this manner.

structural relief Difference in elevation of parts of deformed stratigraphic horizon.

structure Attitudes of deformed masses of rock.

subduction Act of one tectonic unit's descending under another (commonly slab of *lithosphere*).

subduction zone Elongate region along which lithospheric block descends relative to another lithospheric block.

sublimation Process by which solid material passes into gaseous state without first becoming liquid.

subsequent stream Tributary stream flowing along beds of less erosional resistance and parallel to beds of greater resistance. Course determined subsequent to uplift that brought more resistant beds within sphere of erosion.

subsurface water Water below ground surface. Also referred to as *underground water* and *subterranean water*.

subterranean water Water below ground surface. Also referred to as *underground water* and *subsurface water*.

sulfate mineral Mineral formed by combination of complex ion (SO₄)²⁻ with positive ion. Common example: gypsum, CaSO₄ · 2H₂O.

sulfide mineral Mineral formed by direct union of element with sulfur. Examples: argentite, chalcocite, galena, sphalerite, pyrite, and cinnabar.

superposition Law by which, if series of sedimentary rocks has not been overturned, topmost layer is always youngest, and lowermost always oldest.

surface wave Wave that travels along free surface of medium. Earthquake surface waves sometimes represented by symbol *L*.

surge Applied to glaciers, rapid and sometimes catastrophic advance of ice.

suspended water Underground water held in *zone of aeration* by molecular attraction exerted on water by rock and earth materials and by attraction exerted by water particles on one another.

suspension Process by which material is buoyed up in air or water and moved about without making contact with surface while in transit. Contrasts with *traction*.

symmetrical fold *Fold* in which axial plane essentially vertical. Limbs dip at similar angles.

syncline A configuration of folded, stratified rocks in which rocks dip downward from opposite directions to come together in a trough. Reverse of *anticline*.

tablemount See *guyot*.

tabular Shape with large area relative to thickness.

taconite Unleached iron formation of Lake Superior District. Consists of chert with hematite, magnetite, siderite, and hydrous iron silicates. Ore of iron, averaging 25 percent iron, but natural leaching turns it into ore with 50 to 60 percent iron.

talc Silicate of magnesium common among metamorphic minerals. Crystalline structure based on tetrahedra arranged in sheets; greasy and extremely soft. Sometimes known as *soapstone*.

talus Slope established by accumulation of rock fragments at foot of cliff or ridge. Rock fragments that form talus may be rock waste, slide rock, or pieces broken by frost action. Actually, term "talus" widely used to mean rock debris itself.

tarn Lake formed in bottom of *cirque* after glacier ice has disappeared.

tectonic change of sea level Change in sea level produced by land movement.

tectosilicate Mineral with crystal structure containing *silicon-oxygen tetrahedra* arranged in three-dimensional frameworks.

temporary base level Nonpermanent *base level*, such as that formed by lake.

tension Stretching stress that tends to increase volume of a material.

terminal moraine Ridge or belt of *till* marking farthest advance of a glacier. Sometimes called *end moraine*.

terminal velocity Constant rate of fall eventually attained by grain or body when acceleration caused by influence of gravity is balanced by resistance of fluid through which grain falls or air through which body falls.

terrace Nearly level surface, relatively narrow, bordering a stream or body of water and terminating in a steep bank. Commonly term is modified to indicate origin, as in *stream* terrace and *wave-cut* terrace.

terrae Light-toned highlands of moon.

terrigenous deposit Material derived from above sea level and deposited in deep ocean. Example: volcanic ash.

tetrahedron A four-sided solid. Used commonly in describing silicate minerals as shortened reference to *silicon-oxygen tetrahedron*.

texture General physical appearance of rock, as shown by size, shape, and arrangement of particles that make it up.

thermal gradient In earth rate at which temperature increases with depth below surface.

thermal pollution Increase in normal temperatures of natural waters caused by intervention of human activities.

thermal spring Spring that brings warm or hot water to surface. Temperature usually 6.5°C or more above mean air temperature. Sometimes called *warm spring*, or *hot spring*.

thermoremanent magnetism Magnetism acquired by igneous rock as it cools below Curie temperatures of magnetic minerals in it. Abbreviation, TRM.

thin section Slice of rock ground so thin as to be translucent.

tholeiite Group of basalts primarily composed of *plagioclase* (approximately An$_{50}$), *pyroxene*, and iron oxides as *phenocrysts* in glassy groundmass of *quartz* and alkali *feldspar*; little or no *olivine* present.

thrust fault Fault in which *hanging wall* appears to have moved upward relative to *footwall*; opposite of *gravity*, or *normal*, *fault*. Also called *reverse fault*.

tidal current Water current generated by tide-producing forces of sun and moon.

tidal inlet Waterway from open water into a lagoon.

tidal wave Popular but incorrect designation for *tsunami*.

tide Alternate rising and falling of surface of ocean, other bodies of water, or earth itself in response to forces resulting from motion of earth, moon, and sun relative to each other.

till Unstratified, unsorted glacial *drift* deposited directly by glacier ice.

tillite Rock formed by lithification of *till*.

time-distance graph Graph of travel time against distance.

tombolo Sand bar connecting an island to mainland or joining two islands.

topographic deserts Deserts deficient in rainfall because they are either located far from oceans toward center of continents or cut off from rain-bearing winds by high mountains.

topography Shape and physical features of land.

topset bed Layer of sediment constituting surface of *delta*. Usually nearly horizontal and covers edges of inclined *foreset beds*.

tourmaline Silicate mineral of boron and aluminum with sodium, calcium, fluorine, iron, lithium, or magnesium. Formed at high temperatures and pressures through agency of fluids carrying boron and fluorine. Particularly associated with *pegmatites*.

township and range See Appendix D.

traction Process of carrying material along bottom of a stream. Traction includes movement by saltation, rolling, or sliding.

transcurrent fault *Strike-slip fault.*

transducer Device that picks up relative motion between mass of seismograph and ground and converts this into form that can be recorded.

transform fault Point at which strike-slip displacements stop and another structural feature, such as a ridge, develops.

transition element Element in series in which inner shell is being filled with electrons after outer shell has been started. All transition elements metallic in free state.

transpiration Process by which water vapor escapes from a living plant and enters atmosphere.

transverse dune Dune formed in areas of scanty vegetation and in which sand has moved in ridge at right angles to wind. Exhibits gentle windward slope and steep leeward slope characteristic of other dunes.

transverse wave *Shear*, or *shake*, *wave*.

trap rock Popular synonym for *basalt*.

travel time Total elapsed time for wave to travel from source to designated point.

travertine Form of calcium carbonate, $CaCO_3$, which forms stalactites, stalagmites, and other deposits in limestone caves or incrustations around mouths of hot and cold calcareous springs. Sometimes known as *tufa*, or *dripstone*.

trellis pattern Roughly rectilinear arrangement of stream courses in pattern reminiscent of garden trellis, developed in region where rocks of differing resistance to erosion have been folded, beveled, and uplifted.

trenches See *island-arc deeps*.

TRM See *thermoremanent magnetism*.

tropical deserts Deserts lying between 5° to 30° north and south of equator.

truncated spur Beveled end of divide between two tributary valleys where they join a main valley that has been glaciated. Glacier of main valley has worn off end of divide.

tsunami Large wave in ocean generated at time of earthquake. Popularly but incorrectly known as *tidal wave*. Sometimes called *seismic sea wave*.

tufa Calcium carbonate, $CaCO_3$, formed in stalactites, stalagmites, and other deposits in limestone caves, as incrustations around mouths of hot and cold calcareous springs, or along streams carrying large amounts of calcium carbonate in solution. Sometimes known as *travertine*, or *dripstone*.

tuff Rock consolidated from volcanic ash.

tundra Stretch of Arctic swampland developed on top of permanently frozen ground. Extensive tundra regions have developed in parts of North America, Europe, and Asia.

turbidites Sedimentary deposits settled out of turbid water carrying particles of widely varying grade size. Characteristically display *graded bedding*.

turbidity current Current in which limited volume of turbid or muddy water moves relative to surrounding water because of greater density.

turbulent flow Mechanism by which fluid (such as water) moves over or past a rough surface. Fluid not in contact with irregular boundary outruns that slowed by friction or deflected by uneven surface. Fluid particles move in series of eddies or whirls. Most stream flow is turbulent; turbulent flow is important in both erosion and transportation. Contrast with *laminar flow*.

ultimate base level Sea level, lowest possible *base level* for a stream.

unconformity Buried erosion surface separating two rock masses, older exposed to erosion for long interval of time before deposition of younger. If older rocks were deformed and not horizontal at time of subsequent deposition, surface of separation is *angular unconformity*. If older rocks remained essentially horizontal during erosion, surface separating them from younger rocks is called *disconformity*. Unconformity that develops between massive igneous rocks exposed to erosion and then covered by sedimentary rocks is called *nonconformity*.

underground water Water below ground surface. Also referred to as *subsurface water* and *subterranean water*.

uneven fracture Mineral habit of breaking along rough, irregular surfaces.

uniformitarianism Concept that present is key to past. This means that processes now operating to modify earth's surface have also operated in geologic past, that there is uniformity of processes past and present.

unit cell In crystalline structure of mineral three-dimensional grouping of atoms arbitrarily selected so that mineral's structure represented by periodic repetition of this unit in *space lattice*.

unpaired terrace A *terrace* formed when an eroding stream, swinging back and forth across a valley, encounters resistant rock beneath unconsolidated alluvium and is deflected, leaving behind single terrace with no corresponding terrace on other side of stream.

valley glacier Glacier confined to stream valley. Usually fed from cirque. Sometimes called *Alpine glacier* or *mountain glacier*.

valley train Gently sloping plain underlain by glacial outwash and confined by valley walls.

varve Pair of thin sedimentary beds, one coarse, one fine. This couplet has been interpreted as representing a cycle of 1 year or interval of thaw followed by interval of freezing in lakes fringing a glacier.

ventifact Pebble, cobble, or boulder that has had its shape or surface modified by wind-driven sand.

vertical exaggeration See Appendix D.

vesicle Small cavity in aphanitic or glassy igneous rock, formed by expansion of bubble of gas or steam during solidification of rock.

virtual geomagnetic pole Pole consistent with magnetic field as measured at any one locality. Refers to magnetic-field direction of single point, in contrast to *geomagnetic pole*, which refers to best fit of geocentric dipole for entire earth's field. Most *paleomagnetic* readings expressed as virtual geomagnetic poles.

viscosity An internal property of rock that offers resistance to flow. Ratio of deforming force to rate at which changes in shape are produced.

volatile components Materials in magma, such as water, carbon dioxide, and certain acids, whose vapor pressures are high enough to cause them to become concentrated in any gaseous phase that forms.

volcanic ash *Dust-sized pyroclastic* particle: volume equal to, or less than, that of sphere with diameter of 0.06 mm.

volcanic block Angular mass of newly congealed *magma* blown out in eruption. Contrast with *volcanic bomb*.

volcanic bomb Rounded mass of newly congealed *magma* blown out in eruption. Contrast with *volcanic block*.

volcanic breccia Rock formed from relatively large blocks of congealed lava embedded in mass of *ash*.

volcanic dust *Pyroclastic* detritus consisting of particles of *dust* size.

volcanic earthquakes Earthquakes caused by movements of *magma* or explosions of gases during volcanic activity.

volcanic eruption Explosive or quiet emission of *lava*, *pyroclastics*, or volcanic gases at earth's surface, usually from volcano but rarely from fissures.

volcanic mountains Mountains built up from extrusion of *lava* and *pyroclastic debris*.

volcanic neck Solidified material filling vent, or pipe, of dead volcano.

volcanic tremor Continuous shaking of ground associated with certain phase of volcanic eruption.

volcano Landform developed by accumulation of magmatic products near central vent.

vug Small unfilled cavity in rock, usually lined with crystalline layer of different composition from surrounding rock.

wadi See *arroyo*.

warm glacier Reaches melting temperature throughout thickness during summer season.

warm spring *Thermal spring* that brings warm water to surface. Temperature usually 6.5°C or more above mean air temperature.

warp Large section of continent composed of horizontal strata gently bent upward or downward.

wash See *arroyo*.

water gap Gap cut through resistant ridge by superimposed or antecedent stream.

water table Upper surface of *zone of saturation* for underground water. An irregular surface with slope or shape determined by quantity of groundwater and permeability of earth materials. In general, highest beneath hills and lowest beneath valleys.

wave Configuration of matter that transmits energy from one point to another.

weathering Response of materials once in equilibrium within earth's crust to new conditions at or near contact with water, air, or living matter. See also *chemical weathering* and *mechanical weathering*.

wrinkle ridges Ridges found on surfaces of lunar *maria* and flooded craters. May be caused by uplift due to volcanism or to compression.

xenolith Rock fragment foreign to igneous rock in which it occurs. Commonly inclusion of country rock intruded by igneous rock.

yazoo-type river Tributary unable to enter main stream because of *natural levees* along main stream. Flows along back-swamp zone parallel to main stream.

yield point Maximum stress that solid can withstand without undergoing permanent deformation, either by plastic flow or by rupture.

zone of aeration Zone immediately below ground surface, in which openings partially filled with air and partially with water trapped by molecular attraction. Subdivided into (1) *belt of soil moisture*. (2) *intermediate belt*, and (3) *capillary fringe*.

zone of saturation Underground region within which all openings filled with water. Top of zone of saturation is called *water table*. Water contained within zone of saturation is called *groundwater*.

zones of regional metamorphism High-grade, above 700°C; middle-grade, 400 to 700°C; low-grade, 150 to 400°C.

INDEX

The following numbers and exponents (and prefixes) are essentially an outline of the metric system:

$$1{,}000{,}000{,}000{,}000 = 10^{12} \text{ (tera-)}$$
$$1{,}000{,}000{,}000 = 10^{9} \text{ (giga-)}$$
$$1{,}000{,}000 = 10^{6} \text{ (mega-)}$$
$$1{,}000 = 10^{3} \text{ (kilo-)}$$
$$100 = 10^{2} \text{ (hecto-)}$$
$$10 = 10^{1} \text{ (deka-)}$$
$$1 = 10^{0} \text{ [unit]}$$
$$0.1 = 10^{-1} \text{ (deci-)}$$
$$0.01 = 10^{-2} \text{ (centi-)}$$
$$0.001 = 10^{-3} \text{ (milli-)}$$
$$0.000001 = 10^{-6} \text{ (micro-)}$$
$$0.000000001 = 10^{-9} \text{ (nano-)}$$

The fundamental unit of length is the meter, originally defined as one ten millionth of the distance from the equator to the North Pole, later defined as the distance between two marks inscribed on the standard meter bar in Paris, and most recently specified in terms of the wavelength of krypton. So:

$$1 \text{ km (kilometer)} = 10^{3} \text{ m (meter)}$$
$$1 \text{ cm (centimeter)} = 10^{-2} \text{ m}$$
$$1 \text{ mm (millimeter)} = 10^{-3} \text{ m}$$
$$1 \text{ } \mu\text{m (micrometer)} = 10^{-6} \text{ m}$$

The official definition of an inch is based on the length of a meter:

$$1 \text{ in (inch)} = 2.54 \times 10^{-2} \text{ m}$$

Temperature is measured in degrees Celsius, a scale in which the interval between the freezing and boiling points of water is divided into 100 degrees, with 0° representing the freezing point and 100° the boiling point. So:

$$\frac{5}{9} \, ^\circ\text{C (Celsius)} = 1\,^\circ\text{F (Fahrenheit)} \quad \text{or}$$

$$^\circ\text{C} = \frac{(^\circ\text{F} - 32^\circ)}{1.8} \quad \text{or} \quad ^\circ\text{F} = (^\circ\text{C} \times 1.8) + 32^\circ$$

Conversion of mass

	Grams, g	Kilograms, kg	Pounds, lb	Ounces, oz
g	1	1,000	453.6	28.35
kg	0.001	1	0.4536	2.835×10^{-2}
lb	2.205×10^{-3}	2.205	1	6.25×10^{-2}
oz	3.527×10^{-2}	35.27	16	1

Multiply units in the column heads by the figures in the table to convert to units at left. (For example, to convert pounds to kilograms, multiply the number of pounds by 0.4536).

Conversion of volume

	Cubic meters, m³	Cubic yards, yd³	Cubic centimeters, cm³	Cubic inches, in³	Cubic feet, ft³
m³	1	0.7646	1×10^{-6}	1.639×10^{-5}	2.832×10^{-2}
yd³	1.308	1	1.308×10^{-6}	2.143×10^{-5}	3.704×10^{-2}
cm³	1×10^{6}	7.646×10^{5}	1	16.39	2.832×10^{4}
in³	6.102×10^{4}	46,656	6.102×10^{-2}	1	1,728
ft³	35.31	27	3.531×10^{-5}	5.787×10^{-4}	1

Multiply units in the column heads by the figures in the table to convert to units at left. (For example, to convert cubic inches to cubic centimeters, multiply the number of cubic inches by 16.39).